214/05006068/30/3/06 *

20 057 356 39

AF616083

Neuronal Nicotinic Receptors

Neuronal Nicotinic Receptors

Pharmacology and Therapeutic Opportunities

Edited by

Stephen P. Arneric
DuPont/Merck
Wilmington, DE

Jorge D. Brioni
Abbott Laboratories
Abbott Park, IL

A John Wiley & Sons, Inc., Publication

NEW YORK • CHICHESTER • WEINHEIM • BRISBANE • SINGAPORE • TORONTO

NORTHUMBRIA UNIVERSITY
LIBRARY
ITEM No. 200573563 9
CLASS No. 615.78 NEU

This book is printed on acid-free paper. ♾

Copyright © 1999 by Wiley-Liss, Inc. All rights reserved.

Published simultaneously in Canada.

No part of this publication may be reproduced, stored in a retrieval system or transmitted in any form or by any means, electronic, mechanical, photocopying, recording, scanning or otherwise, except as permitted under Sections 107 or 108 of the 1976 United States Copyright Act, without either the prior written permission of the Publisher, or authorization through payment of the appropriate per-copy fee to the Copyright Clearance Center, 222 Rosewood Drive, Danvers, MA 01923, (978) 750-8400, fax (978) 750-4744. Requests to the Publisher for permission should be addressed to the Permissions Department, John Wiley & Sons, Inc., 605 Third Avenue, New York, NY 10158-0012, (212) 850-6011, fax (212) 850-6008, E-Mail: PERMREQ@WILEY.COM.

While the authors, editors and publisher believe that drug selection and dosage and the specification and usage of equipment and devices, as set forth in this book, are in accord with current recommendations and practice at the time of publication, they accept no legal responsibility for any errors or omissions, and make no warranty, expressed or implied, with respect to material contained herein. In view of ongoing research, equipment modifications, changes in governmental regulations and the constant flow of information relating to drug therapy, drug reactions, and the use of equipment and devices, the reader is urged to review and evaluate the information provided in the package insert or instructions for each drug, piece of equipment, or device for, among other things, any changes in instructions or indication of dosage or usage and for added warnings and precautions.

Library of Congress Cataloging-in-Publication Data:

Neuronal nicotinic receptors : pharmacology and therapeutic
opportunities / edited by Stephen P. Arneric, Jorge D. Brioni.
p. cm.
Includes index.
"A Wiley-Liss publication."
ISBN 0-471-24743-X (cloth : alk. paper)
1. Nicotinic receptors. 2. Nicotine—Therapeutic use.
3. Nicotine—Physiological effect. I. Arneric, Stephen Peter.
II. Brioni, Jorge D., 1956– .
[DNLM: 1. Receptors, Nicotinic—physiology. 2. Nicotine—
pharmacology. 3. Nicotine—therapeutic use. WL 102.8 N49414
1999]
QP364.7.N443 1999
615'.78—dc21
DNLM/DLC
for Library of Congress 98-17280

10 9 8 7 6 5 4 3 2 1

Contents

Preface

The field of nicotinic acetylcholine receptor (nAChR) research has focused for several decades on understanding the actions of nicotine on smokers as well as the development of an efficacious smoking cessation treatment. Although several beneficial actions of nicotine were known from anecdotal evidence as well as from documented scientific literature (e.g., analgesia, memory enhancement, anxiolysis, etc.), these positive effects were generally overlooked as a result of the negative consequences of smoking, and the perceived inability to separate the beneficial and detrimental aspects of nicotine. A pervasive mindset has been: How could something good come from an addictive drug? As a result, medicinal chemistry efforts and drug discovery focus in this area has been slow to emerge. This is despite the fact that acetylcholine and its postsynaptic receptor was the first substance to be recognized as a bonafide neutransmitter and the first receptor to be cloned. This book highlights comments on why this has been the case and why there has been a dramatic movement toward the development of novel therapeutics from this rich area of pharmacology.

Curiously, this conservative view of nAChR pharmacology can be contrasted with other, highly productive areas of drug discovery where potent toxins and hallucinogenic substances have been used as the starting point for novel drug programs. A case in point is that of the hallucinogen, lysergic acid diethylamide (LSD), which produces its potent central nervous system effects by interacting with central 5-HT receptors. This latter group of receptors currently consists of at least 15 subtypes that delineate different physiological and therapeutic function. The antidepressant, fluoxetine (Prozac), the antiobesity agent dexfenfluramine, the antinauseant ondansetron, and the antimigraine agent, sumitriptan, all produce their effects on human system function via the direct or indirect modulation of 5-HT receptor function. Yet none of these compounds is viewed as being "LSD-like" as their effects have been delineated to a distinct receptor subtype. Similarly, no one gives a second thought to receiving an injection of a local anesthetic (e.g., lidocaine, procaine) at the dentist, which is a derivative of the first-effect local anesthetic, cocaine. The faith that the addictive properties of cocaine have been removed from the local anesthetics is unquestioned, and the thought of being "hooked" on such agents is nonexistent. So why is this not the case for nicotine? Is it the simplistic reason that nicotine, and the ligand-gated ion channel (LGIC) superfamily with which it shares a name, draws a constant reminder of the negative consequences of tobacco consumption? If this is not the case, would those who suffer from depression and migraine have been equally eager to try an LSD receptor agonist?

More recently there has been a surge of scientific and potential therapeutic advances that may have a profound impact of how this area of research is viewed and how it will affect the medical community in the future. This revolution began with the increased technological advances in the molecular biology and electrophysiology of the neuronal nAChRs. Several homologous gene products related to the different α, β, δ, and γ subunits of the nAChR were identified, and the electrophysiological evaluation of the subunits was carried out in oocytes and in different cell lines. The stable expression of these subunits in mammalian cell lines was also accomplished by several laboratories around the world. The physiological relevancy of these putative nAChR subtypes is still under scrutiny, but we are hopeful that in the near future a specific composition of these subtypes will be determined in peripheral tissue, ganglia, and brain. In the meantime, several novel nAChR ligands such as epibatidine have been discovered and these agents show a unique pharmacological profile that differs from nicotine itself in whole animal studies.

This book was organized to review the most significant developments in the pharmacology of nAChRs and their respective ligands. The chapters represent authoritative updates written by the world's leading researchers in their areas of nAChR biology and chemistry. The reader will find that Parts I, II, and III summarize the latest developments in these exciting fields. Part IV reviews the present understanding of the nicotinic pharmacophore and the impact of the potent nicotinic ligand epibatidine. What clearly delineates this book from others is Part V, which highlights advances leading to future medical treatments. The preclinical pharmacology of the novel nAChR ligands showing facilitation of specific behavioral parameters without exerting the negative effects of nicotine (a nonselective activator of several putative nAChRs) has provided a scientific rationale to advance these agents to the clinical area.

We would like to thank Mike Williams for his guidance and persistent encouragement to pursue this endeavor. He recognized early on that such an undertaking would be a step toward satisfying the unmet need to harmonize and synergize the rapidly evolving information that could form the basis for therapeutic opportunities in this field. The editors also extend their sincere thanks to the authors for their scholarly contributions and patient flexibility. We appreciate the excellent editorial assistance and unyielding professionalism provided by Joe Ingram, Fiona Stevens, and their collaborators at Wiley during the editorial process.

The editors would like to dedicate this work to some very special people in our lives. Jorge Brioni dedicates this book to his daughters Marina Beatriz, Florencia Raquel, and Carla Agostina, well aware that this book is not as exciting as *Amelia Bedelia* or *Henry and Mudge,* but certain that this is the best that Dad can do. Stephen Arneric dedicates this book to his wife, Amy Lou, daughters Katarina Lyn and Elizabeth Leanne, and parents, Inge and Raphael, for their unconditional love, devotion, and patience.

STEPHEN P. ARNERIC
JORGE D. BRIONI

Senior Authors

EDSON X. ALBUQUERQUE, Department of Pharmacology and Experimental Therapeutics, University of Maryland, 655 West Baltimore Street, Baltimore, MD 21201-1559, Ph. (410) 706-7330, Fax (410) 706-3991

STEPHEN P. ARNERIC, DuPont/Merck, CNS Diseases Research, Experimental Station, E400/4416, Rts. 141 & Henry Clay Road, Wilmington, DE 19880-0400, Ph. (302) 695-8745, Fax (302) 695-3730

NEAL BENOWITZ, Division of Clinical Pharmacology and Experimental Therapeutics, University of California, San Francisco, Box 1220, San Francisco, CA 94193-1220, Ph. (415) 206-8324, Fax (415) 206-8949

DARWIN BERG, Department of Biology, University of California, 9500 Gilman Drive, La Jolla, CA 92093-0322, Ph. (619) 534-4680, Fax (619) 534-0301

DANIEL BERTRAND, Department of Physiology, C.M.U., 1, rue, Michel-Servet, CH-1211 Geneva 4, Switzerland, Ph. 41-22.702.53.56, Fax 41-22.702.54.02

PAUL B.S. CLARKE, Department of Pharmacology and Therapeutics, McGill University, McIntire Medical Sciences, 3655 Drummond Street, Montreal, QC H3CG 1Y6, Canada, Ph. (514) 398-3616, Fax (514) 398-6690

FRANCISCO CLEMENTI, Cattedra di Farmacologia, Universita Degli Studi di Milano, 20129 Milano Italy, Ph. 39-2.701.46.262, Fax 39-2.749.90.574

DIANA DONNELLY-ROBERTS, Neurological and Urological Diseases Research, Department 47W, Building AP10-LL, Abbott Laboratories, 100 Abbott Park Road, Abbott Park, IL 60064-3500, Ph. (847) 938-1422, Fax (847) 937-9195

CHRISTOPHER FLORES, Department of Endodontics and Pharmacology, University of Texas Health Science Center, San Antonio, Texas 78284, Ph. (210) 567-3381, Fax (210) 567-3389

RICHARD GLENNON, Department of Pharmacology and Toxicology, MCV Campus, Virginia Commonwealth University, 410 N. 12th Street, PO Box 980613, Richmond, VA 23298-0613, Ph. (804) 828-8487, Fax (804) 828-7625

MARK HOLLADAY, Neurological and Urological Diseases Research, Department 47C, Building AP10-LL, Abbott Laboratories, 100 Abbott Park Road, Abbott Park, IL 60064-3500, Ph. (847) 937-0740, Fax (847) 937-9195

SCOTT LEISCHOW, University of Arizona, Arizona Program for Nicotine and Tobacco Research, 1145 N. Campbell Avenue, PO Box 210228, Tucson, Arizona 85719, Ph. (520) 318-7151, Fax (520) 318-7155

SHERRY S. LEONARD, Department of Psychiatry, University of Colorado Health Sciences Center, C-268-71, 4200 E. 9th Avenue, Denver, CO 80262, Ph. (303) 399-8020, Fax (303) 331-8324

ED LEVIN, Neurobehavioral Research Laboratories, Box 3412, Duke University Medical Center, Durham, NC 27710, Ph. (919) 681-6273, Fax (919) 681-3416

JON LINDSTROM, Institute of Neurological Science, University of Pennsylvania, 217 Stemmler Hall, 36th and Hamilton Walk, Philadelphia, PA 19104-6074, Ph. (215) 573-2859, Fax (215) 573-2015

KEN LLOYD, Vice-President Pharmaceuticals, SIBIA, 505 Coast Boulevard South, Suite 300, La Jolla, CA 92037-4641, Ph. (619) 452-5892, Fax (619) 452-9279

RON LUKAS, Division of Neurobiology, Barrow Neurological Institute, 350 W. Thomas Road, Phoenix, AZ 85013, Ph. (602) 406-3399, Fax (602) 406-4172

MICHAEL MARKS, Institute for Behavioral Genetics, University of Colorado, Campus Box 447, Boulder, CO 80309, Ph. (303) 492-8844, Fax (303) 492-8063

LORNA ROLE, Department of Cell Biology, Center for Neurobiology and Behavior, Columbia Plaza Annex, Room 807, 722 168th Street, New York, NY 10032, Ph. (212) 305-8117, Fax (212) 795-0700

MOHAN SOPORI, Pathophysiology Division, Lovelace Respiratory Research Institute, PO Box 5890, Albuquerque, NM 87185, Ph. (505) 845-1115, Fax (505) 845-1198

VICTOR VILLEMAGNE, Division of Nuclear Medicine, Department of Radiology, Johns Hopkins Medical Institutions, Baltimore, MD 21287, Ph. (410) 955-4138, Fax (410) 955-0691

SUSAN WONNACOTT, Department of Biochemistry, University of Bath, Bath BA2 7AY, United Kingdom, Ph. 44-122.58.26.391, Fax 44-122.58.26.449

Neuronal Nicotinic Receptors

Part I

Molecular Biology and Biochemistry

1

Purification and Cloning of Nicotinic Acetylcholine Receptors

Jon Lindstrom, Ph.D
Medical School of the University of Pennsylvania
Philadelphia, Pennsylvania

This brief introductory chapter is intended to provide a historical context and overview for the subsequent chapters on detailed structure, function, regulation, and pharmacology. It offers a general perspective from which the other specific material may be viewed.

ACETYLCHOLINE RECEPTORS OF MUSCLE AND ELECTRIC ORGANS

Muscle nAChRs

Nicotinic acetylcholine receptors (nAChRs) of muscle became the first and best characterized neurotransmitter receptors in terms of functional role and detailed electrophysiological characterization because skeletal muscle cells provided an easily accessible, uniform source of experimentally tractable tissue. The discovery that cobra and krait venom toxins such as α-bungarotoxin (α-Bgt) bound with very high affinity and specificity to the ACh binding sites of nAChRs permitted the use of labeled toxins to locate, quantitate, and affinity purify nAChRs (Lee et al., 1967; Changeux, 1990). Later, subunit-specific monoclonal antibodies (mAbs) were used in combination first with ^{125}I α-Bgt-labeled native nAChRs and subsequently with expressed cloned nAChR subunits. These provided critical tools for studying the synthesis, con-

Neuronal Nicotinic Receptors: Pharmacology and Therapeutic Opportunities, Edited by S. P. Arneric and J. D. Brioni
ISBN 0-471-24743-x, pages 3–23. Copyright © 1998 by Wiley-Liss, Inc.

formational maturation, and assembly of nAChR subunits and for studying developmental changes in muscle nAChR subunit composition, amount, and localization before innervation of fetal muscle and after denervation of adult muscle (Tzartos and Lindstrom, 1980; Lindstrom, 1985; Merlie et al., 1983b; Blount and Merlie, 1989; Wang et al., 1996).

Electric Organ nAChR Biochemistry

Electric organs, first from electric eels (*Electrophorus electricus*) and later from rays (*Torpedo californica* or *marmorata*), provided rich sources of muscle-type nAChRs. nAChRs from electric organs were the first affinity labeled (Karlin, 1969; Karlin and Couburn, 1973); the first affinity purified (Lindstrom and Patrick, 1974); the first partially protein sequenced (Raftery et al., 1980); the first whose cDNAs were cloned (Ballivet et al., 1982, Noda et al., 1982); and the first, and thus far the only, to be structurally characterized by a diffraction technique (Unwin, 1993, 1995). Thus, electric organ nAChRs have provided the archetype to which we can compare not only neuronal nAChRs but also those receptors in the gene superfamily that share homologous structures with nAChRs ($GABA_A$, glycine, and $5HT_3$ receptors) (Karlin and Akabas, 1995).

Affinity labeling studies by Karlin and co-workers in the 1960s initially identified a nAChR subunit that would later be termed $\alpha 1$ (Karlin and Couburn, 1973). Much later, when the amino acid sequence of α was known, the identification of the site of reaction of these affinity labeling reagents to a unique cysteine pair in positions 192 and 193 localized residues near the ACh binding site (Kao and Karlin, 1986). Subsequent photoaffinity labeling and mutagenesis studies by Changeux and others showed that amino acids from several parts of the N-terminal domains of α subunits also contributed to the formation of this site (Galzi et al., 1990; Bertrand and Changeux, 1995). Recently, Karlin and co-workers have used bifunctional affinity labeling reagents anchored at one end by cysteines 192 and 193 to reach out and label amino acids on adjacent subunits that provide a critical anionic component of the mature ACh binding site now known to be formed at the interface between α and adjacent subunits (Czajkowski and Karlin, 1995; Martin et al., 1996).

The initial affinity purification of electric organ nAChRs in the early 1970s had two immediate effects. One was the realization that nAChRs contained several kinds of subunits (Lindstrom and Patrick, 1974). Ultimately the four kinds found in electric organ nAChRs were termed α, β, γ, and δ in order of their increasing molecular weights (Karlin, 1991; Changeux, 1990). Four subunits were also found in muscle nAChRs. Later it was found that the γ subunits of fetal or denervated muscle were replaced in innervated adult muscle by ϵ subunits (Witzemann et al., 1987), and it was realized that the γ subunits of adult electric organs were more closely related to the ϵ subunits of muscle nAChRs (Nelson et al., 1992). The second immediate effect of purification of milligram amounts of nAChR was the realization that the disease myasthenia gravis (MG) was caused by an autoimmune response to muscle nAChRs (Patrick and Lindstrom, 1973). Immunization of animals with purified electric organ nAChRs resulted in an antibody-mediated autoimmune response that impaired neuromuscular transmission by reducing the number of functional nAChRs. These studies led not only to improved understanding of the pathological mechanisms in MG but also to an immunodiagnostic assay, an emphasis on immunosuppressive therapy, and the development of large libraries of mAbs to nAChR subunits for use both as model autoantibodies and as structural probes (Lindstrom et al., 1988; Lindstrom, 1996; Lindstrom et al., 1999).

The availability of milligram amounts of electric organ nAChR and the developing technology for sequencing small amounts of protein led Raftery and co-workers to sequence the N-termini of the subunits of electric organ nAChRs (Raftery et al., 1980). This confirmed other biochemical data that the subunit stoichiometry was $\alpha_2\beta\gamma\delta$ (Reynolds and Karlin, 1978; Lindstrom et al., 1979) and revealed hints of sequence homologies between the subunits. Finally, the amino acid sequence data provided the critical information necessary for designing oligonucleotide probes when developments in molecular biology techniques shortly thereafter permitted the first cloning of cDNAs for nAChR subunits.

Electric Organ nAChR cDNAs

Numa and co-workers had a revolutionary effect by applying cDNA cloning technology to nAChRs and to several other proteins whose biochemical characterization had matured to the point that they were ripe for application of the new molecular biological techniques (Noda et al., 1982, 1983a,b, 1984, 1986). Initial cloning of cDNAs for subunits of electric organ and muscle nAChR subunits provided dramatic new insights into nAChR structure (Sumikawa et al., 1982; Devillers-Thiery et al., 1983; Claudio et al., 1983; Merlie et al., 1983a). Immediately, this confirmed the homologies between the structures of all of the nAChR subunits. This implied both that all of the subunits had evolved as a gene family from a common precursor and that the nAChR should have a basic pentagonal symmetry. Immediately, the availability of nAChR subunit cDNA sequences resulted in a model for the basic transmembrane orientation of the subunit polypeptide chains, which has largely withstood the test of time (Fig. 1-1). Each subunit was found to have an N-terminal leader sequence that was cleaved cotranslationally. It was proposed that an N-terminal extracellular domain comprised nearly half the subunit (about 210 amino acids of more than 400). This was followed by three closely spaced transmembrane domains (M1–M3), a large cytoplasmic domain, and a fourth transmembrane domain (M4) leading back across the membrane to a short extracellular C-terminal tail. Later, when the cDNA sequences of other neurotransmitter-gated ion channels became available, it was realized that nAChRs were part of a gene superfamily of homologous subunits (Barnard et al., 1987; Betz, 1990; Karlin and Akabas, 1995). Much later, elegant demonstrations of the fundamental homologies between the structures of these receptors were provided by Changeux and co-workers. They showed that the extracellular domain of a nAChR $\alpha 7$ subunit could form a chimera with a $5HT_3$ receptor, resulting in an ACh-gated receptor with the channel properties of a $5HT_3$ receptor (Eisile et al., 1993). They also showed that replacement of as few as three amino acids from the lining of the cation-specific channel lining domain of an nAChR $\alpha 7$ subunit with corresponding amino acids from the anion-specific $GABA_A$ or glycine receptors could convert this excitatory nAChR with a cation-selective channel into an inhibitory nAChR with an anion-selective channel.

Electric Organ nAChR Electron Microscopy

Unwin and co-workers, as well as Stroud and co-workers, have provided much of the information about the basic shape of nAChRs through electron microscopic studies of electric organ nAChRs (Stroud et al., 1990; Unwin, 1993, 1995; Beroukhim and Unwin, 1995). These studies required neither purification nor cloning, only the presence of membrane fragments solidly packed with nAChRs. Amazingly, the nAChRs in some of these vesicles would, during storage over a few weeks, arrange into two-dimensional helical crystalline arrays. These two-dimensional crystalline arrays, coupled with technical and analytical developments that permitted diffraction analysis of their electron micrographs, resulted in electron density maps of nAChRs to a resolution of about 9 Å (Unwin, 1993). This provided a basic picture of the characteristic shape of nAChRs. At this resolution all the receptors in the superfamily would probably look fairly similar. This resolution is insufficient to resolve α helices, much less in-

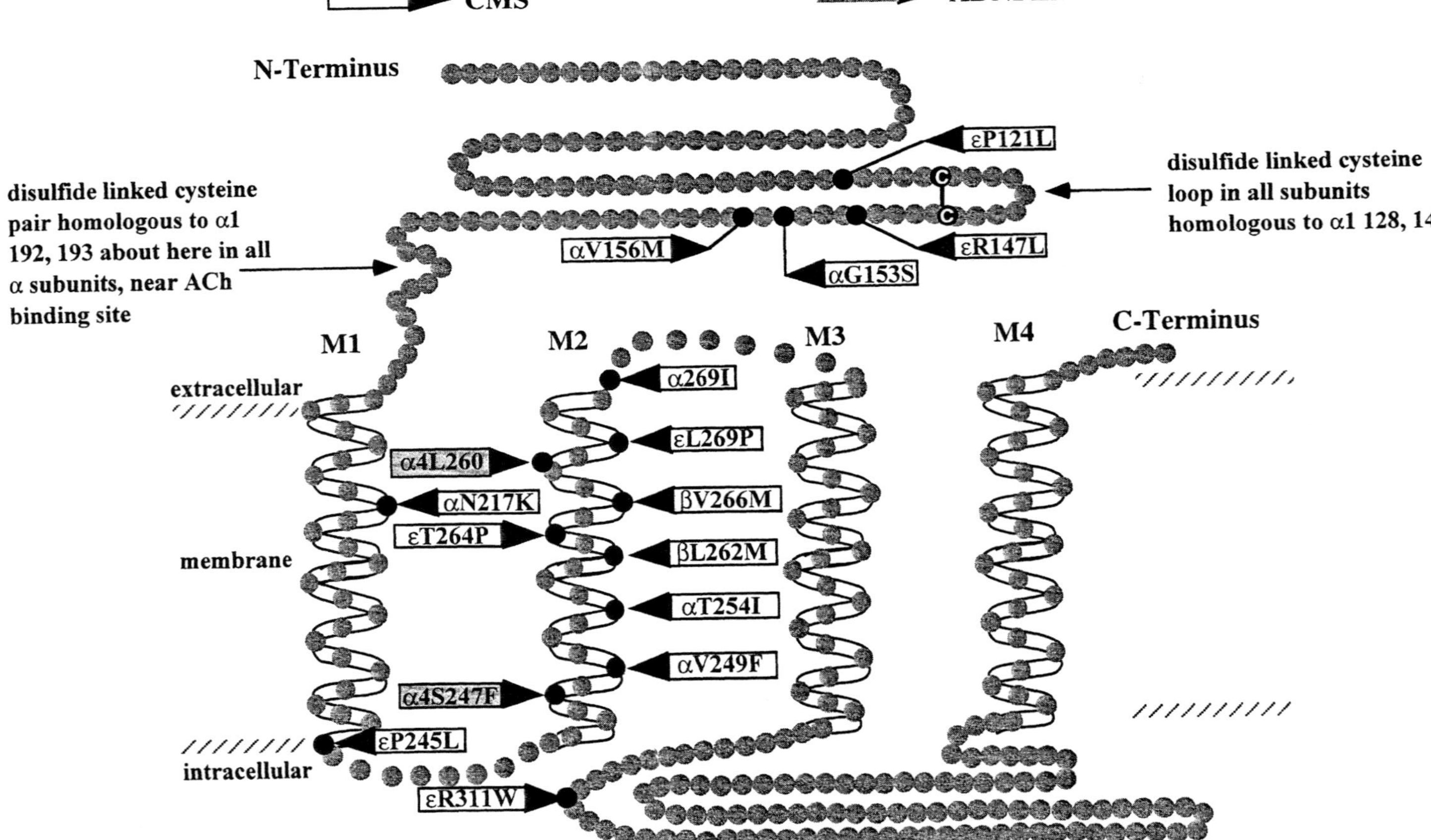
CMS
ADNFLE
N-Terminus
εP121L
disulfide linked cysteine loop in all subunits homologous to α1 128, 142
disulfide linked cysteine pair homologous to α1 192, 193 about here in all α subunits, near ACh binding site
αV156M
εR147L
αG153S
M1
M2
M3
M4
C-Terminus
extracellular
α269I
εL269P
α4L260
βV266M
αN217K
εT264P
βL262M
membrane
αT254I
αV249F
α4S247F
εP245L
intracellular
εR311W

dividual amino acids. It does, however, reveal (when viewed from the top) an 80 Å diameter pentagonal structure surrounding a 25 Å diameter vestibule to the channel that narrows at the level of the membrane. A side view of this cylindrical structure shows the vestibule extending 60 Å above the extracellular surface, another 40 Å through the membrane, and then flaring 15–20 Å on the cytoplasmic surface. This morphological evidence, considered along with cDNA sequence data, affinity labeling studies of the ACh binding site and channel, as well as extensive mutational analysis, has led to a generally accepted model of nAChR structure (Stroud et al., 1990; Unwin, 1993; Changeux, 1990; Karlin and Akabas, 1995). The subunits are thought to be rodlike and organized like barrel staves around the central channel. The extracellular domain forming the channel vestibule and ACh binding sites is thought to be formed from the N-terminal domains of each subunit. It contains glycosylation sites. It also contains a disulfide-linked loop between cysteines homologous to positions 128 and 142 on α subunits. This loop is characteristic of all subunits of receptors in the superfamily, but its function is uncertain. The ion channel is thought to be lined by amino acids from the M2 and top third of the M1 transmembrane domains of each subunit. The M3 and M4 transmembrane domains are thought to contribute to the hydrophobic core of the protein and to interact with the surrounding lipid bilayer. The large cytoplasmic domain formed by the sequence between M3 and M4 is thought to provide regulatory sites for phosphorylation. The cytoplasmic domain also provides anchoring sites for rapsyn, a 43 kDa peripheral membrane protein that links muscle nAChRs to the cytoskeleton as is required for positioning them at their sites of action.

NEURONAL NICOTINIC ACETYLCHOLINE RECEPTORS

Initial Studies

Cloning and purifying neuronal nAChRs proceeded in parallel using reagents initially developed for studies of muscle nAChRs. Low-stringency hybridization with probes for muscle α1 subunit cDNAs led to the identification of a neuronal nAChR subunit cDNA that would ultimately be termed α3 (Boulter et al., 1986). Subsequent waves of low-stringency hybridization screens of libraries from rat and chicken sources led to the recognition of other homologues (Heinemann et al., 1991). Later, several groups cloned human neuronal nAChR subunits (Lindstrom et al., 1996; Gotti et al., 1997). Immunoisolation of chick brain nAChRs using a mAb to the main immunogenic region on α1 subunits led to the purification of nAChRs containing only two kinds of subunits (Whiting and Lindstrom, 1986). Successive libraries of mAbs made against neuronal nAChRs purified from chickens and rats led to mAb probes for a number of nAChR subunits (Whiting and Lindstrom, 1987, 1988; Whiting et al., 1987b, 1991; Lindstrom, 1996). N-terminal amino acid sequencing of the subunits of one subtype of neuronal nAChR

Figure 1.1. Transmembrane orientation of nAChR subunit polypeptide chains. The transmembrane orientation of a generic subunit polypeptide chain is depicted showing a large N-terminal extracellular domain followed by three closely spaced transmembrane domains termed M1, M2, M3, then a large cytoplasmic domain, followed by the fourth transmembrane domain termed M4, and a short C-terminal extracellular domain. The transmembrane domains are depicted as being α helical, but this may not be strictly true. M2 and part of M1 contribute to the lining of the cation channel. Also shown are the approximate locations of some of the mutations in muscle nAChR α1, β, and ε subunits, which have been found to be responsible for various congenital myasthenic syndromes (CMS). In addition, the approximate positions are shown of two mutations in neuronal nAChR α4 subunits responsible for forms of autosomal-dominant nocturnal frontal lobe epilepsy (ADNFLE). The notation for the mutations indicates first the subunit involved, then the native amino acid (in the case of an altered amino acid), then the position of that amino acid in the native subunit, and finally the amino acid present at that position in the mutant subunit.

identified the subunits as corresponding to cDNAs now termed α4 and β2 and identified the actual N-termini of these subunits (Whiting et al., 1987a; Schoepfer et al., 1988). Low-stringency hybridization with oligonucleotide probes based on partial amino acid sequence data from α-Bgt binding nAChRs (Conti-Tronconi et al., 1985) revealed new nAChR subunit homologues from chicken brain (Schoepfer et al., 1990). mAbs made to bacterially expressed putative cytoplasmic domains from these subunits were shown to bind neuronal nAChRs, which bound α-Bgt and are now termed α7 and α8 (Schoepfer et al., 1990; Keyser et al., 1993). The search for an α8 homologue in rats did not find a mammalian α8 but revealed what is now termed α9 (Elgoyhen et al., 1994).

Subunit Nomenclature

Because cloning and purification of neuronal nAChR subunits proceeded rapidly and in parallel in several laboratories, the nomenclature for the various subunits was initially conflicting. The laboratories of Jim Patrick and Steve Heinemann studied rat nAChRs and numbered their cDNAs in order of discovery (Heinemann et al., 1991). Any cDNAs that contained homologues of the cysteine pair homologous to positions 192 and 193 of muscle α subunits, which the pioneering affinity labeling studies of Arthur Karlin had shown were near the ACh binding site, were termed α subunits. Muscle subunits became α1 subunits, and ultimately neuronal subunits α2 through α9 were reported. Homologous neuronal subunits lacking this cysteine pair were termed β subunits because it was initially demonstrated that neuronal β2 subunits could substitute for muscle β subunits in nAChRs expressed in *Xenopus* oocytes (Deneris et al., 1988). Muscle β subunits became β1 subunits, and ultimately neuronal β2 through β4 subunits were identified. Mark Ballivet studied chicken nAChRs and initially termed the subunits for which he cloned cDNAs either α or non-α (Couturier et al., 1990b). My laboratory initially used the convention for naming purified neuronal nAChR subunits that had been employed for naming muscle nAChR subunits: They were termed α and β in order of increasing molecular weight. Also, they were termed "ACh binding subunits" if they could be affinity labeled with Karlin's ACh binding site reagents or "structural subunits" if they could not. This ran into conflict with the cDNA naming scheme when α4 subunits turned out to have a higher molecular weight (75 kDa) than β2 subunits (52 kDa). The N-terminal amino acid sequencing experiments for α4 and β2 subunits sorted this out (Whiting et al., 1987a; Schoepfer et al., 1988). When we initially cloned what are now termed α7 and α8 cDNAs, we termed them "α-Bgt binding proteins 1 and 2" (Schoepfer et al., 1990). Shortly thereafter, Ballivet's lab also cloned α7 and showed that it formed functional homomeric nAChRs (Couturier et al., 1990a). We quickly switched to that nomenclature.

Now a standard nomenclature is in use for the α1, β1, γ, δ, and ϵ subunits of muscle type nAChRs and the α2–α9, β2–β4 subunits of neuronal nAChRs. Often the nAChR gene family is thought to consist of three branches: (1) muscle and electric organ nAChRs (which bind α-Bgt) and are composed of α1, β1, γ or ϵ, and δ subunits; (2) neuronal nAChRs (which do not bind α-Bgt) and are composed of mixtures of α2–α6 subunits with β2–β4 subunits in combinations as simple as α4 and β2 or as complex as α3, β2, β4, and α5; or (3) neuronal nAChRs (which bind α-Bgt) and are composed of α7, α8, or α9 subunits, usually as homomers, though α7α8 heteromers are known and other heteromers may exist. Subunit composition and organization determine all of the properties of an nAChR, except as they might be modified by post-translational events such as glycosylation, covalently attaching lipids, phosphorylation, transport to different membrane regions, or binding to linker proteins associated with the cytoskeleton. The standard nomenclature can conveniently and intuitively convey all that is known about a particular subtype of nAChR by expressing the kinds of subunits present (e.g., α1 nAChR or α1β1εδ nAChR), their stoichiometric ratio (e.g., $(\alpha 1)_2$ β1εδ nAChR), or their stoichiometry and organization around the channel (e.g., α1εα1δβ1 nAChR). This is a vastly more useful nomenclature for defining what is known about nAChR subtypes now or is likely to be learned in the future than is a nomenclature that is used for metabotropic receptors such as muscarinic acetylcholine receptors. Muscarinic receptors have only a single subunit, so the designation M1, M2, . . . tells the defining characteristics of that receptor. A N1, N2, . . . nomenclature would have to number into the thousands to account for all of the combinations and permutations of neurons subunits theoretically possible in neuronal nAChRs, and the names N_{603} or N_{26} would contain no intrinsic information about the nAChRs involved.

nAChR Subtypes

Fortunately, the many potential nAChR subtypes are not found in equal abundance. Thus, muscle-type nAChRs consist of only two subtypes, either fetal α1β1γδ nAChRs or adult α1β1εδ nAChRs. Neuronal nAChRs that do not bind α-Bgt have several subtypes. One set of these subtypes is α3 subunit-containing nAChRs. These are predominant as postsynaptic ganglionic nAChRs in the autonomic nervous system (Vernalis et al., 1993; Heinemann et al., 1991; Sargent, 1993; Role and Berg, 1996). α3 nAChRs are less abundant in the brain than are α4β2 nAChRs. α4β2 nAChRs account for the predominant subtype of brain nAChR with high affinity for (-)-nicotine (Whiting and Lindstrom, 1988; Flores et al., 1992). Neuronal nAChRs that bind α-Bgt consist primarily of α7 subunits in mammals (Del Toro et al., 1994), with α9 being found in limited regions (Elgoyhen et al., 1994). In chickens, α7 predominates in most areas of the brain or autonomic nervous system (Britto et al., 1994; Horch and Sargent, 1995), but α8 predominates in the retina (Keyser et al., 1993; Hamassaki-Britto et al., 1994).

Frequently neurons express a mixture of neuronal nAChR subtypes. For example, ciliary ganglion neurons (Vernalis et al., 1993) and a variety of neuroblastoma cells express both a mixture of α3 nAChR subtypes and α7 nAChRs (Lukas et al., 1993; Gotti et al., 1997). Elegant studies of ciliary ganglia have revealed one of the complexities of neuronal cholinergic transmission. These neurons assemble a mixture of α3-containing nAChRs and arrange them both postsynaptically and perisynaptically (Vernallis et al., 1993; Horch and Sargent, 1995). They also assemble α7 nAChRs and arrange them only perisynaptically. Either the α3 nAChRs or the α7 nAChRs are sufficient to ensure transmission at ciliary ganglion synapses (Zhang et al., 1996; Ullian et al., 1997).

Neuronal nAChRs in the brain frequently appear to have presynaptic roles where they can modulate the release of various transmitters (Wonnacott, 1997; Coggan et al., 1997). No particular subtypes have yet emerged as having exclusively presynaptic or postsynaptic roles. For example, α7 nAChRs have been found in both roles (Gray et al., 1996; Zhang et al., 1996; Albuquerque et al., 1996; Ullian et al., 1997).

nAChRs serve many functional roles. No doubt many of these remain to be determined. In addition to postsynaptic nAChRs (e.g., α1 nAChRs in muscle and α3 nAChRs in ciliary ganglia [Jacob et al., 1984; Horch and Sargent, 1995]), perisynaptic nAChRs (e.g., α7 nAChRs in ciliary ganglia [Jacob and Berg, 1983; Horch and Sargent, 1995; Zhang et al., 1996; Ullian et al., 1997]), and presynaptic nAChRs (e.g., α7 nAChRs in hippocampus (Gray et al., 1996]), there are also extrasynaptic nAChRs (e.g., α7 nAChRs in muscle and tendon transiently during development [Romano et al., 1997a,b] or α3 nAChRs in keratinocytes [Grando et al., 1995, 1996]).

For pharmaceutical purposes some of the less common subtypes of neuronal nAChRs that are expressed in the most limited regions may provide the most precise pharmacological targets. For example, α6 subunits were cloned early on (Heinemann et al., 1991), but remained an orphan devoid of demonstrable function until quite recently. Although very close to $\alpha3$ subunits in sequence, $\alpha6$ subunits result in nAChRs with unique pharmacological properties, at least when cloned nAChRs are expressed in *Xenopus* oocytes (Gerzanich et al., 1997). The real subunit composition of $\alpha6$ nAChRs remains to be determined. $\alpha6$ has been found in only very limited brain regions including the substantia nigra and the ventral tegmental area (LeNovere et al., 1996; Goldner et al., 1997). However, these are intrinsically interesting areas because the dopamine-containing neurons of the substantia nigra are damaged in Parkinson's disease, resulting in nAChR loss (Whitehouse et al., 1988), and also because presynaptic nAChRs are known to facilitate dopamine release (Wonnacott et al., 1997).

Muscle nAChRs provide the model for understanding nAChR subunit stoichiometry, ordering, and conformational changes. Figure 1-2 diagramatically represents a top view of subunit organization around the central channel in major nAChR subtypes. The subunit stoichiometry of muscle type nAChRs is $(\alpha1)_2\ \beta1\gamma\delta$ or $(\alpha1)_2\beta1\epsilon\delta$ (Karlin and Akabas, 1995). The subunits are thought to be oriented around the central cation channel in the order $\alpha1\gamma\alpha1\delta\beta1$. The ACh binding sites formed at the interfaces of one side ("positive") of $\alpha1$ subunits and the other ("negative") side of γ, δ, or ϵ subunits (Tsigelny et al., 1997). $\beta1$ subunits are thought to interact only with the "negative" side of $\alpha1$ subunits. Thus, both $\alpha1$ and γ, δ, or ϵ subunits contribute amino acids to the formation of the ACh binding site. It is thought that, although $\beta1$ subunits do not contribute contact amino acids to the formation of an ACh binding site, they can influence the concentration dependence and kinetics of agonist activation of nAChRs. This is because all of the subunits contribute equally to the lining of the cation channel with their M2 and part of their M1 transmembrane domains, and because all of the subunits probably contribute to the concerted conformational changes involved in nAChR activation and desensitization (Karlin and Akabas, 1995). These agonist-induced conformation changes probably involve subtle iris-like sliding motions of adjacent subunits and transmembrane domains (Karlin and Akabas, 1995; Unwin, 1995). Presumably the location of agonist binding sites at subunit interfaces serves to facilitate conformation changes involved in gating and desensitization, which may involve movement along these interfaces.

Much less is known about the subunit composition and stoichiometry of neuronal nAChRs, which do not bind α-Bgt. The stoichiometry of $\alpha4\beta2$ nAChRs expressed in *Xenopus* oocytes is $(\alpha4)_2(\beta2)_3$ (Anand et al., 1991; Cooper et al., 1991). It is presumed that the order of the subunits is $\alpha4\beta2\alpha4\beta2\beta2$, resulting in two identical ACh binding sites at the appropriate interfaces between $\alpha4$ and $\beta2$ subunits. A small fraction of brain $\alpha4\beta2$ nAChRs also contains an $\alpha5$ subunit that alters the properties of these nAChRs (Conroy et al., 1992; Ramirez-Latorre et al., 1996; Gerzanich et al., 1998). A similar situation occurs with $\alpha3$ nAChRs (Wang et al., 1996). $\alpha3$ can form functional nAChRs in combination with $\beta2$ or $\beta4$ nAChRs. $\alpha5$ cannot form functional nAChRs in combination with $\beta2$ or $\beta4$ nAChRs, but can assemble efficiently with $\alpha3$ and $\beta2$ or $\alpha3$ and $\beta4$. Ciliary ganglion neurons express $\alpha3$, $\alpha5$, $\alpha7$, $\beta2$, and $\beta4$ subunits. These neurons assemble about 80% of their $\alpha3$ nAChRs with the subunit composition $\alpha3\beta4\alpha5$ and about 20% with the subunit composition $\alpha3\beta2\beta4\alpha5$, while assembling $\alpha7$ as homomers (Vernalis et al., 1993). It seems likely that $\alpha5$ subunits assemble in the position equivalent to $\beta1$ subunits such that they do not interface with the "positive" side of $\alpha3$ subunits that can form an ACh binding site. From such a position $\alpha5$ subunits could contribute to the lining of the cation channel and could affect the agonist sensitivity by influencing the concerted conformational changes of the nAChR protein involved in activation and desensitization. In fact, assembly of $\alpha5$ subunits with $\alpha3\beta2$ or $\alpha3\beta4$ nAChRs increases their Ca^{++} permeabilities and rates of desensitization (Wang et al., 1996; Gerzanich et al., 1998). $\alpha5$ increases the ACh sensitivity of $\alpha3\beta2$ nAChRs 50-fold, while having little effect on $\alpha3\beta4$ nAChRs. $\alpha5$ increases the efficacies of (-)-nicotine and DMPP on $\alpha3\beta2$ nAChRs and decreases them on $\alpha3\beta4$

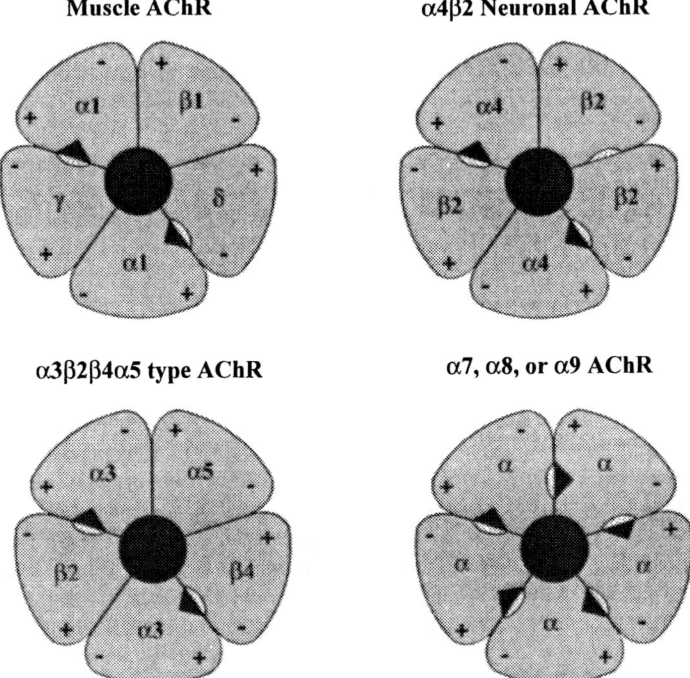

Figure 1.2. Organization of subunits around the central ion channel in various nAChR subtypes. The orientation of subunits in muscle nAChRs is known with the greatest certainty. Each subunit is depicted as having a+ and a− face. The ACh binding sites are depicted as located at subunit interfaces and being comprised of amino acids contributed by both of the adjacent subunits. Neuronal nAChRs can be formed from only one type of subunit (as in the case of α7, α8, or α9 homomers), or of two kinds of subunits (e.g., in the cases of pairs of α2, α3, or α4 with β2 or β4), or of three kinds of subunits (e.g., α5 combined with α 3 and β2 or β4 or with α4 and β2 or β4), or of four kinds of subunits (e.g., the α3β2β4α5 nAChR shown). Because α5 and β3 do not form functional nAChRs in paired combinations with other subunits, it is thought that they may occupy a position similar to that of β1 subunits in muscle nAChRs, which does not directly contribute to the formation of an ACh binding site. It may be that more complex combinations will be found with more than one kind of α subunit contributing to acetylcholine binding sites, for example, α3 β2 α6 β2 β3 nAChRs. In muscle nAChRs, the ACh binding sites formed at α1γ, α1δ, or α1ε interfaces are known to differ in their pharmacological properties, so there is already clear precedent for nAChR molecules containing two different kinds of ACh binding sites. Although not all of the subunits contribute to the formation of ACh binding sites, all of the subunits contribute to the lining of the cation channel and to the concerted conformational changes involved in activation and desensitization. Thus, subunits such as β1, β3, or α5 can contribute to the EC_{50} and other pharmacological properties of nAChRs of which they are a part.

nAChRs (Gerzanich et al., 1998). Addition of α5 subunits to α4β2 nAChRs also affects their permeability (Kuryatov et al., 1997).

A form of human congenital epilepsy is caused by an S247F mutation in the M2 channel lining sequence of the α4 subunit (Steinlein et al., 1995). This mutation impairs α4β2 nAChR function, causing faster desensitization, slower recovery from desensitization, reduced single-channel conductance, and loss of Ca^{++} conductance (Kuryatov et al., 1997). Adding α5 restored the Ca^{++} conductance in mutant α4β2α5 nAChRs. These examples are intended to illustrate that neuronal nAChRs that do not bind α-Bgt can contain as few as two or as many as four kinds of subunits, that all of the subunits contribute to the functional properties of the nAChR, and that each combination of subunits has unique properties. To make things more difficult, a single neuron may express several nAChR subtypes in several discrete regions for functional reasons that may not be obvious (Horch and Sargent, 1995; Zhang et al., 1996; Ullian et al., 1997). Adjacent neurons may express different combinations of nAChR subtypes (Weaver and Chiappinelli, 1996). Thus, trying to identify the nAChR subtype or subtypes in a particular neuron or nucleus or to identify the nAChR subtypes associated with a particular function or use them as a pharmacological target can be a daunting prospect.

The subunit composition and stoichiometry of neuronal nAChRs that bind α-Bgt are more straightforward. α7, α8, and α9 can all function as homomers in expression systems using cloned subunits (Gerzanich et al., 1994; Elgoyhen et al., 1994). In both avians and mammals, α7 nAChRs predominate. α7α8 nAChRs are known to occur in chickens (Keyser et al., 1993). It is not excluded that α7 heteromers exist in mammals. Homomeric α7 nAChRs have been experimentally advantageous for a wide range of mutational studies of nAChR function (Galzi et al., 1992; Eisile et al., 1993). It has even been possible to assemble the extracellular domains of α7 nAChRs into a water-soluble protein with native α7 nAChR binding properties that would be great for structural studies, if a significant amount could be expressed (Wells et al., 1998). The high Ca^{++} permeability of α7, α8, and α9 nAChRs has been one of their allures because this CA^{++} could serve as a second messenger in cells to alter many cellular processes (Seguela et al., 1993; Gerzanich et al., 1994; Elgoyhen et al., 1994). The major limitation on Ca^{++} influx through α7 nAChRs is the speed with which they desensitize. Most neuronal nAChR subtypes exceed the Ca^{++} permeability of muscle nAChRs (Lindstrom et al., 1995; Rathouz et al., 1996). In particular, α3β2α5 nAChRs and α3β4α5 nAChRs have calcium permeabilities near those of α7 nAChRs, but unlike α7 nAChRs they do not rapidly desensitize (Gerzanich et al., 1998). Thus, the sustained net Ca^{++} influx through these nAChR subtypes under some circumstances could exceed that of α7 nAChRs.

Chronic Exposure to Agonists

A general feature of all nAChRs that has long been realized but little understood is that they reversibly desensitize on chronic exposure to agonists (Dani and Heinemann, 1996). This may serve to limit excitotoxicity. On chronic exposure, an agonist such as (-)-nicotine can become a time-averaged antagonist (Hulihan-Giblin et al., 1990a,b). More recently it has become appreciated that after long-term exposure to (-)-nicotine or other agonists nAChRs can be permanently inactivated (Lukas, 1991; Peng et al., 1994; Hsu et al., 1996; Olale et al., 1997). nAChR subtypes differ in their susceptibility to this inactivation (Hsu et al., 1996; Olale et al., 1997; Fenster et al., 1997). This has practical applications in thinking about the actions of nicotine in smokers and may prove relevant to the effects of subtype-specific nicotinic agonists being investigated for pharmaceutical purposes. Exposure of human α4β2 nAChRs or α7 nAChRs to the 0.2 μM concentration of nicotine typical of a tobacco user's serum for 3 hours or more permanently inactivates more than 80% of the functional activity of these nAChRs (Olale et al., 1997), even though chronic exposure can also cause upregulation of the amount of these nAChRs (Peng et al., 1994, 1997).

Net loss of α4β2 nAChR and α7 nAChR function could account for the development of tolerance exhibited by tobacco users to some of the aversive effects of nicotine that are so unpleasant for naive users of tobacco. However, it is generally thought that nicotine addiction is maintained through stimulation of the same dopamine-mediated reward pathways that mediate addiction to cocaine and other abused drugs (Benowitz, 1996; Pich et al., 1997). How could inactivated nAChRs account for this? It turns out that α3 nAChRs are not permanently inactivated by chronic exposure to nicotine (Hsu et al., 1996; Olale et al., 1997; Fenster et al., 1997).

Thus α3 nAChRs, or other subtypes resistant to inactivation by chronic exposure to nicotine, could remain active to respond not only to endogenous ACh but also to the 1 μM boluses of nicotine experienced shortly after inhaling tobacco smoke. Agonist effects on these nAChR subtypes may account for some of the reinforcing and addicting properties of nicotine. Thus, the net effects of nicotine at the concentrations and dose schedules experienced by smokers will (1) have little effect on some nAChR subtypes (e.g., muscle α1 nAChRs), (2) largely inactivate other nAChR subtypes (e.g., α4β2 nAChRs and α7 nAChRs), and (3) activate some subtypes only in a pulsatile fashion during brief concentration peaks after inhaling (e.g., α3 nAChRs).

nAChR Subunit Gene Mutations and Knockouts

Having begun this chapter with a focus on the initial cloning of cDNAs for nAChRs, it seems appropriate to end with a look at the effects of mutating and deleting nAChR genes. Over the past few years, cDNA cloning and expression of intact and mutant nAChR subunits in *Xenopus* oocytes and cell lines has revolutionized understanding of their structure and function. Studies of nAChR mutations in humans and gene knockouts in animals are only beginning to impact our understanding of the functional roles of neuronal nAChRs. The first big realization of gene knockout studies in mice is that mice seem to be able to get along without many nAChR subtypes better than many would have guessed. However, it is worth recalling that loss of dystrophin in the mutant mdx strain of mice is basically asymptomatic, but this mutation in humans causes the ultimately lethal devastation of Duchenne's muscular dystrophy (Engel et al., 1994). Thus, apparent dispensability of an nAChR subtype in mice may not mean that a similar loss in humans would be inconsequential. No doubt further realizations will dawn as knockout mice are studied for more subtle aspects of neuronal function, as conditional knockout mechanisms are employed, and as knocked-out nAChR subunits are replaced with mutant subunits having decreased or increased function. One of the biggest realizations of studies of nAChR mutations in humans and other species is that there may be a lot of mutations out there. Another realization is that the functional properties of at least muscle nAChRs are optimized so that changing any of their parameters very much can have multiple unfortunate effects.

More than 50 mutations of muscle nAChR subunits have been discovered by the elegant studies of Engel and co-workers to cause congenital myasthenic syndromes (CMS) in humans (Engel et al., 1993, 1996a,b, 1997, 1998; Ohno et al., 1995, 1996, 1997; Sine et al., 1995). Frequently, congenital myasthenia was found associated with the presence of two recessive mutations. These observations suggest that there may be many mutations to be found in neuronal nAChRs. However, the physiological effects of neuronal nAChR mutations may be difficult to identify and their histopathological effects will be more difficult to characterize than is the case with nAChRs in muscle biopsies. Excitotoxic effects are caused both by mutations of the muscle nAChR near the ACh binding site, which increase its affinity for ACh and cause prolonged bursts of channel openings, and by mutations in the M2 channel lining domain, which cause prolonged channel opening. These excitotoxic effects probably result from excess calcium influx that disrupts the morphology of the nerve-muscle synapse. The net effect of muscle nAChR mutations, whether they "increase" or decrease nAChR function, is to impair transmission and cause muscle weakness. Some mutations are more damaging than others. More mutations have been found in the ϵ subunits characteristic of the adult form of nAChR than in the other subunits, presumably because many mutations in other subunits are lethal, but at least partial compensation for the loss of ϵ subunits can be provided by expression of the γ subunits characteristic of fetal nAChRs.

The only human neuronal nAChR mutations thus far known to be associated with human disease are two point mutations in α4 subunits that cause a rare form of congenital epilepsy called autosomal-dominant nocturnal frontal lobe epilepsy (ADNFLE) (Steinlein et al., 1995, 1997). These mutations in the M2 channel lining domain reduce α4β2 nAChR function by several mechanisms including preventing Ca^{++} permeability (Kuryatov et al., 1997; Steinlein et al., 1997). This would be especially deleterious if these α4β2 nAChRs were located presynaptically where Ca^{++} influx through them contributed to promoting transmitter release. It has been proposed that the neuronal hyperactivity characteristic of epilepsy could result from reduced α4 nAChR function if these mutant α4 nAChRs were involved presynaptically or postsynaptically in a circuit that released the inhibitory transmitter GABA at the time and places involved in triggering seizures in ADNFLE (Kuryatov et al., 1997). It is interesting that chronic exposure to nicotine in smokers is expected to inactivate a large fraction of α4β2 nAChRs (Olale et al., 1997), yet this does not cause seizures such as ADNFLE. This could result either from the additional effects of nicotine on other nAChR subtypes or indicate that ADNFLE patients have additional risk factors that make them susceptible to loss of α4 nAChR function. Another form of congenital epilepsy has been mapped near the α4 gene (Steinlein et al., 1994), and yet another form has been mapped near the α7 gene (Elmslie et al., 1997), but it remains to be determined if these diseases are actually caused by a mutation in either of these genes. Also, α7 isoforms have been found to be closely linked to an objectively measurable behavioral effect (auditory paired pulse inhibition), which may be a predisposing, but not a directly causative, factor for schizophrenia (Freedman et al., 1997).

Mice with their β2 subunit gene knocked out lack the brain high-affinity nicotine binding sites characteristic of α4β2 nAChRs but otherwise appear grossly normal (Picciotto et al., 1995). β2 knockout mice, not surprisingly, lack a nicotine-induced increase in an avoidance learning behavior found in the wild type. However, strikingly, they perform in this test better than the controls. Thus, apparently, ablating α4β2 nAChR function (as also may occur in smokers [Olale et al., 1997]) is not necessarily altogether a bad thing. However, aged β2 knockout mice do exhibit impaired learning in the Morris water maze test (Zoli et al., 1997). Thus, the effects of the function of various nAChR subtypes may only become apparent under certain conditions.

Mice with their α7 subunit gene knocked out lack brain binding sites for ^{125}I α-Bgt and lack rapidly desensitizing responses to (-)-nicotine in the hippocampal neurons, which normally express large amounts of α7 nAChRs, but α7 knockout mice are grossly behaviorally and anatomically normal (Orr- Urtreger et al., 1997). The last three exons of α7 were deleted, leaving the N-terminus through M1, which is known to be capable of assembling functional α-Bgt binding proteins (Wells et al., in press); however, the altered α7 mRNA was itself unstable and eliminated, resulting in no detectable α7 protein fragments. In tobacco users, results suggest that most of their α7 nAChRs should be chronically inactivated (Olale et al., 1997). Mutant α7-like nAChRs in the nematode worm *Caenorhabditis elegans* (Treinen and Chalfie, 1995) result in excitotoxicity worse than that seen in many congenital myasthenia gravis patients (Engel et al., 1998). It may be that having hyperactive mutant α7 nAChRs or some other mutant nAChR subtypes could be worse than having none at all. More subtle analyses of α7 knockout mice will be required to help determine the functional roles of α7 nAChRs.

Only knockout of α3 genes, among the neuronal nAChR subunits tested, appears to be critical, resulting in runting and death (Xu et al., 1997). Loss of β2 or β4 or α5 was not lethal. It is easy to imagine some compensation for the loss of β2 or β4 or α5 by substitution with one of the others. The critical role of α3 may result in large part from its fundamental postsynaptic role in the autonomic nervous system. Even though at ciliary ganglia synapses blockage of postsynaptic α3 nAChRs still permits transmission through perisynaptic α7 nAChRs (Ullian et al., 1997), α7 does not seem to be able to fully compensate for the loss of α3 nAChRs throughout the nervous system. Even at those synapses at which α7 was also present, it might be expected that at high rates of stimulation the rapid desensitization of α7 might result in failure of transmission in the absence of α3 nAChRs.

CONCLUDING REMARKS

nAChRs have been long studied at many levels and are in many ways the best characterized of neurotransmitter receptors. However, even though most or all of the nAChR subunits have been cloned, we still know a lot more about the properties of expressed cloned neuronal nAChR subunit combinations than we do about the functional roles of these nAChRs in the nervous system. nAChRs are the principal excitatory neurotransmitter receptor in the mammalian peripheral nervous system, but in the brain their amount is tiny as compared to the amount of excitatory glutamate receptors. However, nAChRs are very widely spread in the central nervous system. Many of these are likely to be in presynaptic locations where they can modulate the release of many transmitters. This gives them a lot of leverage and may make them interesting drug targets. The leverage arises from the fact that Ca^{++} flux through very few presynaptic nAChRs would be required to significantly facilitate transmitter release because of the small volume of nerve endings and because voltage-sensitive calcium channels can amplify the effects of these nAChRs (Soliakov and Wonnacott, 1996). nAChRs that modulated signaling on neuronal pathways might have virtues as targets as compared to receptors that when activated turned on spurious new signaling or when turned off blocked signaling altogether.

nAChRs have been found to be involved in many diseases and medically relevant processes (Lindstrom, 1997). They are centrally involved in autoimmune (Lindstrom et al., 1988) and congenital (Engel et al., 1998) myasthenia gravis as well as ADNFLE (Steinlein et al., 1995, 1997). The amount of nAChR is reduced in Alzheimer's disease, Parkinson's disease (Whitehouse et al., 1988), and schizophrenia (Freedman et al., 1995, 1997). Tourette's syndrome (Silver et al., 1996) and other diseases may benefit from nicotine therapy. nAChRs may be a sensitive target for some general anesthetics (Evers et al., 1997; Flood et al., 1997; Violet et al., 1997). Epibatidine is a potent nicotinic agonist 200-fold more potent as an antinociceptive than is morphine (Daly, 1995; Gerzanich et al., 1995). Recently developed less toxic selective nicotinic agonists show great preclinical promise for relief of pain (Holladay et al., 1997). The number of nAChR subtypes possible, the difficulties in identifying them in neurons, the complex combinations in which they can occur, and the many and complex functional roles that they can serve mean that there is still a long way to go from cloning nAChR subunit cDNAs to identifying the normal functional roles of nAChR subtypes. There may be even further to go before understanding the roles of these subtypes in disease and efficiently targeting them for drug therapies. However, nAChR subtype-selective drug candidates may also prove to be valuable reagents for helping figure out the normal functional roles of neuronal nAChRs.

ACKNOWLEDGMENTS

Dr. Lindstrom's laboratory is supported by grants from the National Institutes of Health, Muscular Dystrophy Association, Smokeless Tobacco Research Council, Inc., and the Council for Tobacco Research–USA, Inc.

REFERENCES

Albuquerque E, Alkondon M, Pereira E, Castro N, Schrattenholz A, Barbosa C, Bonfante-Cabarcas R, Aracava Y, Eisenberg H, Maelicke A (1996): Properties of neuronal nicotinic acetylcholine receptors: pharmacological characterization and modulation of synaptic function. J. Pharmacol. Exp. Ther. 280: 1117–1136.

Anand R, Conroy W, Schoepfer R, Whiting P, Lindstrom J (1991): Chicken neuronal nicotinic acetylcholine receptors expressed in *Xenopus* oocytes have a pentameric quaternary structure. J. Biol. Chem. 266: 11,191–11,198.

Ballivet M, Patrick J, Lee J, Heinemann S (1982): Molecular cloning of cDNA for the γ subunit of *Torpedo* acetylcholine receptor. Proc. Natl. Acad. Sci. USA 79: 4466–4470.

Barnard E, Darlison D, Seeburg P (1987): Molecular biology of the $GABA_A$ receptor: the receptor/channel superfamily. Trends Neurosci. 10: 502–509.

Benowitz N (1996): Pharmacology of nicotinic: addiction and therapeutics. Annu. Rev. Pharmacol. Toxicol. 36: 597–613.

Beroukhim R, Unwin N (1995): Three dimensional location of the main immunogenic region of the acetylcholine receptor. Neuron 15: 323–331.

Bertrand D, Changeux J (1995): Nicotine receptor: an allosteric protein specialized for intercellular communication. The Neurosciences 7: 75–90.

Betz H (1990): Homology and analogy in transmembrane channel design: lessons from synaptic membrane proteins. Biochemistry 29: 3591–3599.

Blount P, Merlie J (1989): Molecular basis of the two nonequivalent ligand binding sites of the muscle nicotinic acetylcholine receptor. Neuron 3: 349–357.

Boulter J, Evans K, Goldman D, Martin G, Treco D, Heinemann S, Patrick J (1986): Isolation of a cDNA clone coding for a possible neural nicotinic acetylcholine receptor α subunit. Nature 319: 368–374.

Britto L, Keyser K, Lindstrom J, Karten H (1992): Immunohistochemical localization of nicotinic acetylcholine receptor subunits in the mesencephalon and diencephalon of the chick (*Gallus*). J. Comp. Neurol. 317: 325–340.

Changeux J (1990): Functional archechitecture and dynamics of the nicotinic acetylcholine receptor: an allosteric ligand-gated ion channel. In: 1988–1989 Fidia Research Foundation: Neuroscience Award Lectures, vol. 4. New York: Raven Press, pp. 21–168.

Claudio T, Ballivet M, Patrick J, Heinemann S (1983): *Torpedo californica* acetylcholine receptor 60,000 dalton subunit: nucleotide sequence of cloned cDNA deduced amino acid sequence, subunit structural predictions. Proc. Natl. Acad. Sci. USA 80: 1111–1115.

Coggan J, Paysan J, Conroy W, Berg D (1997): Direct recording of nicotinic responses in presynaptic nerve terminals. J. Neurosci. 17: 5798–5806.

Conroy W, Vernallis A, Berg D (1992): The α5 gene product assembles with multiple acetylcholine receptor subunits to form distinctive receptor subtypes in brain. Neuron 9: 1–20.

Conti-Tronconi B, Dunn S, Barnard E, Dolly J, Lai F, Ray N, Raftery M (1985): Brain and muscle nicotinic acetylcholine receptors are different but homologous proteins. Proc. Natl. Acad. Sci. USA 82: 5208–5212.

Cooper E, Couturier S, Ballivet M (1991): Pentameric structure and subunit stoichiometry of a neuronal nicotinic acetylcholine receptor. Nature 350: 235–238.

Couturier S, Bertrand D, Matter D, Hernandez J, Bertrand S, Millar N, Valera S, Barkas T, Ballivet M (1990a): A neuronal nicotinic acetylcholine receptor subunit (α7) is developmentally regulated and forms a homomeric channel blocked by α-bungarotoxin. Neuron 5: 847–856.

Couturier S, Erkman L, Valera S, Rungger D, Bertrand S, Boulter J, Ballivet M, Bertrand S (1990b): α5, α3, and non α3. Three clustered avian genes encoding neuronal nicotinic acetylcholine receptor related subunits. J. Biol. Chem. 265: 17,560–17,567.

Czajkowski C, Karlin A (1995): Structure of the nicotinic receptor acetylcholine binding site. J. Biol. Chem. 270: 3160–3164.

Daly J (1995): The chemistry of poisons in amphibian skin. Proc. Natl. Acad. Sci. USA 92: 9–13.

Dani J, Heinemann S (1996): Molecular and cellular aspects of nicotine abuse. Neuron 16: 905–908.

Del Toro E, Juiz J, Peng X, Lindstrom J, Criado M (1994): Immunocytochemical localization of the α7 subunit of the nicotinic acetylcholine receptor in the rat central nervous system. J. Comp. Neurol. 349: 325–342.

Deneris E, Connolly J, Boulter J, Wada E, Wada K, Swanson L, Patrick J, Heinemann S (1988): Primary structure and expression of β2: A novel subunit of neuronal nicotinic receptors. Neuron 1: 45–54.

Devillers-Theiry A, Giraudat J, Bentaboulet M, Changeux J (1983): Complete mRNA coding sequence of the acetylcholine binding α subunit of *Torpedo marmorata* acetylcholine receptor: a model for the transmembrane organization of the polypeptide chain. Proc. Natl. Acad. Sci. USA 80: 2067–2071.

Eisile J, Bertrand S, Galzi J, Devillers-Theiry A, Changeux J, Bertrand D (1993): Chimaeric nicotinic-serotonergic receptor combines distinct ligand binding and channel specificities. Nature 366: 479–483.

Elgoyhen A, Johnson D, Boulter J, Vetter D, Heinemann S (1994) α9: An acetylcholine receptor with novel pharmacological properties expressed in rat cochlear hair cells. Cell 79: 705–715.

Elmslie F, Rees M, Willamson M, Kerr M, Kjeldsen M, Pang K, Sundquist A, Friis M, Chadwick D, Richens A, Covavis A, Santos M, Whitehouse W, Gardiner R (1997): Genetic mapping of a major susceptibility locus for juvenile myoclonic epilepsy on chromosome 15q. Hum. Mol. Genet. 6: 1329–1334.

Engel A, Uchitel O, Walls T, Nagel A, Harper C, Bodensteiner J (1993): Newly recognized congenital myasthenic syndrome associated with high conductance and fast closure of the acetylcholine receptor channel. Ann. Neurol. 34: 38–47.

Engel A, Yamamoto M, Fischbeck K (1994): Dystrophinopathies. In: Engel A, Franzini-Armstrong C, editors. Myology, 2nd ed., vol. 2. New York: McGraw-Hill, pp. 1133–1187.

Engel A, Ohno K, Bouzat C, Sine S, Griggs R (1996a): End plate acetylcholine receptor deficiency due to nonsense mutations in the ϵ subunit. Ann. Neurol. 40: 810–817.

Engel A, Ohno K, Milone M, Wang H, Nakano S, Bouzat C, Pruitt J, Hutchinson D, Brengman J, Bren N, Sieb J, Sine S (1996b): New mutations in acetylcholine receptor subunit genes reveal heterogeneity in the slow channel congenital myasthenic syndrome. Hum. Mol. Genetics. 5: 1217–1227.

Engel A, Ohno K, Milone M, Sine S (1997): Congenital myasthenic syndromes caused by mutations in acetylcholine receptor genes. Neurology 48 (suppl. 5): S28–S35.

Engel A, Ohno K, Wang H, Milone M, Sine S (1998): The molecular basis of congenital myasthenic syndromes: mutations in the acetylcholine receptor. NeuroScientist. 4: 185–194

Evers A, Steinbach J (1997): Supersensitive sites in the central nervous system. Anesthesiology 86: 760–762.

Fenster C, Rains M, Noerager B, Quick M, Lester R (1997): Influence of subunit composition on desensitization of neuronal acetylcholine receptors at low concentrations of nicotine. J. Neurosci. 17: 5747–5759.

Flood P, Ramirez-Latorre J, Role L (1997): α4β2 neuronal nicotinic acetylcholine receptors in the central nervous system are inhibited by isoflurane and propofol, but α7-type nicotinic acetylcholine receptors are unaffected. Anesthesiology 86: 859–865.

Flores C, Rogers S, Pabreza L, Wolfe B, Kellar K (1992): A subtype of nicotinic cholinergic receptor in rat brain is composed of α4 and β2 subunits and is upregulated by chronic nicotine treatment. Mol. Pharmacol. 41: 31–37.

Forsayeth J, Kobrin E (1997): Formation of oligomers containing the β3 and β4 subunits of the rat nicotinic receptor. J. Neurosci. 17: 1531–1538.

Freedman R, Hall M, Adler L, Leonard S (1995): Evidence in postmortem brain tissue for decreased numbers of hippocampal nicotinic receptors in schizophrenia. Biopsychiatry 38: 22–33.

Freedman R, Coon H, Myles-Worsley M, Orr-Urtreger A, Olincy A (1997): Linkage of a neuropharmacological deficit in schizophrenia to a chromosome 15 locus. Proc. Natl. Acad. Sci. USA 94: 587–592.

Galzi J, Revah F, Black D, Goeldner M, Hirth C, Changeux J (1990): Identification of a novel amino acid α tyrosine 93 within the cholinergic ligands-binding sites of the acetylcholine receptor by photoaffinity labeling. J. Biol. Chem. 265: 10,430–10,437.

Galzi J, Devillers-Thiery A, Hussy N, Bertrand S, Changeux J, Bertrand D (1992): Mutations in the channel domain of a neuronal nicotinic receptor convert ion selectivity from cationic to anionic. Nature 359: 500–505.

Gerzanich V, Anand R, Lindstrom J (1994): Homomers of α8 subunits of nicotinic receptors functionally expressed in Xenopus oocytes exhibit similar channel but contrasting binding site properties compared to α7 homomers. Mol. Pharmacol. 45: 212–220.

Gerzanich V, Peng X, Wang F, Wells G, Anand R, Fletcher S, Lindstrom J (1995): Comparative pharmacology of epibatidine—a potent agonist for neuronal nicotinic acetylcholine receptors. Mol. Pharmacol. 48: 774–782.

Gerzanich V, Kuryatov A, Anand R, Lindstrom J (1997): "Orphan" α6 nicotinic nAChR subunit can form a functional heteromeric acetylcholine receptor. Mol. Pharmacol. 51: 320–327.

Gerzanich V, Wang F, Kuryatov A, Lindstrom J (1998): α5 subunit alters desensitization, pharmacology, Ca^{++} modulation, and Ca^{++} permeability of human neuronal α3 nicotinic receptors. J. Pharmacol. Exp. Ther. 266: 311–320.

Goldner F, Dineley K, Patrick J (1997): Immunohistochemical localization of the nicotinic acetylcholine receptor subunit α6 to dopaminergic neurons in the substantia nigra and ventral tegmental area. Neuroreport 8: 2739–2742.

Gotti C, Fornasari D, Clementi F (1997): Human neuronal nicotinic receptors. Prog. Neurobiol. 53: 199–237.

Grando S, Horton R, Pereira E, Diethelmokita B, George P, Albuquerque E, Conti-Fine B (1995): A nicotinic acetylcholine receptor regulating cell adhesion and motility is expressed in human keratinocytes. J. Invest. Dermatol. 105: 774–781.

Grando S, Horton R, Mauro T, Kist D, Lee T, Dahl M (1996): Activation of keratinocytes nicotinic cholinergic receptors stimulates calcium influx and enhances cell differentiation. J. Invest. Dermatol. 107: 412–418.

Gray R, Rajan A, Radcliffe K, Yakahiro M, Dani J (1996): Hippocampal synaptic transmission enhanced by low concentrations of nicotine. Nature 383: 713–716.

Hamassaki-Britto D, Gardino P, Hokoc J, Keyser K, Karten H, Lindstrom J, Britto L (1994): Differential development of α bungarotoxin-sensitive and α bungarotoxin-insensitive nicotinic acetylcholine receptors in the chick retina. J. Comp. Neurol. 347: 161–170.

Heinemann S, Boulter J, Connolly J, Deneris E, Duvoisin R, Hartley M, Hermans-Borgmeyer I, Hollman M, O'Shea-Greenfield A, Papke R, Rogers S, Patrick J (1991): The nicotinic receptor genes. Clin. Neuropharmacol. 14: S45–S61.

Holladay M, Wasicak J, Lin N, He Y, Ryther K, Bannon A, Buckley M, Kim D, Decker M, Anderson D, Campbell J, Donnelly-Roberts D, Briggs C, McKenna D, Niforatos W, Piattoni-Kaplan M, Williams M, Arneric S (1997): Identification, initial pharmacological evaluation, and structure-activity relationships of ABT-594 as a potent, orally active analgesic agent acting via neuronal nicotinic acetylcholine receptors. Neuroscience Society Meeting Abstract 477.6, p. 1198 New Orleans.

Horch, H, Sargent P (1995): Perisynaptic surface distribution of multiple classes of nicotinic acetylcholine receptors on neurons in the chicken ciliary ganglion. J. Neurosci. 15: 7778–7795.

Hsu, Y, Amin J, Weiss D, Wecker L (1996): Sustained nicotine exposure differentially affects α3β2 and α4β2 neuronal nicotinic receptors expressed in *Xenopus* oocytes. J. Neurochem. 66: 667–675.

Hulihan-Giblin B, Lumpkin M, Kellar K (1990a): Effects of chronic administration of nicotine on prolactin release in the rat: inactivation of prolactin response by repeated injections of nicotine. J. Pharmacol. Exp. Ther. 252: 21–25.

Hulihan-Giblin B, Lumpkin M, Kellar K (1990b): Effects of chronic administration of nicotine on prolactin release in the rat: inactivation of prolactin response by repeated injections of nicotine. J. Pharmacol. Exp. Ther. 252: 21–25.

Jacob M, Berg D (1983): The ultrastructural localization of α bungarotoxin binding sites in relation to synapses on chick ciliary ganglion neurons. J. Neurosci. 3: 260–271.

Jacob M, Berg D, Lindstrom J (1984): A shared antigenic determinant between the *Electrophorus* acetylcholine receptor and a synaptic component on chick ciliary ganglion neurons. Proc. Natl. Acad. Sci. USA 81: 3223–3227.

Kao P, Karlin A (1986): Acetylcholine receptor binding site contains a disulfide crosslink between adjacent half-cystinyl residues. J. Biol. Chem. 261: 8085–8088.

Karlin A (1969): Chemical modification of the active site of the acetylcholine receptor. J. Gen. Phsyiol. 54: 245–264.

Karlin A (1991): Explorations of the nicotinic acetylcholine receptor. Harvey Lecture Series 85: 71–107.

Karlin, A, Akabas M (1995): Toward a structural basis for the function of nicotinic acetylcholine receptors and their cousins. Neuron 15: 1231–1244.

Karlin A, Couburn D (1973): The affinity-labeling of partially purified acetylcholine receptor from electric tissue of *Electrophorus*. Proc. Natl. Acad. Sci. USA 70: 3636–3640.

Keyser K, Britto L, Schoepfer R, Whiting P, Cooper J, Conroy W, Brozozowska-Prechtl A, Karten H, Lindstrom J (1993): Three subtypes of α bungarotoxin-sensitive nicotinic acetylcholine receptors are expressed in chick retina. J. Neurosci. 13: 442–454.

Kuryatov A, Gerzanich V, Nelson M, Olale F, Lindstrom J (1997): Mutation causing autosomal dominant nocturnal frontal lobe epilepsy alters Ca^{++} permeability, conductance, and gating of human $\alpha4\beta2$ nicotine acetylcholine receptors. J. Neurosci. 17: 9035–9047.

Lee C, Tseng L, Chiu T (1967): Influence of denervation on localization of neurotoxins from clapid venoms in rat diaphragm. Nature 215: 1177–1178.

LeNovere N, Changeux J (1995): Molecular evolution of the nicotinic acetylcholine receptor: an example of a mutigene family in excitable cells. J. Mol. Evol. 40: 155–172.

LeNovere N, Zoli M, Changeux J (1996): Neuronal nicotinic receptor α6 subunit mRNA is selectively concentrated in catecholominergic nuclei of the rat brain. Eur. J. Neurosci. 8: 2428–2439.

Lindstrom J (1985): Nicotinic acetylcholine receptors: use of monoclonal antibodies to study synthesis, structure, function, and autoimmune response. In: Venter J, Fraser C, Lindstrom J, editors. Monoclonal and anti-idiotypic antibodies: probes for receptor structure and function. New York: Alan R. Liss, Inc., pp. 21–57.

Lindstrom J (1996): Neuronal nicotinic acetylcholine receptors. In: Narahashi T, editor. Ion channels, vol. 4. New York: Plenum Press, pp. 377–450.

Lindstrom J (1997): Nicotinic acetylcholine receptors in health and disease. Mol. Neurobiol. 15: 193–222.

Lindstrom J, Patrick J (1974): Purification of the acetylcholine receptor by affinity chromatography. In: Bennet MVL, editor. Synaptic transmission and neuronal interaction. New York: Raven Press, pp. 191–216.

Lindstrom J, Merlie J, Yogeeswaran B (1979): Biochemical properties of acetylcholine receptor subunits from *Torpedo californica*. Biochemistry 18: 4465–4470.

Lindstrom J, Shelton G, Fujii Y (1988): Myasthenia gravis. Adv. Immunol. 42: 233–284.

Lindstrom J, Anand R, Peng X, Gerzanich V, Wang F, Li Y (1995): Neuronal nicotinic receptor subtypes. Ann. N.Y. Acad. Sci. 757: 100–116.

Lindstrom J, Peng X, Kuryatov A, Lee E, Anand R, Gerzanich V, Wang F, Wells G, Nelson M (1998): Molecular and antigenic structure of nicotinic acetylcholine receptors. Ann. N.Y. Acad. Sci., Proceedings of IXth International Conference of Myasthenia Gravis and Related Disorders 841: 71–86.

Lukas R (1991): The effects of chronic nicotinic ligand exposure on functional activity of nicotinic acetylcholine receptors expressed by cells of the PC12 rat pheochromocytoma or the TE671/RD human clonal line. J. Neurochem. 56: 1134–1145.

Lukas R, Norman S, Lucero L (1993): Characterization of nicotinic acetylcholine receptors expressed by cells of the SH-SY5Y human neuroblastoma clonal line. Mol. Cell. Neurosci. 4: 1–12.

Martin M, Czajkowski C, Karlin A (1996): The contributions of aspartyl residues in the acetylcholine receptor γ and δ subunits to the binding of agonists and competitive antagonists. J. Biol. Chem. 271: 13,497–13,503.

Merlie J, Sebbane R, Gardner S, Lindstrom J (1983a): Regulation of acetylcholine receptor gene expression: molecular cloning of a cDNA specific for α subunit of the receptor from the mouse muscle cell line BC3H-1. Proc. Natl. Acad. Sci. USA 80: 3845–3849.

Merlie J, Sebbane R, Gardner S, Olson E, Lindstrom J (1983b): The regulation of acetylcholine receptor expression in mammalian muscle. Cold Spring Harbor Symposium on Quantitative Biology XLVIII: 135–146.

Nelson S, Shelton G, Lei S, Lindstrom J, Conti-Tronconi B (1992): Epitope mapping of monoclonal antibodies to *Torpedo* acetylcholine receptor γ subunits, which specifically recognize the ϵ subunit of mammalian muscle acetylcholine receptor. J. Neuroimmunol. 36: 13–27.

Noda M, Takahashi H, Tanabe T, Toyosato M, Furutani Y, Hirose T, Asai M, Inayama S, Miyata T, Numa S (1982): Primary structure of α subunit precursor of *Torpedo californica* acetylcholine receptor deduced from cDNA sequence. Nature 299: 793–797.

Noda M, Furutani Y, Takahashi H, Toyosato M, Tanabe T, Shimizu S, Kikyotani S, Kayano T, Hirose T, Inayama S, Numa S (1983a): Cloning and sequence analysis of calf cDNA and human genomic DNA encoding α subunit precursor of muscle acetylcholine receptor. Nature 305: 818–823.

Noda M, Takahashi H, Tanabe T, Toyosato M, Kikyotani S, Furutani Y, Hirose T, Takashima H, Inayama S, Miyata T, Numa S (1983b): Structural homology of *Torpedo californica* acetylcholine receptor subunits. Nature 302: 528–532.

Noda M, Shimizu S, Tanabe T, Takai T, Koyano T, Ikeda T, Tokahashi H, Nakoyama H, Kanoboa Y, Minamino N, Kangawa K, Matsuo H, Raftery M, Hirose T, Inayama S, Hayashido H, Miyata T, Numa S (1984): Primary structure of *Electrophorus electricus* sodium channel deduced from cDNA sequence. Nature 312: 121–127.

Noda M, Ikeda T, Suzuki H, Takeshima H, Takakashi T, Kuno M, Numa S (1986): Expression of functional sodium channels from cloned cDNA. Nature 322: 826–828.

Ohno K, Hutchinson D, Milone D, Brengman J, Bouzat C, Sine S, Engel A (1995): Congenital myasthenia syndrome caused by prolonged acetylcholine receptor channel openings due to a mutation in the M2 domain of the ϵ subunit. Proc. Natl. Acad. Sci. USA 92: 758–762.

Ohno K, Wang H, Milone M, Bren N, Brengman J, Nakano S, Quiran P, Pruitt J, Sine S, Engel A (1996): Congenital myasthenic syndrome caused by decreased agonist binding affinity due to a mutation in the acetylcholine receptor ϵ subunit. Nature 17: 157–170.

Ohno K, Quiram P, Milone M, Wang H, Harper M, Pruitt J, Brengman J, Pao L, Fischbeck K, Crawford T, Sine S, Engel A (1997): Congenital myasthenic syndromes due to heteroallelic nonsense-missense mutations in the acetylcholine receptor ϵ subunit gene: identification and functional characterization of six new mutations. Hum. Mol. Gene. 6: 753–766.

Olale F, Gerzanich V, Kuryatov A, Wang F, Lindstrom J (1997): Chronic nicotine exposure differentially affects the function of human α3, α4, and α7 neuronal nicotinic receptor subtypes. J. Pharmacol. Exp. Ther. 283: 675–683.

Orr-Urtreger A, Goldner F, Saeki M, Lorenzo I, Goldberg L, DeBiasi M, Dani J, Patrick J, Beaudet A (1997): Mice deficient in the α7 neuronal nicotinic acetylcholine receptor lack α-bungarotoxin binding sites and hippocampal fast nicotinic currents. J. Neurosci. 17: 9165–9171.

Papke R (1993): The kinetic properties of neuronal nicotinic receptor: genetic basis of functional diversity. Prog. Neurobiol. 41: 509–531.

Patrick J, Lindstrom J (1973): Autoimmune response to acetylcholine receptor. Science 180: 871–872.

Peng X, Anand R, Whiting P, Lindstrom J (1994): Nicotine-induced upregulation of neuronal nicotinic receptors results from a decrease in the rate of turnover. Mol. Pharmacol. 46: 523–530.

Peng X, Gerzanich V, Anand R, Wang F, Lindstrom J (1997): Chronic nicotine treatment upregulates α3 nAChRs and α7 nAChRs expressed by the human neuroblastoma cell line SH-SY5Y. Mol. Pharmacol. 51: 776–784.

Picciotto M, Zoll M, Lena C, Bessis A, Lallemand Y, LeNovere N, Vincent P, Pich M, Brulet P, Changeux J (1995): Abnormal avoidance learning in mice lacking functional high affinity nicotine receptor in the brain. Nature 374: 65–67.

Pich E, Paglusi S, Tessari M, Talabot-Ayer D, Huijsduynen R, Chiamulua C (1997): Common neural substrates for the addictive properties of nicotine and cocaine. Science 275: 83–86.

Raftery M, Hunkapillar M, Strader C, Hood L (1980): Acetylcholine receptor: complex of homologous subunits. Science 208: 1454–1457.

Ramirez-Latorre J, Yu C, Qu X, Perin F, Karlin A, Role L (1996): Functional contribution of α5 subunit to neuronal acetylcholine receptor channels. Nature 380: 347–351.

Rathouz M, Vijayaraghavan S, Berg D (1996): Elevation of intracellular calcium levels in neurons by nicotinic acetylcholine receptors. Mol. Neurobiol. 12: 117–131.

Reynolds J, Karlin A (1978): Molecular weight in detergent solution of acetylcholine receptor from *Torpedo californica.* Biochemistry 17: 2035–2038.

Role L, Berg D (1996): Nicotinic receptors in the development and modulation of CNS synapses. Neuron 16: 1077–1085.

Romano S, Coriveau R, Schwarz R, Berg D (1997a): Expression of the nicotinic receptor α7 gene in tendon and periosteum during early development. J. Neurochem. 68: 640–648.

Romano S, Pugh P, McIntosh J, Berg D (1997b): Neuronal-type acetylcholine receptors and regulation of α7 gene expression in vertebrate skeletal muscle. J. Neurobiol. 32: 69–80.

Sargent P (1993): The diversity of neuronal nicotinic acetylcholine receptors. Ann. Rev. Neurosci. 16: 403–443.

Schoepfer R, Whiting P, Esch F, Blacher R, Shimasaki S, Lindstrom J (1988): cDNA clones coding for the structural subunit of a chicken brain nicotinic acetylcholine receptor. Neuron 1: 241–248.

Schoepfer R, Conroy W, Whiting P, Fore M, Lindstrom J (1990): Brain α-bungarotoxin binding protein cDNAs and mAbs reveal subtypes of this branch of the ligand-gated ion channel superfamily. Neuron 5: 35–48.

Seguela P, Wadiche J, Dinelly-Miller K, Dani J, Patrick J (1993): Molecular cloning, functional properties, and distribution of rat brain α7: a nicotinic cation channel highly permeable to calcium. J. Neurosci. 13: 596–604.

Silver A, Shytle R, Philipp M, Sanberg P (1996): Case study: long term potentiation of neuroleptics with transdermal nicotine in Tourette's syndrome. J. Am. Acad. Child Adolesc. Psychiatry 35: 1631–1636.

Sine S, Ohno K, Bouzat C, Auerbach A, Milone M, Pruitt J, Engel A (1995): Mutation of the acetylcholine receptor α subunit causes a slow channel myasthenic syndrome by enhancing agonist binding affinity. Neuron 15:229–239.

Soliakov L, Wonnacott S (1996): Voltage-sensitive Ca^{++} channels involved in nicotinic receptor-mediated [^{3}H] dopamine release from rat striatal synaptosomes. J. Neurochem. 67: 163–170.

Steinlein O, Smigrodzki R, Lindstrom J, Anand R, Kihler M, Tocharoentanophol C, Vogel F (1994): Refinement of the localization of the gene for neuronal nicotinic acetylcholine receptor α4 subunit (CHRNA4) to human chromosome 20q 13.2-q13.3. Genomics 22: 493–495.

Steinlein O, Mulley J, Propping P, Wallace R, Phillips H, Sutherland G, Scheffer I, Nerkovic S (1995): A missense mutation in the neuronal nicotinic acetylcholine receptor α4 subunit is associated with autosomal dominant nocturnal frontal lobe epilepsy. Nature Genetics 11: 201–203.

Steinlein O, Magnusson A, Stoodt J, Bertrand S, Weiland S, Berkovic S, Nakhem K, Propping P, Bertrand D (1997): An insertion mutation of the CHRNA4 gene in a family with autosomal dominant nocturnal frontal lobe epilepsy. Hum. Mol. Genetics 6: 943–947.

Stroud R, McCarthy M, Shiester M (1990): Nicotinic acetylcholine receptor superfamily of ligand-gated ion channels. Biochemistry 29: 11,009–11,023.

Sumikawa K, Houghton M, Smith J, Bell L, Richards B, Barnard E (1982): The molecular cloning and characterization of cDNA coding for the α subunit of the acetylcholine receptor. Nucl. Acids Res. 10: 5809–5822.

Treinen M, Chalfie M (1995): A mutated acetylcholine receptor subunit causes neuronal degeneration in *C. elegans*. Neuron 14: 871–877.

Tsigelny I, Sugiyama N, Sine S, Taylor P (1997): A model of the nicotinic receptor extracellular domain based on sequence identity and residue location. Biophys. J. 73: 52–66.

Tzartos S, Lindstrom J (1980): Monoclonal antibodies used to probe acetylcholine receptor structure: localization of the main immunogenic region and detection of similarities between subunits. Proc. Natl. Acad. Sci. USA 77: 755–759.

Ullian E, McIntosh J, Sargent P (1997): Rapid synaptic transmission in the avian ciliary ganglion is mediated by two distinct classes of nicotinic receptors. J. Neurosci. 17: 7210–7219.

Unwin N (1993): Nicotinic acetylcholine receptor at 9 Å resolution. J. Mol. Biol. 229: 1101–1124.

Unwin N (1995): Acetylcholine receptor channel imaged in the open state. Nature 373: 37–43.

Vernallis A, Conroy W, Berg D (1993): Neurons assemble acetylcholine receptors with as many as three kinds of subunits while maintaining subunit segregation among receptor subtypes. Neuron 10: 451–464.

Violet J, Downie D, Nakisa R, Lieb W, Franks N (1997): Differential sensitivities of mammalian neuronal and muscle nicotinic acetylcholine receptors to general anesthetics. Anesthesiology 86: 866–874.

Wang Z, Fuhrer C, Shtrom S, Sugiyama I, Ferns M, Hall Z (1996a): The nicotinic acetylcholine receptor at the neuromuscular junction: assembly and tyrosine phosphorylation. Cold Spring Harbor Symposium on Quantitative Biology LXI: 363–371.

Wang F, Gerzanich V, Wells G, Anand R, Peng X, Keyser K, Lindstrom J (1996b): Assembly of human neuronal nicotinic receptor α5 subunits with α3, β2, and β4 subunits. J. Biol. Chem. 271: 17, 656–17,665.

Weaver, W, Chiappinelli V (1996): Single channel recording in brain slices reveals heterogeneity of nicotinic receptors on individual neurons within the chick lateral spiriform nucleus. Brain Res. 724: 95–105.

Wells G, Anand R, Wang F, Lindstrom J (1998): Water-soluble nicotinic acetylcholine receptor formed by α7 subunit extracellular domains. J. Biol. Chem. 273: 964–973.

Whitehouse P, Martino A, Marcus K, Zweig R, Singer H, Prince D, Kellar K (1988): Reductions in acetylcholine and nicotine binding in several degenerative diseases. Arch. Neurol. 45: 722–724.

Whiting P, Lindstrom J (1986): Purification and characterization of a nicotinic acetylcholine receptor from chick brain. Biochemistry 25: 2082–2093.

Whiting P, Lindstrom J (1987): Purification and characterization of a nicotinic acetylcholine receptor from rat brain. Proc. Natl. Acad. Sci. USA 84: 595–599.

Whiting P, Lindstrom J (1988): Characterization of bovine and human neuronal nicotinic acetylcholine receptors using monoclonal antibodies. J. Neurosci. 8: 3395–3404.

Whiting P, Esch F, Shimasaki S, Lindstrom J (1987a): Neuronal nicotinic acetylcholine receptor β subunit is coded for by the cDNA clone α4. FEBS Lett. 219: 459–463.

Whiting P, Liu R, Morley B, Lindstrom J (1987b): Structurally different neuronal nicotinic acetylcholine receptor subtypes purified and characterized using monoclonal antibodies. J. Neurosci. 7: 4005–4016.

Whiting P, Schoepfer R, Conroy W, Gore M, Keyser K, Shimasaki S, Esch F, Lindstrom J (1991): Differential expression of nicotinic acetylcholine receptor subtypes in brain and retina. Mol. Brain Res. 10: 61–70.

Witzemann V, Barg B, Nishikawa Y, Sakmann B, Numa S (1987): Differential regulation of muscle acetylcholine receptor γ and ε subunit mRNAs. FEBS Lett. 223: 104–112.

Wonnacott S (1997): Presynaptic nicotinic ACh receptors. Trends Neurosci. 20: 92–98.

Xu W, Sutcliffe C, Lorenzo I, Goldberg L, Dang H, Patrick J, Beaudet A, Orr-Urtreger A (1997): Gene targeting of the $\alpha7$ and $\beta2$ subunits and the $\beta4$, $\alpha3$, and $\alpha5$ cluster of neuronal nicotinic acetylcholine receptors. Neuroscience Society Meeting Abstract 157.8 Washington, D.C.

Zhang Z, Coggan J, Berg D (1996): Synaptic currents generated by neuronal acetylcholine receptors to α bungarotoxin. Neuron 17: 1231–1240.

Zoli M, Picciotto M, Ferrari R, Changeux J (1997): A spatial learning deficit in aged mice lacking high affinity nicotinic receptors. Neuroscience Society Meeting Abstract 266.3 Washington, D.C.

2

Transcriptional Regulation of Neuronal Nicotinic Receptor Subunit Genes

Diego Fornasari, MD, PhD; E. Battaglioli, PhD; S. Terzano, PhD; and Francesco Clementi, MD

CNR Cellular and Molecular Pharmacology Center
Department of Medical Pharmacology
University of Milan
Milano, Italy

The vertebrate nervous system contains a large number of functionally distinct classes of neurons that have specific regional identities, unique phenotypes, and precise connectivities. This cellular heterogeneity is the consequence of distinct differentiation processes that mainly rely on transcriptional control mechanisms mediated by transcription factors that show discrete temporal patterns of region- and cell-type-specific expression (reviewed in Bang and Goulding, 1996; Quinn, 1996).

The endpoint of every differentiation process is the generation of a cell that expresses a restricted subset of genes, each of which performs highly specific functions. These genes determine the cell phenotype in such a way that an understanding of how their expression is regulated often offers insights into the genetic regulatory mechanisms underlying the cell phenotype itself. It is also likely that transcriptional changes in gene expression have important functions in many aspects of the plasticity of the nervous system, such as learning and memory.

Identification of the genetic mechanisms underlying the expression of neural genes is thus rapidly becoming an invaluable approach for developing our understanding of brain physiology. Neuronal nicotinic acetylcholine receptors (nAChRs) consist of subunits whose anatomi-

Neuronal Nicotinic Receptors: Pharmacology and Therapeutic Opportunities, Edited by S. P. Arneric and J. D. Brioni
ISBN 0-471-24743-x, pages 25–42. Copyright © 1998 by Wiley-Liss, Inc.

cal distribution and developmental regulation represent an attractive model for addressing some crucial questions concerning the generation and maintenance of distinct neural phenotypes (reviewed in McGehee and Role, 1995; Gotti et al., 1997). nAChR subunits may be expressed in very restricted areas of the brain (as in the case of the α2 subunit) or widely distributed throughout the central and peripheral nervous systems (as in the case of the β2 subunit). In rats, the expression of some subunits occurs very early during development, mainly at E11, a stage that coincides with the initial steps of neuronal differentiation (Zoli et al., 1995). This expression may remain stable throughout pre- and perinatal development, or may be transient, as in the case of the α3 subunit, whose widespread expression in the very early phases of central nervous system (CNS) development is rapidly followed by repression in the majority of the examined brain structures.

Our knowledge of the transcriptional mechanisms governing the expression of neuronal nAChR subunits is still in its infancy. The gene regulatory regions of some subunits still need to be characterized, and the already available data relating to the other members of this gene family are sometimes initial, and generally refer to a single species, with the exception of the α3 and α7 promoters. More precisely, information is available concerning the promoters of the chick α2 (Daubas et al., 1993; Bessis et al., 1993; Milton et al., 1995), α7 (Matter-Sadzinski et al., 1992), and β3 subunits (Hernandez et al., 1995; Matter et al., 1995); the mouse β2 subunit (Bessis et al., 1995, 1997); the rat α3 (Duvoisin and Heinemann, 1993; Yang et al., 1994, 1995, 1997; Boyd, 1994, 1996; McDonough and Deneris, 1997) and β4 subunits (Hu et al., 1994, 1995; Bigger et al., 1996; Du et al., 1997); the bovine α7 subunit (Criado et al., 1997); and the human α3 subunit (Fornasari et al., 1997). Although it is clear that much work still remains to be done, many important data have been already published, which are summarized and discussed in this review.

STRUCTURAL AND FUNCTIONAL CHARACTERIZATION OF NICOTINIC SUBUNIT GENE "CORE" PROMOTERS

On the basis of their expression and regulability profiles genes can be roughly classified into two main categories: "housekeeping" and tissue-specific genes. Housekeeping genes are expressed in every cell type because they provide basic functions that are essential for the cell survival; generally, their level of expression is also scarcely regulable. Tissue-specific genes are specifically expressed in one or very few cell types and their level of expression is generally modulated in response to very different requirements; although intermediate categories and exceptions exist, this classification is still in use.

Until a few years ago, it was generally accepted that GC-rich, TATA-less promoters were typical of "housekeeping" genes. As a consequence of the absence of a TATA box and its functional analog, the Initiator (reviewed in Smale, 1997), transcription initiation occurs at multiple nonadjacent sites that can span a region of 80–100 nucleotides. Generally, one or more GC boxes (the DNA elements capable of binding the ubiquitous transcription factor Sp1) are located upstream of the initiation sites and often represent the only recognizable *cis*-acting elements in the minimal promoter of these genes. In the absence of a TATA box, which is specifically recognized by TBP (transcription factor IID binding protein, reviewed in Buratowski, 1994), Sp1 seems to play a fundamental role in the recruitment of the general transcription factors that form the preinitiation complex (Pugh and Tjian, 1990, 1991).

On the contrary, the transcription of tissue-specific genes was thought to be exclusively driven by promoters with a TATA box located approximately 30 bp upstream of a unique initiation site. These promoters have a normal content of G and C nucleotides and often contain consensus sequences for distinct transcription factors that confer flexibility to gene expression. Among these consensus sequences, the CAAT box, which is recognized by various transcription factors, is conserved in the regulatory region of several tissue-specific genes.

Neuronal genes often do not follow this rule, since their promoters more closely resemble those of the housekeeping genes. With the exception of the chicken β3 subunit, this also holds true for the nAChR promoters that have been analyzed so far: They have a high GC content (with a maximum of 83% of G and C nucleotides in the chick α7 promoter), no TATA box, no CAAT box, multiple transcription start sites, and one or more Sp1 binding sites.

Sp1 Regulator of nAChR Subunit Gene Promoters

It has recently been shown that Sp1 is the prototype of a multigene family of zinc finger transcription factors, which includes Sp2, Sp3, and Sp4 (Kingsley and Winoto, 1992; Hagen et al., 1992). These four proteins are capable of recognizing GC or GT boxes, although Sp2 has a lower level of affinity for these DNA motifs. Sp1, Sp2, and Sp3 are ubiquitously distributed, but Sp4 is predominantly expressed in the brain, where it seems to play a role in reproductive behavior, as suggested by knock-out experiments (Supp et al., 1996). Interestingly, Sp1 and Sp3 have different effects on gene expression: Sp3 behaves as a repressor of Sp1-mediated transcription (Majello et al., 1994; Dennig et al., 1996; Kumar and Butler, 1997) and is thus an inhibitory member of this family of regulatory factors. Although Sp1 is ubiquitously present, the level of its expression varies in different cell types and during development, with differences of even 100-fold (Saffer et al., 1991); moreover, its DNA binding activity is regulated by phosphorylation (Daniel et al., 1996; Armstrong et al., 1997), also in response to external stimuli. Taken as a whole, this information clearly indicates that Sp1 can no longer be considered a housekeeping transcription factor involved in the constitutive expression of ubiquitous genes; rather, it is a dynamic regulator of transcription, whose activity is becoming more frequently associated with tissue-specific expression (Baker et al., 1996; Patterson et al., 1997).

In the nAChR field, the involvement of Sp1 has been directly evaluated for the rat α3 and β4 subunit promoters. In the α3 promoter (Yang et al., 1995), Sp1 binds to a GA motif a few nucleotides upstream of the transcription start site region (Fig. 2-1), but electromobility gel shift assay (EMSA) and supershift experiments have shown that in nuclear extracts of PC12 cells there are also Sp1-related proteins that binds this GA motif. Whether or not these proteins correspond to Sp2, Sp3, or Sp4 remains to be elucidated. To determine the functional role of the GA motif, transient transfection experiments have been carried out in which the wild-type promoter (−1647/+47) and a similar construct bearing a mutated α3 Sp1 binding site were analyzed for their ability to drive the expression of the luciferase in PC12 cells: The mutated construct showed a 4-fold decrease in the expression of the reporter gene, thus indicating that the GA motif is critical for the proper activity of the α3 promoter. Similar results have also been obtained with the rat β4 subunit promoter (Bigger et al., 1996), in which a CA box, a few nucleotides upstream of the transcription start sites (Fig. 2-1), binds Sp1 and, probably, Sp1-related proteins. Mutation of this Sp1 binding site leads to an approximately 80% reduction in luciferase activity in the SN17 neuronal cell line.

The α3 and β4 subunits are coexpressed in postganglionic neurons (in which they assemble to form the ganglionic-type nAChR) and in the medial habenula, the only brain structure in which the two subunits are coexpressed at a high level (Zoli et al., 1995). The fact that the activities of the α3 and β4 basal promoters are both strongly dependent on Sp1 may have the functional significance of promoting the coordinated expression of the two genes.

Until now, the only neuronal nicotinic gene regulatory region that has been characterized in humans is that of the α3 subunit (Fornasari et al., 1997). The human α3 promoter has all of

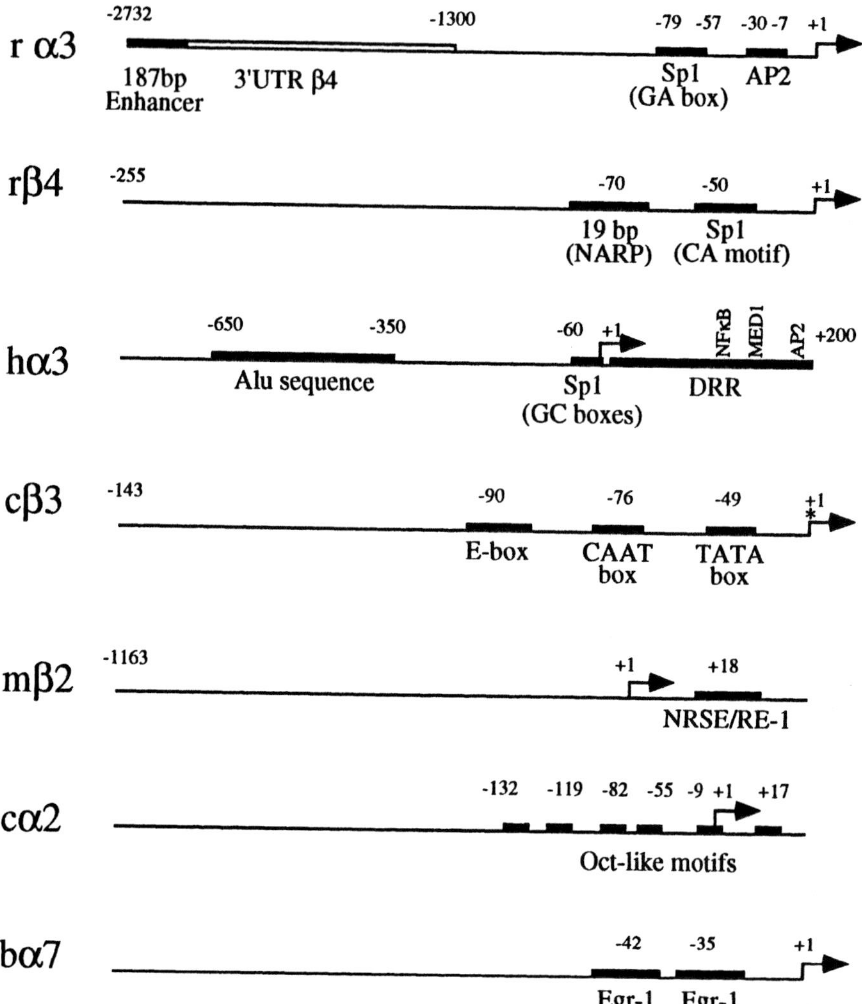

Figure 2.1. Promoter structure of neuronal nAChR subunit genes. Gene names are preceded by a letter indicating the species (r = rat, h = human, c = chicken, m = mouse, b = bovine). The arrows indicate the major transcriptional start site and the direction of the transcription of each gene; positive and negative numbering starts from this site. The asterisk indicates the presence of a unique transcription initiation site. Note that the lengths of the different regulatory regions are not to scale. The elements of each of the promoters discussed in the text are indicated.

the general features of its rat counterpart, with a 63% identity in a stretch of 178 nucleotides that roughly correspond to the basal promoters. Outside this region, we failed to align the two sequences, even though some scattered stretches of 6 to 7 nucleotides are identical in approximately corresponding positions of the two regulatory regions. Whether these conserved sequences are actually *cis*-acting elements remains to be demonstrated. The human α3 DNA region that lies immediately upstream of the cluster of transcription start sites contains four GC boxes that overlap with the same number of putative AP2 binding sites (Fig. 2-1). The functional role of the Sp1 transcription factor in the activity of the human α3 promoter has not been so finely characterized as in its rat counterpart. However, a 60 bp DNA fragment that apparently contains only the above-mentioned putative *cis*-acting elements is still capable of driving the expression of the luciferase reporter gene when transfected in human neuroblastoma cell lines (Terzano, unpublished results); furthermore, several distinct constructs that consistently promote the expression of the reporter gene become transcriptionally inactive when the 60 bp fragment is removed. Taken together, these observations strongly support the idea that the 60 bp fragment is essential for the proper function of the human α3 promoter and, by inference, that Sp1 may also play a pivotal role in the expression of this nAChR subunit gene.

Additional *Cis*-Acting Element in the Regulation of Basal Promoters

The rat β4 minimal promoter contains an additional 19 bp regulatory element (Du et al., 1997) immediately upstream of the Sp1 binding site (Fig. 2-1). Site-directed mutagenesis of this 19 bp sequence leads to an approximately 60% decrease in promoter activity. EMSAs using nuclear extracts from SN17 cells and different rat tissues show that the DNA element is capable of binding a nuclear protein with an apparent molecular weight of 50 kDa, as shown by crosslinking experiments, which is mainly but not exclusively expressed in the brain as well as in the SN17 cell line. This protein, which has been called NARP (neuronal ACh receptor promoter binding protein), has been purified from bovine brain by means of sequential chromatographic steps followed by final affinity chromatography on a column with covalently linked oligonucleotides containing several copies of the 19 bp element (Du et al., 1997). Analysis of the proteins eluted from this column revealed the presence of four distinct species with molecular masses of 31, 43, 65, and 114 kDa. The 43 kDa protein (NARP 43) has been microsequenced and corresponds to the transcription factor Purα. Purified Purα is actually capable of binding the 19 bp element, but it is not known whether it corresponds to the 50 kDa protein identified in the rat brain by crosslinking experiments. The identities and properties of the other three proteins that copurify with NARP 43 remain to be elucidated, as does the question of whether they directly interact with DNA or establish protein-protein interactions with NARP 43.

Purα was originally described as a sequence-specific single-stranded DNA binding protein, with high affinity for the purine-rich motif (GGN)n (Bergemann et al., 1992), and since it is capable of binding to some stably bent DNA regions involved in DNA replication, it was initially envisaged as an initiator of replication. However, its additional functions in the regulation of transcription have rapidly emerged with the discovery that Purα is involved in the gene expression regulation of the myelin basic protein in oligodendrocytes (Haas et al., 1995). Interestingly, Purα has also been shown to interact with the retinoblastoma protein Rb (Johnson et al., 1995). It has therefore been speculated that, in the context of the β4 basal promoter, this factor may establish a direct contact with the nearby Sp1, a protein that has also been shown to dimerize with a number of transcriptional activators, facilitating its interaction with the CA box.

The rat α3 basal promoter does not seem to contain any additional regulatory element, since the mutation of an AP2 binding site, located within the transcription start site region (Fig. 2-1), has very little effect on the activity of the promoter in PC12 (Yang et al. 1995). However, these authors did not find any AP2 DNA binding activity in the nuclear extract of PC12, even though AP2 is widely expressed in cells that derive from the neural crest. It is thus also possible that the lack of function of AP2 mainly reflects its absence or very low expression in that particular PC12 subclone, rather than an intrinsic property of the rat α3 promoter.

TISSUE- AND CELL-TYPE-SPECIFIC EXPRESSION OF NICOTINIC SUBUNIT GENES

The Promoter of the Chicken β3 Subunit

The chicken β3 subunit represents a sort of exception in the field of nAChRs because, as mentioned previously, its promoter is unique in having an appropriately spaced TATA box and a CAAT box (Fig. 2-1), a moderate GC content (49%), and only one transcription start site (Hernandez et al., 1995). The chicken β3 subunit is expressed in many developing sensory ganglia, but in the CNS its expression is confined to the retina, specifically to the ganglion cell layer and inner half of the inner nuclear layer (Hernandez et al., 1995). Very interestingly, a 143 bp DNA fragment containing the above-mentioned boxes is sufficient to direct the expression of a reporter gene in the specific subsets of retinal cells (Hernandez et al., 1995; Matter et al., 1995). This 143 bp promoter also possesses an E-box (Fig. 2-1), which may play an essential role in the proper neuronal and cell-type-specific expression of the β3 subunit. The E-box was originally described in muscle-specific genes as a CANNTG motif capable of binding members of the bHLH (basic helix-loop-helix) family of transcription factors, such as MyoD, that are essential for the muscle differentiation and development (reviewed in Olson and Klein, 1994). In the nervous system, the bHLH transcription factors were originally discovered in *Drosophila,* in which Achete-Scute complex (AS-C) and *atonal* work as "proneural genes" that induce the cells to assume a neural fate (reviewed in Jan and Jan, 1994; Campos-Ortega, 1993); subsequently, a number of bHLH proteins have also been identified in vertebrates that share homology with their *Drosophila* counterparts (Johnson et al., 1990; Jasoni et al., 1994; Lee et al., 1995).

Although the specific role of the E-box remains to be ascertained, it has been speculated that the 143 bp β3 subunit promoter may properly work with a limited number of genetic elements because the spatial and temporal expression patterns of the gene are rather simple. Indeed, the β3 promoter starts to be active at E4 and remains functional, from the beginning, only in a very restricted subset of retinal cells. On the contrary, other nAChR subunits display a wider distribution in the CNS and a more complex pattern of developmental regulation. For instance, the α7 subunit is expressed in many distinct areas of the CNS (Britto et al., 1992; Dominguez del Toro et al., 1994; Rubboli et al., 1994) but also in the autonomic nervous system (Zhang et al., 1996) as well as in some selected non-neuronal tissues (Tarroni et al., 1992; Romano et al., 1997a,b). At the beginning of the development of chicken brain, its promoter appears to be widely active in undifferentiated tissues and only subsequently acquires specificity for certain neuronal subtypes (Matter-Sadzinski et al., 1992). It is interesting to note that in chicken retina both β3 and α7 subunits are expressed in ganglion cells. However, the structures of the two promoters are quite different and the developmental strategies that are followed to obtain this cell-type-specific expression seem to be distinct and independent.

Cis-Acting Elements That Participate in Neuron-Specific Expression of the Human and Rat α3 Subunit Genes

Complex patterns of gene expression are likely to rely on the combination of positive and negative regulation mechanisms that restrict the expression of a certain gene to neuronal tissues and select the subpopulations of neuronal cells in which this expression has to be maintained. Relevant aspects of the neuro-specific expression of the human α3 nicotinic subunit have emerged from the study of the DNA region specifying the 5′ UTR of the mRNA, which lies immediately downstream of the previously described Sp1 site-containing fragment (Fornasari et al., 1997). The presence of this downstream regulatory region (DRR) (Fig. 2-1), which is located in the Acc III-Nco I fragment, quadruples the expression of the reporter gene in neuronal cells and its activity seems to rely more on transcriptional than post-transcriptional mechanisms. Indeed, the DRR contains putative *cis*-acting elements for the transcription a factors NFκB, AP2, and MED-1: The mutation of one of these sites leads to a 50% decrease in the activity of the region, whereas a double mutation in both the NFκB and AP2 sites completely abolishes the positive effect of the DRR. However, all the functional properties of the DRR cannot be easily explained only on the basis of the presence of these *cis*-acting elements. The activity of the DRR seems to be dependent on the cellular and promoter contexts in which it is placed. Indeed, the DRR does not affect the expression of the reporter gene in non-neuronal cells, and only modestly increases the expression of luciferase in neuronal cells, when placed downstream of the viral SV40 early promoter. A restricted interaction between a tissue-specific regulatory element and the natural core promoter has already been demonstrated for the human myoglobin gene, in which the muscle-specific enhancer is able to work in conjunction with the core promoter elements of the myoglobin gene but not in combination with the SV40 early promoter (Wefald et al., 1990). The activity of the DRR has also been investigated in a non-neuronal cell type that physiologically expresses the α3 transcript: the T lymphocyte (Mihovilovic and Roses, 1993). In these cells, the expression of the gene is driven by the same promoter as that operating in neurons, but the DRR is inactive (Battaglioli et al., 1998). Thus, the DRR is a neuron-specific element with positive effects on the expression of the α3 gene.

A distinct 187 bp DNA region that seems to participate in the neuro-specific expression of the α3 subunit has also been identified in the rat gene (Fig. 2-1), in the β4 3′ untranslated exon, approximately 2.5 Kb upstream of the basal promoter (McDonough and Deneris, 1997). As known, the β4, α3, and α5 subunit genes are located on the same chromosome, forming a genomic cluster that has been conserved during evolution. The 187 bp region contains two almost perfect 37 bp direct repeats that are separated from each other by 6 bp, and a nearly perfect 14 bp palindrome that begins 7 bp downstream of the second repeat. The presence of this 187 bp region in different reporter constructs increases the expression of the luciferase gene by approximately four times in PC12 cells. This region displays the functional properties of an enhancer, since it works in a position-, orientation-, and distance-independent fashion with regard to the α3 minimal promoter. It is noteworthy that the 187 bp fragment is also capable of stimulating the transcriptional activity of the β4 promoter. This strongly suggests that this region may also be implicated in the coordinated gene expression regulation of both the α3 and β4 subunit genes, in those districts in which the two subunits are actually coexpressed. As previously mentioned, this DNA region has a tissue-specific profile of activity, since it stimulates transcription in PC12 cell but not in non-neuronal cell lines.

The dissection of the 187 bp fragment has revealed that the functional properties of the entire region actually rely on the two 37 bp direct repeats. By EMSAs, the first repeat has been shown to bind a nuclear factor (or factors), which is differentially expressed in PC12; this binding can be challenged by the second repeat, thus suggesting that the two elements probably interact with the same protein(s). Altogether, these data clearly indicate that the two repeats form a tissue-specific enhancer that participates in the neuro-specific expression of the rat α3 subunit and probably also influences the expression of the upstream β4 subunit.

An in vivo study of the tissue- and cell-type-specific profile of the entire β4-α3 intergenic region, including the β4 3′ untranslated exon that contains the previously described enhancer, has been carried out in transgenic mice (Yang et al., 1997). The *LacZ* reporter gene was not expressed in non-neuronal tissues and in the CNS the expression of the transgene was mainly located in the nuclei that strongly transcribe the endogenous α3 gene. These data suggest that some DNA elements that restrict the expression of the α3 subunit to specific neuronal cell subsets should exist in the tested genomic fragment. However, surprisingly, no expression of *LacZ* was detected in autonomic ganglia, in which the α3 subunit is very abundantly expressed; this is also in contrast with what is observed in PC12 cells, in which the same regulatory region shows consistent and cell-type restricted activity. It has been suggested that a *cis*-acting element that is specifically active in postganglionic neurons may be missing from the tested genomic fragment. Since the luciferase reporter gene is much more sensitive than *LacZ*, the absence of this hypothetical element would become evident in transgenic mice rather than in PC12 cells that have been transfected with luciferase constructs. Alternatively, the lacking element might be essential to the α3 expression in postganglionic neurons of the autonomic nervous system but not in undifferentiated PC12, which, although closely related to autonomic neurons, is actually a pheochromocytoma cell line, derived from the endocrine cells of the adrenal medulla. Finally, the activity of the transgene in postganglionic neurons may be heavily dependent on the integration site in the mouse chromosome.

NRSF/REST: Negative Regulation in Non-Neuronal Cells

A *trans*-acting factor that represses the expression of neuronal genes in non-neuronal cells or undifferentiated neural progenitors has recently been identified and named neuron-restrictive silencer factor (NRSF) (Schoenherr and Anderson, 1995) or repressor element-1 silencing transcription factor (REST) (Chong et al., 1995), according to the two labs that cloned this protein. NRSF/REST contains nine zinc fingers related to the consensus sequence for zinc finger proteins in the Gli-Kruppel family, which allow the factor to bind a conserved DNA element (NRSE/RE1 = neural-restrictive silencer element/repressor element-1) located in the regulatory region of many neuronal genes (Schoenherr et al., 1996). Since NRSF/REST is generally not expressed, or only scarcely expressed in neurons, its target genes undergo negative regulation only in non-neuronal tissues (reviewed in Schoenherr and Anderson, 1995b).

Among the neuronal genes that are regulated by NRSF/REST, there is also the mouse β2 nicotinic subunit (Bessis et al., 1995). In this gene, a DNA element that binds NRSF/REST is located in the DNA region specifying the 5′ UTR of the mRNA (Fig. 2-1). The functional role of this element has been studied in both in vitro and in vivo models. In tissue culture models, an intact 1163 bp β2 promoter was highly active in a human neuroblastoma cell line (157 times over the background) but did not work at all in fibroblasts (Bessis et al., 1995). A double mutation in the DNA element NRSE/RE-1 led to a 3-fold increase in the expression of the reporter gene in neuronal cells but especially determined the appearance of a spectacular transcriptional activity in fibroblast cells that was comparable with that of the wild-type promoter in neurons (Bessis et al., 1995). This clearly suggested that the neural-restrictive DNA element was necessary and sufficient to confine the expression of the β2 subunit to neuronal cells, but the data have not fully confirmed this matter in transgenic mice: Whereas the wild-type 1.2 kb regulatory region appropriately drove the expression of the β-galactosidase reporter gene in many brain structures, but not in non-neuronal tissues (Bessis et al., 1995), the mutation of the NRSE/RE-1 had dramatic effects on the expression profile of the reporter gene in the CNS but did not produce any ectopic appearance of β-galactosidase activity in non-neuronal cells (Bessis et al., 1997). To reconcile this discrepancy, it has been speculated that additional neuro-specific elements may restrict the expression of the β2 subunit to the nervous system.

However, this hypothetical element(s) does not seem to be functional in the fibroblast cell line, as the NRSE mutation is sufficient to convert the silent β2 promoter to a very efficient transcriptional activator. Among the many differences that distinguish tissue culture cell line and transgenic mice experiments, the physical state of the tested construct may sometimes be critical. In transient transfection assays the construct remains episomal, whereas in transgenic animals it is integrated into the host chromosome and associated with chromatin, which is known to participate in the molecular mechanisms of gene expression. It is therefore possible that the hypothesized additional β2 promoter neuro-restrictive element(s) can properly work only in transgenic models.

Studies of the β2 NRSE/RE-1 have also produced new insights into the regulatory mechanisms that rely on this DNA element and its specific transcription factor. As previously mentioned, the mutation of this regulatory sequence in the β2 promoter also has effects on the expression of the reporter gene in the neuroblastoma cell line, leading to the relief of partial inhibition. This is in line with the observation that REST is also expressed at a low level in undifferentiated neurons, such as neuroblastoma cells (Nishimura et al., 1996). An even more dramatic effect has been observed in transgenic mice studied at E 13.5, in which NRSE mutation switched on the β2 promoter in the cortex. The same mutation in the same animals had the effect of switching off the expression of the reporter gene in the majority of the brain structures positive for the wild-type promoter, thus suggesting that NRSE/RE-1 may work as an enhancer in these districts (Bessis et al., 1997). These complex effects of NRSE/RE-1 on the activity of the β2 promoter in the brain of transgenic mice may be due to the presence of additional NRSE binding proteins with distinct functional properties, which may participate in the regulation of the cell-type-specific neuronal gene expression.

In parallel with these findings, it has also been shown that positive or negative effects of NRSE/RE-1 on gene expression in neuroblastoma cells may depend on the distance of this DNA element from the basal promoter. Indeed, NRSE/RE-1 becomes an enhancer when it is located close to the TATA box of the SV40 promoter (Bessis et al., 1997). It is still unclear if these results mainly reflect an intrinsic and autonomous property of NRSE-RE-1 or also depend on the particular promoter that has been used in these experiments. Actually, there is a discrepancy with what has been observed in the same neuroblastoma cells with the natural β2 promoter, in which a "close" NRSE/RE-1 works as a silencer rather than an enhancer (Bessis et al., 1995).

Positive and Negative Regulation of Gene Expression by POU-Domain Proteins

A few years ago, the almost simultaneous identification of four transcription factors (the ubiquitous Oct-1, the B-cell-restricted Oct-2, the pituitary-specific Pit-1/GHF-1, and Unc-86 from *Caenorhabditis Elegans*) and the comparison of their primary structures revealed the existence of a large region of sequence similarity that was called the POU (Pit, Oct, and Unc) domain. This is a bipartite structure consisting of two highly conserved regions separated by a hypervariable linker segment. The amino-terminal region has peculiar features that are exclusive to this family of proteins and, for this reason, has been called POU-specific domain; on the contrary, the carboxyl-terminal region is homologous to the homeodomain of several developmental regulators first identified in *Drosophila* and has therefore been called the POU homeodomain. Approximately 40 POU proteins have so far been identified and divided into six distinct classes on the basis of sequence similarities in the linker region and POU-specific domain (reviewed in Herr and Cleary, 1995; Ryan and Rosenfeld, 1997). POU-domain proteins have been shown to exert key developmental functions in early embryogenesis, as well as in cell-type-specific differentiation processes; several members of this family are also expressed in the nervous system (reviewed in McEvilly and Rosenfeld, 1997). There is evidence that two nAChR subunit genes might be regulated by POU-domain proteins: the chicken α2 and the rat α3 subunits.

The Expression of the Chicken α2 Subunit Gene Regulated by Class IV Members of the POU-Domain Protein Family

The α2 subunit has a highly restricted expression pattern in both rodents (Wada et al., 1989) and chicken (Daubas et al., 1990; Morris et al., 1990), a pattern that also seems to rely on negative mechanisms of transcriptional regulation. Indeed, the 5′-flanking region of the chicken gene contains a six-time repeated 11-nucleotide motif, which is similar to the Oct binding site (Fig. 2-1). When the DNA fragment containing the six repeats has been cloned upstream of a viral heterologous promoter, fused to the luciferase reporter gene, and transfected in different cell lines, inhibition of reporter gene expression has been observed in all of the tested cell lines, including neurons, with maximal effect in fibroblasts and PC12 cells (respectively 25% and 33% of the activity of the SV40 viral promoter alone) (Bessis et al., 1992). Interestingly, the silencing effect of the α2 DNA region was found to be completely dependent on the presence of all six octamer-like elements, since the presence of only one, two, or four elements enhanced the activity of the viral promoter. To try to explain this paradoxical situation, in which the sum of six enhancer elements produces a silencer effect, it can be speculated that the silencing activity of the intact region relies on a highly organized structure of DNA-binding proteins and corepressors. Changes in the number of Oct-like elements may modify the geometry of the complex and thus promote the access of different DNA-binding proteins or recruit coactivators instead of corepressors. The octamer-like motif is capable of binding a nuclear factor extracted from fibroblasts, as demonstrated by EMSA; however, since an oligonucleotide bearing the canonical octamer sequence was unable to compete with the α2 octamer-like element for binding with the fibroblast nuclear factor, this factor is distinct from the ubiquitous POU-domain protein Oct-1 (Bessis et al., 1992).

In order to evaluate whether POU-domain proteins could actually interact with the α2 silencer region, functional experiments have been carried out in which the previously described reporter construct was cotransfected with plasmids driving the expression of Brn-3a, Brn-3b, and Brn-3c (Milton et al., 1995). These transcription factors belong to class IV of the POU-domain protein family and are expressed in discrete regions of the midbrain, hindbrain, and spinal cord. In these experiments, the reporter construct was strongly transactivated by Brn-3b, whereas the other two transcription factors had no effect.

Brn-3b-mediated enhancement of gene expression was observed only when all six Oct-like motifs were present, whereas a reporter construct containing one, two, or four copies of the element did not respond. Interestingly, these deleted reporter constructs became positively regulated by Brn-3c, thus confirming the functional complexity of the chicken α2 regulatory region and the high degree of functional flexibility of the members of the POU-domain protein family.

Altogether, these findings seem to indicate that a still unidentified member of the POU-protein family may repress the expression of the α2 subunit gene in non-neuronal cells and in non-expressing neurons, whereas Brn-3b could be one of the positive regulators of the α2 promoter in a subset of nuclei in the CNS. Because of the difficulties of isolating the very few α2-expressing neurons, this hypothesis remains to be directly demonstrated. Interestingly, Brn-3b null mice have already been generated that survive through adulthood and are fertile. These animals have a severe loss of retinal ganglion cells that begins in postnatal life, whereas the majority of the brain structures seem to be normal (Erkman et al., 1996; Gan et al., 1996), and so they may represent a useful model for investigating (for instance, by in situ hybridization) whether the destruction of the Brn-3b gene has any effect on α2 nicotinic subunit expression.

Expression regulation by class IV members of the POU-domain protein family has also been investigated in other nicotinic subunit gene promoters. By means of cotransfection experiments, it has been shown that the rat α3 promoter is activated by Brn-3a but repressed by Brn-3b and Brn-3c, whereas the chicken α7 and the rat β4 subunit promoters are not affected by any of these transcription factors (Milton et al., 1996).

The Rat α3 Promoter Regulated by Oct-6/Scip/Tst-1

The rat α3 promoter has also been shown to be regulated by Oct-6/Scip/Tst-1 (Yang et al., 1994), which is a class III member of the POU-domain protein family. Oct-6/Scip/Tst-1 was originally described in oligodendrocyte precursor and developing Schwann cells, and considered a developmental regulator of myelination as knockout experiments subsequently demonstrated (Bermingham et al., 1996; Jaegle et al., 1996). However, Oct-6/Scip/Tst-1 is also expressed in several nuclei of the developing and adult brain (He et al., 1989; Zwart et al., 1996), including the medial habenula (Yang et al., 1994) in which the α3 nicotinic subunit is abundantly expressed. The rat α3 regulatory region contains putative DNA binding sites for this POU-domain transcription factor and DNase footprint analyses show that bacterially expressed Oct-6/Scip/Tst-1 protein is actually capable of binding these elements. Nevertheless, Oct-6/Scip/Tst-1 does not need to bind the DNA to transactivate the rat α3 promoter as cotransfection of a minimal α3 reporter construct, lacking any DNA binding site for this transcription factor, can still fully respond to Oct-6/Scip/Tst-1 in PC12 cells (Yang et al., 1994). Molecular dissection of this transcription factor has also shown that the POU-specific domain alone is sufficient to transactivate the α3 promoter (Fyodorov and Deneris, 1996).

Interestingly, the effects of Oct-6/Scip/Tst-1 are promoter specific, since little or no response is observed with different viral promoters, and tissue specific, since a strong transactivation of the α3 promoter has only been detected in PC12 (Yang et al., 1994). Since the binding of Oct-6/Scip/Tst-1 to DNA is completely dispensable for the transactivation of the rat α3 promoter, it becomes evident that protein-protein interactions should play a pivotal role in its activity. Oct-6/Scip/Tst-1 may interact with a transcription activator that directly contacts the DNA: One obvious candidate would be the transcription factor Sp1, whose function in the context of the rat α3 promoter has been previously discussed. However, specific experiments designed to test this hypothesis produced negative results (Fyodorov and Deneris, 1996). Oct-6/Scip/Tst-1 may also interact with a transcription activator that binds elements located in the DNA region specifying the 5′ UTR of the mRNA. As previously discussed, this region has peculiar properties in the human α3 promoter but has not yet been investigated in the rat counterpart.

Although direct interaction with a general transcription factor of the preinitiation complex cannot be ruled out, an alternative intriguing possibility is that Oct-6/Scip/Tst-1 actually titrates out a corepressor protein that, in its turn, might be bound to some factor in the basal complex. As for the other members of the POU-domain protein family, no physiological correlation between Oct-6/Scip/Tst-1 and α3 expression has yet been found. It is worth noting that the α3 gene is very abundantly transcribed in PC12, as may be expected from a cell line derived from the adrenal medulla, whereas Oct-6/Scip/Tst-1 is not expressed at all. It is therefore likely that this transcription factor works in these cells by mimicking the activity of an endogenous POU-domain protein, although the effects of Oct-6/Scip/Tst-1 are rather specific since other POU-domain proteins, such as Oct-1, Oct-2, and Brn-4 produce minimal effects on the activity of the rat α3 promoter (Yang et al., 1994).

In the CNS, the role of Oct-6/Scip/Tst-1 on the activity of the α3 promoter is not known, although the fact that the transcription factor and the nAChR subunit are coexpressed in the medial habenula is suggestive. Also, in this case, the use of Oct-6/Scip/Tst-1 null mice, some of which can survive several weeks after birth (Jaegle et al., 1996), may help to understand this issue.

GENE EXPRESSION REGULATION OF nAChR SUBUNIT GENE PROMOTERS BY EXTERNAL STIMULI AND SECOND MESSENGER SYSTEMS

The ability of neural cells to modify their pattern of gene expression in response to external stimuli is not only essential for the realization of specific differentiation programs, leading to the generation of different neuronal phenotypes, but also underlies most, if not all, of the adaptive processes known as neuronal plasticity, including learning and memory. The stimulation of membrane receptors by growth factors, cytokines, and neurotransmitters induces the activation of second messenger systems, which, by means of different pathways, can promote a very rapid and protein synthesis–independent transcription of a class of regulatory genes known as immediate early genes (IEGs) (reviewed in Hughes and Dragunow, 1995). IEGs encode transcription factors that modulate the expression of several target genes responsible for the phenotypical features of the cell, often referred to as "late-response genes." Among the different IEGs, Egr-1 is the prototype of a class of closely related transcription factors that also includes Egr-2, Egr-3, and Egr-4; these are expressed at basal level in many districts of the nervous system but are rapidly induced by various extracellular stimuli (reviewed in Beckmann and Wilce, 1997).

It has recently been shown that the bovine α7 subunit gene promoter is regulated by Egr-1 in chromaffin cells (Criado et al., 1997). Two Egr-1 DNA-binding sites have been identified in the minimal promoter (Fig. 2-1), located a few nucleotides upstream of the major transcription start point, and these bind the transcription factor with different affinities. Cotransfection experiments with a plasmid-expressing Egr-1 showed that different α7 reporter constructs were stimulated up to seven times by the transcription factor in chromaffin cells and in a human neuroblastoma cell line.

Since Egr-1 is expressed at a low level in bovine chromaffin cells, it probably participates in the regulation of the basal activity of the α7 promoter. However, the highly probable functional role of Egr-1 with regard to the α7 subunit is to transactivate its promoter in order to enhance the expression of the α7 gene in response to external cues. The identities of these cues are still unknown. Glucocorticoids released by the nearby cortex have been taken into account, but to the best of our knowledge no evidence of regulation of the Egr-1 gene by these steroids yet exists. It is worth noting that nAChRs have been shown to induce the expression of the IEG, c-fos, in differentiated PC12 cells, by increasing intracellular calcium concentration (Greenberg et al., 1986). Moreover, in vivo, the IP injection of nicotine induces rapid Egr-1 expression in rat superior cervical ganglion (Koistinaho et al., 1993). Thus, a potential trigger of Egr-1 induction in chromaffin cells may be the increase of Ca^{2+} ions, initiated by nAChR activation. A positive feedback in which an enhancement of splanchnic nerve activity is followed, after a few hours, by an increase in α7 expression may mediate a long-lasting modification (potentiation) in chromaffin cell activity.

The β4 subunit gene promoter has been shown to be regulated by NGF (Hu et al., 1994). In at least some subclones of PC12 cells, treatment with NGF induces a 4- to 5-fold increase in the expression of the two β4 endogenous transcripts. This increase mainly relies on transcriptional mechanisms, since different β4 promoter reporter constructs are consistently activated upon NGF treatment. The DNA elements that allow the β4 promoter to respond to this growth factor have not yet been investigated but were sufficient to confer NGF responsiveness approximately 100 bp upstream of the major transcription start site.

The pharmacological manipulation of second messenger systems has been exploited in order to study the regulability of α3 subunit expression in PC12 (Boyd, 1996). Increased cAMP levels due to forskolin administration, or reduced protein kinase C activity caused by chronic treatment with Phorbol Myristate Acetate (PMA), have been shown to increase the expression of the α3 transcripts as well as to activate different α3 promoter constructs, previously transfected in PC12.

Altogether, these data suggest that the transcription of neuronal nAChR subunit genes undergoes modifications in response to environmental signals, and thus strengthen the hypothesis that this class of receptors is involved in some adaptive functions of the nervous system.

SPECIES-SPECIFIC REGULATION OF GENE EXPRESSION

Differences in the regulatory mechanisms underlying neuronal nicotinic subunit gene expression in different species might be suspected on the basis of the not completely overlapping expression pattern of certain subunits in equivalent brain structures. It is conceivable that, during evolution, the acquisition or even loss of *cis*-acting elements in the transcriptional regulatory region may allow a nAChR subunit gene to be expressed in a new neuronal circuit or even to become regulated by additional external signals. One indication that this can occur has been given by the study of the human α3 nAChR subunit gene, whose transcriptional regulatory region, upstream and close to the basal promoter, contains an Alu repeat (Fig. 2-1) (Fornasari et al., 1997). This repeat shows 85% identity to the modern, primate-specific *Alu* consensus sequence described by Britten (Britten, 1994). Alu sequences are transposable elements, which, by moving into the proximity of gene promoters, are capable of modulating gene expression by means of different mechanisms, thus adding novel properties to the regulatory region. Therefore, Alu sequences are believed to be important sources of evolutionary changes (reviewed in McDonald, 1993; Britten, 1996).

The nearly 300-bp-long α3 Alu repeat is embedded in a DNA region of 600 bp that, when entirely removed, causes a 3- to 4-fold increase in the transcriptional activity of the downstream promoter. This indicates that negative regulatory elements are located in the 600 bp deleted fragment. Further dissections of this region have shown that the *Alu* sequence contains positive and negative elements that modulate the expression of the downstream α3 promoter, thus confirming its functional role in the expression of this gene. The specific DNA elements that, in the context of the Alu sequence, regulate α3 expression are currently under study.

CONCLUSION AND PERSPECTIVE

As previously discussed, many aspects of gene expression regulation of the neuronal nAChR subunit genes remain to be elucidated. Nevertheless, the data summarized here are very promising and, for their related implications, seem to anticipate even more relevant discoveries. The analysis and comparison in different species of the transcription regulatory regions controlling the expression of a certain subunit will help to define the conserved regulatory mechanisms responsible for tissue specificity or responsiveness to external cues, as well as to unravel the more recently evolved elements that may confer further regulability and flexibility to gene expression.

The study of the regulation of the promoter activity of nAChR subunit genes by extracellular signals will provide new insights into the role of these receptors in neuronal plasticity. The characterization of human promoters may be useful in the light of the possible involvement of nAChRs in brain pathologies.

An altered expression of neuronal nAChRs has been documented in Alzheimer's disease, Parkinson's disease, Lewy's body dementia, and schizophrenia, and the involvement of the nAChR system in Tourette's syndrome has been suspected on the basis of the beneficial effects of nicotine on disease symptoms (see Gotti et al., 1997, for an overview). In some of these pathological conditions the defect in the expression of the receptors may be directly related to alterations in the promoter activity of specific nAChR subunit genes. The role of promoters in the pathogenesis of brain diseases has recently received a great deal of attention since the documentation of an association between anxiety-related traits and polymorphism of the serotonin transporter regulatory region (Lesch et al., 1996).

The data summarized in this review clearly indicate that such studies are not only useful for acquiring a greater understanding of the physiological significance of this class of receptor molecules in the nervous system but are also likely to have a more general impact on our knowledge of transcription regulation in neural cells.

REFERENCES

Armstrong SA, Barry DA, Leggett RW, Mueller CR (1997): Casein kinase II-mediated phosphorylation of the C terminus of Sp1 decreases its DNA binding activity. J. Biol. Chem. 272: 13,489–13,495.

Baker DL, Dave V, Reed T, Periasamy M (1996): Multiple Sp1 binding sites in the cardiac/slow twitch muscle sarcoplasmatic reticulum Ca^{2+}-ATPase gene promoter are required for expression in Sol8 muscle cells. J. Biol. Chem. 271: 5921–5928.

Bang AG, Goulding MD (1996): Regulation of vertebrate neural cell fate by transcription factors. Curr. Opin. Neurobiol. 6: 25–32.

Beckmann AM, Wilce PA (1997): Egr transcription factors in the nervous system. Neurochem. Int. 31: 477–510.

Bergemann AD, Ma ZW, Johnson EM (1992): Sequence of cDNA comprising the human *pur* gene and sequence-specific single-stranded-DNA-binding properties of the encoded protein. Mol. Cell Biol. 12: 5673–5682.

Bermingham JR, Scherer SS, O'Connell S, Arroyo E, Kalla KA, Powell FL, Rosenfeld MG (1996): Tst-1/Oct-6/SCIP regulates a unique step in peripheral myelination and is required for normal respiration. Genes Dev. 10: 1751–1762.

Bessis A, Savatier N, Devillers-Thiery A, Bejanin S, Changeux JP (1993): Negative regulatory elements upstream of a novel exon of the neuronal nicotinic acetylcholine receptor α2 subunit gene. Nucleic Acid Res. 21: 2185–2192.

Bessis A, Salmon AM, Zoli M, Le Novere N, Picciotto M, Changeux JP (1995): Promoter elements conferring neuron-specific expression of the β2 subunit of the neuronal nicotinic acetylcholine receptor studied in vitro and in transgenic mice. Neuroscience 69: 807–819.

Bessis A, Champtiaux N, Chatelin L, Changeux JP (1997): The neuron-restrictive silencer element: a dual enhancer/silencer crucial for patterned expression of a nicotinic receptor gene in the brain. Proc. Natl. Acad. Sci. USA 94: 5906–5911.

Bigger CB, Casanova EA, Gardner PD (1996): Transcriptional regulation of neuronal nicotinic acetylcholine receptor genes. J. Biol. Chem. 271: 32,842–32,848.

Boyd RT (1994): Sequencing and promoter analysis of the genomic region between the rat neuronal nicotinic acetylcholine receptor β4 and α3 genes. J. Neurobiol. 25: 960–973.

Boyd RT (1996): Transcriptional regulation and cell specificity determinants of the rat nicotinic acetylcholine receptor α3 gene. Neurosci. Lett. 208: 73–76.

Britten RJ (1994): Evolutionary selection against change in many Alu repeat sequences interspersed through primate genomes. Proc. Natl. Acad. Sci. USA 91: 5992–5996.

Britten RJ (1996): DNA sequence insertion and evolutionary variation in gene regulation. Proc. Natl. Acad. Sci. USA 93: 9374–9377.

Britto LRG, Keyser KT, Lindstrom JM, Karten HJ (1992): Immunohistochemical localization of nicotinic acetylcholine receptor subunits in the mesencephalon and diencephalon of the chick (*Gallus gallus*). J. Comp. Neurol. 317: 325–340.

Buratowski S (1994): The basics of basal transcription by RNA polymerase II. Cell 77: 1–3.

Campos-Ortega JA (1993): Mechanisms of early neurogenesis in *Drosophila melanogaster.* J. Neurobiol. 24: 1305–1327.

Chong JA, Tapia-Ramirez J, Kim S, Toledo-Aral JJ, Zheng Y, Boutros MC, Altshuller YM, Frohman MA, Krnaer SD, Mandel G (1995): REST: a mammalian silencer protein that restricts sodium channel expression to neurons. Cell 80: 949–957.

Criado M, Dominguez del Toro E, Carrasco-Serrano C, Smillie FI, Juiz JM, Viniegra S, Ballesta JJ (1997): Differential expression of α-bungarotoxin-sensitive neuronal nicotinic receptors in adrenergic chromaffin cells: a role for transcription factor Egr-1. J. Neurosci. 17: 6554–6564.

Daniel S, Zhang S, DePaoli-Roach AA, Kim KH (1996): Dephosphorylation of Sp1 by protein phosphatase 1 is involved in the glucose-mediated activation of the acetyl-CoA carboxylase gene. J. Biol. Chem. 271: 14,692–14,697.

Daubas P, Devillers-Thiery A, Geoffroy B, Martinez S, Bessis A, Changeux JP (1990): Differential expression of the neuronal acetylcholine receptor α2 subunit gene during chick brain development. Neuron 5: 49–60.

Daubas P, Salmon AM, Zoli M, Geoffroy B, Devillers-Thiery A, Bessis A, Medevielle F, Changeux JP (1993): Chicken neuronal acetylcholine receptor α2-subunit gene exhibits neuron-specific expression in the brain and spinal cord of transgenic mice. Proc. Natl. Acad. Sci. USA 90: 2237–2241.

Dennig J, Beato M, Suske G (1996): An inhibitor domain in SP3 regulates its glutamine-rich activation domains. EMBO J. 15: 5659–5667.

Dominguez del Toro E, Juiz JM, Peng X, Lindstrom J, Criado M (1994): Immunocytochemical localization of the α7 subunit of the nicotinic acetylcholine receptor in the rat central nervous system. J. Comp. Neurol. 349: 325–342.

Du Q, Tomkinson AE, Gardner PD (1997): Transcriptional regulation of neuronal nicotinic acetylcholine receptor genes. J. Biol. Chem. 272: 14,990–14,995.

Duvoisin RM, Heinemann S (1993): Transcription control elements of the rat neuronal nicotinic acetylcholine receptor subunit α3. Braz. J. Med. Biol. Res. 26: 137–150.

Erkman L, McEvilly RJ, Luo L, Ryan AK, Hooshmand F, O'Connell SM, Keithley EM, Rapaport DH, Ryan AF, Rosenfeld MG (1996): Role of transcription factors Brn-3.1 and Brn-3.2 in auditory and visual system development. Nature 381: 603–606.

Fornasari D, Battaglioli E, Flora A, Terzano S, Clementi F (1997): Structural and functional characterization of the human α3 nicotinic subunit gene promoter. Mol. Pharmacol. 51: 250–261.

Fyodorov D, Deneris E (1996): The POU domain of SCIP/Tst-1/Oct-6 is sufficient for activation of an acetylcholine receptor promoter. Mol. Cell Biol. 16: 5004–5014.

Gan L, Xiang M, Zhou L, Wagner DS, Klein WH, Nathans J (1996): POU domain factor Brn-3b is required for the development of a large set of retinal ganglion cells. Proc. Natl. Acad. Sci. USA 93: 3920–3925.

Gotti C, Fornasari D, Clementi F (1997): Human neuronal nicotinic receptors. Prog. Neurobiol. 53: 199–237.

Greenberg ME, Ziff EB, Greene LA (1986): Stimulation of neuronal acetylcholine nicotinic receptors induces rapid gene transcription. Science 234: 80–83.

Haas S, Thatikunta P, Steplewski A, Johnson E, Khalili K, Amini S (1995): A 39-kD DNA-binding protein from mouse brain stimulates transcription of myelin basic protein gene in oligodendrocytic cells. J. Cell Biol. 130: 1171–1179.

Hagen G, Muller S, Beato M, Suske G (1992): Cloning by recognition site screening of two novel GT box binding protein: a family of Sp1 related genes. Nucleic Acids Res. 20: 5519–5525.

He X, Treacy MN, Simmons DM, Ingraham HA, Swanson LW, Rosenfeld MG (1989): Expression of a large family of POU-domain regulatory genes in mammalian brain development. Nature 340:35–41.

Hernandez MC, Erkman L, Matter-Sadzinski L, Roztocil T, Ballivet M, Matter JM (1995): Characterization of the nicotinic acetylcholine receptor β3 gene. J. Biol. Chem. 270: 3224–3233.

Herr W, Cleary MA (1995): The POU domain: versatility in transcriptional regulation by a flexible two-in-one DNA-binding domain. Genes Dev. 9: 1679–1693.

Hu M, Whiting Theobald NL, Gardner PD (1994): Nerve growth factor increases the transcriptional activity of the rat neuronal nicotinic acetylcholine receptor β4 subunit promoter in transfected PC12 cells. J. Neurochem. 62: 392–395.

Hu M, Bigger CB, Gardner PD (1995): A novel regulatory element of a nicotinic acetylcholine receptor gene interacts with a DNA binding activity enriched in rat brain. J. Biol. Chem. 270: 4497–4502.

Hughes P, Dragunow M (1995): Induction of immediate-early genes and the control of neurotransmitter-regulated gene expression within the nervous system. Pharmacol. Rev. 47: 133–178.

Jaegle M, Mandemakers W, Broos L, Zwart R, Karis A, Visser P, Grosveld F, Meijer D (1996): The POU factor Oct-6 and Schwann cell differentiation. Science 273: 507–510.

Jan YN, Jan LY (1994): Genetic control of cell fate specification in *Drosophila* peripheral nervous system. Ann. Rev. Genetics 28: 373–393.

Jasoni CL, Walker MB, Morris MD, Reh TA (1994): A chicken achaete-scute homolog (CASH-1) is expressed in a temporally and spatially discrete manner in the developing nervous system. Development 120: 769–783.

Johnson EM, Chen PL, Krachmarov CP, Barr SM, Kanowsky M, Ma ZW, Lee WH (1995): Association of human Purα with the retinoblastoma protein; Rb, regulates binding to the single-stranded DNA Purα recognition element. J. Biol. Chem. 270: 24,352–24,360.

Johnson JE, Birren SJ, Anderson DJ (1990): Two rat homologues of *Drosophila* achete-scute specifically expressed in neuronal precursor. Nature 346: 858–861.

Kingsley C, Winoto A (1992): Cloning of GT box-binding proteins: a novel Sp1 multigene family regulating T-cell receptor gene expression. Mol. Cell Biol. 12: 4251–4261.

Koistinaho J, Pelto-Huikko M, Sagar SM, Dagerlind A, Roivainen R, Hokfelt T (1993): Differential expression of immediate-early genes in the superior cervical ganglion after nicotine treatment. Neuroscience 56: 729–739.

Kumar AP, Butler AP (1997): Transcription factor Sp3 antagonizes activation of decarboxylase promoter by Sp1. Nucleic Acids Res. 25: 2012–2019.

Lee JE, Hollemberg SM, Snider L. Turner DL, Lipnick N, Weintraub H (1995): Conversion of *Xenopous* ectoderm into neurons by NeuroD, a basic helix-loop-helix protein. Science 268: 836–844.

Lesch KP, Bengel D, Heils A, Sabol SZ, Greenberg BD, Petri S, Benjamin, J, Muller CR, Hamer DH, Murphy DL (1996): Association of anxiety-related traits with a polymorphism in the serotonin transporter gene regulatory region. Science 274: 1527–1531.

Majello B, DeLuca P, Hagen G, Suske G, Lania L (1994): Different members of the Sp1 multigene family exert opposite transcriptional regulation of the long terminal repeat of HIV-1. Nucleic Acids Res. 22: 4914–4921.

Matter JM, Matter-Sadzinski L, Ballivet M (1995): Activity of the β3 nicotinic receptor promoter is a marker of neuron fate determination during retina development. J. Neurosci. 15: 5919–5928.

Matter-Sadzinski L, Hernandez MC, Roztocil T, Ballivet M, Matter JM (1992): Neuronal specifity of the α7 nicotinic acetylcholine receptor promoter develops during morphogenesis of the central nervous system. EMBO J. 11: 4529–4538.

McDonald JF (1993): Evolution and consequences of transposable elements. Curr. Opin. Genet. Dev. 3: 855–864.

McDonough J, Deneris E (1997): β43′: enhancer displaying neural-restricted activity is located in the 3′-untranslated exon of the rat nicotinic acetylcholine receptor β4 gene. J. Neurosci. 17: 2273–2283.

McEvilly RJ, Rosenfeld MG (1997): Genetically defined roles of class I, class III, and class IV POU domain factors in the development of the mammalian endocrine and nervous systems. Curr. Opin. Endocrinol. Diabetes 4: 172–183.

McGehee DS, Role L (1995): Physiological diversity of nicotinic acetylcholine receptors expressed by vertebrate neurons. Ann. Rev. Physiol. 57: 521–546.

Mihovilovic M, Roses AD (1993): Expression of α-3, α-5, and β-4 neuronal acetylcholine receptor subunit transcripts in normal and myastenia gravis thymus. J. Imunol. 151: 6517–6524.

Milton NG, Bessis A, Changeux JP, Latchman DS (1995): The neuronal nicotinic acetylcholine receptor α2 subunit gene promoter is activated by the Brn-3b POU family transcription factor and not by Brn-3a or Brn-3c. J. Biol. Chem. 270: 15,143–15,147.

Milton NGN, Bessis A, Changeux JP, Latchman DS (1996): Differential regulation of neuronal nicotinic acetylcholine receptor subunit gene promoters by Brn-3 POU family transcription factors. Biochem. J. 317: 419–423.

Morris BJ, Hicks AA, Wisden W, Darlison MG, Hunt SP, Barnard EA (1990): Distinct regional expression of nicotinic acetylcholine receptor genes in chick brain. Mol. Brain Res. 7: 305–315.

Nishimura E, Sasaki K, Maruyama K, Tsukada T, Yamaguchi K (1996): Decrease in neuron-restrictive silencer factor (NRSF) mRNA levels during differentiation of cultured neuroblastoma cells. Neurosci. Lett. 211: 101–104.

Olson EN, Klein WH (1994): bHLH factors in muscle development: dead lines and commitments, what to leave in and what to leave out. Genes Dev. 8: 1–8.

Patterson C, Wu Y, Lee ME, DeVault JD, Runge MS, Haber E (1997): Nuclear protein interactions with the human KDR/flk-1 promoter in vivo. J. Biol. Chem. 272: 8410–8416.

Pugh BF, Tjian R (1990): Mechanisms of transcriptional activation by Sp1: evidence for coactivators. Cell 61: 1187–1197.

Pugh BF, Tjian R (1991): Transcription from a TATA-less promoter requires a multisubunit TFIID complex. Gene Dev. 5: 1935–1945.

Quinn JP (1996): Neuronal-specific gene expression. The interaction of both positive and negative transcriptional regulators. Prog. Neurobiol. 50: 363–379.

Romano SJ, Corriveau RA, Schwarz RI, Berg DK (1997a): Expression of the nicotinic receptor α7 gene in tendon and periosteum during early development. J. Neurochem. 68: 640–648.

Romano SJ, Pugh PC, McIntosh JM, Berg DK (1997b): Neuronal-type acetylcholine receptors and regulation of α7 gene expression in vertebrate skeletal muscle. J. Neurobiol. 32: 69–80.

Rubboli F, Court JA, Sala C, Morris C, Chini B, Perry E, Clementi F (1994): Distribution of nicotinic receptors in human hippocampus and thalamus. Eur. J. Neurosci. 6: 1596–1604.

Ryan AK, Rosenfeld MG (1997): POU domain family values: flexibility, partnerships, and developmental codes. Genes Dev. 11: 1207–1225.

Saffer JD, Jackson SP, Annarella MB (1991): Developmental expression of Sp1 in the mouse. Mol. Cell Biol. 11: 2189–2199.

Schoenherr CJ, Anderson DJ (1995a): The neuron-restrictive silencer factor (NRSF): a coordinate repressor of multiple neuron-specific genes. Science 267: 1360–1363.

Schoenherr CJ, Anderson DJ (1995b): Silencing is golden: negative regulation in the control of neuronal gene transcription. Curr. Opin. Neurobiol. 5: 566–571.

Schoenherr CJ, Paquette AJ, Anderson DJ (1996): Identification of potential target genes for the neuron-restrictive silencer factor. Proc. Natl. Acad. Sci. USA 93: 9881–9886.

Smale ST (1997): Transcription initiation from TATA-less promoters within eukariotic protein-coding genes. Biochim. Biophys. Acta 1351: 73–88.

Supp DM, Witte DP, Branford WW, Smith EP, Potter SS (1996): Sp4, a member of the Sp1-family of zinc finger transcription factors, is required for normal murine growth, viability and male fertility. Dev. Biol. 176: 284–299.

Tarroni P, Rubboli F, Chini B, Zwart R, Oortgiesen M, Sher E, Clementi F. (1992): Neuronal-type nicotinic receptors in human neuroblastoma and small-cell lung carcinoma cell lines. FEBS Letters 312: 66–70.

Wada E, Wada K, Boulter J, Deneris E, Heinemann S, Patrick J, Swanson LW (1989): Distribution of α2, α3, α4 and β2 neuronal nicotinic receptor subunit mRNA in the central nervous system: a hybridization histochemical study in the rat. J. Comp. Neurol. 284: 314–335.

Wefald FC, Devlin BH, William RS (1990): Functional hetereogeneity of mammalian TATA-box sequences revealed by interaction with a cell-specific enhancer. Nature 344: 260–262.

Yang X, McDonough J, Fyodorov D, Morris M, Wang F, Deneris ES (1994): Characterization of an acetylcholine receptor α3 gene promoter and its activation by the POU domain factor Scip/Tst-1. J. Biol. Chem. 269: 10,252–10,264.

Yang X, Fyodorov D, Deneris ES (1995): Transcriptional analysis of acetylcholine receptor α3 gene promoter motifs that bind Sp1 and AP-2. J. Biol. Chem. 270: 8514–8520.

Yang X, Yang F, Fyodorov D, Wang F, McDonough J, Herrup K, Deneris E (1997): Elements between the protein-coding regions of the adjacent β4 and α3 acetylcholine receptor genes direct neuron-specific expression in the central nervous system. J. Neurobiol. 32: 311–324.

Zhang Z, Coggan JS, Berg DK (1996): Synaptic currents generated by neuronal acetylcholine receptors sensitive to α-bungarotoxin. Neuron 17: 1231–1240.

Zoli M, Le Novere N, Hill JA, Changeux JP (1995): Developmental regulation of nicotinic Ach receptor subunit mRNA in the rat central and peripheral nervous system. J. Neurosci. 15: 1912–1939.

Zwart R, Broos L, Grosveld G. Meijer D (1996): The restricted expression pattern of the POU factor Oct-6 during early development of the mouse nervous system. Mech. Dev. 54: 185–194.

3

Molecular Composition and Biophysical Characteristics of Nicotinic Receptors

J. Ramirez-Latorre, PhD
Department of Biology
Temple University
Philadelphia, Pennsylvania

G. Crabtree, BS, Jennifer Turner, MA, and Lorna Role, PhD
Columbia College of Physicians and Surgeons
Department of Anatomy and Cell Biology in the Center for Neurobiology and Behavior
New York, New York

Nicotine has been demonstrated to have diverse effects on the nervous system. In addition to nicotine's well-known potency as an addictive agent, administration of nicotine has been shown to enhance "working" memory, enhance attention, increase arousal and alertness, and decrease anxiety (for reviews, see Clarke et al., 1995; Decker et al., 1995; Dehaene and Changeux, 1995; James and Nordberg, 1995; Schroder et al., 1995; Vidal, 1996). Indeed, even if smoking were not addictive, the reportedly positive effects of nicotine on cognition and mood might well support its continued (ab)use.

Many of these effects of nicotine—to which most smokers can attest—have now been

Neuronal Nicotinic Receptors: Pharmacology and Therapeutic Opportunities, Edited by S. P. Arneric and J. D. Brioni
ISBN 0-471-24743-x, pages 43–64. Copyright © 1998 by Wiley-Liss, Inc.

traced to specific changes in neural excitability within discreet regions of the nervous system. These many effects are mediated by nicotine's interaction with particular cellular receptors—those that normally bind the neurotransmitter acetylcholine (ACh). Nicotine's diverse and complex activities in the nervous system are due to its ability to qualitatively mimic an important endogenous neurotransmitter by interacting with a subset of the receptor sites for ACh.

As you will recall, there are three broad classes of ACh receptors: muscarinic receptors, muscle-type nicotinic receptors, and neuronal nicotinic ACh receptors (nAChRs). Muscle-type nicotinic receptors bind nicotine relatively poorly, and although there are many muscarinic receptors expressed by neurons, smooth muscle, and gland cells, nicotine does not interact with muscarinic sites. Neuronal nAChRs are nicotine's major target in the body—and the subject of this chapter.

We focus our discussion on recent work addressing the subunit composition and consequent biophysical properties of nAChRs. Although at first glance the reader may think the topic is rather dry, the structure and functional features of nAChRs are actually the key physiological determinants of nicotine's diverse biological activities. Specifically, we consider the physiology of recombinant nAChRs of known composition and the molecular determinants of native nAChRs, native nAChR subtypes, and targeting of nAChRs to subcellular domains.

KEY BIOPHYSICAL AND PHARMACOLOGICAL PROPERTIES OF NICOTINE-GATED CHANNELS

The net effects of nicotine are determined (in part) by the particular subtype(s) of nAChR(s) with which they interact. The term *nAChR subtypes* refers to the fact that nAChRs can be composed of a variety of subunits and that these differences in subunit composition manifest in meaningful differences in nAChR properties. Thus, nAChRs that differ in subunit composition can be distinguished from each other by distinct pharmacological and biophysical profiles (Fig. 3-1 and references therein). In addition, it is widely thought, though not yet proven, that differences in nAChR composition may subserve differential susceptibility to modulation by post-translational mechanisms. Post-translational modifications such as phosphorylation and glycosylation can modulate nAChR assembly and/or the functional properties of the resultant nAChR complex.

Functional Distinctions Among nAChR Subtypes

As the variety of neuronal nAChR complexes is due, in large part, to differences in subunit composition, a brief review of the subunits that (currently) comprise the nAChR gene family is appropriate. Since the first cloning of a neuronal nicotine receptor subunit gene in 1986 by Heinemann, Patrick, and their colleagues (Boulter et al., 1986), 10 additional sequences have been identified (for reviews, see Le Novre and Changeux, 1995; Ortells and Lunt, 1995). The family now includes α2, α3, α4, α5, α6, α7, α8, α9, β2, β3, and β4. The first round of functional characterization of each of these subunit genes has employed heterologous expression of the subunit cRNA (or cDNA), both alone and in combination with other α or β subunits. This approach has been enormously informative and has yielded the bulk of what we know today about the various homomeric and heteromeric nAChR subunit combinations that *can* form (for reviews, see Patrick et al., 1993; Sargent, 1993; Lindstrom et al., 1995; McGehee and Role, 1995; Changeux et al., 1996; Colquhoun and Patrick, 1997; also: Gross et al., 1991; Bertrand et al., 1993; Gerzanich et al., 1994; Hussy et al., 1994; Cohen et al., 1995; Galzi et al., 1996;

AChR Composition	α3β4	α3β2	α4β2*	α4β2α5*	α7*	α7α5β2/4
Channel Profile	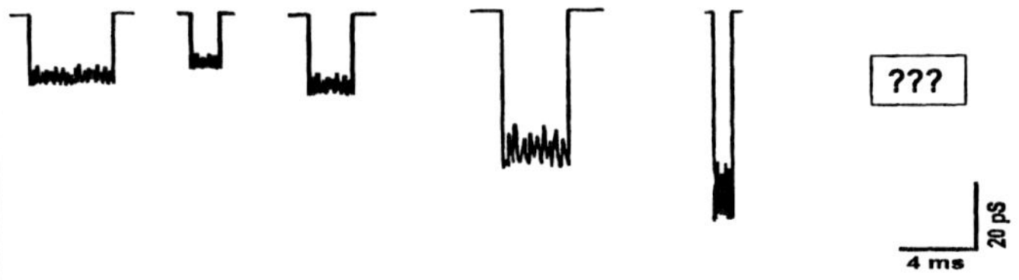					??? (4 ms; 20 pS)
P_{Ca}/P_{Na}	~1	~1	~1	ND	12	ND
ACh EC_{50} (μM)	30	200	1	150	~200	~5000
ACh Affinity (nM) ^	800	50	60	ND	ND	ND
Nicotine EC_{50} (μM)	2	ND	~1	~5	~2	ND
Nicotine Affinity (nM)^	400	30	8	ND	ND	ND
nBgTx (τ_{off} min)	< 0.1	~120	<0.1	ND	ND	ND
DHβE (K_i, μM) ^^	30	0.2	0.4	ND	ND	ND
αBgTx block	no	no	no	no	yes < 1 nM	yes (>10 nM)
IC_{50} MLA	>1 μM	>1 μM	>1 μM	>1 μM	~100 pM	(~10 pM)

Figure 3.1. *All data are from the heterologous expression of rat nAChRs in* Xenopus *oocytes unless otherwise noted. Recent reviews: McGehee and Role, 1995; Lindstrom, 1996; Role and Berg, 1996; Colquhoun and Patrick, 1997. Primary references: Ballivet et al., 1988; Papke et al., 1989, 1997; Revah et al., 1991; Charnet et al., 1992; Bertrand et al., 1993b; Lester and Dani, 1993; Luetje et al., 1993; Covernton et al., 1994; Cohen et al., 1995; Harvey et al., 1996; Harvey and Luetje, 1996; Ramirez-Latorre et al., 1996; Wang et al., 1996; Ragozzino et al., 1997; Parker et al., 1997; Luetje et al., 1997. Definition of symbols: * indicates data from studies of chick nAChRs. ^ indicates data from Luetje and colleagues assessing relative agonist affinity by competition with binding of 3H-epibatadine: Parker and Luetje, 1997. ^^ indicates data from Luetje et al., personal communication. Data in parentheses are estimates from preliminary studies by the authors.*

Gopalakrishan et al., 1996; Rogers and Dani, 1996; Wang et al., 1996; Fenster et al., 1997; Rogazzino et al., 1997; Sivilotti et al., 1997). One must, however, keep in mind the emphasis on *can* form, as an inherent limitation of this sort of analysis. From these studies we know that combinations of α2, α3, and α4 with either β2 or β4 can form functional complexes and, furthermore, that these channels are distinct from each other in their pharmacological and biophysical profiles. Likewise, heterologous expression studies of α7, α8, and α9 have demonstrated that each of these subunits can form homomeric complexes with particular pharmacological and biophysical characteristics that distinguish them both from each other and from their α/β siblings (Couturier et al., 1990; Anand et al., 1993; Bertrand et al., 1993; Seguela et al., 1993; Elgoyhen et al., 1994; Gerzanich et al., 1994). These and other key features of specific nAChR complexes, as deduced from heterologous expression studies, are summarized in Figure 3-1. Highlights of recent studies of the structural and functional properties of various nAChR subtypes are discussed in more detail below.

One note before we plunge into the *Xenopus* oocyte and heterologous expression. The ultimate goal of subjecting recombinant nAChRs to detailed pharmacological scrutiny and structure/function analyses is, of course, to better understand how *native* nAChRs actually work. More immediate goals are to discern key physiological properties or "fingerprints" of the nAChR subtypes and ascertain which pharmacological probes might facilitate the identification and study of specific nAChR subtypes in vivo. This cataloguing of characteristics, though perhaps less glamorous than in vivo work, is a necessary prerequisite to being able to reliably discern *which* nAChR subtype(s) are involved in *each* component of nicotine's broad spectrum of biological activities. In addition, a detailed understanding of nAChR structure and function is essential to the development of nAChR-subtype-specific compounds that are potentially useful in smoking cessation programs or in the treatment of various neurological disorders (see Part V). To date, such efforts have used fairly classic pharmacological approaches and have focused on the identification of compounds with nAChR agonist activity (e.g., Decker et al., 1993; Arneric et al., 1994; Papke et al., 1996, 1997; Schrattenholtz et al., 1996; Briggs et al., 1997; Damaj et al., 1997; Sullivan et al., 1997). In view of the potential role of nAChR desensitization in the net effects of nicotine, one might propose that subtype-specific *antagonists* may be an even more fruitful line of investigation in the development of nAChR-related drugs (see Chapter 4 in this book; and Dani and Heinneman, 1996, for a particularly lucid discussion of this issue). The latter idea underscores the fundamental importance of studies (described below) that attempt to delineate the structure determinants of antagonist, as well as agonist, sensitivity (e.g., Bertrand et al., 1993; Dani et al., 1993; Luetje et al., 1993; Patrick et al., 1993; Harvey et al., 1996a,b; 1997).

Pharmacology of Recombinant nAChRs

One of the fundamental characteristics of nAChR subtypes, extensively studied by heterologous expression of nAChRs in *Xenopus* and in cell lines, is the diversity in receptor pharmacology. Distinctions in the agonist pharmacology of specific α/β combinations were initially assessed by examining the relative potency of different agonists in the activation of macroscopic currents (reviewed in Dani et al., 1993; Patrick et al., 1993; Clarke et al., 1995; McGehee and Role, 1995; Role and Berg, 1996). Recent studies directly quantitate agonist-binding affinity for the six "major" α/β pairs, in competitive-binding assays with ^{3}H-epibatidine (e.g., Harvey et al., 1996a,b; 1997; Parker and Luetje, 1997; and see Fig. 3-1). Specific nAChR subtypes differ in the rank order of potency and in the apparent affinity and efficacy of ACh versus nicotine. In particular, the apparent affinity of $\alpha2\beta2$ vs. $\alpha3\beta4$ complexes for ACh differs by more than two orders of magnitude (~ nM vs. 800 nM, respectively; Parker and Luetje, 1997). With respect to relative affinity for nicotine versus ACh, the $\alpha4\beta2$ complex is the most disparate of those studied to date, as the K_{app} for nicotine is nearly 10-fold lower than that for ACh (Parker and Luetje, 1997; Fig. 3-1).

Studies of recombinant α/β nAChRs expressed in *Xenopus* oocytes and in cell lines have also delineated important differences amongst the nAChR subtypes in antagonist pharmacology (for reviews, see Dani et al., 1993; Patrick et al., 1993; Colquhoun and Patrick, 1997; also: Hussy et al., 1994; Johnson et al., 1995; Stetzer et al., 1996; Harvey et al., 1996a,b; 1997). These studies began with detailed assessment of the interactions of neuronal bungarotoxin with nAChRs; it is now clear that the kinetics as well as the apparent affinity for neuronal bungarotoxin are nAChR-subtype-specific. The distinctions have been mapped to particular domains of both the α and β subunit sequences in an elegant series of experiments of Luetje and colleagues (e.g., Luetje et al., 1990, 1991, 1993; Harvey et al., 1996a,b). For example, these investigators find that Thr59 of the $\beta2$ subunit and that glycosylation of asn141 of the $\alpha3$ subunit are the key determinants of the particularly high-affinity, rapid-on, slow-off block by nBgTx characteristic of the $\alpha3\beta2$ complex (Luetje et al., 1990; Papke et al., 1993; Wheeler et al., 1993; Colquhoun and Patrick, 1997).

The importance of these findings extends beyond a precise understanding of nAChR pharmacology to unveil fundamental aspects of nAChR structure. They indicate that the agonist-binding pocket of neuronal nicotinic receptors, such as that in the muscle AChR, must reside between adjacent subunits. Work on the structural determinants of the ligand-binding domains of different nAChR subtypes is certain to reveal crucial functional domains that are the logical targets of modern drug design. A new set of neuronal nAChR-active toxins that have recently emerged from the work of McIntosh, Olivera, and their colleagues (e.g., McIntosh et al., 1994; Johnson et al., 1995; Pereira et al., 1996; Harvey et al., 1997) constitute powerful probes for dissecting the structural domains essential to nAChR function. Initial studies suggest that the blocking activity of some of these toxins is highly subtype—indeed, highly *residue*—specific, such as recent reports demonstrating that sensitivity to α-conotoxin-MII is determined by three specific domains within the β2 N-terminal sequence and identifying the critical residue as thr59 (Harvey et al., 1997). The bottom line: If this level of resolution is brought to bear on each of the native nAChR-like subunit combinations, a detailed understanding of the structure-function determinants of nicotine action should emerge in short order!

Complexes comprised of α7, α8, and α9 subunits share susceptibility to the classic neuromuscular AChR antagonist, αBgTx, and oddly, are also potently blocked by strychnine (Elgoyhen et al., 1994). Homomeric α7 complexes are selectively blocked by subnanomolar concentrations of methyllycaconitine (MLA) (see Macallan et al., 1988; Ward et al., 1990; Alkondon et al., 1992; Yum et al., 1996; Alkondon et al., 1997; Orr-Urtreger et al., 1997) and, according to a recent study, selectively activated by choline (Papke et al., 1996; Alkondon et al., 1997a,b). Although the pharmacology of native α7-containing nAChRs is qualitatively the same as recombinant, homomeric α7 nAChRs, more quantitative analyses reveal interesting and potentially important distinctions. In fact, comparison of the properties of the nAChR complexes studied to date in heterologous expression systems with those of native nAChRs reveals only partial overlap, as discussed in more detail below.

Novel nAChRs?

Recent heterologous expression studies have resurrected several subunits that were previously designated as "nonfunctional" based on assays that tested only pairwise combinations of α and β subunits. Thus, work by several groups (including our laboratory) has now established that α5, α6, and β3 can participate in functional nAChR complexes (Ramirez-Latorre et al., 1996; Wang et al., 1996; Forsayeth and Kobrin, 1997; Gerzanich et al., 1997; Sivilotti and Colquhoun, 1997). Neuronal nicotinic AChR complexes containing the α5 subunit with either α3 and β4 or α4 and β2 also have significantly higher affinity for ACh than their non-α5-containing equivalents. It is particularly interesting that although the inclusion of α5 in nAChR complexes significantly decreases ACh affinity, it has little effect on either nicotine affinity or potency. Thus, the EC50 for ACh-gated currents in oocytes expressing $\alpha5\alpha4\beta2$ nAChRs is more than 100 times that of $\alpha4\beta2$, but the concentration-response profile to nicotine is similar, whether or not α5 is included in the nAChR complex (Ramirez-Latorre et al., 1996). Such findings are likely important to understanding the differential efficacy of nicotine and ACh in the central nervous system (CNS). Clearly, the relationship between the apparent affinity, efficacy, and potency of activation by ACh as opposed to nicotine must be examined in detail for each of the native nAChR complexes in order to understand the actions and interactions of the endogenous and exogenous ligands in vivo.

Further pharmacological and functional diversity of nAChRs is suggested by reports that several versions of nonhomomeric, α7-containing nAChRs can be expressed in heterologous systems (Crabtree et al., Soc. Neurosci. abstract, 1997). For example, our preliminary experiments reveal that coexpression of α7 with a cysteine-tagged α5 and β2 or β4 yields functional (and pharmacologically distinct) nicotine-gated channels. The inclusion of all three subunits in these complexes is confirmed by probing for the cysteine-tagged subunit(s) using the SCAM technique as described by Akabas and Karlin (Akabas et al., 1992; Ramirez-Latorre et al., 1996; GC, JR-L, and LR, unpublished). We hasten to note, however, that the possibility that α7 participates in heteromeric nAChRs complexes in vivo is viewed with considerable skepticism in the field. On the "pro-heteromeric-α7-nAChR" side of the debate is evidence for α7-containing and/or α-Bgt-sensitive nAChRs in vivo that are physiologically distinct from the α7 homomeric channels in heterologous expression studies (Gerzanich et al., 1994; McGehee et al., 1995; Gray et al., 1996; Yu and Role, 1998a). In this regard, Lindstrom and colleagues have provided the most compelling evidence for native receptors that are heteromeric complexes of α7/α8 and are the first to suggest that native α7/α8 heteromers may be further diversified by the inclusion of other, as yet unidentified, subunits. Also consistent with the possibility of native α7-containing heteromeric channels are biochemical studies demonstrating co-immunoprecipitation of α-BgtT-binding sites with other (non-α7-type) subunits. Finally, the existence of heteromeric, α7-containing nAChRs in vivo is further supported by studies of native nAChRs that are relatively (or completely) insensitive to α-BgtT and/or α-MLA but are functionally deleted by α7-specific antisense oligonucleotides (McGehee et al., 1996; Alkondon et al., 1997a,b; Liang and Vizi, 1997; Yu and Role, 1998a,b). On the other hand, recent work demonstrates that, at least in rat brain, all of the high-affinity α-BgtT binding sites are homomeric α7 nAChRs (Chen et al., 1997). One should note the stipulation of *high-affinity* α-BgtT binding, as the latter study examined the composition of receptors with sufficiently high α-BgtT affinity as to be immunopurified as a stable complex with the toxin. Thus, the cumulative data suggest that homomeric α7 nAChRs may be the sole high-affinity α-BgtT sites in CNS. However, *if* α7 can coassemble in heteromeric complexes *then* we would predict that such nAChRs would bind α-BgtT relatively poorly. In fact, this appears to be the case: Preliminary studies of the α7/α5/β2 or β4 nAChRs that can be formed (as described above) have lower affinity for α-BgTx than α7 homomeric channels (GC, JR-L, and LR, unpublished observations).

Nicotine-Gated Channels and Calcium

In addition to differing in the characteristics of their agonist interactions, nAChR subtypes show a complex regulation by Ca^{2+} (Mulle et al., 1992a,b; Vernino et al., 1992; Bertrand et al., 1993; Seguela et al., 1993; Vernino et al., 1994; Albuquerque et al., 1995; Rogers and Dani, 1995; Zwart et al., 1995; Galzi et al., 1996; Wilkie et al., 1996; Fenster et al., 1997). As early as 1979, it was shown that the flux of multiple ionic species through pores could not be assumed to be independent, due to ionic interactions with, and competition for, sites within the pore (Lewis, 1979). Divalent cation flux through nicotinic nAChRs is surely complex, as these ionic species—in particular Ca^{2+}—can both permeate and modulate the open channel.

Studies in *Xenopus* oocytes have demonstrated that the relative permeability for Ca^{2+} versus Na^+ ($P_{Ca/Na}$) of neuronal heteromeric α/β nAChRs ranges from 1 to 1.5, compared with less than 0.2 for muscle nAChRs and about 7 for glutamate receptors activated by N-methyl-D-aspartate (NMDA). The current champion is the α7 homomeric channel with a $P_{Ca/Na}$ value of about 20 (Vernino et al., 1992, 1994; Bertrand et al., 1993; Sequela et al., 1993; Rogers and Dani, 1995). Although the permeability properties of these channels are such that ion permeation through the pore deviates from the predictions of Goldman–Hodgkin/Katz, the percentage of total current carried by Ca^{2+} is still significantly higher than for muscle nAChR channels and is comparable to NMDA receptor channels if assessed under identical conditions (e.g., Rogers and Dani, 1995). Approximately 5% of the total current through α/β combinations is carried by Ca^{2+}, compared to 2% via muscle nAChRs (Vernino et al., 1994). Ca^{2+} permeation through NMDA receptor channels is about 12% (Rogers and Dani, 1995) and is estimated at about 10% for α7 homomeric nAChRs. α7 homomeric nAChRs may be an important determinant of intracellular Ca^{2+} in neurons as they constitute a major route of Ca^{2+} entry at membrane potentials equal to or hyperpolarized from rest potential (e.g., Castro and Albuquerque, 1995; Lindstrom et al., 1995; Rogers and Dani, 1995). Under these conditions—where voltage-dependent Mg^{2+} block of NMDA receptors is greatest—α7-containing nAChRs do not rectify. Other high-calcium-permeability nAChR subtypes such as α9 homomers, however, display strong rectification at hyperpolarized potentials and as such are less likely to mediate significant Ca^{2+} influx at rest potential (Elgoyhen et al., 1994).

In addition to the divalent cation permeability of nAChRs, a number of studies have highlighted the effects of divalent cations in general, and of calcium in particular, on nAChR channel conductance and opening probability. In general, divalent ions affect conductance of neuronal ACh channels profoundly. The physical localization of the divalent cation site(s) responsible for changes in nAChR conductance is not known. In PC12 cells Mg^{+2} ions affect the single-channel conductance asymmetrically, reducing both inward (external Mg^{+2}) and outward (internal Mg^{+2}) ion fluxes, with the latter effect being more profound (Neuhaus and Cachelin, 1990). In general, divalent ions reduce single-channel conductance. Extracellular Ca^{+2} reduces inward current (e.g., Mulle et al., 1992a,b; Rogers and Dani, 1995). Other divalent cations, including Ba^{+2} and Sr^{+2}, have similar effects on nAChRs, and recent studies suggest that such effects may be subtype specific (Sands et al., 1993; Zwart et al., 1995; Galzi et al., 1996).

In adrenal chromaffin cells, which express α3, β4, and/or β2, external Ca^{+2} increases the opening frequency of nAChRs 4-fold. The burst length is increased by a factor of 1.7 (Amador and Dani, 1995). In rat medial habenula, the presence of Ca^{+2} did not change the open time of the channel (Mulle et al., 1992), suggesting that the closing rate α (see Eq. 3-1) did not change, but rather that the opening rate β increased by a factor of 15. Other reports, however, suggest that habenula neurons express multiple nAChR subtypes, so it is possible that effects of external calcium on the opening and/or closing rate of a particular nAChR subtype may have been overlooked (Connolly et al., 1995; Brussaard et al., 1994). The following model is presented to provide the reader with a more substantive perspective on potential mechanisms underlying nAChR modulation by Ca^{2+}. A typical (although surely oversimplified) kinetic scheme for ACh binding to and opening nAChRs is as follows:

$$A + R \underset{k_{-1}}{\overset{2k_1}{\Leftrightarrow}} AR + A \underset{2k_{-2}}{\overset{k_2}{\Leftrightarrow}} ARA \underset{\alpha}{\overset{\beta}{\Leftrightarrow}} O^{*} \tag{3-1}$$

where R, AR, and ARA denote the unliganded, single-liganded, and double-liganded receptor, respectively, and O* is the receptor in the open state.

The first two reactions are assumed to be fast ($k_1 \cong k_2 \cong 10^8 M^{-1} sec^{-1}$) compared to the forward and reverse reactions for channel opening (i.e., $1/\alpha \sim 10^{-3}$ sec). As the ligand-receptor reactions proceed much faster than the conformational changes, examination of the population of nAChRs in the time range of $1/\alpha$ would reveal receptors distributed amongst the R, AR, and ARA states according to a quasi-steady state. That is, if $w^* = O^*$, $w = ARA$, $z = AR$, and $y = R$, $[ACh] = x$

$$\frac{dw^*}{dt} = w\beta - \alpha w^*$$

$$\frac{dw}{dt} = k_2 xy - 2k_2 w - \beta w + \alpha w^*$$

$$\frac{dz}{dt} = 2k_2 w - zxk_2 + 2k_1 xy - k_{-1} z$$

$$\frac{dy}{dt} = -2k_1 xy + k_{-1} z$$

or $z = 2K_1 xy$, $w = zxK_2/2$, and hence $w = K_2 K_1 x^2 y$. Given that the total number of receptors (i.e., the sum in all states) is constant, $C = y + z + w + w^*$, we get

$$y = \frac{C - w^*}{1 + 2K_1 X + K_1 K_2 X}$$

and by setting $\frac{dw^*}{dt} = 0$ we obtain $I_p = \frac{w^*}{C}$ for the peak current, and

$$\frac{I}{I^O} = \frac{\left(\frac{\beta}{\alpha}\right) K_1 K_2 X^2}{1 + 2K_1 X + K_1 K_2 X^2 \left(1 + \frac{\beta}{a}\right)} \tag{3-2}$$

Ratios of β/α equal to ~ 10 have been suggested for the muscle-type nAChRs. With the above considerations in mind, the suggestion that external Ca^{+2} increases the magnitude of nAChR-mediated macroscopic currents solely by altering channel opening rate requires that the initial β/α ratio be relatively low. Likewise this proposal assumes that the reduction in the single-channel conductance by calcium involves a distinct modulatory site.

Although the locus of calcium-dependent modulation of nAChR gating is not known, the short cytoplasmic loop between TM1 and TM2 loop appears to be well suited to coordinate Ca^{+2} ion binding. The presence of charges in this region of each of the five subunits that surround the pore suggests a plethora of possible interactions with divalent ions. In particular, preliminary attempts to identify key sites related to nAChR modulation by changes in internal Ca^{2+} suggest that one or more cysteine residues within the TM1-TM2 cytoplasmic loop are functionally "close" to a Ca^{+2} regulatory site (JR-L, GC, and LR, unpublished observations). Significantly, chemical modification of such cysteine residues with MTS reagents mimics the effects of increased $[Ca]_{int}$, whereas site-directed mutagenesis of the TM1-TM2 cysteines to serine residues ablates nAChR modulation by internal Ca^{2+}.

In view of the complexity of dissecting the effects and sites of calcium modulation in native nAChR channels, we have pursued this issue by studying the subunit combinations thought to underlie native (chick) habenula nAChRs, using heterologous expression in *Xenopus* oocytes (Ramirez-Latorre et al., 1996). Indeed, comparison of recombinant $\alpha 4\beta 2$ and $\alpha 4\alpha 5\beta 2$ nAChRs reveals distinct effects of altering intracellular and extracellular calcium. Perhaps most intriguing are the effects of changes in internal calcium on nAChR opening probability and on nAChR open and closed time durations. Thus, increasing calcium on the cytoplasmic face of excised patches of α4β2 nAChRs (from nominally zero to approximately 1 μM Ca) markedly increases opening probability and open duration while decreasing interburst interval (Ramirez-Latorre et al., 1996; Soc. Neurosci. abstract; Crabtree et al., 1997, Soc. Neurosci. abstract). In contrast, preliminary study of α5+ α4β2 nAChRs reveals that identical changes in cytoplasmic calcium have precisely the opposite effect. Thus, the inclusion (or lack thereof) of the α5 subunit critically regulates the extent of nAChR activation. As these changes in calcium are likely to occur within activated synaptic terminals, one might predict that the inclusion of α5 in presynaptic nAChR complexes would be an important determinant of both the duration and extent of synaptic facilitation by nicotine (see Fig. 3-2 and below).

The differential modulation of α4β2 ± α5 nAChR complexes may be due to the presence of two cysteine residues within the cytoplasmic loop between TM1-TM2 of both α4 and β2; the same region of the α5 subunit includes only one cysteine residue. These two cysteines are tandemly conserved in α2, α3, α4, α6, β2, and β4 while both cysteines are absent from the α7 and β3 sequences. The *absence* of cysteine residues in the comparable region of the α7 sequence may contribute to the high calcium permeation and, we would predict, minimal susceptibility to modulation by internal Ca^{2+}. This analysis also suggests that the binding and modulation of heteromeric nAChRs by Ca^{2+} may underlie the relatively low calcium permeability of these receptors compared with α7 nAChRs. Additional support for the functional significance of the TM1-TM2 cytoplasmic loop in Ca^{2+}–nAChR interactions comes from a recent report describing a point mutation in this region of the α4 subunit (Kuryatov et al., 1997). This mutation diminishes the Ca^{2+} permeability of α4β2 nAChRs, which can be restored by inclusion of the α5 subunit in the complex. Thus, further studies directed at dissecting the calcium permeability and modulation of α5- and α7-containing complexes will prove to be of great interest.

THE DIVERSITY OF NATIVE NICOTINIC RECEPTORS: SUBTYPES AND SUBCELLULAR DOMAINS

Heterologous Expression of Specific nAChR Subunits Doth Not a Native nAChR Make

The previous discussion may prompt one to ask why there has been so much experimental fuss over determining which nAChR complexes *can* be generated in heterologous expression systems. Of course, one assumption underlying these studies has been that examination of recombinant receptors of known subunit composition will help us determine the composition of native nAChRs. Implicit to this idea is the notion that heterologous expression of nAChRs yields receptors that are reasonable facsimiles of native nAChRs. Unfortunately, the news on this front is less than encouraging (for a recent review, see Colquhoun and Patrick, 1997). Exhaustive studies comparing native nAChRs from rat ganglionic neurons with both duplex and triplex combinations of recombinant α3, α5, and β4 reveal that the properties of those channels do not match (Sivilotti and Colquhoun, 1997). For example, these investigators find coexpression of only two subunits (e.g., α3 and β4) yields multiple nAChR subtypes. Although these channels are distinct from each other in their conductance and/or kinetic properties, the oocyte-produced subtypes do not recapitulate the native array of nAChRs. One interpretation is that the oocyte permits assembly of nAChRs that differ in subunit stoichiometry—i.e., 2(α3) + 3(β4) and 3(α3) + 2(β4)—whereas such promiscuity in assembly ratios is not supported in neurons. Another possibility is that the nAChRs produced and assembled in oocytes may be

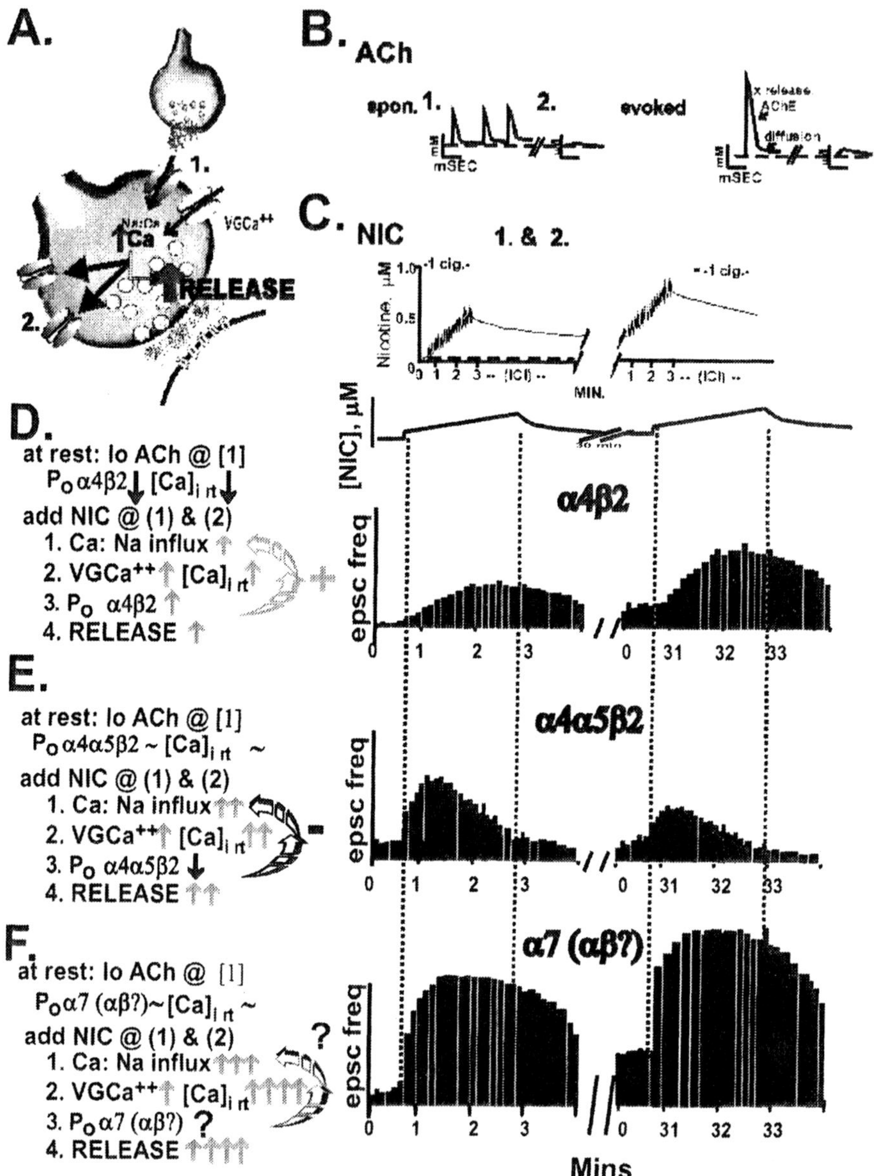

Figure 3.2. Consideration of the potential impact of distinct nAChR subtypes as well as distinct patterns of delivery of ACh versus nicotine on the time course of nAChR-mediated presynaptic facilitation in vivo. ***A:*** *Diagram of an axo-axonic synapse. Cholinergic axon terminal (upper right) releases ACh directly onto a second synaptic terminal (lower left) at an axo-axonic synapse (1_1) The second terminal is studded with nAChRs, so that activation leads to the facilitation of transmitter release.* ***B:*** *Theoretical profile of the concentration of ACh as a function of time at sites (1_1) and (2) under conditions of spontaneous (spon.) and action potential evoked release from the cholinergic bouton. The concentration at site (2_1) would depend on the distance of site (2_1) from (1_1), the local concentration of AChE, and the amount of ACh released at (1).* ***C:*** *Theoretical profile of the concentration of nicotine as a function of time at sites (1) and (2) under conditions of smoking two cigarettes, with an intercigarette interval of several minutes.* ***D, E,*** *and* ***F*** *(left-hand side): Proposed changes in intraterminal Ca and consequent modulation of presynaptic nAChRs under conditions where the presynaptic nAChRs at sites (1) and (2) are heteromeric complexes of $\alpha 4 + \beta 2$ (**D**) $\alpha 4 + \alpha 5 + \beta 2$ (**E**), or $\alpha 71$ $\alpha\beta$-type subunits.* ***D,***

subject to an array of post-translational modifications to which they are not exposed in neurons. In any case, these results underscore that one cannot be too cautious in interpreting heterologous expression data, as channels formed may not be reliable predictors of native nAChR composition. Nevertheless, the usefulness of heterologous expression as a venue for dissecting how one nAChR differs from another should not be underestimated. In fact, such a reductionist approach is a necessary component of any search for subtype-specific pharmaceutical agents and is an important first step toward understanding the actions and interactions of nicotine and ACh on cholinoceptive neurons.

Assessing the Composition of Native nAChRs

So, how do we determine what subsets of the many theoretically possible combinations of nAChR subunits are actually assembled into bona fide nAChR complexes and used by CNS and PNS (peripheral nervous system) neurons? This is a difficult but important problem that is under attack by many investigators using the full armamentarium of modern molecular genetic, biochemical, and biophysical techniques (for reviews, see Dani et al., 1993; Patrick et al., 1993; McGehee and Role, 1995; Lindstrom, 1996; Colquhoun and Patrick, 1997; see also Deneris et al., 1989; Wada et al., 1989; Heinemann et al., 1990; Clarke 1993; Sargent, 1993).

Immunohistochemical, autoradiographic, and in situ hybridization studies have yielded the crucial database of information: Such approaches tabulate the distribution of receptor subunit mRNA and protein throughout the nervous system (see Chapter 7). Unfortunately, it is difficult to obtain antibodies that are truly selective between the 11 (relatively homologous) nAChR subunit proteins. Although several of the desired panel of subunit-specific antibodies have been isolated and characterized (see Sargent, 1993, for review), the current subunit-by-subunit protein map awaits the development of additional immunological probes. Comparable mapping of the subunit mRNAs is available. However, in view of the important role of nAChRs targeted to axon terminal domains (see below and Chapter 8), in situ hybridization analyses may be misleading in attempting to deduce which of the subunits expressed in the cell body might be destined to participate in the presynaptic nAChRs. Of course, coexpression of multiple subunit mRNAs or proteins in an individual neuron does not necessarily imply coassembly. As such, both CNS and PNS nAChRs have been further scrutinized in experiments using antagonist binding or immunopurification to isolate nAChR complexes, followed by biochemical and immunological assays of subunit composition. These studies indicate that about half of the total pool of CNS nAChRs include the α4 and β2 subunits; the other half include α-BgTx-binding/α7 subunits (see Lindstrom, 1996, for review). In contrast α3-containing complexes are a detectable but relatively small component of the CNS nAChR. Assay of the number of CNS nAChRs that include the α5 subunit suggests that such complexes are present at about 10–20% of the levels of α4/β2 nAChR (but NB: the latter figure likely overestimates the contribution of α5-containing complexes to the total CNS pool; see Forsayeth et al., 1997; J. Lindstrom, personal communication). The overall picture in PNS neurons is quite different, with α3/β4 (as opposed to α4β2) nAChRs and α-BgTx-binding/α7 nAChRs constituting the bulk of the total nAChR population expressed by chick ciliary ganglia (the biochemistry of nAChRs has been analyzed in most detail in this system (e.g., Pugh et al., 1995; Chapter 10 in this book).

E, and F (right-hand side): Proposed profile of changes in synaptic facilitation (theoretical plots of EPSC frequency vs time) in response to nicotine administration (as shown in C). The proposed changes in synaptic facilitation illustrated in D, E, and F assume the conditions detailed to the left of each panel. (i.e., response to exogenous nicotine with ongoing spontaneous release of ACh and presynaptic nAChRs comprised of α4 + β2 (D), α4 + α5 + β2 (E), or α71 αβ (F) type subunits). See text for assumptions, disclaimers, and discussion of this highly speculative proposal.

Also, unlike in CNS, a significant percentage of the total nAChR pool in ciliary ganglion neurons also includes the α5 subunit (e.g., Conroy et al., 1992, 1995; Corriveau and Berg, 1993; Vernallis et al., 1993; Levey et al., 1994).

The strength of such biochemical analyses rests, per force, on the sensitivity of the assays employed. As the recovery of receptor protein and the specificity of the antibodies obtained have improved, the resolution of these analyses has evolved. The current technology is impressive: Receptor complexes that comprise as little as 5–10% of the total population are reliably detected. Nevertheless, even with the development of such high-resolution techniques, biochemical analyses of nAChR composition are still constrained by the fact that one must assay the *total* nAChR pool. There are several reasons why assay of the total nAChR pool may confound an analysis of the physiological contribution of a particular subunit.

First, analysis of nAChRs in neurons in vitro reveals that at least 50% of nAChRs are within the *intra*cellular pool; the ratio of intracellular to surface nAChRs in vivo is not well defined (J. Lindstrom, personal communication). A second limitation of biochemical assay of nAChR composition, even if confined to the surface nAChRs, is revealed by studies demonstrating that only a subset of the surface nAChR sites are actually functional (i.e., activatable) receptor/channels (Margiotta et al., 1987). The existence of spare receptors and/or functional receptors that do not normally contribute to transmitter-induced responses is not at all uncommon. Few studies have addressed whether or not nAChRs—like other transmitter receptors—include such a spare receptor pool. Finally, how do we know whether the profile of nAChR subtypes on the cell surface is the same as that of the total nAChR pool? A particular subtype may be disproportionately represented in one or the other pool, so an estimate of the percent of total receptors that include (for example) α5 may grossly overestimate (or underestimate) the proportion of surface nAChRs that include this subunit.

The reader may be inclined to view this litany of reservations in interpreting the hard-fought battles over nAChR biochemistry as petty or even irrelevant concerns, relative to the overall goal of dissecting the compositions of native nAChRs. Nevertheless (and admitting our inevitable biophysical bias), we propose that understanding the actions and interactions of nicotine and ACh in the nervous system must begin with understanding the composition of the *physiologically relevant* nAChR pool. That is, it is essential to determine the subunit composition of the receptors that constitute the actual targets of endogenous and exogenous cholinergic agents. This is no small task, as even in the "best-case scenario"—i.e., one in which *all* surface nAChRs are physiologically relevant—the estimated number of such nAChRs is very low.

Studies aimed at dissecting the functional contribution of specific subunits and specific nAChR subtypes have typically combined molecular genetic and electrophysiological approaches. Many attempts to assess receptor composition by comparing the properties of nAChR channels of known subunit composition with those of the native nAChRs have been frustrated by the generally poor match between these obviously diverse experimental systems. In some cases, however, the properties of native receptors have been reasonably well recapitulated by heterologous expression in particular cell lines and, in a few cases, in *Xenopus* oocytes. For example, examination of the properties of native nAChRs in chick retina and comparison with the properties of heterologously expressed α7 and/or α8 subunits strongly suggest that the native retinal nAChRs include a homomeric α7 channel as well as a heteromeric α7/α8 channel (Gerzanich et al., 1994). Likewise, the properties of two of the four predominant subtypes of neuronal nAChRs expressed in chick habenula are reasonably well matched by heterologous coexpression of α4β2 and α4α5β2. Specifically, the conductance and kinetic profiles as well as the time course of nicotine-induced desensitization of the 25 and 50 pS native channels are strikingly similar to those of the nAChR channels produced by coexpression of α4β2 and α4α5β2, respectively (Brussaard et al., 1994; Ramirez-Latorre et al., 1996). Examination of nAChRs expressed in "vanilla" receptor background cell lines has provided still better matches with native nAChR profiles. For example, study of nAChRs produced co-expression of the α4 and β2 or α3 and β4 subunits in HEek293 cells, defined the pharmacological, physiological, and regulatory properties of these receptors, and demonstrated that the properties of the transfected cells match those of native CNS and PNS nAChRs (Storch et al., 1996; Golpalakrishan et al., 1996).

The ability to delete individual nAChR subunits selectively and analyze the impact of such deletions on functions ranging from cognition and behavior to the biophysical minutiae of nAChR gating promises to be the most powerful approach to understanding how nicotine and nAChRs actually work. In fact, mice genetically engineered to lack each of the neuronal nicotinic subunit genes are being developed by several laboratories, and experiments delineating the impact of gene-specific knockouts on nAChR functions are just emerging (Picciotto et al., 1995; Orr-Urtreger et al., 1997). The first examination of nAChR function in genetically modified mice was reported by Changeux and colleagues (Picciotto et al., 1995), who targeted the β2 subunit for deletion. Their study revealed an important role for β2-containing nAChRs in some forms of associative memory. A recent report documents some of the consequences of selective deletion of the α7 subunit (Orr-Urtreger et al., 1997). The α7-*l*-mice are disappointingly healthy, having no gross neuroanatomical or neurophysiological abnormalities, but CNS α-Bgt-binding sites are not detectable, and the rapidly activating and inactivating currents that are a hallmark of α7-containing nAChRs in wild type (WT) hippocampal neurons are abolished in the homozygous α7 nulls (Orr-Urtreger et al., 1997). We anxiously await the behavioral analyses of α7(-*l*-)mice.

The use of genetically engineered animals to study the physiological role and functional properties of nAChRs has its own complications, however. Despite initial assays demonstrating that knockout of β2 was not accompanied by a complimentary upregulation of α4—the prime suspect if compensation were to occur—subsequent studies revealed increased expression of α7, the physiological affect of which is still under analysis (Picciotto, 1996, Soc. Neurosci. abstract). It is interesting to note the similarities between results obtained with the "poor persons" knockout techniques (i.e., subunit-targeted antisense oligonucleotides) and those obtained from study of genetically modified mice. Preliminary studies of α5(-*l*-) mice reveal that the deletion of α5 removes one of the three nAChR channel subtypes normally expressed by SCG. In addition, α5(-*l*-) mice express an array of novel "mutant" channels, distinct from any detected in WT animals. These findings are strikingly reminiscent of those obtained when the contribution of α5 to native nAChRs is probed by treatment of primary sympathetic neurons in vitro with α5-specific antisense oligonucleotides (Yu and Role, 1998a,b). In fact, compensatory changes in the profile of nAChR channels following subunit-specific knockout may be the rule rather than the exception, even within the relatively brief time periods required to knock down subunit gene expression by using antisense oligonucleotides. This observation underscores the difficulties involved in dissecting the composition and function of native nAChRs, even with the powerful and presumably selective approach of targeted gene knockout. If compensation is a common corollary of subunit deletion, biophysical analysis of the functional contribution of an individual subunit will be unavoidably confounded by the generation of "mutant" channels. At this point at least one of the authors (LR) is willing to concede that biochemical analysis of subunit composition (even of a minuscule number of nAChRs) may not be such a forbidding task after all!

Are Particular Subtypes of nAChRs Selectively Targeted to Certain Cellular Domains?

The last 10 years of nAChR research have resulted in significant advances in our understanding of the details of nAChR function. The biophysical and pharmacological profiles of many nAChR subtypes have been defined. Despite the limitations of any individual approach to analyzing the structure, composition, and functional properties of native nAChRs, the combined efforts of many have, in fact, yielded considerable insight into the biology and physiology of nicotine. Perhaps one of the most intriguing prospects unveiled over the last decade stems from the discovery that nAChRs in the CNS may regulate neuronal excitability by modulating—as well as directly mediating—synaptic transmission (for reviews, see McGehee and Role, 1995; Colquhoun and Patrick, 1996; Role and Berg, 1996; Wonnacott, 1997; Chapter 8 of this book; see also Whyte et al., 1986; Rapier et al., 1987, 1988; McGehee et al., 1995; Gray et al., 1996; Wilke et al., 1996; Marshall et al., 1997). This notion, initially suggested by studies by Wonnacott and her colleagues demonstrating that nicotine could enhance the release of dopamine from synaptosomal preparations of ventral striatum, may explain many of the apparent mysteries of nicotine's potent and broad-ranging activity (Rapier et al., 1987, 1988). The conundrum has been this: How could a drug that acts by gating a population of channels that constitute such a small percentage of the transmitter receptors in the brain do so much? There are, of course, many mechanisms by which small primary signals are efficiently amplified in biological systems. The synaptosome results suggested that impedance and short-circuiting were the tricks underlying nicotine's broad and efficacious activity. Receptors that mediate synaptic transmission at cell body and dendritic synapses are expressed at relatively high levels and/or are organized or clustered at synaptic sites. Their conductances are summed and can overcome the relatively low input resistance of the cell body or can control the potential within a dendritic spine. If a receptor is expressed at low levels, however, its influence will be proportionally small unless the receptors are targeted to a high-impedance locale. Effects will be greatest if receptors are targeted to a domain within or close to the output or business end of the neuron—the site of transmitter release. In fact, the activation of only a few such strategically placed receptors is sufficient to influence local changes in membrane potential dramatically.

The preceding description of the mechanisms by which receptors on axon terminals can influence synaptic transmission is both familiar and obvious. Indeed, it is well established that nAChRs are present at both pre- and postsynaptic sites within the CNS (e.g., see Role and Berg, 1996; Wonnacott, 1997, for reviews). Furthermore, nicotine's ability to enhance the release of many neurotransmitters—including GABA, norepinephrine, serotonin, acetylcholine, and glutamate, as well as dopamine—has been demonstrated by numerous laboratories. Less clear is why so many different types of nAChRs would be needed to enhance pre- and/or postsynaptic excitability. Given the low levels of nAChRs, why not just produce one type of postsynaptic nAChR with high conductance and opening probability, and long open duration? On the other hand, nAChRs with relatively high calcium permeability might be best for a receptor population assigned to modulating transmitter release.

Of course, such a simplified scenario is at least naive, and at worst iconoclastic. Instead, we stand firm in the belief that the diverse array of nAChR subtypes expressed in the CNS reflects physiologically important, rather than biologically redundant, phenomena. Therefore, we conclude our discussion with (admittedly speculative) considerations of possible mechanisms by which nAChR diversity is a major determinant of the broad spectrum of behavioral and cognitive effects of nicotine. As committed reductionists (i.e., cellular, as opposed to animal, physiologists), we recognize that our theories are limited by our channel-o-centric view of the brain. The true biologists may be offended by our extrapolations and, as such, we excuse these readers from subjecting themselves to our fanciful finale.

We return to the topic of targeting of nAChR subtypes to specific cellular domains, and propose that the specific subtype of nAChR expressed on an individual neuronal terminal will dictate the magnitude and time course of synaptic facilitation, and hence the net physiological effects, of endogenous neurotransmitter and exogenous nicotine. First, the reader will recall the findings presented above (and summarized in Fig. 3-1)—that nicotinic receptor subtypes have distinct agonist and antagonist profiles, distinct conductance, kinetics and permeability to divalent cations, and, last but not least, are differentially modulated by changes in intracellular calcium. A compelling example of the pharmacological and physiological diversity of nAChRs is provided by comparison of α4β2-type channels with their α5-containing counterparts. Additional appreciation of nAChR diversity derives from the further comparison of these two subtypes with α7-containing nAChRs.

Next, we need to remind ourselves of the vast differences in the temporal and spatial pattern of *delivery* of exogenously administered nicotine versus endogenously released ACh (see Fig. 3-2A). Thus, the spontaneous and action-potential-evoked release of ACh at (presumed) axo-axonic synapses would result in an essentially instantaneous ($t_{1/2}$ rise ~ 1–2 msec), short-duration (~ 5–20 msec) increase in *local* ACh concentration, which at the peak of release could reach the 10–100 mM levels proposed in classic studies of nerve-muscle synapses. The transmitter concentration profile is rapidly returned to resting levels (<1 μM?) due to the efficiency of enzymatic hydrolysis by acetylcholinesterase. In contrast, consider the likely profile of nicotine delivery during and after its administration by inhalation (smoking). Nicotine is delivered in small (submicromolar), and relatively slow (5–10 sec) pulses with each drag on the cigarette. The *interdrag interval* (a new technical term) is characterized by a slow and relatively small decline in the overall nicotine concentration (i.e., throughout the brain). The next nicotine dose (next puff, same cigarette) is presumably additive with the waning concentration. Furthermore, because the concentration of nicotine inhaled increases as the self-administrator puffs through to the end of one delivery vehicle (one commercial cigarette lasts 3–5 minutes [LR, personal communication]), the concentration of nicotine in the brain rises steadily, achieving levels of 0.3–1.0 μM (Benowitz et al., 1990) depending on the particular type of cigarette, the inhalation pattern of the smoker, and the number of cigarettes previously consumed. The contribution of the latter condition will also depend on the time between repeated nicotine doses relative to the time course of nicotine clearance from the CNS. Figure 3-2B illustrates a theoretical concentration profile for nicotine with an intercigarette interval of approximately 30 minutes. In summary, the key differences in delivery of ACh versus nicotine are in the rate of delivery, the largely uniform increase in nicotine concentration as it percolates through the brain, and the additive nature of sequential doses as the drug is only slowly eliminated from the CNS.

The next step is to consider these theoretical concentration-versus-time profiles for ACh and nicotine delivery in the context of a synaptic terminal that releases (say) dopamine and is studded with a single class of nAChRs (subtype to be designated). To simplify the considerations, we presume that the synaptic terminal is at rest (i.e., the dopaminergic neuron is not depolarized) with a low but detectable level of spontaneous release (i.e., the transmitter release and consequent postsynaptic currents are TTX-resistant minis). A final simplifying premise: The cholinergic projections to the dopaminergic (and cholinoceptive) presynaptic terminal are inactive. That is, we set the local concentration of ACh to a (speculative) value of 1 μM, assuming the lack of either spontaneous or stimulus-evoked release of ACh. If nicotine administration is superimposed on this setting, what would be the time course and extent of facilitation (if any) of dopamine release?

Clearly the resultant changes in the profile of spontaneous dopamine release would depend on the particular nAChR subtype on the synaptic terminal. Figure 3-2D, E, and F present our hypothetical profiles of nicotine-induced changes in dopamine release, indicated by changes in epsc frequency (y axis) as a function of time. Each panel shows the hypothetical result that we propose if release were measured from a cholinoceptive terminal with only α4β2 nAChRs [D] or only α4α5β2 nAChRs [E] or α7-containing nAChRs [F]. The predicted profiles of release are based on the relative affinity of each nAChR subtype for ACh versus nicotine, the relative permeability to calcium, the dependence (or lack thereof) of Po on intracellular Ca^+, and the rate of inactivation with continued exposure to low concentrations of ACh and nicotine. Our predictions suggest that presynaptic terminals expressing α4β2 will elicit the smallest increases in release, but because of the increase in Po with increased intracellular calcium, the release from these terminals will be the most sustained and potentially additive from cigarette to cigarette (assuming an intercigarette interval of 30 minutes and an equivalent $t_{1/2}$ for the clearance of nicotine from the brain).

Terminals with α4α5β2-type nAChRs are expected to produce an initial, more robust release of transmitter because of the higher calcium permeability of these channels. However, because of the decrease in Po induced by increased intracellular calcium and the faster rate of receptor desensitization of α5-containing nAChRs, these terminals are subject to an intrinsic "negative feedback." These characteristics will shorten the duration of the peak synaptic facilitation and decrease the extent to which release is continuously elevated in the interval between nicotine administrations. Last but not least are the α7-containing nAChRs. Although α7 homomeric channels are renowned for their rapid inactivation in the continued presence of agonists, recent studies of low ACh- and nicotine-elicited responses reveal that α7 nAChRs desensitize very slowly with prolonged, low-level agonist exposure (Fenster et al., 1997). Thus, because of the high calcium permeability of α7 nAChRs, we predict a rapid and robust increase in release with nicotine administration and a sustained increase in activity between nicotine doses.

CONCLUSION

This discussion is, of course, rife with assumptions and, in some aspects, based on sheer guesses rather than on any available data (e.g., ACh levels in the vicinity of cholinergic projections?). Nevertheless, the thought experiment is a useful one and at least provides testable hypotheses that may facilitate the analysis of presynaptic nAChR subtypes. Finally, as the overall goal of this chapter is to stimulate interest in the molecular composition and biophysical properties of nicotinic receptors, what better way to entice the uninitiated than with a slightly tongue-in-cheek consideration of possible mechanisms underlying "your brain on nicotine"?

REFERENCES

Akabas MH, Stauffer DA, et al. (1992): Acetylcholine receptor channel structure probed in cysteine-substitution mutants. Science 258(5080): 307–310.

Albuquerque EX, Pereira EF, et al. (1995): Nicotinic receptor function in the mammalian central nervous system. Ann. NY Acad. Sci. 757: 153–168.

Albuquerque EX, Pereira EF, et al. (1997): Nicotinic acetylcholine receptors on hippocampal neurons: distribution on the neuronal surface and modulation of receptor activity. J. Recept. Signal Transduct. Res. 17(1–3): 243–266.

Alkondon M, Pereira EF, et al. (1992): Blockade of nicotinic currents in hippocampal neurons defines methyllycaconitine as a potent and specific receptor antagonist. Mol. Pharmacol. 41(4): 802–808.

Alkondon MP, Pereira EF, Barbosa CT, Albuquerque EX (1997a): Neuronal nicotinic acetylcholine receptor activation modulates GABA release from CA1 neurons of rat hippocampal slices. J. Pharmacol. Exp. Ther. 283(3): 1396–1411.

Alkondon M, Pereira E, Cortes WE, Maelicke A, Albuquerque EX (1997b): Choline is a selective agonist of α7 nicotinic adetylcholine receptors in rat brain neurons. Neurosciences

Amador M, Dani JA (1995): Mechanism for modulation of nicotinic acetylcholine receptors that can influence synaptic transmission. J. Neurosci. 15: 4525–4532.

Anand R, Peng X (1993): Pharmacological characterization of α-bungarotoxin-sensitive acetylcholine receptors immunoisolated from chick retina: contrasting properties of α7 and α8 subunit-containing subtypes. Mol. Pharmacol. 44(5): 1046–1050.

Arendash GW, Sanberg PR, et al. (1995): Nicotine enhances the learning and memory of aged rats. Pharmacol. Biochem. Behav. 52(3): 517–523.

Arneric SP, Sullivan JP (1994): (S)-3-methyl-5-(1-methyl-2-pyrrolidinyl) isoxazole (ABT 418): a novel cholinergic ligand with cognition-enhancing and anxiolytic activities: I. In vitro characterization. J. Pharmacol. Exp. Ther. 270(1): 310–318.

Ballivet MP, Nef P, et al. (1988): Electrophysiology of a chick neuronal nicotinic acetylcholine receptor expressed in Xenopus oocytes after cDNA injection. Neuron 1: 847–852.

Benowitz NL, Porchet H, et al. (1990): In Wonnacott S, Russell MAH, Stolerman IP, editors. Nicotine pharmacology: molecular, cellular, and behavioral aspects. Oxford: Oxford University Press, pp. 112–157.

Bertrand D, Galzi JL, et al. (1993): Mutations at two distinct sites within the channel domain M2 alter calcium permeability of neuronal α7 nicotinic receptor. Proc. Natl. Acad. Sci. USA 90: 6971–6975.

Boulter J, Evans, K, et al. (1986): Isolation of a cDNA clone coding for a possible neural nicotinic acetylcholine receptor α-subunit. Nature 319: 368–374.

Boyd RT, Jacob MH, et al. (1988): Expression and regulation of neuronal acetylcholine receptor mRNA in chick ciliary ganglia. Neuron 1: 495–502.

Briggs CA, Anderson DJ, et al. (1997): Functional characterization of the novel neuronal nicotinic acetylcholine receptor ligand GTS-21 in vitro and in vivo. Pharmacol. Biochem. Behav. 57(1–2): 231–241.

Brussaard AB, Yang, X, Doyle JP, et al. (1994): Developmental regulation of multiple nicotinic nAChR channel subtypes in embryonic chick habenula neurons: contributions of both the α2 and α4 subunit genes. Pflugers Archiv./Eur. J. Physiol. 429: 27–43.

Castro NG, Albuquerque EX (1995a): The α-bungarotoxin-sensitive hippocampal nicotinic receptor channel has a high calcium permeability. Biophys. J. 68: 516–524.

Castro NG, Albuquerque EX (1995b): The α-bungarotoxin-sensitive hippocampal nicotinic receptor channel has a high calcium permeability. Biophys. J. 68: 516–524.

Changeux JP, Bessis A, et al. (1996): Nicotinic receptors and brain plasticity. Cold Spring Harbor Symposium on Quantitative Biology LXI: 343–362.

Charnet P, Labarca C, et al. (1992): Pharmacological and kinetic properties of α4β2 neuronal nicotinic acetylcholine receptors expressed in Xenopus oocytes. J. Physiol. 450: 375–394.

Chen D, Patrick JW (1997): The alpha-bungarotoxin-binding nicotinic acetylcholine receptor from rat brain contains only the alpha7 subunit. J. Biol. Chem. 272(38): 24,024–24,029.

Clarke PBS (1992): The fall and rise of neuronal alpha-bungarotoxin binding proteins. Trends Pharmacol. Sci. 13: 407–413.

Clarke PBS (1993): Nicotinic receptors in mammalian brain: localization and relation to cholinergic innervation. Progr. Brain Res. 98: 77–83.

Clarke PBS, Pert A (1985): Autoradiographic evidence for nicotine receptors on nigrostriatal and mesolimbic dopaminergic neurons. Brain Res. 348: 355–358.

Clarke BS, Schwartz RD, et al. (1985): Nicotinic binding in rat brain: autoradiographic comparison of [^{3}H]acetylcholine, [^{3}H]nicotine, and [^{125}I]-alpha-bungarotoxin. J. Neurosci. 5: 1307–1315.

Clarke P, Quik M, et al., Eds. (1995): Effects of nicotine on biological systems. Vol. 2: Advances in pharmacological sciences. Basel, Switzerland, Birkhauser Verlag Press.

Coggan JS, Paysan J, et al. (1997): Direct recording of nicotinic responses in presynaptic nerve terminals. J. Neurosci. 17(15): 5798–5806.

Cohen BN, Figl A, et al. (1995): Regions of beta 2 and beta 4 responsible for differences between the steady state dose-response relationships of the alpha 3 beta 2 and alpha 3 beta 4 neuronal nicotinic receptors. J. Gen. Physiol. 105(6): 745–764.

Colquhoun LM, Patrick JW (1997): Pharmacology of neuronal nicotinic acetylcholine receptor subtypes. Adv. Pharmacol. 39: 191–220.

Connolly JG, Gibb AJ, et al. (1995): Heterogeneity of neuronal nicotinic acetylcholine receptors in thin slices of rat medial habenula. J. Physiol. 484: 87–105.

Conroy WG, Berg DK (1995): Neurons can maintain multiple classes of nicotinic acetylcholine receptors distinguished by different subunit compositions. J. Biol. Chem. 270(9): 4424–4431.

Conroy WG, Vernallis AB, et al. (1992): The α5 gene product assembles with multiple acetylcholine receptor subunits to form distinctive receptor subtypes in brain. Neuron 9: 679–691.

Corriveau RA, Berg DK (1993): Coexpression of multiple acetylcholine receptor genes in neurons: quantification of transcripts during development. J. Neurosci. 13(6): 2662–2671.

Couturier S, Bertrand D, et al. (1990): A neuronal nicotinic acetylcholine receptor subunit (α7) is developmentally regulated and forms a homo-oligomeric channel blocked by α-Btx. Neuron 5: 847–856.

Covernton PJ, Kojima H, et al. (1994): Comparison of neuronal nicotinic receptors in rat sympathetic neurons with subunit pairs expressed in oocytes. J. Physiol. 481: 27–34.

Damaj JI, Creasy KR, et al. (1995): Comparative pharmacology of nicotine and ABT-418, a new nicotinic agonist. Psychopharmacology (Berl) 120(4): 483–490.

Damaj MI, Glassco W, et al. (1997): Pharmacological investigation of (+)- and (−)-cis-2,3,3a,4,5,9b-hexahydro-1-methyl-1H-pyrrolo-[3,2-h]isoquinoline, a bridged-nicotine analog. J. Pharmacol. Exp. Ther. 282(3): 1425–1434.

Dani JA (1993): Structure, diversity, and ionic permeability of neuronal and muscle acetylcholine receptors. Exs 66: 47–59.

Dani JA, Heinemann S (1996): Molecular and cellular aspects of nicotine abuse. Neuron 16(5): 905–908.

Decker MW, Majchrzak MJ, et al. (1993): Effects of lobeline, a nicotinic receptor agonist, on learning and memory. Pharmacol. Biochem. Behav. 45(3): 571–576.

Decker MW, Brioni JD, et al. (1995): Diversity of neuronal nicotinic acetylcholine receptors: lessons from behavior and implications for CNS therapeutics. Life Sci. 56(8): 545–570.

Dehaene S, Changeux JP (1995): Neuronal models of prefrontal cortical functions. Ann NY Acad. Sci. 769: 305–319.

Deneris ES, Boulter J, et al. (1989): Beta 3: a new member of nicotinic acetylcholine receptor gene family is expressed in brain. J. Biol. Chem. 264(11): 6268–6272.

Dominguez del Toro E, Juiz JM, et al. (1997): Expression of alpha 7 neuronal nicotinic receptors during postnatal development of the rate cerebellum. Brain Res. Dev. Brain Res. 98(1): 125–133.

Eisele JL, Bertrand S, et al. (1993): Chimaeric nicotinic-serotonergic receptor combines distinct ligand binding and channel specificities. Nature 366: 479–483.

Elgoyhen AB, Johnson ES, et al. (1994): α9: an acetylcholine receptor with novel pharmacological properties expressed in rat cochlear hair cells. Cell 79: 705–715.

Fenster CP, Rains MF, et al. (1997): Influence of subunit composition on desensitization of neuronal acetylcholine receptors at low concentrations of nicotine. J. Neurosci. 17(15): 5747–5759.

Flood P, Ramirez-Latorre J, et al. (1997): α4β2 neuronal nicotinic acetylcholine receptors in the central nervous system are potently inhibited by isoflurane and propanol, but α7-type nicotinic acetylcholine receptors are unaffected. Anesthesiology 86(4): 859–865.

Forsayeth JR, Kobrin E (1997): Formation of oligomers containing the beta3 and beta4 subunits of the rat nicotinic receptor. J. Neurosci. 17(5): 1531–1538.

Galzi J-L, Devillers-Thiery A, et al. (1992): Mutations in the channel domain of a neuronal nicotinic receptor convert ion selectivity from cationic to anionic. Nature 359: 500–505.

Galzi JL, Bertrand S, et al. (1996): Identification of calcium binding sites that regulate potentiation of a neuronal nicotinic acetylcholine receptor. Embo. J. 15(21): 5824–5832.

Gerzanich V, Anand R, et al. (1994): Homomers of α8 and α7 subunits of nicotinic receptors exhibit similar channel but contrasting binding site properties. Mol. Pharmacol. 45: 212–220.

Gerzanich V, Kuryatov A, et al. (1997): "Orphan" alpha6 nicotinic AChR subunit can form a functional heteromeric acetylcholine receptor. Mol. Pharmacol. 51(2): 320–327.

Gopalakrishnan M, Monteggia LM, et al. (1996): Stable expression, pharmacologic properties and regulation of the human neuronal nicotinic acetylcholine alpha 4 beta 2 receptor. J. Pharmacol. Exp. Ther. 276(1): 289–297.

Gray R, Rajan AS, et al. (1996): Hippocampal synaptic transmission enhanced by low concentrations of nicotine [see comments]. Nature 383(6602): 713–716.

Harvey SC, Luetje CW (1996): Determinants of competitive antagonist sensitivity on neuronal nicotinic receptor beta subunits. J. Neurosci. 16(12): 3798–3806.

Harvey SC, Maddox FN, et al. (1996): Multiple determinants of dihydro-beta-erythroidine sensitivity on rat neuronal nicotinic receptor alpha subunits. J. Neurochem 67(5): 1953–1959.

Harvey SC, McIntosh JM, et al. (1997): Determinants of specificity for alpha-conotoxin MII on alpha3beta2 neuronal nicotinic receptors. Mol. Pharmacol. 51(2): 336–342.

Heinemann S, Boulter J, et al. (1990): The brain nicotinic acetylcholine receptor gene family. Progr. Brain Res. 86: 195–203.

Hussy N, Ballivet M, et al. (1994): Agonist and antagonist effects of nicotine on chick neuronal nicotinic receptors are defined by alpha and beta subunits. J. Neurophysiol. 72(3): 1317–1326.

Ifune CK, Steinbach, JH (1991): Voltage dependent block by magnesium of neuronal nicotinic receptor channels in rat phaeochromocytoma cells. J. Physiol. 443: 683–701.

James JR, Nordberg A (1995): Genetic and environmental aspects of the role of nicotinic receptors in neurodegenerative disorders: emphasis on Alzheimer's disease and Parkinson's disease. Behav. Genet. 25: 149–159.

Johnson DS, Martinez J, et al. (1995): Alpha-conotoxin lml exhibits subtype-specific nicotinic acetylcholine receptor blockade: preferential inhibition of homomeric alpha 7 and alpha 9 receptors. Mol. Pharmacol. 48(2): 194–199.

Konishi J, Magata Y, et al. (1993): Studies on the effects of smoking on cerebral perfusion and metabolism: basic studies with in vitro tracer methods. Yakubutsu Seishin Kodo 13(3): 167–174.

Kuryatov A, Gerzanich V, et al. (1997): Mutation causing autosomal dominant nocturnal frontal lobe epilepsy alters Ca^{2+} permeability, conductance, and gating of human alpha4beta2 nicotinic acetylcholine receptors [in process citation]. J. Neurosci. 17(23): 9035–9047.

Lena C, Changeux JP (1997): Role of Ca^{2+} ions in nicotinic facilitation of GABA release in mouse thalamus [in process citation]. J. Neurosci. 17(2): 576–585.

Lena C, Changeux JP, et al. (1993): Evidence for "preterminal" nicotinic receptors on GABAergic axons in the rat interpeduncular nucleus. J. Neurosci. 13: 2680–2688.

Le Novere N, Changeux JP (1995): Molecular evolution of the nicotinic acetylcholine receptor: an example of multigene family in excitable cells. J. Mol. Evol. 40: 155–172.

Levey MS, Brumwell C, et al. (1994): Innervation and target tissue interactions differentially regulate acetylcholine receptor subunit transcript levels in developing neurons in situ. Neuron 14(1): 153–162.

Lewis CA (1979): Ion-concentration dependence of the reversal potential and the single channel conductance of ion channels at the frog neuromuscular junction. J. Physiol. 286: 417–445.

Lindstrom J (1996): Neuronal nicotinic acetylcholine receptors. Ion Channels 4: 377–450.

Lindstrom JR, Anand R, et al. (1995): Neuronal nicotinic receptor subtypes. Ann. NY Acad. Sci. 757: 100–116.

Listerud M, Brussard AB, et al. (1991): Functional contribution of neuronal AChR subunits by antisense oligonucleotides. Science 254: 1518–1521.

Luetje CW, Patrick J. (1991): Both α- and β-subunits contribute to the agonist sensitivity of neuronal nicotinic acetylcholine receptors. J. Neurosci. 11: 837–845.

Luetje CW, Wada K, et al. (1990): Neurotoxins distinguish between different neuronal nicotinic acetylcholine receptor subunit combinations. J. Neurochem 55(2): 632–640.

Luetje CW, Piattoni M, et al. (1993): Mapping of ligand binding sites of neuronal nicotinic acetylcholine receptors using chimeric alpha subunits. Mol. Pharmacol. 44(3): 657–666.

Macallan DR, Lunt GG, et al. (1988): Methyllycaconitine and (+)-anatoxin-α differentiate between nicotinic receptors in vertebrate and invertebrate nervous systems. FEBS Lett. 226(2): 357–363.

Margiotta JF, Berg DK, et al. (1987): Cyclic AMP regulates the proportion of functional acetylcholine receptors on chicken ciliary ganglion neurons. Proc. Natl. Acad. Sci. USA 84: 8155–8159.

Marshall DL, Redfern PH, et al. (1997): Presynaptic nicotinic modulation of dopamine release in the three ascending pathways studied in vivo microdialysis: comparison of naive and chronic nicotine-treated rats. J. Neurochem. 68(4): 1511–1519.

Mathie A, Colquhoun D, et al. (1990): Rectification of currents activated by nicotinic acetylcholine receptors in rat sympathetic ganglion neurons. J. Physiol. 427: 625–655.

McGehee DS, Role LW (1995): Physiological diversity of nicotinic acetylcholine receptors expressed by vertebrate neurons. Annu. Rev. Physiol. 57: 521–546.

McGehee D, Heath M, et al. (1995): Nicotine enhancement of fast excitatory synaptic transmission in CNS by presynaptic receptors. Science 269: 1692–1697.

McIntosh JM, Yoshikami D, et al. (1994): A nicotinic acetylcholine receptor ligand of unique specificity, alpha-conotoxin lml. J. Biol. Chem. 269(24): 16,733–16,739.

Mulle C, Choquet D, et al. (1992a): Calcium influx through nicotinic receptor in rat central neurons: its relevance to cellular regulation. Neuron 8: 135–143.

Mulle C, Lena C, et al. (1992b): Potentiation of nicotinic receptor response by external calcium in rat central neurons. Neuron 8: 937–945.

Neuhaus R, Cachelin AB (1990): Changes in the conductance of the neuronal nicotinic acetylcholine receptor channel induced by manesium. Proc. R. Soc. Lond. B241: 78–84.

Nordberg A (1994): Human nicotinic receptors—their role in aging and dementia. Neurochem. Int. 25(1): 93–97.

Nutter TJ, Nutter ADJ (1995): Monovalent and divalent cation permeability and block of neuronal nicotinic receptor channels in rat parasympathetic ganglia. J. Gen. Physiol. 105: 701–723.

Olale F, Gerzanich V, et al. (1997): Chronic nicotine exposure differentially affects the function of human alpha3, alpha4, and alpha7 neuronal nicotinic receptor subtypes [in process citation]. J. Pharmacol. Exp. Ther. 283(2): 675–683.

Orr-Urtreger A, Goldner FM, et al. (1997): Mice deficient in the alpha7 neuronal nicotinic acetylcholine receptor lack alpha-bungarotoxin binding sites and hippocampal fast nicotinic currents [in process citation]. J. Neurosci. 17(23): 9165–9171.

Ortells MO, Lunt GC (1995): Evolutionary history of the ligand-gated ion-channel superfamily of receptors. TINS 18: 121–127.

Papke RL, Boulter J, et al. (1989): Single-channel currents of rat neuronal nicotinic acetylcholine receptors expressed in Xenopus oocytes. Neuron 3: 589–596.

Papke RL, Duvoisin RM, et al. (1993): The amino terminal half of the nicotinic β-subunit extracellular domain regulates the kinetics of inhibition by neuronal bungarotoxin. Proc. R. Soc. Lond. Biol. 252(1334): 141–148.

Papke RL, Bencherif M, et al. (1996): An evaluation of neuronal nicotinic acetylcholine receptor activation by quaternary nitrogen compounds indicates that choline is selective for the alpha 7 subtype. Neurosci. Lett. 213(3): 201–204.

Papke RL, Thinschmidt JS, et al. (1997): Activation and inhibition of rat neuronal nicotinic receptors by ABT-418. Br. J. Pharmacol. 120(3): 429–438.

Parker MJL, Parker CW (1997): [3]H Epibatidine affinity for neuronal nicotinic receptors is primarily determined by beta subunits. Soc. Neurosci. Abstr. 23(1): 385.

Patrick J, Seguela P, et al. (1993): Functional diversity of neuronal nicotinic acetylcholine receptors. Prog. Brain Res. 98: 113–120.

Pereira EF, Alkondon M, et al. (1996): Alpha-conotoxin-Iml: a competitive antagonist at alpha-bungarotoxin-sensitive neuronal nicotinic receptors in hippocampal neurons. J. Pharmacol. Exp. Ther. 278(3): 1472–1483.

Picciotto MR, Zoli M, et al. (1995): Abnormal avoidance learning in mice lacking functional high-affinity nicotine receptor in the brain. Nature 374(6517): 65–67.

Pugh PC, Corriveau RA, et al. (1995): A novel subpopulation of neuronal acetylcholine receptors among those binding α-bungarotoxin. Mol. Pharmacol. 47: 171–225.

Ragozzino D, Fucile S, et al. (1997): Functional properties of neuronal nicotinic acetylcholine receptor channels expressed in transfected human cells. Eur. J. Neurosci. 9(3): 480–488.

Ramirez-Latorre J, Yu CR, et al. (1996): Functional contributions of alpha5 subunit to neuronal acetylcholine receptor channels. Nature 380(6572): 347–351.

Rapier C, Wonnacott S, et al. (1987): The neurotoxin histrionicotoxin interacts with the putative ion channel of the nicotinic acetylcholine receptors in the central nervous system. FEBS Lett. 212(2): 292–296.

Rapier C, Lunt GG, et al. (1988): Stereoselective nicotine-induced release of dopamine from striatal synaptosomes: concentration dependence and repetitive stimulation. J. Neurochem. 50: 1123–1130.

Revah F, Bertrand D, et al. (1991): Mutations in the channel domain alter desensitization of a neuronal nicotinic receptor. Nature 353: 846–849.

Rogers M, Dani JA (1995): Comparison of quantitative calcium flux through NMDA, ATP, and ACh receptor channels. Biophys. J. 68: 501–506.

Sands SB, Barish ME (1991): Calcium permeability of neuronal nicotinic receptor channels in PC12 cells. Brain Res. 560: 38–42.

Sands SB, Costa AC, et al. (1993): Barium permeability of neuronal nicotinic receptor alpha 7 expressed in Xenopus oocytes. Biophys. J. 65(6): 2614–2621.

Sargent PB (1993): The diversity of neuronal nicotinic acetylcholine receptors. Ann. Rev. Neurosci. 16: 403–433.

Schrattenholz A, Pereira EF, et al. (1996): Agonist responses of neuronal nicotinic acetylcholine receptors are potentiated by a novel class of allosterically acting ligands. Mol. Pharmacol. 49(1): 1–6.

Schroder H, Giacobini E, et al. (1995): Nicotinic receptors in Alzheimer's disease. Brain Imaging of Nicotine and Tobacco Smoking, ed. Edward F. Domino. Ann Arbor, NPP Books, 1995, pp. 73–93.

Sequela P, Wadiche J, et al. (1993): Molecular cloning, functional properties and distribution of rat brain $\alpha 7$: a nicotinic cation channel highly permeable to calcium. J. Neurosci. 13: 596–604.

Sivilotti LG, McNeil DK, et al. (1997): Recombinant nicotinic receptors, expressed in Xenopus oocytes, do not resemble native rat sympathetic ganglion receptors in single-channel behaviour. J. Physiol. (London) 500(Pt. 1): 123–138.

Stetzer E, Ebbinghaus U, et al. (1996): Stable expression in HEK-293 cells of the rate alpha3/beta4 subtype of neuronal nicotinic acetylcholine receptor. FEBS Lett. 397(1): 39–44.

Sullivan JP, Donnelly-Roberts D, et al. (1997): ABT-089 [2-methyl-3-(2-(S)-pyrrolidinylmethoxy)pyridine]: I. A potent and selective cholinergic channel modulator with neuroprotective properties. J. Pharmacol. Exp. Ther. 283(1): 235–246.

Vernallis AB, Conroy WG, et al. (1993): Neurons assemble acetylcholine receptors with as many as three kinds of subunits while maintaining subunit segregation among receptor subtypes. Neuron 10: 451–464.

Vernino S, Amador M, et al. (1992): Calcium modulation and high calcium permeability of neuronal nicotinic acetylcholine receptors. Neuron 8: 127–134.

Vernino S, Rogers M, et al. (1994): Quantitative measurement of calcium flux through muscle and neuronal nicotinic acetylcholine receptors. J. Neurosci. 14(9): 5514–5524.

Vidal C (1996): Nicotinic receptors in the brain. Molecular biology, function, and therapeutics. Mol. Chem. Neuropathol. 28(1–3): 3–11.

Vidal C, Changeux JP (1989): Pharmacological profile of nicotinic acetylcholine receptors in the rat prefrontal cortex: an electrophysiological study in a slice preparation. Neuroscience 29: 261–270.

Wada E, Wada K, et al. (1989): Distribution of α2, α3, α4, and β2 neuronal nicotinic receptor subunit mRNAs in the central nervous system: a hybridization histochemical study in the rat. J. Comp. Neurol. 284: 314–335.

Wang F, Gerzanich V, et al. (1996): Assembly of human neuronal nicotinic receptor alpha5 subunits with alpha3, beta2, and beta4 subunits. J. Biol. Chem. 271(30): 17,656–17,665.

Ward JM, Cockcroft VB, et al. (1990): Methyllycaconitine: a selective probe for neuronal alpha-bungarotoxin binding sites. FEBS Lett. 270(1–2): 45–48.

Wheeler SV, Chad JE, et al. (1993): Residues 1 to 80 of the N-terminal domain of the beta subunit confer neuronal bungarotoxin sensitivity and agonist selectivity on neuronal nicotinic receptors. FEBS Lett. 332(1–2): 139–142.

Whyte J, Harrison R, et al. (1986): Subcellular fractionation and distribution of cholinergic binding sites in fetal human brain. Neurochem. Res. 11(7): 1011–1123.

Wilkie GI, Hutson P, et al. (1996): Pharmacological characterization of a nicotinic autoreceptor in rat hippocampal synaptosomes. Neurochem. Res. 21(9): 1141–1148.

Wonnacott S, Swanson KL, et al. (1992): Homoanatoxin: a potent analogue of anatoxin-A. Biochem. Pharmacol. 43(3): 419–423.

Yu C, Role L (1998a): Functional contributions of the α7 subunit to native nitotinic receptors. J. Physiol. 509(3): 651–665.

Yu C, Role L (1998b): Functional contributions of the α5 subunit to native nicotinic receptors. J. Physiol. 509(3): 667–681.

Yum L, Wolf KM, et al. (1996): Nicotinic acetylcholine receptors in separate brain regions exhibit different affinities for methyllycaconitine. Neuroscience 72(2): 545–555.

Zwart R, Van Kleef RG, et al. (1995): Potentiation and inhibition of subtypes of neuronal nicotinic acetylcholine receptors by Pb^{2+}. Eur. J. Pharmacol. 291(3): 399–406.

4

Desensitization and the Regulation of Neuronal Nicotinic Receptors

Michael J. Marks, Ph.D
Institute for Behavioral Genetics
University of Colorado
Boulder, Colorado

It is generally recognized that nicotine is the pharmacologically active alkaloid that is responsible for the physiological and behavioral effects of tobacco and is likely the active substance leading to dependence. Therefore, the consequences of chronic nicotine exposure on physiological, behavioral, and biochemical measures have been widely investigated in an effort to understand the basis for tobacco use.

The development of ligand binding assays that could measure sites with distinct nicotinic pharmacology made possible the study of the effects of chronic nicotine exposure on putative nicotinic receptors in the brain. Subsequently, the identification of the molecular composition of the receptors and the development of assays for nicotinic receptor function has expanded the investigation of the development of tolerance and dependence on the effects of nicotine to a more biochemical level. Initially binding assays permitted the measurement of two distinct types of nicotinic sites: those labeled by the high-affinity binding of nicotinic agonists such as acetylcholine (ACh) (Schwartz et al., 1992), nicotine (Romano and Goldstein, 1980), cytisine (Pabreza et al., 1991), and methylcarbachol (Abood and Grassi, 1986; Boska and Quirion, 1987) and those labeled by the binding of alpha-bungarotoxin (α-Bgt) (Oswald and Freeman, 1981). Through the use of molecular biological and immunochemical techniques it has now been established that most of the high-affinity agonist binding sites measurable with nicotine

Neuronal Nicotinic Receptors: Pharmacology and Therapeutic Opportunities, Edited by S. P. Arneric and J. D. Brioni
ISBN 0-471-24743-x, pages 65–80. Copyright © 1998 by Wiley-Liss, Inc.

and cytisine are comprised of receptors assembled from α4 and β2 subunits (Flores et al., 1992; Whiting and Lindstrom, 1988) while α-Bgt binding sites contain α7 subunits (Schoepfer et al., 1990; Couturier et al., 1990; Seguela et al., 1993).

This review focuses on three aspects of nicotinic receptor regulation in response to agonist exposure: (1) receptor number and function after chronic exposure of animals or cells to nicotine, (2) desensitization of neuronal nicotinic receptor function in response to acute exposure to nicotinic agonists, and (3) possible role of receptor desensitization in the subsequent regulation of receptor number and function following chronic agonist treatment.

EFFECTS OF CHRONIC NICOTINE TREATMENT ON NICOTINIC RECEPTORS IN THE CENTRAL NERVOUS SYSTEM

High-Affinity Agonist Binding Sites

Chronic Treatment with Nicotine. Initially surprising results demonstrated that the number of high-affinity agonist binding sites in rat (Schwartz and Kellar, 1983) and mouse (Marks et al., 1983) brain increased following chronic nicotine treatment. This unusual response to chronic agonist treatment has subsequently been replicated frequently (see Wonnacott, 1990, for review). Furthermore, it has been shown that this is not an oddity of the use of experimental animals since high-affinity nicotine binding sites are higher in human smokers (Benwell et al., 1988; Breese et al., 1997). Immunoprecipitation of high-affinity cytisine binding sites from rats chronically treated with nicotine demonstrated that the agonist binding sites are the α4/β2 subtype, which is also the predominant agonist binding subtype in drug-naive rats (Flores et al., 1991).

Changes in the number of these binding sites are dependent on the time of nicotine treatment. Maximal increases in high-affinity binding in rat brain following twice daily injections were observed following 10 days of treatment (Schwartz and Kellar, 1985) or following 4 days of intravenous infusion of nicotine in mice (Marks et al., 1985) and rats (Collins et al., 1990). More prolonged treatment of either rats or mice did not elicit more extensive changes, indicating that at a given dose the increase in the number of binding sites attains steady state. The nicotine-induced increases in the number of binding sites are reversible. A time-dependent return of the numbers of binding site to control levels occurs with withdrawal of treatment in rats (Schwartz and Kellar, 1985) and mice (Marks et al., 1985). Furthermore, the number of nicotine binding sites in former smokers did not differ from that in nonsmokers (Breese et al., 1997).

The increase in number of binding sites is dose-dependent. For example, a concentration-dependent increase in nicotine binding that reached maximal levels after treatment with 2 mg/kg/hr was observed in mice that had been infused intravenously with six different doses of nicotine (0–6 mg/kg/hr), indicating an upper limit to the extent of change elicited by chronic nicotine treatment (Marks et al., 1991).

The doses of nicotine required to elicit increases in the number of binding sites varies with species. In rats, for example, nicotinic binding sites increase substantially following relatively mild treatments such as twice daily injections (Schwartz and Kellar, 1983; Collins et al., 1988) or fairly low constant infusions (Collins et al., 1990; Fung and Lau, 1988). In mice, however, three daily injections had no measurable effect on the number of binding sites (Pauly et al., 1992) and continual treatment doses are much higher than those required for rats. Whether these species differences arise from variations in receptor interactions, pharmacokinetic parameters, and/or metabolic clearance rates has not yet been established.

Chronic nicotine treatment does not affect nicotinic receptor binding uniformly throughout the brain. Receptor levels in cerebral cortex and hippocampus increased substantially with treatment, while areas such as thalamus showed a smaller response (Marks et al., 1983, 1991; Pauly et al., 1991; Sanderson et al., 1993; Breese et al., 1997). Quantitative autoradiographic analyses of mouse brains after chronic treatment confirms the variation in response (Pauly et al., 1991; Marks et al., 1992). However, the changes observed with all regions showed similar dose-response patterns (Pauly et al., 1991), suggesting that the concentration dependence for the response is similar even if the magnitude of the change is variable.

Chronic Treatment with Other Agonists. Nicotine is not unique in eliciting changes in nicotinic receptor binding following chronic treatment. Increases in high-affinity agonist binding have been observed following chronic intraperitoneal injection of cytisine (Schwartz and Kellar, 1985), chronic subcutaneous infusion of anatoxin-a (Rowell and Wonnacott, 1990), and chronic intraventricular injection of methylcarbachol to rats (Yang and Buccafusco, 1994) as well as chronic infusion of anabasine in mice (Bhat et al., 1991).

αBgt BINDING SITES IN BRAIN

Chronic nicotine treatment tends to increase the number of nicotinic binding sites in brain measured with α-Bgt (Marks et al., 1983, 1991); however, the increases are smaller and fewer brain regions have significant increases. The increases observed with chronic treatment are dose dependent, but the doses required to achieve effects are higher than those for the high-affinity agonist binding sites. The changes are also time dependent, but the time course for both increase with treatment and decrease with withdrawal are faster than that for the agonist binding receptors (Marks et al., 1985).

NICOTINIC RECEPTOR FUNCTION FOLLOWING CHRONIC TREATMENT

With the development of biochemical assays for nicotinic receptor function, the effects of chronic nicotine treatment on function can be evaluated. In contrast to the universal observation of agonist-induced increases in ligand binding, changes in functional responses have been more variable. Nicotinic-receptor-mediated neurotransmitter release has been observed using several different species and methods (see Wonnacott, 1997, for review). The effects of chronic nicotine treatment on several of these systems have been investigated.

Responses to chronic agonist treatment depend on the response measured, dosing schedule, choice of agonist, and species. Decreased responses (tolerance) have been reported in brain tissue from chronically treated animals for nicotine-stimulated ACh release (Lapchak et al., 1989), Dopamine (DA) release (Marks et al., 1993b; Grady et al., 1997; Benwell et al., 1995), tetraphenylphosphonium uptake (Hillard and Pounds, 1993), and Rb efflux (Marks et al., 1993b). However, preliminary results for Rb efflux indicate the magnitude of the effect and the response across brain regions vary markedly (Collins et al., 1996). For some nicotine-stimulated responses, such as prolactin release (Hulihan-Giblin, 1990), chronic nicotine injection causes a complete loss of responsiveness that requires many days to recover. These results suggest tolerance to the effects of nicotine even though ligand binding sites increased. Note, however, that the nicotinic receptor subtypes regulating each response have not been definitively determined, although the results suggest that nicotine-stimulated Rb efflux is mediated by the α4/β2 receptor subtype (Marks et al., 1993a).

Chronic nicotine treatment has also been reported to result in an increase in functional responsiveness that roughly parallels the increase in ligand binding sites. For example, chronic nicotine injections in rats increased the nicotine-stimulated release of DA and 5-hydroxytryptamine (5HT) (but not ACh) in rat striatal slices (Yu and Wecker, 1994). Intermittent injections also increased the in vivo DA release following a challenge dose. This enhanced response was abolished when the rats were also continuously treated with a low nicotine dose (Benwell et al., 1995). These results emphasize the importance of dosing schedule in influencing the development of tolerance or supersensitivity to nicotine. However, constant infusion of the potent nicotinic agonist, anatoxin-a, resulted in increased nicotine-stimulated DA release (Rowell and Wonnacott, 1990). The differential functional response observed following continuous infusion of nicotine and anatoxin-a suggests that subtleties in agonist action can profoundly affect receptor function following chronic treatment. Such differences may be extremely important when evaluating the therapeutic potential of nicotinic agonists.

CHRONIC NICOTINE TREATMENT IN CELLS AND CELL LINES

Both clonal cell lines expressing endogenous nicotinic receptors and lines transfected with defined receptor subtypes have been examined for their responses to chronic exposure to nicotinic drugs. Rat pheochromocytoma cells, PC12, which express receptors related to those in adrenal chromaffin cells, had decreased function following chronic treatment with carbachol (Robinson and McGee, 1985). The loss of function upon chronic treatment with carbachol and nicotine is both time and dose dependent with a maximal decrease in response of approximately 75% occurring after 1 hr of treatment. The concentrations of agonists required to affect functional loss are comparable to those that elicit acute functional response. The loss of function with time is biphasic in that a readily reversible decrease ascribed to receptor desensitization is observed with short times of treatment while a nearly irreversible decrease representing inactivation occurs with longer times of treatment (Simasko et al., 1986; Boyd, 1987; Lukas, 1991). The effect of treatments on receptor numbers has not been evaluated.

The TE671 cell line that expresses neuromuscular junction-type nicotinic receptors also shows a time- and concentration-dependent loss of function upon chronic treatment with agonists that is observed at concentrations lower than those that activate functional response (Lukas, 1991). Similar concentrations of chronic nicotine elicit increases in receptor number measured by α-Bgt binding (Siegel and Lukas, 1988). Thus, in this cell line apparent receptor upregulation (increase in binding) is accompanied by functional downregulation.

Cells that contain α7 nicotinic receptors that bind α-BGT also respond to chronic nicotine treatment. α-Bgt binding in primary cultures of rat hippocampal neurons increased 40% following treatment with nicotine (Barrantes et al., 1995). The EC50 for this response (1.7 μM) was substantially lower than the EC50 for activation of this receptor in cultured hippocampal cells (27 μM) (Alkondon and Albuquerque, 1993). A 40% increase in α-BGT binding was also observed in SH-SY5Y cells after chronic nicotine treatment (Peng et al., 1997). This increase was complete in 24 hr and occurred with an EC50 for nicotine of 56 μM.

Chronic nicotine treatment also affects α4/β2 nicotinic receptors in cells in culture. Chronic nicotine treatment of either *Xenopus* oocytes expressing chick α4/β2 nicotinic receptors or of fibroblasts transfected with this subtype (M10 cells) elicits a concentration-dependent (EC50 = 0.21 μM, maximal change about 100%) increase in nicotine binding (Peng et al., 1994). Functional response decreases with chronic treatment (about 50% and 90% following treatment with 1 and 10 μM nicotine, respectively). A confirmation of the effect of chronic nicotine treatment on M10 cells as well as a demonstration of a nicotine-induced increase in nicotine binding in primary cortical cell cultures has been reported (Bencherif et al., 1995). In addition, HEK cells transfected with human α4/β2 receptors display a very large (14-fold) increase in cytisine binding sites with chronic nicotine treatment (EC50 = 0.56 μM) (Gopalakrishnan et al., 1996a). These cells also demonstrate a 40% increase in nicotine-stimulated efflux of ^{86}Rb without effect on the EC50 for activation (about 0.6 μM) following treatment with 0.1 μM nicotine, a dose at which a 4-fold increase in cytisine binding was observed.

The concentrations of nicotine required to induce increases in the numbers of ligand binding sites are considerably lower for the α4/β2 receptors than for any other subtype studied. Comparison of the relative functional response for control cells with that for drug-treated cells indicates that the functional response per unit binding site observed when increases in receptor number are maximal is reduced by 75% (Peng et al., 1994) or by 65% (Gopalakrishnan et al., 1996a, 1997). This apparent change in specific activity of this receptor subtype is comparable to the 30% decrease in Rb efflux per nicotine binding site observed for nicotine-stimulated efflux in synaptosomes of chronically treated mice (Marks et al., 1993b; Collins et al., 1996). Whether this loss of activity is a uniform reduction of each receptor or the presence of a pool of active and inactive receptors in unknown. Decreased function could also arise because the fraction of the receptors in the plasma membrane is reduced; however, measurement of the receptors immunologically suggests that the fraction of receptors on the cell surface is similar in control and treated cells (Peng et al., 1994). The nicotine-induced increase in α/β2 receptors occurs in cell lines that do not endogenously express this receptor subtype, indicating that regulation of this subtype may be an intrinsic property of the receptor protein itself.

POSSIBLE MECHANISM OF NICOTINIC RECEPTOR REGULATION: THE DESENSITIZATION HYPOTHESIS

In order to explain the apparent anomalous upregulation of nicotinic receptors following chronic nicotine treatment, it has been postulated that the property of nicotinic receptors to lose function or desensitize following prolonged or repeated exposure to nicotine or other nicotinic agonists is the driving force for the upregulation. Nicotine-induced paralysis at the ganglia was observed a century ago (Langely and Dickenson, 1889). Furthermore, the observation that prolonged exposure to concentrations of nicotinic agonists that do not activate also reduce receptor function was made 40 years ago (Katz and Thesleff, 1957). At that time these authors advanced a cyclical model postulating interconvertible high- and low-affinity agonist binding sites to explain the desensitization observed following exposure to nonstimulating and stimulating concentrations of agonists, respectively. This simple model is a useful framework within which to consider the properties of desensitization and the possible role of desensitization in the subsequent regulation of nicotinic receptors occurring with chronic agonist exposure.

The model allows for two paths to desensitization. One route follows receptor activation during which the liganded receptor undergoes a conformational change leading to channel opening and an increase in conductance. The liganded receptor can also undergo a conformational change leading to a loss of function and the accumulation of a liganded desensitized form of the receptor that is not only inactive but has higher affinity for the agonist. A second route to the accumulation of liganded desensitized receptor can occur at low agonist concentrations at which binding to the unliganded, desensitized form of the receptor occurs. Both routes to desensitization are made possible by the property that the desensitized conformation of the receptor has higher affinity for the agonist than the ground state receptor and that the unliganded ground state and desensitized state of the receptor are interconvertible. Therefore, following exposure to either stimulating or substimulating concentrations of agonist for sufficient time, a significant portion of the receptors are refractory to activation owing to the accumulation of the liganded desensitized nonfunctional receptor.

It should be noted that the basic cyclical model may be an oversimplification. Results obtained with Torpedo receptors led to the advancement of more complex models including two ligand binding sites (Dilger and Liu, 1992) or the existence of an inactivated state in addition to ground and desensitized states (Changeaux, 1990; Lena and Changeaux, 1993). Receptor desensitization has been studied in detail in Torpedo and reviews of this work are available (Ochoa et al., 1989; Lena and Changeaux, 1993), so these results will not be discussed here. Instead, experiments involving neuronal nicotinic receptors either expressed endogenously in brain or cells or expressed heterologously in *Xenopus* oocytes or cell lines will be discussed.

DESENSITIZATION OF NEURONAL NICOTINIC RECEPTORS

Heterologously Expressed Receptors

Inasmuch as conformational change is generally regarded as the initial step for both receptor activation and receptor desensitization the subunit composition of the receptor in question may markedly affect the properties of agonist-induced desensitization. Functional receptors obtained by the injection of mRNA encoding either α2, α3, or α4 with mRNA encoding either the β2 or β4 differ in their response to stimulation with nicotinic agonists, and both α and β subunits affect the response (Luetje and Patrick, 1991). The pharmacology of α7 homomers heterologously expressed in *Xenopus* oocytes also differs from that of the other receptor subtypes in agonist sensitivity (Couturier et al., 1990) and in the difference in sensitivity to inhibition by α-Bgt or methyllycaconitine (Seguela et al., 1993). Furthermore, homomeric α7 nicotinic receptors desensitize rapidly in the continued presence of either nicotine or ACh. These studies confirm that subunit composition dramatically influences both the pharmacology and kinetics of neuronal nicotinic receptors.

Although it has been recognized that the desensitization properties of nicotinic receptors can be dramatically influenced by receptor subtype, relatively few studies have directly assessed desensitization properties of cloned receptors of defined subunit composition. The concentration dependence of the activation and desensitization of chick α3 and α4 subunits coexpressed with chick non-alpha-1 (nα1) (analogous to β2 for mammalian receptors) has been evaluated (Gross et al., 1991). α4/nα1 receptors have higher affinity for ACh and slower desensitization than α3/nα1 receptors. Construction of chimeric receptors in which the extracellular domain of one subtype was exchanged for that of another strongly suggested that the structure of the extracellular domain dramatically affects agonist affinity, while the structure of the transmembrane and cytoplasmic domains affects the desensitization rate. Results with rat nicotinic subunits α2, α3, and α4 in combination with β2 confirmed that the affinity of nicotine and ACh, the rate of desensitization observed with prolonged exposure to agonist, and the rate of recovery from desensitization varies with α subunit composition (Vibat et al., 1995). Desensitization is also influenced by the β subunit. For example, rat α3/β2 receptors heterologously expressed in *Xenopus* oocytes desensitized much more quickly than did α3/β4 receptors (Cachelin and Jaggi, 1991).

The studies described immediately above evaluated desensitization observed with prolonged exposure to activating concentrations of either nicotine of ACh. In normal neuronal function, this type of desensitization would be expected to predominate. However, when nicotinic agonists may be present at relatively low concentrations and for relatively long times as is encountered in tobacco use or during the therapeutic application of nicotinic agonists, significant desensitization may occur with little or no receptor activation and may persist as long as low concentrations remain in the body.

Desensitization of cloned receptors heterologously expressed in *Xenopus* oocytes following exposure to either activating or subactivating nicotine concentrations has been reported (Fenster et al., 1997). α4-containing receptors were activated with lower nicotine concentrations than α3-containing receptors when expressed with either β2 or β4 structural subunits. Each nicotinic receptor subtype examined displayed a time- and concentration-dependent decrease in responsiveness following exposure to nicotine concentrations below those required for activation. Inhibition was virtually complete for every subtype, except α3/β2. α7 receptors were also desensitized by exposure to subactivating nicotine concentrations, but the concentrations required to achieve desensitization were higher than those for receptors composed of α and β subunits. The nicotine concentrations required for activation were at least 50-fold higher than those required for desensitization for all five receptor subtypes studied. Rates of recovery from desensitization also varied among subtypes: β4-containing receptors generally recovered more slowly than β2-containing receptors and α4-containing receptors recovered more slowly than α3-containing receptors. The rates of desensitization and recovery for α7 homomers were more rapid than any other subtype.

These experiments confirm that neuronal nicotinic receptors desensitize following prolonged exposure to both stimulating and substimulating concentrations of nicotine in a pattern generally consistent with the Katz–Thesleff model. Furthermore, the experiments demonstrate that the subunit composition of the receptors markedly affect both activation and desensitization. These results obtained using cloned receptors also provide a reference of comparison to the response measured in cells or tissues. Note, however, that the studies described above employed receptors assembled from either one or two subunits. Those receptor subtypes containing additional subunits such as α5 (Vernallis et al., 1993; Ramirez-Latorre et al., 1996) or α6 (Gerzanich et al., 1997) also exist and may have distinct patterns of activation and desensitization.

Desensitization of Responses in Brain Preparations

In order for nicotinic receptor desensitization to be physiologically important, it must be demonstrated in preparations from brain or other nervous tissue. Several studies confirm that nicotinic responses measured in samples isolated from brain desensitize following exposure to either activating or subactivating concentrations of nicotinic agonists.

Whole cell patch clamp of acutely dissociated medial habenular neurons has been used to demonstrate that the response to activating concentrations of nicotine diminished rapidly with time and the rate of desensitization increased with increasing nicotine concentration (Lester and Dani, 1994). Furthermore, receptor function also decreased following exposure to concentrations significantly lower than those required for activation. However, the onset of this inhibition was considerably slower than that observed following receptor activation. Function returned to control levels following termination of nicotine treatment.

Nicotine-stimulated DA release from mouse striatal synaptosomes also desensitizes following exposure to either activating or subactivating concentrations of nicotine (Grady et al., 1994). The onset of desensitization was slower following treatment with low concentrations. Desensitization was incomplete following exposure to either stimulating or nonstimulating conditions, such that a substantial residual release (about 20% of maximal) persisted. A similar desensitization by low nicotine doses was observed in rat brain synaptosomes as well (Rowell and Hillebrand, 1994; Rowell, 1995). Agonists in addition to nicotine also induced desensitization at both stimulating and substimulating doses (Grady et al., 1997). The IC50 values measured for desensitization at substimulating concentrations of agonists were highly correlated with the affinity of these agonists for the high-affinity nicotine binding site.

Desensitization of response has also been observed using nicotinic-receptor-mediated Rb efflux from mouse thalamic synaptosomes (Marks et al., 1994). Nicotine induced a concentration-dependent decrease in response following either treatment with stimulating or substimulating doses of the drug. The onset of desensitization is slower at substimulating doses and is at least partially reversible. Other nicotinic agonists also induce desensitization at both stimulating or substimulating doses (Marks et al., 1996). As with DA release, the concentration of agonist required to induce desensitization with little activation was closely correlated with the affinity of these agonists for the high-affinity nicotine binding site.

The studies discussed above demonstrate that neural nicotinic receptors do indeed desensitize when exposed to agonists. Desensitization can occur with little or no receptor activation, in general agreement with the model of Katz and Thesleff (1957). The fact that neural nicotinic receptors desensitize fulfills the minimum requirement for the desensitization hypothesis advanced to explain the anomalous upregulation of nicotinic binding sites observed with chronic nicotine treatment.

POSSIBLE MECHANISMS FOR MODULATION OF NICOTINIC RECEPTORS AFTER CHRONIC TREATMENT

A cohesive and universal explanation of the regulation of nicotinic receptor number and function following chronic nicotine treatment is unavailable. In fact, the widely disparate functional responses observed for the various nicotinic subtypes may make a universal explanation impossible. In this section, the various steps that may be involved in receptor regulation in response to chronic agonist or antagonist treatment are outlined and specific examples are provided for various nicotinic subtypes where possible. In general, the following questions are addressed: (1) the role of receptor desensitization and activation, (2) the immediate consequences of desensitization, including receptor inactivation, and (3) potential cellular responses to long-term receptor desensitization or inactivation that lead to changes in the number and function of the receptors.

DESENSITIZATION IN THE REGULATION OF NICOTINIC RECEPTOR FUNCTION

When the initial observations that chronic treatment with the agonist, nicotine, resulted in an increase in the number of nicotinic receptor binding sites, it was apparent that the adaptive response to nicotine was quite different than that observed following treatment with other neurotransmitter receptor agonists (Creese and Sibley, 1980). Since nicotinic receptors in muscle were known to desensitize after exposure to agonists, the hypothesis was advanced that a similar desensitization of brain nicotinic receptors also occurred (Marks et al., 1983; Schwartz and Kellar, 1985; Morrow et al., 1985). As a consequence, chronic treatment with nicotine would be the functional equivalent to chronic treatment with an antagonist (because the receptors were less active rather than overstimulated). The general response of the cells to chronic treatment with an antagonist is to increase the number of receptors in an attempt to restore homeostasis. The observation that nicotinic receptors in neural tissue desensitize with exposure to either stimulating or substimulating concentrations of nicotinic agonists meets the minimum requirement that the formation of the desensitized state of the receptor with chronic agonist treatment is the initial signal for the increase in the receptors with chronic treatment.

If receptor desensitization is the initial signal eventually resulting in the regulation of receptor levels, nicotinic receptors with different affinities for agonists should respond to chronic drug treatment at doses or concentrations reflecting these differences. Several observations are in general accord with this prediction. For example, treatment doses that increase agonist binding sites, which have a relatively high affinity for nicotine (Ki about 5 nM), are lower than those that increase the numbers of α-Bgt binding sites (Ki about 500 nM) (Marks et al., 1991, Pauly et al., 1991). The concentrations of nicotine to half maximally induce increases in nicotinic binding sties for α4/β2, α3-containing, and α7-containing subtypes (Peng et al., 1994, 1997) are qualitatively similar to the concentrations required for desensitization of rat receptors expressed in *Xenopus* oocytes in which α4/β2 receptors are 50-fold more sensitive to desensitization by nicotine than either α3- or α7-containing receptors (Fenster et al., 1997). It should be emphasized that the receptors in question are from different species, so quantitative comparisons may be misleading. The nicotine dose-response curves required to induce functional desensitization in rat α4/β2 and α3/β2 receptors expressed in *Xenopus* oocytes also demonstrated a large difference in nicotine concentration required for loss of function (Yu and Wecker, 1996). Preliminary reports of the effects of prolonged nicotine exposure on human α4/β2 and human α7 receptors transfected in HEK cells indicate that the latter cells require 100-fold higher concentrations (Gopalakrishnan et al., 1996b, 1997). In general, then, differences in agonist doses required to induce binding and/or functional changes in nicotinic receptors with chronic treatment roughly parallel doses required to achieve acute receptor desensitization. This observation is generally consistent with the desensitization hypothesis for modulation of nicotine receptors with chronic treatment.

If desensitization resulting from conformational change is the only receptor alteration induced by chronic treatment and leads to changes in receptor number, then the desensitization should be readily reversible. Consequently, functional receptor response would be expected to parallel the change in receptor number and increased functional response should be observed upon withdrawal of nicotine. A preliminary report suggests that this is the case for human α7 receptor expressed in HEK cells. These cells demonstrate markedly higher functional responses following withdrawal of nicotine that parallels the increase in the number of α-Bgt binding sites (Gopalakrishnan et al., 1996b). It would seem that for this receptor subtype the change in receptor turnover following reversible desensitization may be sufficient to explain the increase in the number of binding sites following agonist treatment. It remains possible that other readily reversible alterations in receptor conformation are also present.

However, for most nicotinic receptors the model invoking receptor desensitization is probably overly simple. Even though the rank order of concentration required to elicit desensitization roughly parallels that for changes observed following chronic treatment, it should be noted that a very robust correlation between desensitizing and activating agonist concentrations has been reported for many nicotinic agonists for both DA release and Rb efflux from mouse synaptosomes, although the concentrations required for these two processes are at least 10-fold different (Marks et al., 1996; Grady et al., 1997). In addition, the absolute concentrations of agonists required to elicit changes with chronic treatment tend to be higher than those that inhibit ligand binding (Peng et al., 1994; Gopalakrishnan et al., 1997; but contrast Hsu et al., 1996). The relationships between agonist affinity for the desensitized state of the receptor and the concentration required to induce changes in the number and function of that receptor does not prove that receptor desensitization is the major driving force behind the regulation. Detailed analyses of dose-response relationships for the effects of nicotine and other nicotinic drugs on the number of binding sites and also on receptor function will be required to completely evaluate the role of this initial drug-receptor interaction in modulating receptor numbers and function. Such a study has been described for HEK cells transfected with human α4/β2 receptors where the potency of several nicotinic agonists as inhibitors of cytisine binding was closely correlated to the concentration needed to increase receptor numbers (Gopalakrishnan et al., 1997).

However, the concentrations for inhibition of ligand binding were several orders of magnitude lower than those that induced changes in receptor number.

LONG-LASTING CHANGES IN RECEPTOR FUNCTION FOLLOWING DESENSITIZATION

A fundamental property of nicotinic receptor desensitization induced by a conformational change of the receptor is that the loss of function should be reversible. Indeed, under many circumstances desensitization is rapidly reversible, but molecular composition of the receptor affects recovery. For example, α7-containing nicotinic receptors undergo a very rapid desensitization, but after removal of the agonist these receptors rapidly and completely recover. In contrast, α4/β2 receptors desensitize quite slowly and also recover from desensitization slowly (Fenster et al., 1997).

The extent of recovery from desensitization is dependent on the length of exposure to the agonist. Although functional recovery can be complete after short exposure to agonists, if exposure time is longer a nearly irreversible loss of function can occur (Boyd, 1987; Shimasko et al., 1986). This functional inactivation can have profound consequences following chronic nicotine treatment such that after several days, a time scale comparable to that required for new protein synthesis, are required to regain functional responsiveness of α4/β2 receptors (Peng et al., 1994). Functional inactivation of this receptor subtype following chronic treatment is also more pronounced as nicotine concentration exceeds that where changes in ligand binding are already maximal (Peng et al., 1994; Gopalakrishnan et al., 1997), suggesting that the biochemical signals for these two processes are not identical. Therefore, exposure of α4/β2 receptors to nicotine concentrations higher than those that induce maximal increases in the number of receptors causes substantial reduction in both absolute and specific functional activity. However, the concentrations required to achieve this profound loss of function are considerably higher than those present in smokers or nicotine-treated animals. Nevertheless, partial functional inactivation is probably an important component in the regulation of receptor responsiveness of both α3 and α4 nicotinic receptors.

BIOCHEMICAL BASIS FOR RECEPTOR REGULATION

Modulation of Receptor Function by Phosphorylation

The precise mechanisms regulating either receptor desensitization or inactivation with prolonged exposure to agonists are not yet known. A simple conformational change to a desensitized state is probably adequate to induce short-term, reversible loss of function. The longer-lasting, virtually irreversible inactivation of receptor function observed with prolonged exposure to agonists probably involves additional mechanisms. Receptor phosphorylation may be one such mechanism. Phosphorylation of the muscle nicotinic receptor alters the desensitization of this receptor (Swope et al., 1992) and receptor phosphorylation is increased upon activation (Miles et al., 1994). Although much less is known about the phosphorylation of neuronal nicotinic receptors, the intracellular domains of the neuronal receptors contain consensus phosphorylation sites (Role, 1992) and direct phosphorylation of intracellular domains of α4 (Nakayama et al., 1993) and α7 (Moss et al., 1996) subunits in vitro and of α3 subunits in chick ciliary ganglia (Vijayaraghavan et al., 1990) has been demonstrated. Pharmacological studies suggest that alteration of phosphorylation affects the expression of nicotinic receptors in ganglia (Haselbeck and Berg, 1996) and cells transfected with α4 and β2 subunits (Gopalakrishnan et al., 1997). In the latter study treatment of the cells with activators of either protein kinase A or C results in an increase in the number of active receptors. The effects of nicotine and the activators were synergistic. Therefore, receptor phosphorylation is a potential mechanism by which chronic nicotinic treatment could regulate both receptor number and function. Whether receptor phosphorylation can subsequently alter receptor clustering and stability in a manner similar to that at the neuromuscular junction remains to be determined.

Receptor Localization

A critical question about the regulation of nicotinic receptor numbers and function following chronic nicotine treatment concerns the cellular location of these receptors. Several studies have addressed this issue and the effects of treatment on receptor distribution are dependent on the subtype. The α4/β2 receptors expressed in transfected cells increase markedly with chronic nicotine treatment (Peng et al., 1994; Bencherif et al., 1995; Gopalakrishnan et al., 1996, 1997) and the proportion of receptors on the cell surface after chronic nicotine treatment does not appear to differ from that of untreated cells (Peng et al., 1994; Gopalakrishnan et al., 1997). Therefore, the number of receptors on the cell surface increases with chronic treatment. An interesting consequence of this observation is that the average functional activity of each receptor must be substantially lower following chronic treatment than it is in untreated cells. In contrast, α3-containing receptors in SH-SY5Y cells, which increase dramatically following chronic nicotine treatment, are located primarily in intracellular pools with very little increase in the number of receptors on the cell surface (Peng et al., 1997). Therefore, effects of treatment on cellular distribution of various nicotinic receptor subtypes are not universal.

Effects on mRNA Levels

One possible explanation for the nicotine-induced increase in receptor number is that the synthesis of the receptors is increased as a result of increase in the amount of mRNA encoding the subunits. This mechanism is unlikely. Quantitative *in situ* hybridization of brains of mice chronically treated with nicotine demonstrated that mRNA encoding α2, α3, α4, α5, and β2 receptor subunits was unaffected (Marks et al., 1992). Similarly, α4 and β2 mRNA in primary rat cortical cells in culture was unchanged by chronic nicotine treatment (Bencherif et al., 1995). However, a small increase in one of the three α4 transcripts was observed in a single brain area after five nicotine injections, suggesting that chronic nicotine injections may have a modest selective effect on mRNA levels (Yu et al., 1996). No change in the amount of mRNA encoding either α3 or α7 was observed after chronic exposure of SH-SY5Y cells to nicotine for 4 days, a time of maximal increase in receptor numbers (Peng et al., 1997). In general, therefore, no changes in steady-state mRNA levels occur with chronic nicotine treatment, although questions about the turnover rates of mRNA have not been addressed.

Effects on Protein Turnover

Increases in steady-state levels of nicotinic receptors occurring as a result of chronic nicotine treatment can arise from a decrease in the rate of protein degradation and/or an increase in the rate of protein synthesis. Several studies have addressed this question. In M10 cells expressing α4/β2 receptors the rate of degradation of nicotine binding sites following blockade of protein synthesis was markedly decreased in the presence of nicotine, indicating that the drug-receptor complex was resistant to proteolysis (Peng et al., 1994). However, a different study using this same cell line found no change in degradation rate in the presence of nicotine, leading to the proposal that nicotine promotes the assembly of measurable receptors from a precursor pool (Bencherif et al., 1995).

Blockade of protein synthesis in HEK cells transfected with human α4/β2 prevented the robust increase in cytisine binding observed in the presence of nicotine (Gopalakrishnan et al., 1997). However, the rate of receptor degradation was not measured and cells treated with nicotine plus cycohexamide displayed higher cytisine binding than cells treated with cyclohexamide alone. This study indicates that protein synthesis is required for maximal increase in receptor number following chronic agonist treatment. In the absence of a change in the amount of mRNA, this result suggests an improved efficiency of protein translation or assembly in drug-treated cells. It was also noted that blockade of protein synthesis prevented the increase in α3-containing receptors in SH-SY5Y cells (Peng et al., 1997). It was postulated that continued protein synthesis was required to stabilize internalized receptors from degradation.

At present, it appears that each nicotinic receptor subtype may be uniquely regulated at the level of protein synthesis and degradation. The mechanisms of agonist-induced receptor stabilization or alterations of synthesis remain to be elucidated. Full receptor function does not seem to be required for the regulation of any of the subtypes measured, since simultaneous antagonist treatment does not usually prevent the increases in receptor number.

SUMMARY

Although the levels of most nicotinic receptor subtypes increase with chronic nicotine treatment, the distinct mechanisms that regulate this increase are not yet well defined. Several observations apply to all of the receptors studied to date: (1) Each receptor subtype desensitizes with prolonged exposure to agonists, although the concentrations required for activation vary widely. (2) Desensitization can also be induced by exposure to agonist concentrations much lower than those required for activation. (3) Agonist concentrations required to elicit increase in the number of receptors roughly parallel the relative affinity of the subtype for the agonist. (4) Chronic agonist exposure has little effect on the levels of mRNA encoding the receptor subunits. Other details of receptor responses to chronic drug treatments vary markedly among the subtypes. As a result, overall change in receptor number and function may increase, decrease, or remain essentially unchanged depending on subunit composition of the particular receptor, treatment dose, dosing scheme,and the brain region or cell type examined. Although receptor desensitization may be important in the initial stages of receptor regulation, it is certainly not the only factor prompting increases in receptor number. Outstanding questions about the changes in receptor structure, both reversible and irreversible, that lead progressively to desensitization, inactivation, and long-term regulation remain to be answered. The answers obtained for each receptor subtype will be to some extent unique.

ACKNOWLEDGMENTS

Thanks to Dr. Sharon Grady for a critical reading of the manuscript. Thanks to Dr. Allan Collins for many years of collaboration. This work was supported in part by grants DA03194 and DA10156 from the National Institute on Drug Abuse.

REFERENCES

Abood LG, Grassi S (1986): [^{3}H]Methylcarbamylcholine, a new radioligand for studying brain nicotinic receptors. Biochem. Pharmacol. 35: 4199–4202.

Alkonbdon M, Albuquerque EX (1993): Diversity of nicotinic acetylcholine receptors in rat hippocampal neurons. I. Pharmacological and functional evidence for distinct structural subtypes. J. Pharmacol. Exp. Ther. 265: 1455–1473.

Barrantes GE, Rogers AT, Lindstrom J, Wonnacott S (1995): α-Bungarotoxin binding sites in rat hippocampal and cortical cultures: initial characterization, colocalisation with α-7 subunits and up-regulation by chronic nicotine treatment. Brain Res. 672: 228–236.

Bencherif M, Fowler K, Lukas RJ, Lippiello PM (1995): Mechanisms of upregulation or neuronal nicotinic acetylcholine receptors in clonal cell lines and primary cultures of fetal rat brain. J. Pharmacol. Exp. Ther. 275: 987–994.

Benwell MEM, Balfour DKJ, Anderson JM (1988): Evidence that tobacco smoking increases the density of (−)-[^{3}H]nicotine binding sites in human brain. J. Neurochem 50: 1243–1247.

Benwell MEM, Balfour DKJ, Birrell CE (1995): Desensitization of the nicotine-induced mesolimbic dopamine responses during constant infusion with nicotine. Br. J. Pharmacol. 114: 454–460.

Bhat RV, Turner SL, Selvaag SR, Marks MJ, Collins AC (1991): Regulation of brain nicotinic receptors by chronic agonist infusion. J. Neurochem. 56: 1932–1939.

Boska P, Quirion R (1987): [^{3}H]N-Methylcarbamylcholine, a new radioligand specific for nicotinic acetylcholine receptors in brain. Eur. J. Pharmacol. 139: 323–333.

Boyd ND (1987): Two distinct phases of desensitization of acetylcholine receptors in clonal rat PC12 cells. J. Physiol. (London) 389: 45–67.

Breese CR, Marks, MJ, Logel J, Adams CE, Sullivan B, Collins AC, Leonard S (1997): Effect of smoking history on [^{3}H]nicotine binding in human postmortem brain. J. Pharmacol. Exp. Ther. 282: 7–13.

Cachelin AB, Jaggi R (1991): Beta subunits determine the time course for desensitization of rat alpha3 neuronal nicotinic acetylcholine receptors. Pflugers Archiv. 419: 579–582.

Cachelin AB, Rust G (1995): Beta-subunits codetermine the sensitivity of rat neuronal nicotinic receptors to antagonists. Pflugers Archiv. 429: 449–451.

Changeaux JP (1990): Functional architecture and dynamics of the nicotinic acetylcholine receptor: An allosteric ligand-gated ion channel. Fidia Res. Found. Neurosci. Award Lect. 4: 21–168.

Cohen BN, Figl A, Quick MW, Labarca C, Davidson N, Lester HA (1995): Regions of β2 and β4 responsible for differences in the steady-state dose-response relationships of the α3β2 and α3β4 neuronal nicotinic receptors. J. Gen. Physiol. 105: 745–764.

Collins AC, Grady SR, Booker TK, Robinson SF, Bullock AE, Stitzel JA, Clark AL, Marks MJ (1996): Differential effects of chronic nicotine treatment on nicotine-stimulated rubidium efflux in various mouse brain areas. Soc. Neurosci. Abstr. 22: 269.

Collins AC, Romm E, Wehner JM (1988): Nicotine tolerance: An analysis of the time course of its development and loss in the rat. Psychopharmacology 96: 7–14.

Collins AC, Romm E, Wehner JM (1990): Dissociation of the apparent relationship between nicotine tolerance and up-regulation of nicotinic receptors. Brain Res. Bull. 25: 373–379.

Couturier S, Bertrand D, Matter J, Hernandez M, Bertrand S, Millar N, Valera S, Barkas T, Ballivet M (1990): A neuronal nicotinic acetylcholine receptor subunit (α7) is developmentally regulated and forms an homoligomeric channel blocked by α-bungarotoxin. Neuron 5: 847–856.

Creese I, Sibley DR (1980): Receptor adaptation to centrally acting drugs. Ann. Rev. Pharmacol. Toxicol. 21: 357–391.

Dilger JP, Liu Y (1992): Desensitization of acetylcholine receptors in BC3H-1 cells. Pflugers Archiv. 420: 479–485.

Fenster CP, Rains, MF, Noerager B, Quick MW, Lester RAJ (1997): Influence of subunit composition on desensitization of neuronal acetylcholine receptors at low concentrations of nicotine. J. Neurosci. 17: 5747–5759.

Flores CM, Rogers SW, Pabreza LA, Wolfe BB, Kellar KJ (1992): A subtype of nicotinic cholinergic receptor in rat brain is composed of α4 and β2 subunits and is up-regulated by chronic nicotine treatment. Mol. Pharmacol. 41: 31–37.

Fung YK, Lau YS (1988): Receptor mechanisms of nicotine-induced locomotor hyperactivity in chronic nicotine-treated rats. Eur. J. Pharmacol. 152: 263–271.

Gerzanich V, Kuryatаov A, Anand R, Lindstrom J (1997): "Orphan" α6 nicotinic AChR subunit can form a functional heteromeric acetylcholine receptor. Mol. Pharmacol. 51: 320–327.

Gopalakrishnan M, Monteggia LM, Anderson DJ, Molinari EJ, Piattoni-Kaplan M, Donnelly-Roberts D, Arneric SP, Sullivan JP (1996a): Stable expression, pharmacological properties and regulation of the human neuronal nicotinic acetylcholine α4 β2 receptor. J. Pharmacol. Exp. Ther. 276: 289–297.

Gopalakrishnan M, Delbono O, Molinari EJ, Renganathan M, Messi L, Arneric SP, Sullivan JP (1996b): Regulation of recombinant human α7 nicotinic receptors by activator and antagonist ligands. Soc. Neurosci. Abstr. 22: 1527.

Gopalakrishnan M, Molinari EJ, Sullivan JP (1997): Regulation of human α4 β2 neuronal nicotinic receptors by cholinergic channel ligands and second messenger pathways. Mol. Pharmacol. 52: 524–534.

Grady SR, Marks MJ, Collins AC (1994): Desensitization of nicotine-stimulated [^{3}H]dopamine release from mouse striatal synaptosomes. J. Neurochem. 62: 1390–1398.

Grady SR, Grun EU, Marks MJ, Collins AC (1997): Pharmacological comparison of transient and persistent [^{3}H]dopamine release from mouse striatal synaptosomes and response to chronic L-nicotine treatment. J. Pharmacol. Exp. Ther. 282: 32–43.

Gross A, Ballivet M, Rungger D, Bertrand D (1991): Neuronal nicotinic acetylcholine receptors expressed in Xenopus oocytes: Role of the α subunit in agonist sensitivity and desensitization. Pflugers Archiv. 419: 545–551.

Haselbeck RL, Berg DK (1996): Tyrosine kinase inhibitors alter composition of nicotinic receptors on neurons. J. Neurobiol. 31: 404–414.

Hillard CJ, Pounds JJ (1993): Effects of chronic nicotine treatment on the accumulation of [^{3}H]tetraphenylphosphonium by cerebral cortical synaptosomes. J. Neurochem. 60: 687–695.

Hsu Y-N, Amin J, Weiss DS, Wecker L (1996): Sustained nicotine exposure differentially affects α3 β2 and α4 β2 neuronal nicotinic receptors expressed in Xenopus oocytes. J. Neurochem. 66: 667–675.

Hulihan-Giblin BA, Lumpkin MA, Kellar KJ (1990): Effects of chronic administration of nicotine on prolactin release in the rat: Inactivation of prolactin response by repeated injections of nicotine. J. Pharmacol. Exp. Ther. 252: 21–25.

Hussy N, Ballivet M, Bertrand D (1994): Agonist and antagonist effects of nicotine on chick neuronal nicotinic receptors are determined by α and β subunits. J. Neurophysiol. 72: 1317–1326.

Katz B, Thesleff S (1957): A study of the "desensitization" produced by acetylcholine at the motor endplate. J. Physiol. (London) 138: 63–80.

Langely JN, Dickenson WL (1889): On the local paralysis of peripheral ganglia and on the connection of different classes of nerve fibers with them. Proc. R. Soc. London B Biol. Sci. 46: 423–431.

Lapchak PA, Araujo DM, Quirion R, Collier B (1989): Effect of chronic treatment on nicotinic autoreceptor function and N-[^{3}H]methylcarbamylcholine binding sites in rat brain. J. Neurochem. 52: 483–491.

Lena C, Changeux J-P (1993): Allosteric modulations of the nicotinic acetylcholine receptor. TINS 16: 181–186.

Lester RAJ, Dani JA (1995): Acetylcholine receptor desensitization induced by nicotine in rat medial habenula neurons. J. Neurophysiol. 74: 195–206.

Luetje CW, Patrick J (1991): Both α- and β-subunits contribute to the agonist sensitivity of neuronal nicotinic acetylcholine receptors. J. Neurosci. 11: 837–845.

Lukas RJ (1991): Effects of chronic nicotinic ligand exposure on functional activity of nicotinic acetylcholine receptors expressed by cells of the PC12 rat pheochromocytoma or TE671/RD human clonal line. J. Neurochem. 56: 1134–1145.

Marks MJ, Burch JB, Collins AC (1983): Effects of chronic nicotine infusion on tolerance development and cholinergic receptors. J. Pharmacol. Exp. Ther. 226: 806–816.

Marks MJ, Stitzel JA, Collins AC (1985): Time course study of the effects of chronic nicotine infusion on drug response and brain receptors. J. Pharmacol. Exp. Ther. 235: 619–628.

Marks MJ, Campbell SM, Romm E, Collins AC (1991): Genotype influences the development of tolerance to nicotine in the mouse. J. Pharmacol. Exp. Ther. 259: 392–402.

Marks MJ, Pauly JR, Gross SD, Deneris ES, Hermans-Borgmeyer I, Heinemann SF, Collins AC (1992): Nicotine binding and nicotinic receptor subunit RNA after chronic nicotine treatment. J. Neurosci. 12: 2765–2784.

Marks MJ, Farnham DA, Grady SR, Collins AC (1993a): Nicotinic receptor function determined by stimulation of rubidium efflux from mouse brain synaptosomes. J. Pharmacol. Exp. Ther. 264: 542–552.

Marks MJ, Grady SR, Collins AC (1993b): Downregulation of nicotinic receptor function after chronic nicotine infusion. J. Pharmacol. Exp. Ther. 266: 1268–1275.

Marks MJ, Grady SR, Yang J-M, Lipiello PM, Collins AC (1994): Desensitization of nicotine-stimulated $^{86}Rb^+$ efflux from mouse brain synaptosomes. J. Neurochem. 63: 2125–2135.

Marks MJ, Robinson SF, Collins AC (1996): Nicotinic agonists differ in activation and desensitization of $^{82}Rb^+$ efflux from mouse thalamic synaptosomes. J. Pharmacol. Exp. Ther. 277: 1383–1396.

Marshall DL, Redfern PH, Wonnacott S (1997): Presynaptic nicotinic modulation of dopamine release in the three ascending pathways studied by in vivo mictodialysis: Comparison of naive and chronic nicotine-treated rats. J. Neurochem. 86: 1511–1519.

Miles K, Audigier SS, Greengard P, Huganir RL (1994): Autoregulation of phosphorylation of the nicotinic acetylcholine receptor. J. Neurosci. 14: 3271–3279.

Morrow AL, Loy R, Creese I (1985): Alteration of nicotinic cholinergic agonist binding sites in hippocampus after fimbria transection. Brain Res. 334: 309–314.

Moss SJ, McDonald BJ, Rudhard Y, Schoepfer R (1996): Phosphorylation of the predicted major intracellular domains of the rat and chick neuronal nicotinic acetylcholine receptor $\alpha 7$ by cAMP dependent protein kinase. Neuropharmacology 35: 1023–1028.

Nakayama H, Okuda H, Nakashima T (1993): Phosphorylation of rat brain nicotinic acetylcholine receptor by cAMP dependent protein kinase in vitro. Mol. Brain Res. 20: 171–177.

Ochoa EL, Chattopadhyay A, McNamee MG (1989): Desensitization of the nicotinic acetylcholine receptor: Molecular mechanisms and effects of modulators. Cell Mol. Neurobiol. 9: 141–178.

Oswald RE, Freeman JA (1981): α-Bungarotoxin binding and central nervous system nicotinic receptors. Neuroscience 6: 1–14.

Pabreza LA, Dhawan S, Kellar KJ (1991): [^{3}H]Cytisine binding to nicotinic cholinergic receptors in brain. Mol. Pharmacol. 39: 9–12.

Pauly JR, Marks MJ, Gross SD, Collins AC (1991): An autoradiographic analysis of cholinergic receptors in mouse brain after chronic nicotine treatment. J. Pharmacol. Exp. Ther. 258: 1127–1136.

Pauly JR, Grun EU, Collins AC (1992): Tolerance to nicotine following chronic treatment by injections: A potential role for cortecosterone. Psychopharmacology 108: 33–39.

Peng X, Gerzanicih V, Anand R, Whiting PJ, Lindstrom J (1994): Nicotine-induced increase in neuronal nicotinic receptors results from a decrease in the rate of receptor turnover. Mol. Pharmacol. 46: 523–530.

Peng X, Gerzanich V, Anand R, Wang F, Lindstrom J (1997): Chronic nicotine treatment up-regulates $\alpha 3$ and $\alpha 7$ acetylcholine receptor subtypes expressed by human neuroblastoma cell line SH-SY5Y. Mol. Pharmacol. 51: 776–784.

Ramirez-Latorre J, Yu CR, Qu X, Perin F, Karlin A, Role L (1996): Functional contributions of $\alpha 5$ subunit to neuronal acetylcholine receptor channels. Nature 380: 547–551.

Robinson D, McGee R (1985): Agonist-induced regulation of neuronal nicotinic receptor of PC12 cells. Mol. Pharmacol. 27: 409–417.

Role LW (1992): Diversity in primary structure and function of neuronal nicotinic acetylcholine receptor channels. Curr. Opin. Neurobiol. 2: 254–262.

Romano C, Goldstein A (1980): Stereospecific nicotine receptors on rat brain membranes. Science 210: 647–649.

Rowell PP, Hillebrand JA (1994): Characterization of nicotine-induced desensitization of evoked dopamine release from rat striatal synaptosomes. J. Neurochem. 63: 561–569.

Rowell PP, Winkler DL (1984): Nicotine stimulation of [^{3}H]acetylcholine release from mouse cerebral cortical synaptosomes. J. Neurochem. 43: 1593–1598.

Rowell PP, Wonnacott S (1990): Evidence for functional activity of up-regulated nicotine binding sites in rat striatal synaptosomes. J. Neurochem 55: 2105–2110.

Sanderson EM, Drasdo AL, McCrea K, Wonnacott S (1993): Upregulation of nicotinic receptors following continuous infusion of nicotine is brain-region-specific. Brain Res. 617: 349–352.

Schoepfer R, Conroy WG, Whiting P, Gore M, Lindstrom J (1990): Brain α-bungarotoxin binding protein cDNAs and mAbs reveal subtypes of this branch of the ligand-gated ion channel superfamily. Neuron 5: 35–48.

Schwartz RD, Kellar KJ (1983): Nicotinic cholinergic receptor binding sites in brain: regulation in vivo. Science 220: 214–216.

Schwartz RD, Kellar KJ (1985): In vivo regulation of [^{3}H]acetylcholine recognition sites in brain by nicotinic cholinergic drugs. J. Neurochem. 45: 427–433.

Schwartz RD, McGee R, Kellar KJ (1982): Nicotinic cholinergic receptors labeled by [^{3}H]acetylcholine in rat brain. Mol. Pharmacol. 22: 56–62.

Sequela P, Wadiche J, Dinely-Miller K, Dani JA, Patrick J (1993): Molecular cloning, functional properties, and distribution of rat brain α7: a nicotinic channel highly permiable to calcium. J. Neurosci. 13: 596–604.

Siegel HN, Lukas RJ (1988): Nicotinic agonists regulate α-bungarotoxin binding sites of TE671 human medulloblastoma cells. J. Neurochem. 50: 1272–1278.

Simasko SM, Soares JR, Weiland GA (1986): Two components of carbamylcholine-induced loss of nicotinic acetylcholine receptor function in the neuronal cell line PC12. Mol. Pharmacol. 30: 6–12.

Swope SL, Moss SJ, Blackstone CD, Huganir RL (1992): Phosphorylation of ligand-gated ion channels: A possible mode of synaptic plasticity. FASEB J. 6: 2514–2523.

Vernallis AB, Conroy WG, Berg DK (1993): Neurons assemble AChRs with as many as 3 kinds of subunits while maintaining subunit segregation among subtypes. Neuron 10: 451–464.

Vibat CRT, Lasalde JA, McNamee MG, Ochoa ELM (1995): Differential desensitization of rat neuronal nicotinic acetylcholine receptor subunit combinations expressed in *Xenopus laevis* oocytes. Cell Mol. Neurobiol. 15: 411–425.

Vijaaraghavan S, Schmid HA, Halvorson SW, Berg DK (1990): Cyclic AMP-dependent phosphorylation of a neuronal acetylcholine receptor α-type subunit. J. Neurosci. 10: 3255–3262.

Whiting PJ, Lindstrom J (1988): Characterization of bovine and human neuronal nicotinic acetylcholine receptors using monoclonal antibodies. J. Neurosci. 8: 3395–3404.

Wonnacott S (1990): The paradox of nicotinic receptor upregulation by nicotine. TIPS 11: 216–219.

Wonnacott S (1997): Presynaptic nicotinic ACh receptors. TINS 20: 92–98.

Yang X, Buccafusco JJ (1994): Effect of chronic central treatment with acetylcholine analog methylcarbamylcholine on cortical nicotinic receptors: Correlation between changes and behavioral function. J. Pharmacol. Exp. Ther. 271: 651–659.

Yu ZJ, Wecker L (1994): Chronic nicotine administration differentially affects neurotransmitter release from rat striatal slices. J. Neurochem. 63: 186–194.

Yu ZJ, Morgan DG, Wecker L (1996): Distribution of three nicotinic α4 mRNA transcripts in rat brain: Selective regulation by nicotine administration. J. Neurochem. 66: 1326–1329.

5

Cell Lines as Models for Studies of Nicotinic Acetylcholine Receptors

Ronald J. Lukas, Ph.D
Division of Neurobiology
Barrow Neurological Institute
Phoenix, Arizona

Other chapters in this volume and recent reviews (Lukas and Bencherif, 1992; Sargent, 1993; Galzi and Changeux, 1994; Lindstrom, 1996; Lukas, 1998) address diversity in nicotinic acetylcholine receptor (nAChR) subtypes and subunits. The wisdom of exploiting as many research tools as possible in studying the widely distributed and physiologically important members of the nAChR family should be evident (Table 5-1). We have championed the use of cell lines as experimental models for studies of nAChRs or, for that matter, as models for studies of many cellular and molecular processes in neurobiology (Lukas, 1988, 1995; Lukas and Bencherif, 1992). Presented here is our rationale, both in practical and theoretical terms, for using cell line models. Also elaborated as a literature review are examples of how cell lines have been used reliably not only to elucidate the structure and function of diverse nAChR subtypes but also to illuminate cellular and molecular mechanisms involved in the regulation of nAChR expression (i.e., numbers and function of nAChRs). The chapter ends with a perspective on recent advances in the use of cell lines as hosts for studies of heterologously expressed nAChRs.

It is important to us to acknowledge investigators whose work established the utility of cell lines as models and inspired our own efforts. Seminal studies that also concerned nAChRs were conducted by Schubert et al. (1974a,b; BC_3H-1 mouse muscle cell line and several central neu-

Neuronal Nicotinic Receptors: Pharmacology and Therapeutic Opportunities, Edited by S. P. Arneric and J. D. Brioni
ISBN 0-471-24743-x, pages 81–97. Copyright © 1998 by Wiley-Liss, Inc.

TABLE 5.1. Experimental Models and Their Uses

	Model					
Modality	Intact Animal	Tissue Slices/ Primary Culture	Tissue Extracts*	Cell Culture	Transfected Cell	Oocyte
Behavior	✓					
Systems physiology	✓	~				
Tracts, connections	✓	~				
Pharmacology	✓	✓	✓	✓	✓	✓
Electrophysiology	✓	✓		✓	✓	✓
Imaging (Ca^{2+}, molecular)	~	✓		✓	✓	✓
Metabolism	~	~	~	✓	✓	
Subcellular fractionation			✓	✓	✓	~
Membrane protein studies		~	✓	✓	✓	~
Protein/immunochemistry		~	✓	✓	✓	~
Nucleic acid chemistry		✓	✓	✓	~	
Regulation of expression	~	~		✓	✓	~
Mutagenesis	~	~	~	~	✓	✓

Several experimental model systems have been used in studies of nAChRs and responsiveness to nicotinic cholinergic ligands. Examples of these models range from intact, behaving animals to tissue extracts* (including membranes, RNA, synaptosomes, etc.). Many different means or experimentally measureable "modalities" have been used to assess activity of nicotinic ligands and nAChRs, and examples are presented. Some of these modalities are quite broad (e.g., "pharmacology" of responses to drug treatment can be assessed using many experimental models, "metabolism" includes assays ranging from pharmacokinetic measures of drug distribution to assessment of enzyme activity to evaluation of neurotransmitter uptake/release) and can be assessed in several ways across model systems. Entries illustrating modalities accessible through the use of selected experimental models, including clonal cell lines and transfected cells, are not presented with magistral intent but rather to prompt thought and debate (e.g., is it clear that regulation of gene expression can be studied in cell lines or transfected cells [signified by ✓], but perhaps also [signified by ~] in intact animals [using knockout or antisense strategies], in slices or primary cultures derived from those animals, and in appropriately engineered oocytes, although not in tissue extracts?).

ronal lines), Greene and Rein (1977; PC12 rat autonomic neuronlike pheochromocytoma), Patrick and Stallcup (1977a,b; PC12), and Syapin et al. (1982; TE671/RD human clone expressing muscle-type nAChR). Others who developed the cell lines that we have used are McAllister et al. (1969, 1977; TE671/RD), Tumilowicz et al. (1970; IMR-32 human autonomic neuron-like peripheral neuroblastoma), Greene and Tischler (1976; PC12), Biedler and colleagues (Ross et al., 1983; SH-SY5Y human peripheral neuroblastoma), and Chikaraishi and co-workers (Suri et al., 1994; CATH.a mouse central catecholaminergic neuronal line). Patrick et al. (1978) provided an excellent earlier perspective on utility of cell lines as models, also emphasizing and reviewing earlier studies of nAChRs.

RATIONALE FOR THE USE OF CLONAL CELL LINES AS MODELS

In practical terms, our use of cell lines as models was initially driven by limitations in other preparations. For example, initial studies of radiolabeled α-bungarotoxin (Bgt) binding sites in membrane and detergent-solubilized preparations from mammalian brain or neural crest-derived tissues clearly showed that Bgt binding sites had physical, chemical, and pharmacological properties expected for a central nervous system (CNS) form of nAChR (Lukas and Bencherif, 1992). However, those preparations could not be used to study function of Bgt binding sites. Electrophysiological and ion flux studies in animals, primary cell cultures, or cell lines (PC12) were initially interpreted as showing Bgt insensitivity of autonomic (e.g., see Patrick and Stallcup, 1977a,b; Margiotta and Berg, 1982) or CNS nAChR responses (reviewed in Lukas and Bencherif, 1992; Albuquerque et al., 1997). However, many of the preparations used for functional studies were not suitable for elucidation of physical and chemical properties of functional nAChRs under study (Table 5-1).

The notion of nAChR diversity, first demonstrated in elegant pharmacological studies done decades ago (Paton and Zaimis, 1952; Dale, 1954), was reborn based on the discovery of nAChR-like radioligand binding site heterogeneity and on cloning of a diverse family of genes encoding distinct nAChR subunits (reviewed in Lukas and Bencherif, 1992; Lindstrom, 1996). Hence, there was impetus to identify model systems amenable to recombinant DNA-based studies of nAChR expression. Finally, given their wide distributions and critical roles in classical and novel forms of chemical signaling throughout the brain and body, nAChRs are ideal targets for the regulation of nervous system function (Lukas, 1995). Thus, it seemed important to develop model systems that would allow studies of how nAChR numbers and function were regulated by exogenous agents, such as nicotine from tobacco products, as well as by natural processes, including those that influence the cytoskeleton, the extracellular matrix, intracellular signaling cascades, and gene expression.

ADVANTAGES OF CELL CULTURE AND CELL LINES

The following list, which is not necessarily exhaustive, includes advantages that cell culture techniques and cell lines offer in studies of complex signaling molecules such as nAChRs:

- Limitless quantities of material (immortalized cells)
- Homogeneity of material (similarity or identity of sister cells)
- Ease of manipulation (of cells and environment—medium)
- Cell as a basic functional unit and as a network element in co-cultures

First, to the extent that continuous cell lines have become immortalized and can be maintained through many passages, which distinguishes them from cells in primary cultures, quantities of experimental material are limitless. More cells can always be grown, harvested, and processed or preserved as needed if the sensitivity of a given assay requires this. Second, cells from continuous lines are highly homogeneous, and every cell in a clonal line culture initially isolated from a single cell should have the same characteristics, aside from variations in characteristics as cells traverse the cell cycle. Thus, continuous or clonal cell cultures are simple compared to intact tissue, and their homogeneity also translates into simplification in the range of macromolecules (nAChR subtypes) expressed. Third, clonal cell lines and their immediate environment (cell culture medium) can be manipulated quickly and easily. Fourth, cells are the basic functional units of eukaryotic organisms, but principles of cell-cell interactions can be revealed through studies of co-cultures.

Radioligand Binding and Pharmacological/Physical Characterization

Clearly feasible is the use of cell lines for radioligand binding assays to establish numbers and subcellular distributions of nAChRs as well as their affinities for radioligands based on equilibrium, pseudoequilibrium, or kinetic (rates of association and dissociation) determinations. Variations on these assays can also be used to characterize nAChRs pharmacologically with regard to the abilities (potency, competition efficacy) of drugs to inhibit binding of radioligands to their targets, sometimes revealing microheterogeneity in sites. For example, these approaches have been used to characterize Bgt binding muscle-type nAChRs (BC_3H-1 cells—Sine and Taylor, 1979; TE671/RD cells—Lukas, 1986a; Ward et al., 1990), Bgt binding α7-nAChRs (IMR-32 cells—Clementi et al., 1986; Lukas, 1993; PC12 cells—Lukas, 1986b; SH-SY5Y cells—Lukas et al., 1993), ^3H-labeled acetylcholine binding muscle-type or α3β4 nAChRs (Lukas, 1990, 1993; Lukas et al., 1993), and ^3H-labeled epibatidine binding nAChRs (Gerzanich et al., 1995). Radioligand labeling has also been used to track muscle-type nAChR assembly intermediates and their physical properties in TE671/RD and other cell types (e.g., see Conroy et al., 1990). In addition, immobilized ligands can be used in affinity purification of nAChRs, taking advantage of the theoretically limitless quantities of materials that can be obtained through cell line cultivation (TE671/RD cells—Luther et al., 1989).

Functional Characterization

Access to cultured cells is simple, allowing them to be manipulated in a variety of ways to measure nAChR function. For example, efflux of the isotopic K^+ analogue, $^{86}Rb^+$, from preloaded cells rinsed free of extracellular isotopic ion and challenged with nicotinic ligands has been used to determine whether those agents activate or inhibit ion efflux from cells via nAChR channels (BC_3H-1—Mauger et al., 1978; PC12 or TE671/RD—Stallcup, 1979; Lukas and Cullen, 1988; Lukas, 1989; SH-SY5Y or IMR-32—Lukas, 1993; Lukas et al., 1993). Nicotinic ligands can be applied to cells in medium containing isotopic ions, thereby allowing determination of nAChR function based on influx of ions through nAChR channels (PC12—Stallcup and Cohn, 1976; Robinson and McGee, 1985). Ion flux assays typically are done by integrating channel responses from the entire ensemble of nAChRs in many thousands of cells, thereby providing a quick means to determine average functional properties over millions of receptors. Cells are more fragile than sealed membrane vesicles, which limits the ability to use stopped-flow approaches to characterize the most rapid events in nAChR channel function.

However, ion flux assays can be done expeditiously for many conditions of time and dose of drug exposure for many combinations of drugs tested alone or in combination with a proven agonist, offering a convenient and high-throughput option for screening for nAChR ligand agonist or antagonist activity (Bencherif et al., 1996). Assays can be easily designed to test for competitive or noncompetitive mechanisms of block for ligands as diverse as forskolin (McHugh and McGee, 1986), local anesthetic-like ligands (Bencherif et al., 1995a), steroids (Ke and Lukas, 1996), and substance P (Lukas and Eisenhour, 1996) across several nAChR subtypes. Effects of nAChR covalent modification can also be assessed (e.g., at *cys* residues—Leprince, 1983; at his residues—Bouzat et al., 1993). Culture medium can also be manipulated to allow determination of effects of cell membrane potential and *trans*-membrane ion distributions on nAChR function (DeLorme and McGee, 1988). In addition, by using isotopes such as $^{22}Na^+$ as a Na^+ analogue, $^{86}Rb^+$ as a K^+ or Na^+ analogue, or $^{45}Ca^{2+}$ to monitor Ca^{2+}, permeability of nAChRs to specific ions and sensitivity to different classes of channel blockers can be determined (Stallcup, 1979; Donnelly-Roberts et al., 1995). Indeed, it can be argued that ion flux assays measure the most elementary process in nAChR ion channel function.

In cases where epiphenomena to these elementary ion flux events are to be measured, cell lines still offer advantages. For example, cultured cells can be subjected to whole cell current recording to determine the electrical status of the cell in response to exposure of the entire ensemble of nAChRs on the cell surface to nicotinic ligands (Oortigiesen and Vijverberg, 1989; Ifune and Steinbach, 1990; Siara et al., 1990; Gould et al., 1992). Cells can also be manipulated to extract information on nAChR single-channel responses to nicotinic ligands (Oswald et al., 1989; Ifune and Steinbach, 1990). Electrical recording also offers means to establish mechanisms of nAChR functional block (local anesthetics and phencyclidine—Papke and Oswald,

1989; chlorpromazine—Benoit and Changeux, 1993). Ready accessibility of cells in culture to recording electrodes and rapid perfusion or drug application systems can be used to advantage to determine the sometimes extremely rapid kinetics of nAChR function, providing an advantage over ion flux assays applied to intact, but fragile, cells (Niu and Hess, 1993). Moreover, the small size of cloned cells (compared to frog oocytes; Costa et al., 1994) makes it easier to apply drugs rapidly and to the entire ensemble of receptors on the cell, helping to ensure that receptors are all activated nearly simultaneously.

Several other kinds of functional assays can be done using cultured cells. For example, they can be used to assess effects of nAChR activation on neurotransmitter release (norepinephrine release from PC12 cells—Greene and Rein, 1977; from SH-SY5Y cells—Murphy et al., 1991). Similarly, release of peptide (dynorphin, PC12—Margioris et al., 1992), hormone (prolactin, rat GH_3 pituitary cells—Coleman and Bancroft, 1995), or growth factor (transforming growth factor β from SiHa cervical cancer cells; Rokowicz-Szulczynska et al., 1994) can be assessed. Effects of nicotinic ligands on small cell carcinoma growth (Schuller, 1989; Quik et al., 1994) identify novel nicotinic cholinoceptive targets potentially involved in tobacco-use-related pathology. These studies indicate how nicotinic ligands might affect complex processes far downstream from nAChR activation, perhaps through release of growth-promoting substances (Cattaneao et al., 1993).

Studies using PC12 cells show nicotine-induced rescue from growth factor deprivation-induced neuronlike cell death (Yamashita and Nakamura, 1996) and promotion of β amyloid precursor protein release (Kim et al. 1997). Studies using IMR-32 cells show nicotinic protection against cytotoxic challenges (Donnelly-Roberts et al., 1996). These findings also indicate that complex processes could be influenced by nicotinic ligands, perhaps out of therapeutic intent. nAChR function can also be detected through activation of gene transcription (*zif*268—Abu-Shakra et al., 1994; *fos, jun*—see Lukas, 1995) or based on changes in enzyme activity and/or gene expression (tyrosine hydroxylase in PC12 cells—Nose et al., 1985). Cultured cells can also be used in suspension or in situ to assess effects of nAChR activation, directly or indirectly, on accumulation of intracellular Ca^{2+} measured fluorimetrically as cell suspensions, in fluorescence microplate readers, or during microscopic examination (e.g., see Quik et al., 1997). Just as is the case for radioligand binding assays, any of these functional assays can be used to determine the pharmacology of nAChRs mediating these functional responses with regard to agonist potencies and efficacies as well as antagonist potencies and mechanisms of action.

Immunological Characterization

Cell lines are amenable to a variety of immunologically based assays of nAChR expression (Patrick and Stallcup, 1977b; Gotti et al., 1989; Luther et al., 1989; Lindstrom, 1996) and to assessment of anti-nAChR activity in disease (Maricq et al., 1985; Lang et al., 1988; Kennel et al., 1993; Lindstrom, 1996). They also can be used to define specificity and features of anti-nAChR subunit antibodies. Immunocytochemical techniques can be perfected using cell lines and to establish features of specific antibodies to then be used in studies of more heterogeneous pools of tissue, such as sections from brain. Western blot analyses can take advantage of the potentially limitless quantities of nAChRs of known composition from cell lines to ensure that enough antigen is presented to be detectable and to check for cross-subunit reactivity of antibodies. In fact, the potential availability of limitless quantities of purified native nAChRs from cell culture systems could aid in generation of novel kinds of antibodies. Cell lines can also be used in immunoprecipitation of radioligand-receptor complexes or in immunoisolation of unliganded receptors.

Identification of nAChR Subunit Genes

Cell lines have demonstrated usefulness as sources for identification and cloning of nAChR subunit genes and their isoforms (BC_3H-1 γ—Yu et al., 1986; TE671/RD human α1—Schopefer et al., 1988; TE671/RD human δ—Luther et al., 1989; PC12 rat α5, β4—Boulter et al., 1990; IMR-32 human α3—Fornasari et al., 1990; C2 mouse muscle cell γ long and short isoforms—Mileo et al., 1990; IMR-32 human β2—Tarroni et al., 1992; SH-SY5Y human α7—Peng et al., 1994a; SH-SY5Y or IMR-32 human β3, β4—Groot Kormelink and Luyten, 1997). Studies using clonal cells also have facilitated or confirmed identification of subunits expressed not only in those cells (op. cit.; Lukas et al., 1993) but also in their non-neoplastic analogues, in some cases also helping to establish origins of clonal cells (e.g., TE671/RD cells in expression of muscle-type rather than a CNS nAChR).

Studies of Human nAChRs

Another value of clonal culture systems is their utility in identification and development of nicotinic drugs to treat diseases and to combat habitual use of tobacco products. Ultimately, such efforts will require an understanding of the selectivity/specificity of nicotinic agents for different human nAChR subtypes. Consequently, human clonal lines naturally expressing defined nAChR subtypes have great practical value, particularly given limited availability of other experimental models of human origin.

COMPARISONS OF ADVANTAGES TO LIMITATIONS

Cell lines have limitations like any other model experimental system (Table 5-1). Proponents of cell lines as models must repeatedly evaluate suitability of these systems for their intended uses and demonstrate that findings obtained using cell lines can be extrapolated to systems in vivo. Studies to date, however, discount erroneous, but frequently perceived, limitations of cell line culture. For example, although most cell lines are derived from tumors and have, by definition, become immortalized, they do not seem to simply represent entirely aberrant freaks of nature.

Receptor and Cell Identification

Table 5-2 lists criteria that should and seem to be met to demonstrate similarities between cell lines and their non-neoplastic analogues, at least with regard to studies of nAChRs. First, the types of nAChRs expressed by a given cell line should be the same as those expressed by cells in the tissue of tumor origin. Pheochromocytoma or peripheral neuroblastoma cells are ana-

TABLE 5-2. Relationships to Non-Neoplastic Tissues

1. Clonal cell lines express the came nAChRs as are found in analogous non-neoplastic tissues.
Muscle—rhabdomyosarcoma, muscle cell lines—α1β1(γ/ε)δ nAChRs
Autonomic neurons—pheochromocytoma, peripheral neuroblastoma—α3β4 nAChRs and α7 nAChRs
Central neurons—CNS neuron-like tumors (medulloblastoma?)—α4β2 nAChRs and α7 nAChRs?—other, novel subtypes?

2. Clonal cell lines express nAChRs that are regulated as they are in analogous, non-neoplastic tissue.
Sensitivity to second messenger modulation, phosphorylation, growth factors, nicotine, extracellular matrix or cytoskeletal interactions, transcriptional regulators?

logues of autonomic neurons, reflecting common origins in adrenoneuroblasts of the neural crest. Just as is the case for characterized nAChRs of autonomic ganglia, cells of the PC12 rat pheochromocytoma and the IMR-32 or SH-SY5Y human neuroblastomas express nAChR $\alpha 3$, $\alpha 5$, $\alpha 7$, $\beta 2$, and $\beta 4$ subunit genes and at least two ganglionic nAChR subtypes containing $\alpha 7$ and $\alpha 3$ plus $\beta 4$ subunits, respectively (reviewed in Lindstrom, 1996, Lukas, 1998). Just as is the case for nAChRs in vertebrate muscle, nAChRs of the TE671/RD human clonal line and of the BC_3H-1 mouse muscle cell lines are composed of $\alpha 1$, $\beta 1$, γ, and δ subunits (Luther et al., 1989). Although there has been some debate as to the precise origins of both of these cell lines (TE671 cells were thought to have derived from a cerebellar medulloblastoma, but they now seem to be derived from the RD rhabdomyosarcoma; McAllister et al., 1969, 1977; Stratton et al., 1989; BC_3H-1 cells were isolated from the CNS but seem to represent a smooth muscle cell line; Schubert et al., 1974a), there is no uncertainty about the origins of the C2 (Kaffe and Saxel, 1977), G8 (Noble et al., 1978; Morris et al., 1989), L6 (Stallcup and Cohn, 1976), and other rat or mouse muscle cell lines, each of which expresses muscle-type nAChR.

From one perspective, once it is established that a given cell type expresses specific nAChRs, there seems to be no reason not to use that cell in studies of at least basic properties of those nAChRs. On the other hand, does expression of the same type of nAChRs provide insights into shared lineages of clonal lines and their non-neoplastic analogues? For example, TE671/RD cells express muscarinic acetylcholine receptors thought to be expressed in neurons but not in muscle cells (Bencherif and Lukas, 1992), and there is a tumor type called a medullomyoblastoma that expresses diverse properties. Do these observations imply that these cells have adopted features of a multipotent precursor to both muscle and neuronal cells during immortalization?

Regulation of nAChR Expression

Another criterion for demonstrating utility of clonal cell lines as models for studies of nAChRs is whether those nAChRs are regulated in the same way as they are in analogous, non-neoplastic cells. This area of study is still in the early stages, but where effects can be compared they seem to indicate close similarities in effects and mechanisms of regulation. In practical terms, it would be very difficult to elucidate supramolecular mechanisms involved in the control of nAChR expression using reduced preparations, intact animals, or many heterologous expression systems (Table 5-1). Cell lines are perhaps most useful for such studies, at least in helping to narrow the realm of mechanistic possibilities for regulation of nAChR expression so that other studies can be designed more incisively.

For example, clonal cells possess intact second messenger signaling systems, allowing assessment of how nAChR numbers and function can be affected by agents altering cyclic nucleotide production, phospholipid turnover, protein kinase/phosphatase activities, and/or intracellular Ca^{2+} accumulation. Studies with cell lines suggest that their nAChRs are affected by second messenger signaling as are nAChRs in analogous non-neoplastic cells (Simantov and Sachs, 1973; McGee and Liepe, 1984; Mulle et al., 1988; Smith et al., 1989; Giovannelli et al., 1990; Bencherif and Lukas, 1991; Rothhut et al., 1996). Growth factors are known to affect nAChR expression in autonomic ganglia, and studies of autonomic neuronlike clonal lines also reveal growth factor effects on the same class(es) of nAChRs (Amy and Bennett, 1983; Mitsuka and Hatanaka, 1983; Rogers et al., 1992; Henderson et al., 1994).

Chronic exposure to nicotine is known to induce increases in numbers of nAChRs in the brains of intact animals. Functional responsiveness to acute challenges with nicotine is diminished in animals previously and chronically exposed to nicotine. Studies using clonal lines expressing ganglionic or muscle-type nAChRs also show the same kinds of effects of chronic nicotine exposure on nAChR numbers and function (for examples and reviews of earlier work, see Lukas, 1991, 1995; Peng et al., 1994b, 1997; Lukas et al., 1996).

Interactions of nAChRs with the extracellular matrix can be elucidated using cell lines (see C2 cell studies of Mook-Jung and Gordon, 1995). Cell lines are well suited to establish how changes in nAChR expression affect or are affected by changes in cytoskeletal status (Bencherif and Lukas, 1993). Cell lines could also be foci of research into requirements for proper assembly of nAChR subunits, including possible roles of chaperones in that process.

In cell lines, nAChR subunit gene expression is stable and under natural promoter control, providing excellent models for studies of regulation of subunit transcription coordinately (Gotti et al., 1987; Lukas, 1995) or differentially (different $\alpha 1$ isoforms in TE671/RD—Morris et al., 1991; induction of subunit switches—Shepherd and Brehm, 1994). Cells can be genetically altered by treatment with antisense oligonucleotides to delineate whether and how signaling cascades, cytoskeletal changes, post-translational processing, and transcriptional factors control nAChR expression (e.g., see C2 H/N-ras studies of Olson et al., 1987; PC12 studies of Henderson et al., 1994).

nAChRs are ideal targets for a variety of therapeutic and recreational drugs, and cell culture systems are attractive models to establish whether agents such as hallucinogens, hormones, peptides, and potential therapeutic agents act directly at nAChRs and/or via effects on second messenger, cytoskeletal, assembly, or nuclear signals.

Studies of effects of chronic nicotine exposure on nAChR expression provide particularly strong support for use of cell lines as models to provide mechanistic insights not easily obtained using other experimental systems. For example, it had been unclear from the literature whether functional responses affected by chronic nicotine treatment occurred far downstream from nAChR activation in whole animals or preparations composed of heterogeneous cell populations. It was unknown whether the same or different populations of nAChRs were being upregulated/functionally inactivated on chronic nicotine exposure. Mechanisms underlying these effects had not been delineated. Studies with cell lines demonstrate unequivocally that chronic nicotine treatment does produce functional losses immediately at the nAChR level in a single population of cells in which numbers of the same population of nAChRs are increased but without sustained or transient changes in nAChR subunit mRNA levels (see Siegel and Lukas, 1988; Lukas, 1991, 1995; Peng et al., 1994b, 1997; Lukas et al., 1996; Bencherif et al., 1995b).

Early studies of chronic nicotine effects were considered to be counterdogmatic and paradoxical because of the apparent upregulation of nAChRs on chronic agonist exposure and because of the apparent uncoupling between effects on nAChR numbers, function, and subunit gene transcription. Recent studies now strongly indicate that upregulation and persistent inactivation are post-transcriptionally mediated, and possibly distinct, processes (op. cit.). Chronic nicotine exposure has different levels of effects on neuromuscular, autonomic, and central functions in vivo, but bases for these differential effects were unknown. Recent studies suggest that chronic nicotine can induce persistent inactivation and upregulation of all nAChR subtypes, but they also show differences in doses of nicotine required to induce persistent inactivation or upregulation for a given nAChR subtype and across nAChR subtypes (op. cit.). Thus, these in vitro studies illuminate potential bases for effects of nicotine in vivo.

The Problem of the CNS Neuronal Cell Line

A substantial limitation in use of clonal cell lines is that those identified may not express all of the important nAChR subtypes that mediate nicotine's effects in vivo. For example, neuron-like cell lines that express nAChR subtypes found in the CNS are rare. If one assumes that neoplastic transformation can only occur in dividing cells, upon realization that dividing CNS neurons are predominantly restricted to periventricular zones early in embryonic development, neither the rarity of clonal neuronal cell lines expressing CNS nAChR subtypes nor the rarity of CNS neuronlike tumors is surprising. Tumorigenesis in periventricular neuroblasts most likely would occlude cerebral spinal fluid flow and/or adversely impact development of the surrounding brain, thereby compromising fetal survival, long before any tumor could be detected in an experimental animal or a human fetus.

Most CNS tumors occurring postnatally are of metastatic or glial origin. The exception is cerebellar medulloblastoma, which is thought to derive from primitive neuroepithelial cells that have migrated to a superficial surface of the cerebellum, forming the external germinal layer. These cells continue to divide postnatally and are the sources for cerebellar granule, basket, and stellate neurons. Human children present with these not necessarily lethal tumors at 2–7 years of age due to compromised brain stem or cerebellar function.

Aside from early studies of the misidentified TE671/RD cell lines, it remains surprising that no other medulloblastomas have been screened for expression of CNS nAChRs. CNS neuronlike cell lines induced by carcinogen treatment were suggested to express nAChRs (Schubert et al., 1994a), but further studies did not follow this observation. CNS neuronlike cell lines developed through fusions of primary CNS neuronal cells with peripheral neuroblastomas or gliomas (Minna et al., 1972; Klee and Nirenberg, 1974), but they have not been widely screened for nAChRs, perhaps with good reason. Particularly for such cell lines derived by fusion with peripheral neuroblastomas, there could be uncertainty as to how the nAChR subunits expressed in neuroblastoma cells (having autonomic neuronal character) would contribute to or suppress expression of nAChR subunits truly originating from the CNS neuronal cells. Use of virally transformed (Hashimaru et al., 1976) or otherwise immortalized (H19-7 hippocampal cells—Komourian and Quik, 1996; CATH.a catecholaminergic cells—Lucero and Lukas, 1996) neuronal cells promises to be further exploited in studies of nAChRs. Similarly, success realized in studies of nAChRs from pluripotent cell lines such as the P19 embryonal carcinoma (Cauley et al., 1996) suggests that immortalized stem cells (Bulloch et al., 1977; Gritti et al., 1996; Johe et al., 1996) could also be models for studies of neuronal nAChR from specific cell types and brain regions.

CLONAL CELL LINES AS HOSTS FOR HETEROLOGOUS EXPRESSION OF RECOMBINANT nAChRs

As a way to combat the rarity of cell lines expressing important nAChR subtypes, and as a way to substantially augment studies of recombinant nAChRs (frequently done using the oocyte expression system), cell lines transfected with nAChR subunit genes to exogenously express diverse nAChR subtypes have been developed recently. Initial success was achieved in stable or transient expression of hetero-oligomeric muscle-type nAChRs from *Torpedo* (Claudio et al., 1987; Forsayeth et al., 1990; Gu et al., 1990) or mouse (Blount and Merlie 1990; Blount et al., 1990; Sine and Claudio, 1991; Chen et al., 1995). Stably expressed chick $\alpha4\beta2$ in AChRs in mouse fibroblasts (M10 cells—Whiting et al., 1991) show modest levels of functional activity (Court et al., 1994; Peng et al., 1994b). Functional human $\alpha4\beta2$ nAChRs have been exogenously expressed (HEK 293 human embryonic kidney cells—Gopalakrishnan et al., 1996, 1997; Buisson et al., 1996).

Although success has been spotty in introducing functional, recombinant $\alpha7$ nAChRs into cell types that do not naturally express nAChRs (Cooper and Millar, 1997), neuronlike cells have been excellent models for expression de novo or overexpression of recombinant nAChRs (rat $\alpha7$ in SH-SY5Y cells—Puchacz et al., 1994; rat $\alpha7$ in rat pituitary cells—Quik et al., 1996;

chick α7 in SH-SY5Y cells—Puchacz et al., 1995). This success has recently been extended to create native nAChR null human cells that can express functional, recombinant human α7 nAChRs at very high levels and, in some cases, under inducible promoter control (in HEK 293 cells—Gopalakrishnan et al., 1995; in SH-EP1 human epithelial cells—Puchacz et al., 1995; Peng and Lukas, 1988; Peng et al., 1998). The nature of the cellular host or even subclones of those cells, the design of the transfection vector, and species matches/mismatches between the host cell and the introduced DNA all could influence success of expression (Chen et al., 1995; Quik et al., 1996; Cooper and Millar, 1997), and creation of stable cell lines expressing recombinant nAChRs can be labor intensive.

However, once established, transfected cell lines can be used like any cell lines to study nAChRs. They are especially valuable in site-directed mutagenesis and receptor chimera studies to help elucidate basic principles of nAChR structure and function (e.g., muscle-type nAChRs—Blount and Merlie, 1990; α7 nAChRs—Puchacz et al., 1995). Moreover transfected cell models can be used to examine regulation of nAChR expression, whether by nicotine or other agents (M10 cells—Peng et al., 1994b; Bencherif et al., 1995b; Zhang et al., 1995; Rothhut et al., 1996; α7 nAChR-expressing cells—Lukas et al., 1996). Transfected cell lines also may be useful in identifying features of nAChR subunits that vary due to gene polymorphisms or mutations that could contribute to neurological, psychiatric, or other kinds of disease and that could point to novel therapeutic interventions (Steinlein et al., 1997).

ACKNOWLEDGMENTS

Work in the author's laboratory has been supported, at different stages, by grants from the National Institute on Drug Abuse (DA07319), the National Institute of Neurological Disorders and Stroke (NS16821), the Smokeless Tobacco Research Council (0277-01), the Arizona Disease Control Research Commission (82-1-098, 9615, and 9730), the American Parkinson Disease Association, and the Council for Tobacco Research—U.S.A. (1683A and 4366M), by Epi-Hab Phoenix, Inc., and by faculty endowment and laboratory capitalization funds from the Men's and Women's Boards of the Barrow Neurological Foundation. The contents of this report are solely the responsibility of the authors and do not necessarily represent the views of the aforementioned awarding agencies.

REFERENCES

It is inevitable and unintentional that many relevant reports will not be mentioned in this necessarily nonexhaustive overview. Bibliographies in the articles and reviews cited here provide broader access to the literature.

Abu-Shakra SR, Cole AJ, Adams RN, Drachman DB (1994): Cholinergic stimulation of skeletal muscle cells induces rapid immediate early gene expression: Role of intracellular calcium. Molec. Brain Res. 26: 55–60.

Albuquerque EX, Alkondon M, Pereira EFR, Castro NG, Schrattenholz A, Barbosa CTF, Bonfante-Cabarcas R, Aracava Y, Eisenberg HM, Maelicke A (1997): Properties of neuronal nicotinic acetylcholine receptors: Pharmacological characterization and modulation of synaptic function. J. Pharmacol. Exp. Ther. 280: 1117–1136.

Amy CM, Bennett EL (1983): Increased sodium ion conductance through nicotinic acetylcholine receptor channels in PC12 cells exposed to nerve growth factor. J. Neurosci. 3: 1547–1553.

Bencherif M, Lukas RJ (1991): Differential regulation of nicotinic acetylcholine receptor expression by human TE671/RD cells following second messenger modulation and sodium butyrate treatments. Molec. Cell. Neurosci. 2: 52–65.

Bencherif M, Lukas RJ (1992): Ligand binding and functional characterization of muscarinic acetylcholine receptors on the TE671/RD human cell line. J. Pharmacol. Exp. Ther. 257: 946–953.

Bencherif M, Lukas RJ (1993): Cytochalasin modulation of nicotinic cholinergic receptor expression and muscarinic receptor function in human TE671/RD cells: a possible functional role of the cytoskeleton. J. Neurochem 61: 852–864.

Bencherif M, Eisenhour CM, Prince RJ, Lippiello PM, Lukas RJ (1995a): The "calcium antagonist" TMB-8 [3,4,5-trimethoxy benzoic acid 8-(diethylamino)octyl ester] is a potent, non-competitive, functional antagonist at diverse nicotinic acetylcholine receptor subtypes. J. Pharmacol. Exp. Ther. 275: 1418–1426.

Bencherif M, Fowler K, Lukas RJ, Lippiello PM (1995b): Mechanisms of upregulation of neuronal nicotinic acetylcholine receptors in clonal cell lines and primary cultures of fetal rat brain. J. Pharmacol. Exp. Ther. 275: 987–994.

Bencherif M, Lovette ME, Fowler KW, Arrington S, Reeves L, Caldwell WS, Lippiello PM (1996): RJR-2403: a nicotinic agonist with CNS selectivity. I. *In vitro* characterization. J. Pharmacol. Exp. Ther. 279: 1413–1421.

Benoit P, Changeux JP (1993): Voltage dependence of the effects of chlorpromazine on the nicotinic receptor channel from mouse muscle cell line So18. Neurosci. Lett. 160: 81–84.

Blount P, Merlie JP (1990): Mutational analysis of muscle nicotinic acetylcholine receptor subunit assembly. J. Cell Biol. 111: 2613–2622.

Blount P, Smith MM, Merlie JP (1990): Assembly intermediates of the mouse muscle nicotinic acetylcholine receptor in stably transfected fibroblasts. J. Cell Biol. 111: 2601–2611.

Boulter J, O'Shea-Greenfield A, Duvoisin RM, Connolly JG, Wada E, Jensen A, Gardner PD, Ballivet M, Deneris ES, McKinnon D, Patrick J, Heinemann S (1990): Alpha 3, alpha 5, and beta 4: three members of the rat neuronal nicotinic acetylcholine receptor-related gene family form a gene cluster. J. Biol. Chem. 265: 4472–4482.

Bouzat CB, Lacorazza HD, Biscoglio de Jimenez Bonino M, Barrantes FJ (1993): Effects of chemical modification of extracellular histidyl residues on the channel properties of the nicotinic acetylcholine receptor. Pflug. Arch.—Eur. J. Physiol. 423: 365–371.

Buisson B, Gopalakrishnan M, Arneric SP, Sullivan JP, Bertrand D (1996): Human $\alpha4\beta2$ neuronal nicotinic acetylcholine receptor in HEK 293 cells: A patch-clamp study. J. Neurosci. 16: 7880–7891.

Bulloch K, Stallcup W, Cohn M (1978): The derivation and characterization of neuronal cell lines from rat and mouse brain. Brain Res. 135: 25–36.

Cattaneao MG, Codignola A, Vicenti LM, Clementi F, Sher E (1993): Nicotine stimulates a serotonergic autocrine loop in human small cell lung carcinoma. Cancer Res. 53: 5566–5568.

Cauley K, Marks M, Gahring LC, Rogers SW (1996): Nicotinic receptor subunits $\alpha3$, $\alpha4$, and $\beta2$ and high affinity nicotine binding sites are expressed by P19 embryonal cells. J. Neurobiol. 30: 303–314.

Chen Q, Fletcher GH, Steinbach JH (1995): Selection of stably transfected cells expressing a high level of fetal muscle nicotinic receptors. J. Neurosci. Res. 40: 606–612.

Claudio T, Green WN, Hartman DS, Hayden F, Paulson HL, Sigworth FJ, Sine SM, Swedlund A (1987): Genetic reconstitution of functional acetylcholine receptor channels in mouse fibroblasts. Science 238: 1688–1694.

Clementi F, Cabrini D, Gotti C, Sher E (1986): Pharmacological characterization of cholinergic receptors in a human neuroblastoma cell line. J. Neurochem. 47: 291–297.

Coleman DT, Bancroft C (1995): Nicotine acts directly in pituitary GH3 cells to inhibit prolactin promoter activity. J. Neuroendocrinol. 7: 785–789.

Conroy WG, Saedi MS, Lindstrom J (1990): TE671 cells express an abundance of a partially mature acetylcholine receptor α subunit which has characteristics of an assembly intermediate. J. Biol. Chem. 265: 21,642–21,651.

Cooper ST, Millar NS (1997): Host cell-specific folding and assembly of the neuronal nicotinic acetylcholine receptor $\alpha7$ subunit. J. Neurochem. 68: 2140–2151.

Costa A, Patrick J, Dani J (1994): Improved technique for studying ion channel expression in Xenopus oocytes, including fast perfusion. Biophys. J. 67: 1–7.

Court JA, Perry EK, Spurden D, Lloyd S, Gillespie JI, Whiting P, Barlow R (1994): Comparison of the binding of nicotinic agonists to receptors from human and rat cerebral cortex and from chick brain ($\alpha4\beta2$) transfected into mouse fibroblasts with ion channel activity. Brain Res. 667: 118–122.

Dale HH (1954): The beginnings and the prospects of neuro-humoral transmission. Pharmacol. Rev. 6: 7–13.

DeLorme E, McGee R Jr. (1988): Effects of prolonged depolarization on the nicotinic acetylcholine receptors of PC12 cells. J. Neurochem. 50: 1248–1252.

Donnelly-Roberts DL, Arneric SP, Sullivan JP (1995): Functional modulation of human "ganglionic-like" neuronal nicotinic acetylcholine receptors (nAChRs) by L-type calcium channel antagonists. Biochem. Biophys. Res. Commun. 213: 657–662.

Donnelly-Roberts DL, Xue IC, Arneric SP, Sullivan JP (1996): In vitro neuroprotective properties of the novel cholinergic channel activator (ChCA), ABT-418. Brain Res. 719: 36–44.

Fornasari D, Chini B, Tarroni P, Clementi F (1990): Molecular cloning of human neuronal nicotinic receptor $\alpha3$-subunit. Neurosci. Lett. 111: 351–356.

Forsayeth JR, Franco A Jr., Rossi AB, Lansman JB, Hall ZW (1990): Expression of functional mouse muscle acetylcholine receptors in Chinese hamster ovary cells. J. Neurosci. 10: 2771–2779.

Galzi J-L, Changeux J-P (1994): Ligand-gated ion channels as unconventional allosteric proteins. Curr. Opin. Struct. Biol. 4: 554–565.

Gerzanich V, Peng X, Wang F, Wells G, Anand R, Fletcher S, Lindstrom (1995): Comparative pharmacology of epibatidine: a potent agonist for neuronal nicotinic acetylcholine receptors. Molec. Pharmacol. 48: 774–782.

Giovannelli A, Farini D, Gauzzi MC, Alema S, Eusebi F (1990): Regulation of acetylcholine receptor desensitization in mouse myotubes by cytosolic cyclic AMP. Cell. Signal. 2: 347–352.

Gopalakrishnan M, Buisson B, Touma E, Giordano T, Campbell JE, Hu IC, Donnelly-Roberts D, Arneric SP, Bertrand D, Sullivan JP (1995): Stable expression and pharmacological properties of the human $\alpha7$ nicotinic acetylcholine receptor. Eur. J. Pharmacol. 290: 237–246.

Gopalakrishnan M, Monteggia LM, Anderson DJ, Molinari EJ, Piattoni-Kaplan M, Donnelly-Roberts D, Arneric SP, Sullivan JP (1995): Stable expression, pharmacological properties and regulation of the human neuronal nicotinic acetylcholine $\alpha4\beta2$ receptor. J. Pharmacol. Exp. Ther. 276: 289–297.

Gopalakrishnan M, Molinari EJ, Sullivan JP (1997): Regulation of human $\alpha4\beta2$ neuronal nicotinic acetylcholine receptors by cholinergic channel ligands and second messenger pathways. Molec. Pharmacol. 52: 524–534.

Gotti C, Sher E, Cabrini D, Bondiolotti G, Wanke E, Mancinelli E, Clementi F (1987): Cholinergic receptors, ion channels, neurotransmitter synthesis, and neurite outgrowth are independently regulated during the in vitro differentiation of a human neuroblastoma cell line. Differentiation 34: 144–155.

Gotti C, Ogando AE, Clementi F (1989): The α-bungarotoxin receptor purified from a human neuroblastoma cell line: biochemical and immunological characterization. Neuroscience 32: 759–767.

Gould J, Reeve HL, Vaughan PF, Peers C (1992): Nicotinic acetylcholine receptor in human neuroblastoma (SH-SY5Y) cells. Neurosci. Lett. 145: 201–204.

Greene LA, Rein G (1977): Release of [^{3}H]norepinephrine from a clonal line of pheochromocytoma cells (PC12) by nicotinic cholinergic stimulation. Brain Res. 138: 521–528.

Greene LA, Tischler AS (1976): Establishment of a noradrenergic clonal line of rat pheochromocytoma cells which respond to nerve growth factor. Proc. Natl. Acad. Sci. USA 73: 2424–2428.

Gritti A, Parati EA, Cova L, Frolichstahl P, Galli R, Wanke E, Faravelli L, Morassutti DJ, Roisen F, Nickel DD, Vescovi AL (1996): Multipotential stem cells from the adult mouse brain proliferate and self-renew in response to basic fibroblast growth factor. J. Neurosci. 16: 1091–1100.

Groot Kormelink PJ, Luyten WH (1997): Cloning and sequencing of full-length cDNAs encoding the human neuronal nicotinic acetylcholine receptor (nAChR) subunits β3 and β4 and expression of seven nAChR subunits in the human neuroblastoma cell line SH-SY5Y and/or IMR-32. FEBS Lett. 400: 309–314.

Gu Y, Franco A Jr., Gardner PD, Lansman JB, Forsayeth JR, Hall ZW (1990): Properties of embryonic and adult muscle acetylcholine receptors transiently expressed in COS cells. Neuron 5: 147–157.

Hashimaru M, Ray J, Sah DWY, Gage FH (1996): Differentiation of the immortalized adult neuronal progenitor cell line HC2S2 into neurons by regulatable suppression of the v-*myc* oncogene. Proc. Natl. Acad. Sci. USA 93: 1518–1523.

Henderson LP, Gdovin MJ, Liu C-L, Gardner PD, Maue RA (1994): Nerve growth factor increases nicotinic ACh receptor gene expression and current density in wild-type and protein kinase A-deficient PC12 cells. J. Neurosci. 14: 1153–1163.

Ifune CK, Steinbach JH (1990): Rectification of acetylcholine-elicited currents in PC12 pheochromocytoma cells. Proc. Natl. Acad. Sci. USA 87: 4794–4798.

Johe KK, Hazel TG, Muller T, Dugich-Djordjevic MM, McKay RDG (1996): Single factors direct the differentiation of stem cells from the fetal and adult central nervous system. Genes and Devel. 10: 3129–3140.

Ke L, Lukas RJ (1996): Effects of steroid exposure on ligand binding and functional activities of diverse nicotinic acetylcholine receptor subtypes. J. Neurochem. 67: 1100–1112.

Kennel P, Vilquin JT, Fonteneau P, Tranchant C, Poindron P, Walter JM (1993): Value of TE671 cells in the detection of anti-acetylcholine receptor antibodies. Revue Neurologique 149: 771–775.

Kim SH, Kim YK, Jeong SJ, Hass C, Kim YH, Suh YH (1997): Enhanced release of secreted form of Alzheimer's amyloid precursor protein form PC12 cells by nicotine. Mol. Pharmacol. 52: 430–436.

Klee WA, Nirenberg M (1974): A neuroblastoma X glioma hybrid cell line with morphine receptors. Proc. Natl. Acad. Sci. USA 71: 3474–3477.

Komourian J, Quik M (1996): Characterization of nicotinic receptors in immortalized hippocampal neurons. Brain Res. 718: 37–45.

Lang B, Richardson G, Rees J, Vincent A, Newsome-Davis J (1988): Plasma from myasthenia gravis patients reduces acetylcholine receptor agonist-induced Na^+ flux into TE671 cell line. J. Neuroimmunol. 19: 141–148.

Leprince P (1983): Chemical modification of the nicotinic cholinergic receptor of PC-12 nerve cell. Biochem. 22: 5551–5556.

Lindstrom J (1996): Neuronal nicotinic acetylcholine receptors. In: Narahashi T, editor. Ion channels, vol. 4. New York: Plenum Press, pp. 377–450.

Lucero L, Lukas RJ (1996): A neuronal cell line expressing a unique, CNS nicotinic acetylcholine receptor subtype. Soc. Neurosci. Abstr. 22: 1272.

Lukas RJ (1986a): Characterization of curaremimetic neurotoxin binding sites on membrane fractions derived from the human medulloblastoma clonal line, TE671. J. Neurochem. 46: 1936–1941.

Lukas RJ (1986b): Characterization of curaremimetic neurotoxin binding sites on cellular membrane fragments derived from the rat pheochromocytoma PC12. J. Neurochem. 47: 1768–1773.

Lukas RJ (1988): Nicotinic acetylcholine receptor diversity: agonist binding and functional potency. Prog. Brain Res. 79: 117–127.

Lukas RJ (1989): Pharmacological distinctions between functional nicotinic acetylcholine receptors on the PC12 rat pheochromocytoma and the TE671 human medulloblastoma. J. Pharmacol. Exp. Ther. 251: 175–182.

Lukas RJ (1990): Heterogeneity of high affinity nicotinic ^{3}H-acetylcholine binding sites. J. Pharmacol. Exp. Ther. 253: 51–57.

Lukas RJ (1991): Effects of chronic nicotinic ligand exposure on functional activity of nicotinic acetylcholine receptors expressed by cells of the PC12 rat pheochromocytoma or the TE671/RD human clonal line. J. Neurochem. 56: 134–145.

Lukas RJ (1993): Expression of ganglia-type nicotinic acetylcholine receptors and nicotinic ligand binding sites by cells of the IMR-32 human neuroblastoma clonal line. J. Pharmacol. Exp. Ther. 265: 294–302.

Lukas RJ (1995): Diversity and patterns of regulation of nicotinic receptor subtypes. Ann. NY Acad Sci. 757: 153–168.

Lukas RJ (1998): Neuronal nicotinic acetylcholine receptors. In: Barrantes FJ, editor. The Nicotinic Acetylcholine Receptor: Current Views and Future Trends. New York: Landes Bioscience, pp. 145–173.

Lukas RJ, Bencherif M (1992): Heterogeneity and regulation of nicotinic acetylcholine receptors. Int. Rev. Neurobiol. 34: 25–131.

Lukas RJ, Cullen MJ (1988): An isotopic rubidium ion efflux assay for the functional characterization of nicotinic acetylcholine receptors on clonal cell lines. Analyt. Biochem. 175: 212–218.

Lukas RJ, Eisenhour CM (1996): Interactions between tachykinins and diverse, human nicotinic acetylcholine receptor subtypes. Neurochem. Res. 21: 1245–1257.

Lukas RJ, Norman SA, Lucero L (1993): Characterization of nicotinic acetylcholine receptors expressed by cells of the SH-SY5Y human neuroblastoma clonal line. Molec. Cell Neurosci. 4: 1–12.

Lukas RJ, Ke L, Bencherif M, Eisenhour CM (1996): Regulation by nicotine of its own receptors. Drug Develop. Res. 38: 136–148.

Luther MA, Schoepfer R, Whiting P, Caset B, Blatt Y, Montal MS, Montal M, Lindstrom J (1989): A muscle acetylcholine receptor is expressed in the human cerebellar medulloblastoma cell line TE671. J. Neurosci. 9: 1082–1096.

Margioris AN, Markogiannakis E, Makrigiannakis A, Gravanis A (1992): PC12 rat pheochromocytoma cells synthesize dynorphin. Its secretion is modulated by nicotine and nerve growth factor. Endocrinology 131: 703–709.

Margiotta JF, Berg DK (1982): Functional synapses are established between ciliary ganglion neurons in dissociated cell culture. Nature 296: 152–154.

Maricq AV, Gu Y, Hestrin S, Hall Z (1985): The effects of a myasthenic serum on the acetylcholine receptors of C2 myotubes. II. Functional inactivation of the receptor. J. Neurosci. 5: 1917–1924.

Mauger JP, Moura AM, Worcel M (1978): Pharmacology of the adenoceptors and cholinoceptors of the BC_3H-1 nonfusing muscle cell line. Brit. J. Pharmacol. 64: 29–36.

McAllister RM, Melnyk J, Finkelstein JZ, Adams EC Jr., Gardner MB (1969): Cultivation in vitro of cells derived from a human rhabdomyosarcoma. Cancer 24: 520–526.

McAllister RM, Issacs H, Rougey R, Peet M, Au W, Soukkup SW, Gardner MB (1977): Establishment of a human medulloblastoma cell line. Int. J. Cancer 20: 206–212.

McGee R Jr., Liepe B (1984): Acute elevation of cyclic AMP does not alter the ion-conducting properties of the neuronal nicotinic acetylcholine receptor of PC12 cells. Molec. Pharmacol. 26: 51–56.

McHugh EM, McGee R Jr. (1986): Direct anesthetic-like effects of forskolin on the nicotinic acetylcholine receptors of PC12 cells. J. Biol. Chem. 261: 3103–3106.

Mileo AM, Monaco L, Palma E, Grassi F, Miledi R, Eusebi F (1995): Two forms of acetylcholine receptor gamma subunit in mouse muscle. Proc. Natl. Acad. Sci. USA 92: 2686–2690.

Minna J, Glazer D, Nirenberg M (1972): Genetic dissection of neuronal properties using somatic cell hybrids. Nature 235: 225–230.

Mitsuka M, Hatanaka H (1983): Selective loss of acetylcholine sensitivity in a nerve cell line cultured in hormone-supplemented serum-free medium. J. Neurosci. 3: 1785–1790.

Mook-Jung I, Gordon H (1995): Acetylcholine receptor clustering in C2 muscle cells requires chondroitin sulfate. J. Neurobiol. 28: 482–492.

Morris A, Beeson D, Jacobsen L, Baggi F, Vincent A, Newsome-Davis J (1991): Two isoforms of the muscle acetylcholine receptor α-subunit are translated in the human cell line TE671. FEBS Lett. 295: 116–118.

Morris CE, Montpetit MR, Sigurdson WJ, Iwasa K (1989): Activation by curare of acetylcholine receptor channels in a murine skeletal muscle cell line. Can. J. Physiol. Pharmacol. 67: 152–158.

Mulle C, Benoit P, Pinset C, Roa M, Changeux J-P (1988): Calcitonin gene-related peptide enhances the rate of desensitization of the nicotinic acetylcholine receptor in cultured mouse muscle cells. Proc. Natl. Acad. Sci. USA 85: 5728–5732.

Murphy NP, Ball SG, Vaughan PF (1991): Potassium- and carbachol-evoked release of [^{3}H]noradrenaline from human neuroblastoma cells, SH-SY5Y. J. Neurochem. 56: 1810–1815.

Niu L, Hess GP (1993): An acetylcholine receptor regulatory site in BC_3H-1 cells: Characterized by laser-pulse photolysis in the microsecond-to-millisecond time region. Biochemistry 32: 3831–3835.

Noble MD, Brown TH, Peacock JH (1978): Regulation of acetylcholine receptor levels by a cholinergic agonist in mouse muscle cell cultures. Proc. Natl. Acad. Sci. USA 75: 3488–3492.

Nose PS, Griffith LC, Schulman H (1985): Ca^{2+}-dependent phosphorylation of tyrosine hydroxylase in PC12 cells. J. Cell Biol. 101: 1182–1190.

Olson EN, Spizz G, Tainsky MA (1987): The oncogenic forms of N-ras or H-ras prevent skeletal myoblast differentiation. Molec. Cell Biol. 7: 2104–2111.

Oortgiesen M, Vijverberg HPM (1989): Properties of neuronal type acetylcholine receptors in voltage clamped mouse neuroblastoma cells. Neuroscience 31: 169–179.

Oswald RE, Papke RL, Lukas RJ (1989): Characterization of nicotinic acetylcholine receptor channels on the TE671 human medulloblastoma clonal line. Neurosci. Lett. 96: 207–212.

Papke RL, Oswald RE (1989): Mechanisms of non-competitive inhibition of acetylcholine-induced single channel currents. J. Gen. Physiol. 93: 785–811.

Paton WDM, Zaimis EJ (1952): The methonium compounds. Pharmacol. Rev. 4: 219–253.

Patrick J, Stallcup B (1977a): α-Bungarotoxin binding and cholinergic receptor function on a rat sympathetic nerve line. J. Biol. Chem. 252: 8629–8633.

Patrick J, Stallcup WB (1977b): Immunological distinctions between acetylcholine receptor and the α-bungarotoxin binding component on sympathetic neurons. Proc. Natl. Acad. Sci. USA 74: 4689–4692.

Patrick J, Heinemann S, Schubert D (1978): Biology of cultured nerve and muscle. Annu. Rev. Neurosci. 1: 417–443.

Peng J-H, Lukas RJ (1998): Heterologous expression of α-bungarotoxin- and epibatidine-binding human α7-neuronal nicotinic acetylcholine receptors in a native nicotinic receptor-null human epithelial cell line. Soc. Neurosci. Abst. 24: in press.

Peng J-H, Lucero L, Fryer J, Herl J, Leonard SS, Lukas RJ (1998): Inducible, heterologous expression of human α7-neuronal nicotinic acetylcholine receptors in a native nicotinic receptor-null human clonal line. Naunyn-Schmiedberg's Arch. Pharmacol. 358(52): RS77.

Peng X, Katz M, Gerzanich V, Anand R, Lindstrom J (1994a): Human α7 acetylcholine receptor: cloning of the α7 subunit from the SH-SY5Y cell line and determination of pharmacological properties of native receptors and functional α7 homomers expressed in *Xenopus* oocytes. Molec. Pharmacol. 45: 546–554.

Peng X, Anand R, Whiting P, Lindstrom J (1994b): Nicotine-induced upregulation of neuronal nicotinic receptors results from a decrease in the rate of turnover. Molec. Pharmacol. 46: 523–530.

Peng X, Gerzanich V, Anand R, Wang F, Lindstrom J (1997): Chronic nicotine treatment up-regulates α3 and α7 nicotinic acetylcholine receptor subtypes expressed by the human neuroblastoma cell line SH-SY5Y. Molec. Pharmacol. 51: 776–784.

Puchacz E, Buisson B, Bertrand D, Lukas RJ (1994): Functional expression of nicotinic acetylcholine receptors containing rat α7 subunits in human neuroblastoma cells. FEBS Lett. 354: 155–159.

Puchacz E, Galzi J-L, Buisson B, Bertrand D, Changeux J-P, Lukas RJ (1995): Properties of transgenic nicotinic acetylcholine receptors stably expressed in human cells as homooligomers of wild-type or mutant rat or chick α7 subunits. Soc. Neurosci. Abst. 21: 1335.

Quik M, Chan J, Patrick J (1994): α-Bungarotoxin blocks the nicotinic receptor mediated increase in cell number in a neuroendocrine cell line. Brain Res. 655: 161–167.

Quik M, Choremis J, Komourian J, Lukas RJ, Puchacz E (1996): Similarity between rat brain nicotinic alpha-bungarotoxin receptors and stably expressed alpha-bungarotoxin binding sties. J. Neurochem. 67: 145–154.

Quik M, Philie J, Choremis J (1997): Modulation of α7 nicotinic receptor-mediated calcium influx by nicotinic agonists. Molec. Pharmacol. 51: 499–506.

Rakowicz-Szulcznska EM, McIntosh DG, Smith M (1994): Growth factor–mediated mechanisms of nicotine-dependent carcinogenesis. Carcinogen 15: 1839–1846.

Robinson D, McGee R, Jr. (1985): Agonist-induced regulation of the neuronal nicotinic acetylcholine receptor of PC12 cells. Molec. Pharmacol. 6: 357–382.

Rogers SW, Mandelzys A, Deneris ES, Cooper E, Heinemann S (1992): The expression of nicotinic acetylcholine receptors by PC12 cells treated with NGF. J. Neurosci. 12: 4611–4623.

Ross RA, Spengler BA, Biedler JL (1983): Coordinate morphological and biochemical interconversion of human neuroblastoma cells. J. Natl. Cancer Inst. 71: 741–747.

Rothhut B, Romano SJ, Vijayaraghavan S, Berg DK (1996): Post-translational regulation of neuronal nicotinic acetylcholine receptors stably expressed in a mouse fibroblast cell line. J. Neurobiol. 29: 115–125.

Sargent P (1993): The diversity of neuronal nicotinic acetylcholine receptors. Annu. Rev. Neurosci. 16: 403–443.

Schoepfer R, Luther M, Lindstrom J (1988): The human medulloblastoma cell line TE671 expresses a muscle-like acetylcholine receptor. Cloning of the α1 subunit cDNA. FEBS Lett. 226: 235–240.

Schubert D, Heinemann S, Carlisle W, Tarikas H, Kimes B, Patrick J, Steinbach JH, Culp W, Brandt BL (1974a): Clonal cell lines from the rat central nervous system. Nature 249: 224–227.

Schubert D, Harris AJ, Devine CE, Heinemann S (1974b): Characterization of a unique muscle cell line. J. Cell Biol. 61: 398–413.

Schuller HM (1989): Cell type specific, receptor-mediated modulation of growth kinetics in human lung cancer cell lines by nicotine and tobacco-related nitrosamines. Biochem. Pharmacol. 38: 3439–3442.

Shepherd D, Brehm P (1994): Adult forms of nicotinic acetylcholine receptors are expressed in the absence of nerve during differentiation of a mouse skeletal muscle cell line. Develop. Biol. 162: 549–557.

Siara J, Ruppersberg JP, Rudel R (1990): Human nicotinic acetylcholine receptor: the influence of second messengers on activation and desensitization. Pflug. Arch.—Eur. J. Physiol. 415: 701–706.

Siegel HN, Lukas RJ (1988): Nicotinic agonists regulate α-bungarotoxin binding sites of TE671 human medulloblastoma cells. J. Neurochem. 50: 1272–1278.

Simantov R, Sachs L (1973): Regulation of acetylcholine receptors in relation to acetylcholinesterase in neuroblastoma cells. Proc. Natl. Acad. Sci. USA 70: 2902–2905.

Sine SM, Claudio T (1991): Stable expression of the mouse nicotinic acetylcholine receptor in mouse fibroblasts. Comparison of receptors in native and transfected cells. J. Biol. Chem. 266: 13,679–13,689.

Sine S, Taylor P (1979): Functional consequences of agonist-mediated state transitions in the cholinergic receptor: studies in cultured muscle cells. J. Biol. Chem. 254: 3315–3325.

Smith MM, Merlie JP, Lawrence JC Jr. (1989): Ca^{2+}-dependent and cAMP-dependent control of nicotinic acetylcholine receptor phosphorylation in muscle cells. J. Biol. Chem. 264: 12,813–12,819.

Stallcup WB (1979): Sodium and calcium fluxes in a clonal nerve cell line. J. Physiol. 286: 525–540.

Stallcup WB, Cohn M (1976): Electrical properties of a clonal cell line as determined by measurement of ion fluxes. Exp. Cell Res. 98: 277–284.

Steinelin OK, Mangusson A, Stoodt J, Bertrand S, Weiland S, Berkovic SF, Nakken KO, Propping P, Bertrand D (1997): An insertion mutation of the CHRNA4 gene in a family with autosomal dominant nocturnal frontal lobe epilepsy. Hum. Molec. Genet. 6: 943–947.

Stratton MR, Darling J, Pilkington GJ, Lantos PL, Reeves BR, Cooper CS (1989): Characterization of the human cell line TE671. Carcinogenesis 10: 899–905.

Suri C, Fung B, Tischler AS, Chikaraishi DM (1993): Catecholaminergic cell lines from the brain and adrenal glands of tyrosine hydroxylase-SV40 T antigen transgenic mice. J. Neurosci. 13: 1280–1291.

Syapin PJ, Salvaterra PM, Engelhardt JK (1982): Neuronal-like features of TE671 cells: presence of a functional nicotinic cholinergic receptor. Brain Res. 231: 365–377.

Tarroni P, Rubboli F, Chini B, Zwart R, Oortgiesen M, Sher E, Clementi F (1992): Neuronal-type nicotinic receptors in human neuroblastoma and small-cell lung carcinoma cell lines. FEBS Lett. 312: 66–70.

Tumilowicz JJ, Nichols WW, Cholan JJ, Greene AE (1970): Definition of a continuous human cell line derived from neuroblastoma. Cancer Res. 30: 2110–2118.

Ward JM, Cockroft VB, Lunt GG, Smillie FS, Wonnacott S (1990): Methyllycaconitine: a selective probe for neuronal α-bungarotoxin binding sties. FEBS Lett. 270: 45–48.

Whiting P, Schoepfer R, Lindstrom J, Priestly T (1991): Structural and pharmacological characterization of the major brain nicotinic acetylcholine receptor subtype stably expressed in mouse fibroblasts. Molec. Pharmacol. 40: 463–472.

Yaffe D, Saxel O (1977): Serial passaging and differentiation of myogenic cells isolated from dystrophic mouse muscle. Nature 270: 725–727.

Yamashita H, Nakamura S (1996): Nicotine rescues PC12 cells from death induced by nerve growth factor deprivation. Neurosci. Lett. 213: 145–147.

Yu L, LaPolla RJ, Davidson N (1986): Mouse muscle nicotinic acetylcholine receptor γ subunit: cDNA sequence and gene expression. Nucl. Acids Res. 14: 3539–3555.

Zhang X, Gong ZH, Hellstrom-Lindahl E, Nordberg A (1995): Regulation of α4β2 nicotinic acetylcholine receptors in M10 cells following treatment with nicotinic agents. Neuroreport 6: 313–317.

6

Stable Expression of Human Neuronal Nicotinic Receptors

Bruno Buisson, Ph.D
Department of Physiology
Faculty of Medicine
University of Geneva, Switzerland

Murali Gopalakrishnan, Ph.D
Neurological and Urological Diseases Research (Dept. 47C)
Abbott Laboratories
Abbott Park, Illinois

Daniel Bertrand, Ph.D
Department of Physiology
Faculty of Medicine
University of Geneva, Switzerland

Neuronal nicotinic acetylcholine receptors (nAChRs) belong to the large family of ionotropic receptors (Bertrand and Changeux, 1995; Galzi and Changeux, 1995). Because they constitute both the ion-permeable channel and specific ligand binding site(s), they are frequently designed as ligand-gated channels (LGCs). Fast synaptic transmission depends on the activation of LGCs by the neurotransmitter. Its release in synaptic clefts is very fast and, at the neuromuscular junction, it was shown that free diffusion of ACh in this restricted space limits the raising phase of the ACh-evoked current (Katz and Miledi, 1973). Following binding of ACh on

Neuronal Nicotinic Receptors: Pharmacology and Therapeutic Opportunities, Edited by S. P. Arneric and J. D. Brioni
ISBN 0-471-24743-x, pages 99–124. Copyright © 1998 by Wiley-Liss, Inc.

the nAChR, transition between the close resting state and an open active state is achieved within microseconds (Colquhoun and Sakmann, 1981; Matsubara et al., 1992). Since nAChRs, like other LGCs, let ions flow in their open state, ion flux measurements give access to the functionality of these macromolecules. Ion fluxes can be measured by using indirect biochemical techniques (isotopic rubidium [$^{86}Rb^+$] efflux) or in real time through direct electrical recordings of agonist-evoked currents. This last method includes the so-called voltage-clamp methods. The patch-clamp technique, which represents the most advanced development of voltage-clamp methods, allows for investigation of elementary events at the picoampere (pA) and microsecond (μs) resolutions with the monitoring of a single LGC activity (Colquhoun and Sakmann, 1981).

It is assumed that neuronal nAChRs result from assembly of five subunits belonging to the α- and/or non-α type (Anand et al., 1991; Cooper et al., 1991). Until now, eight α subunits (α2 to α9) and three β subunits (β2, β3, and β4) have been identified in the nervous system (McGehee and Role, 1995). Homo-oligomers can also be formed by the assembly of five identical subunits as in the case of α7, α8 or α9 as shown by several studies (Couturier et al., 1990; Elgoyhen et al., 1994; Peng et al., 1994; Gopalakrishnan et al., 1995). Although it is widely accepted that functional properties of reconstituted homo-oligomeric receptors resemble those of native nAChRs known to contain the α7 subunits, the composition and stoichiometry of these receptors in the nervous system further await characterization. Hetero-oligomers are assumed to be formed by the assembly of α and β subunits (Bertrand and Changeux, 1995). However, more complex arrangements, including, for instance, the α5 subunit, as part of the complex (i.e., α3α5β4 etc.) may be formed in some neurons (Conroy et al., 1992; Role and Berg, 1996). Since immunoprecipitation and in situ hybridization studies, together with immunochemical labeling, have revealed the presence of several nAChR subunits in a single neuronal population (Corriveau and Berg, 1993; Pugh et al., 1995; Ullian and Sargent, 1995; Flores et al., 1996; Horch and Sargent, 1996), it is likely that the native composition of nAChRs would be much more complex. Further, since functional investigations have indicated that neuronal nAChRs could be localized both on pre- and postsynaptic membranes (Role and Berg, 1996; Léna and Changeux, 1997; Wonnacott, 1997), the main challenge is now to determine which subunits are coassembled and what precisely are their cellular distribution, physiological and pharmacological properties.

The identification and cloning of genes coding human neuronal nAChRs has opened new ways to investigate the physiological and pharmacological properties of this class of LGCs (Lindstrom, 1996). Human cDNAs have been cloned from different sources including adult (China et al., 1994) and embryonic (Anand and Lindstrom, 1990; Monteggia et al., 1995) brain tissues and various human neuroblastoma cell lines (Peng et al., 1994). Moreover, alignment of all the sequences coding for the α4 or α7 subunits identified by different laboratories reveals non-negligible variations. Although few differences may be attributed to sequencing mismatches, most of them should be representative of the human genome polymorphism. Until now, physiological and pharmacological investigations were only performed by expression of one cDNA type. Thus, caution should be taken regarding the generalization of these data to brain functions.

Expression of human nAChR subunits in heterologous systems represents the closest physiological approaches for a noninvasive and functional characterization of these receptors. Two alternative approaches can be considered for the reconstitution of functional receptors with: (1) expression of mRNAs or cDNAs in *Xenopus* oocytes and (2) transient or stable transfection in a cell line. The electrophysiological approaches to investigate the properties of LGCs have been recently reviewed elsewhere (Bertrand et al., 1997a). The aim of this work is to describe the stable transfection methods of some of the human nAChR subunits in human embryonic kidney cells (HEK-293) and to review their principal physiological and pharmacological characteristics.

ADVANTAGES OF EXPRESSION IN CELL LINES

The ability to introduce exogenous DNA into cultured mammalian cells (transfection) has provided a powerful tool for studying gene expression, function, and regulation of nAChRs. Expression in cell lines offers distinct advantages compared to other host systems such as amphibian oocytes:

- Reconstitution of diverse nAChR subtypes in cell lines permits analysis of their physiological and pharmacological characteristics in a well-defined system without interference from other nAChR subtypes.
- Events such as post-translational modifications or other steps in the receptor assembly process may differentially occur in different cell types, which may influence the properties of the reconstituted nAChRs leading to possible functional consequences and pharmacological differences.
- Studies with cell lines permit faster drug application and better temporal resolution to investigate the fast kinetics and other properties of the nAChRs.
- Stable cell lines expressing defined nAChR subtypes have, more recently, emerged as a method of choice for high throughput screening of diverse compound libraries, thus offering the potential to yield novel selective agents unlikely to be discovered by traditional methods.

METHODOLOGICAL CONSIDERATIONS

A number of parameters need to be considered for successful expression of nAChRs in cell lines. These are briefly outlined below.

Choice of Transfection Methodology

Various methods to transfer DNA into eukaryotic cells are available. These include the use of calcium phosphate or other divalent cations, polycations, liposomes, retroviruses, microinjection, and electroporation (Roth, 1994). Among these, the cationic liposome-mediated transfection technique (lipofection), a versatile and convenient method to deliver foreign genes into mammalian cells, has been widely employed (Gopalakrishnan et al., 1996, 1997). This procedure involves the use of a mixture of synthetic cationic lipid molecules where a positively charged head group is attached to a lipid backbone by an ester linkage. The positively charged head groups associate with negative charges, such as phosphate groups of nucleic acids, resulting in the formation of lipid–DNA complexes internalized into the cell by fusion with the cell membrane or by endocytosis (Felgner et al., 1987).

Selection of the Expression Vector

A second important consideration is to identify an appropriate plasmid expression vector. Plasmids used for expression studies generally contain the following essential elements: an efficient eukaryotic transcription unit, a prokaryotic origin of replication, a viral-derived origin of replication and a prokaryotic selectable marker besides mRNA processing signals (polyadenylation sequences, etc.), and the polylinker containing multiple restriction endonuclease sites for insertion of foreign DNA.

For stable expression, it is important to use a vector that has a strong promoter for transcribing the gene of interest together with a marker for amplification of the stable transfectants. A good promoter for high-level transcription is the major immediate early promoter of the human cytomegalovirus (CMV; Brewer, 1994) and several vectors that incorporate this promoter are commercially available. Other constitutively active promoters including Rous sarcoma virus (RSV) and simian virus 40 (SV40) may also be employed.

A widely used selectable marker is the aminoglycoside phosphotransferase gene (*neoR* gene), the expression of which in mammalian cells confers resistance to the antibiotic G418. G418 is an aminoglycoside antibiotic produced by *Streptomyces* that induces cytotoxicity by blocking translation. Other commonly used selectable markers are the genes encoding hygromycin B phosphotransferase (HPH), adenosine deaminase (ADA), dihydrofolate reductase (DFHR), thymidine kinase (TK), xanthine-guanine phosphoribosyl transferase (XGPRT), etc. In cases where the expressed protein may be cytotoxic, an expression vector with an inducible promoter may be chosen to regulate the levels of expression. Inducible vectors (such as the mouse mammary tumor virus long terminal repeat or MMTV-LTR, metallothionein, *Lac* repressor) are well suited for controlled expression of proteins that may be potentially cytotoxic or could perhaps impair specific cellular functions.

Choice of the Cell Line

In choosing a cell line for the stable expression of nAChRs, several factors need consideration. First, it is desirable that the host cell line does not express the protein of interest whose gene is to be introduced. This could be assessed by a number of techniques including detection of mRNA levels by reverse transcriptase–polymerase chain reaction (RT–PCR) or Northern analysis, protein levels (radioligands and/or antibodies), and/or receptor function (patch-clamp/ion flux studies). Second, it is important that the cell line chosen can easily be maintained in culture and can be readily transfected using the conditions employed. Cells at low passage numbers that have been regularly passed to ensure that they are in a permanent growth phase are generally preferred. Human embryonic kidney-293 cells or their variants have been widely employed in the stable expression of various nAChRs, although other cell types such as L cells, COS cells, and CHO cells may also be used.

Optimization of Transfection Efficiency

Optimization of transfection method is another critical consideration since cells are generally quite variable with respect to the efficiency of transfection. With the lipofection method, the parameters that have to be optimized for a successful transfection include amount of DNA and lipid needed (charge ratio) and the effect of serum.

For studies using transiently transfected cells, it is necessary to have good and consistent levels of transfection efficiencies. The quality of transient transfection can be assessed by co-expression with a reporter gene such as β-galactosidase, chloramphenicol acetyltransferase (CAT), β-lactamase, firefly luciferase, or green fluorescent protein (GFP). One of the commonly used positive controls for monitoring transfection efficiencies in mammalian cells is the β-galactosidase vector. The *Escherichia coli lacZ* gene that codes bacterial β-galactosidase (β-gal) can be assessed either by a simple in situ staining procedure or by quantitative methods.

Alternatively, cells can be cotransfected with GFP, a 27 kD protein that forms a fluorophore within itself so that when expressed in a cell it will fluoresce (Heim and Tsien, 1996). The advantage of using GFP is that it does not require the addition of exogenous substrate or cofactors for activity, thereby permitting real-time detection without damaging the cells. The use of a fluorescent marker may also provide the opportunity to select positive cells by a fluorescent-activated cell sorter (FACS).

Stable Versus Transient Expression

The choice of the expression system critically depends on the objective of the studies. In some cases, it may be only necessary to show that the nAChR subunit (wild type or mutant) is functionally active or that the objective is to study the influence of additional subunits on physiological properties of those nAChRs that have already been stably expressed in cells; in such instances transient expression of the gene is usually satisfactory. However, for studies involving longer duration or if the transfected cells are required for long-term use, for example, in drug screening efforts, then large and consistent amounts of the protein are required and it is desirable to have a stable cell line.

Although transient transfections may yield results within days instead of several weeks, these suffer from several disadvantages not generally observed with stable cell lines. For example, even after careful optimization, transient transfection efficiencies may vary from day to day, which could result in significant variations in the evaluation of the end results (biochemical or functional). Second, even in the best optimized system, transfection rates are mostly below 20%. Finally, protein expression generally decreases after about 4 days, thus limiting the feasibility of studies that require a longer duration. Thus, a stable cell line is generally desirable. With stable transfection, the transfected DNA is either integrated into the chromosomal DNA or maintained as an episome. Stable transfection involving integration of the DNA is most efficient when linearized plasmid DNA is used, since linearization facilitates recombination of the DNA with host cell chromosomes. However, since this occurs only in a small fraction, stably transfected cells need to be amplified by selection, for example, by drug resistance, which usually takes several weeks, as opposed to transient expression, which is accomplished in as little as 24 to 72 hours.

STABLE EXPRESSION OF nAChRs IN CELL LINES

Considerable effort over the past few years in several laboratories has led to the successful and stable expression of various nAChR subtypes in mammalian cell lines. Table 6-1 summarizes many of the stable cell lines along with their salient properties. Our own studies have resulted in the generation of a library of cell lines stably expressing various nAChR subtypes using CMV-based plasmid vectors (Gopalakrishnan et al., 1996, 1997; Buisson et al., 1996), whose functional and biochemical properties have been maintained in culture for more than 1 to 2 years. To date, three different cell lines obtained by stable transfection of HEK-293 cells have been characterized by our laboratories—K28, K177, and K4-14—which express respectively the following nAChR subunits: $\alpha 7$, $\alpha 4\beta 2$, and $\alpha 3\beta 2$. These studies have employed HEK-293 cells since they do not endogenously express neuronal nAChRs as assessed by radioligand binding and/or functional studies.

Transfections were routinely carried out using the lipofection method using Lipofectamine (GIBCO). In this method, the plasmid samples of DNA (1–1.5 μg) are mixed with Lipofectamine and are used to transfect cells ($1\text{–}2 \times 10^6$) previously grown to approximately 50–60%

Table 6.1. Mammalian Cell Lines Expressing Neuronal Nicotinic Acetylcholine Receptor Subtypes

nAChR Subtype (Species)	Cell Line	Radioligand Binding	Functional Assessment	References
α7 (human)	HEK-293 (K28)	[^{125}I]α-Bungarotoxin K_D = 0.7 nM; B_{max} = 973 fmol/mg	ACh-evoked current (EC_{50} = 155 μM); ACh-evoked Ca^{2+} flux (EC_{50} = 108 μM)	Gopalakrishnan et al., 1995 Delbono et al., 1997
α7 (rat)	Human neuroblastoma SH-SY5Y	[^{125}I]α-Bungarotoxin: 1400 fmol/mg	ACh-evoked current (EC_{50} = 150 μM)	Puchacz et al., 1994
α7 (rat)	Rat pituitary GH3 cells	—	ACh-evoked Ca^{2+} flux (EC_{50} = 17 μM)	Quik et al., 1997
α4β2 (avian)	Mouse fibroblasts Ltk (M10)	[^{3}H]Nicotine: 2300 sites/cell	ACh-evoked current (EC_{50} = 20 μM)	Whiting et al., 1991 Pereira et al., 1994
α4β2 (human)	HEK-293 (K177)	[^{3}H]Cytisine: K_D = 0.2 nM; B_{max} = 1359 fmol/mg)	ACh-evoked $^{86}RB^{+}$ efflux (EC_{50} = 43 μM); ACh-evoked current (EC_{50} = 3 μM)	Gopalakrishnan et al., 1995 Buisson et al., 1996
α4β2 (human)	HEK-293 (A4B2)	—	Nicotine-evoked Ca^{2+} flux (EC_{50} = 4.4 μM)	Sacaan et al., 1997
α2β2 (human)	HEK-293	[^{3}H]Epibatidine: K_D = 28 pM; B_{max} = 756 fmol/mg	—	Kuntzweiler et al., 1997
α3β2 (human)	HEK-293	[^{3}H]Epibatidine: K_D = 26 pM; B_{max} = 176 fmol/mg	ACh-evoked Ca^{2+} flux (EC_{50} = 87 μM) ACh-evoked current (EC_{50} = 70 μM)	Kuntzweiler et al., 1997
α3β2 (human)	HEK-(tsa201) cells	[^{3}H]Epibatidine = 6 × 10^3 sites/cell	—	Weng et al., 1997
α3β2α5 (human)	HEK-293 (83-19-15)	—	ACh-evoked Ca^{2+} flux (EC_{50} = 0.09 μM)	Gillespie et al., 1997
α2β4 (human)	HEK-293	[^{3}H]Epibatidine: K_D = 15 pM; B_{max} = 525 fmol/mg	—	Stauderman et al., 1995
α3β4 (human)	HEK-(tsa201) cells	[^{3}H]Epibatidine = 6 × 10^4 sites/cell	ACh-evoked current (EC_{50} = 79 μM)	Wang et al., 1995
α3β4 (rat)	HEK-293	[^{3}H]Epibatidine Binding K_D = 300 pM; B_{max} = 8640 fmol/mg	Nicotine-evoked current (EC_{50} = 30 μM)	Xiao et al., 1997
α4β4 (human)	HEK-293	[^{3}H]Epibatidine Binding K_D = 202 pM; B_{max} = 8640 fmol/mg	—	Stauderman et al., 1995

confluence. The amount of lipid that results in peak activity of transfection is initially assessed by transfection of a reporter gene such as β-galactosidase. Following transfection, the cells are incubated in an incubator for 2 to 3 days during which period the transient expression of the gene may be assessed. For generating stable transfectants, the cells are split 48 hours post-transfection, diluted, and plated in media containing appropriate antibiotic(s) for selecting resistant clones. The concentration of the antibiotic to be used is determined from a lethality curve generated using untransfected cells. When antibiotic-resistant clusters of cells begin to appear (typically 2–3 weeks), the colonies are chosen at random with the aid of cloning cylinders and cells are further propagated in media containing the appropriate antibiotic(s).

As described, many nAChR subtypes function as pairwise or triplex complexes in native cells. These have been reconstituted by introducing two or three heterologous genes into a single cell line, which is accomplished by cotransfecting the genes under identical promoters but containing different antibiotic selection markers. For example, the human α4β2 nAChRs have been stably expressed by cotransfection with α4 and β2 nAChR DNAs, both cloned into a CMV-based expression vector containing the neomycin and hygromycin resistance genes respectively, followed by appropriate antibiotic selection and screening (K177 cell line; Gopalakrishnan et al., 1996). Alternatively, a two-step selection strategy could be performed by first introducing one set of genes followed by a second round of transfection or using a plasmid with expression cassettes for two different genes and containing a single selectable marker, thereby ensuring a 1:1 stoichiometry at the mRNA level.

It is to be noted that in some instances expression of multisubunit receptors may pose difficulties because of the potential for incorrect assembly of subunits. Artifactual data may result from incomplete or incorrect expression and/or processing of receptor protein from the heterologous gene. In some cases, the expressed protein may be antiproliferative or cytotoxic depending on whether the foreign protein modifies the cell membrane potential or alters intracellular ionic concentrations (for example, Ca^{2+}). This may pose difficulty in generating stable cell lines, especially in the case of receptors that are constitutively active (Bertrand et al., 1997b).

BIOCHEMICAL CHARACTERIZATION OF nAChRs STABLY EXPRESSED IN CELL LINES

A primary goal of molecular biological and expression studies of nAChRs is to identify genes that encode native channels present in normal cells. A variety of analytical approaches is employed to compare the biophysical, physiological, and pharmacological properties of the recombinant nAChR subunit combinations stably expressed in cell lines with those of native nAChRs. These include (1) mRNA level detection, (2) binding techniques using radiolabeled ligands or fluorescently labeled ligands, (3) electrophysiological recordings, (4) immunoassays (Western blot), and (5) radiotracer and/or fluorometric assays.

mRNA Analysis

One of the common methods to monitor gene expression involves techniques such as Northern blotting or nuclease protection assays to quantify the specific mRNAs transcribed from the gene(s) of interest. RT–PCR analysis of total RNA from transfected cells (to indicate the appropriate translation of mRNAs corresponding to the gene of interest) may also be performed and compared, although not quantitatively, with the RNA of untransfected cells).

Radioligand Binding

The specific binding of high-affinity radiolabeled nAChR ligands to stably transfected cells has been successfully employed to assess nAChR expression. Commonly used radioligands to identify clones that express high levels of α3/α4 subunit-containing nAChRs include [^{3}H]cytisine, [^{3}H]nicotine, [^{3}H]acetylcholine, [^{3}H]epibatidine, or [^{3}H]A-85380, whereas [^{125}I]α-bungarotoxin is a high-affinity ligand for the α7 subunit-containing nAChRs. Typically, the specific binding levels are assessed in membrane preparations or in intact cells from the several clones, and highly expressing cell lines are chosen for further studies. Studies using [^{3}H]cytisine with the K177 cell line stably expressing the human α4β2 nAChRs revealed a maximal binding density value of 1.4 pmol/mg protein and a K_D value of 0.2 nM (Fig. 6-1). This affinity is similar to those values observed in human brain and in rodent cortical tissues but with receptor densities about 10- to 20-fold higher than those observed in rodent tissues. [^{125}I]α-bungarotoxin and [^{3}H]epibatidine have also been successfully utilized to investigate stably transfected nAChRs derived from α7 and α3β2 combinations, respectively (Gopalakrishnan et al., 1995; Kuntzweiler et al., 1997; Wang et al., 1997).

Fluorescent Labeling

The uniformity and distribution of the expressed nAChRs can be assessed by using fluorescently labeled ligands. As an illustration, the human α7 nAChRs stably expressed in HEK-293 cells (K28 cell line) were labeled by fluorescein isothiocyanate-conjugated α-bungarotoxin (Fig. 6-2) and more than 80% of the cells were found to be specifically labeled.

Radiotracer Methods

Radiotracers such as $^{86}Rb^+$ have also been successfully employed in the evaluation of functional expression of certain nAChR subtypes that desensitize slowly in the continuous pres-

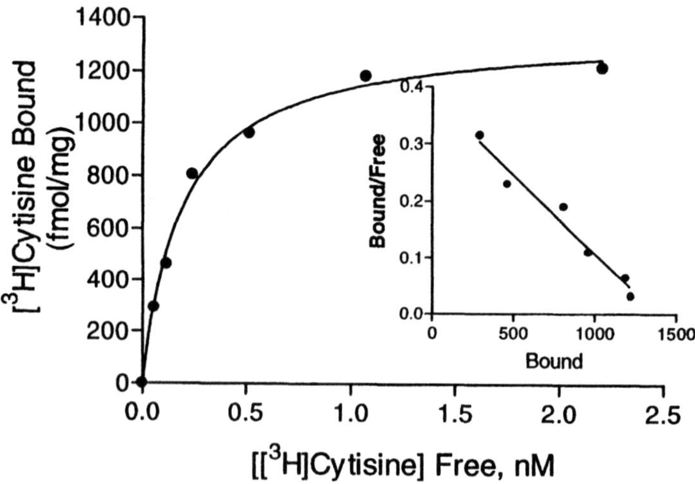

Figure 6.1. Stable expression of human α4β2 nAChRs in HEK-293 cells. Specific binding of [^{3}H]-cytisine to transfected cell membranes (K177 cell line). Inset: Linearization of the data in the form of a Scatchard plot (B_{max} = 1400 fmol/mg protein; K_D = 0.2 nM).

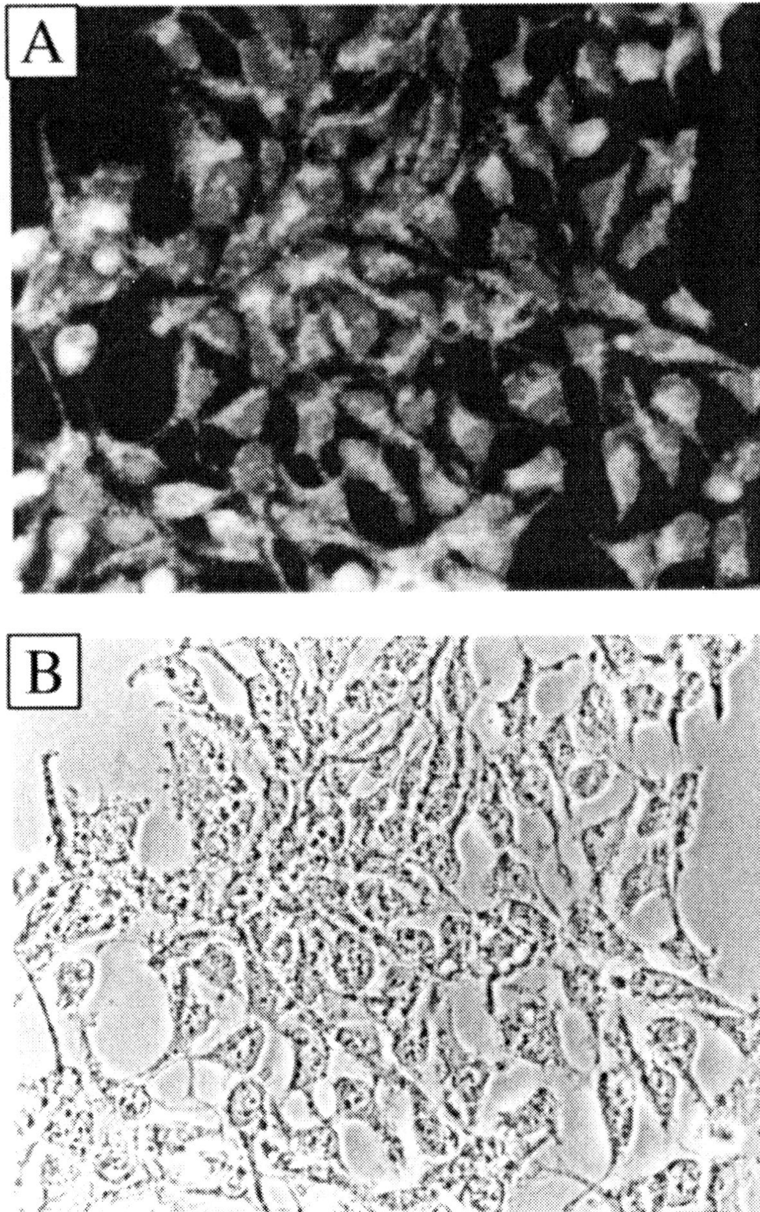

Figure 6.2. *HEK-293 cells stably expressing human $\alpha 7$ nAChRs (K28 cell line). (**A**) Fluorescent labeling by fluorescein isothiocyanate (FITC) conjugated-α-bungarotoxin showing cell surface expression of human $\alpha 7$ nAChRs. (**B**) Phase contrast micrograph of the cells shown in (**A**) (Modified from Gopalakrishnan et al., 1995, with permission.)*

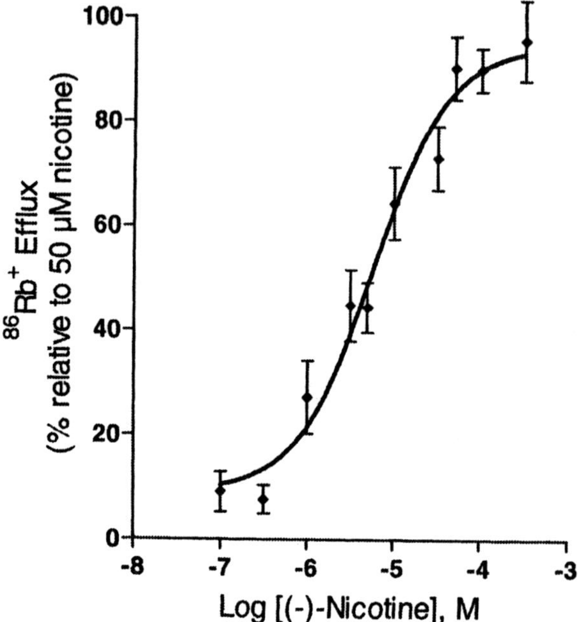

Figure 6.3. *Functional evaluation of human α4β2 nAChRs expressed in K177 cells by cation efflux. Shown is the concentration dependency of $^{86}Rb^+$ efflux response evoked by incubation of cells with increasing concentrations of (−)-nicotine for a period of 5 min.*

ence of the agonist. The $^{86}Rb^+$ flux protocol involves assessing the ability of agonists to stimulate efflux of the radioisotope from preloaded cells (Lukas and Cullen, 1988). In cells stably expressing the α4β2 nAChRs (K177 cell line), cation efflux was stimulated in a concentration-dependent manner by acetylcholine and (−)-nicotine and by other nAChR ligands including (±)-epibatidine, ABT-418, and DMPP (Fig. 6-3). Agonist-stimulated efflux was inhibited by 10 μM mecamylamine or dihydro-β-erythroidine (DHβE). It should be noted that ion efflux studies generally require prolongated incubations (in the minute range) in presence of the agonist that contrast to the rapid exposure (in the ms range) typically employed in patch-clamp studies.

Fluorometric Analysis

Functional assessment of nAChRs that are Ca^{2+} permeable may be assessed by agonist-stimulated Ca^{2+} influx. For example, intracellular calcium responses evoked by activation of human α7 nAChRs expressed in K28 cells have been studied by simultaneous current recording and fast fluorimetric intracellular Ca^{2+} imaging (Delbono et al., 1997). After activation with 100 μM nicotine, these cells showed a significant increase in intracellular calcium that attained maximal levels (3- to 5-fold over basal) when the desensitization of current responses was complete (within 100 ms). An excellent correlation was seen between agonist pharmacological profiles obtained by measuring whole cell currents and calcium responses. More recently, a fluorimetric imaging plate reader (FLIPR) has been developed that can be employed to rapidly detect nAChR-mediated changes in intracellular calcium using fluorescent calcium indicator dyes such as fluo-2 or calcium green (Schroeder and Neagle, 1996; Kuntzweiler et al., 1998). The advantages of this method include (1) rapid (~ 1s) measurement of functional nAChR activation, (2) real-time analysis of signaling events, and (3) simultaneous measurement of 96-well plates permitting high throughput screening capabilities. Agonist-stimulated intracellular Ca^{2+} changes have also been studied in HEK-293 cells stably expressing human α4β2 and α4β4 subtypes (Sacaan et al., 1997).

ELECTROPHYSIOLOGICAL CHARACTERIZATION OF nAChRs

Fast Drug Application and Cell Adhesion

In regard to the natural synaptic function, fast agonist delivery is an indispensable condition for the investigation of LGCs. Of the many systems developed for rapid solution exchanges, the best known are undoubtedly the U-tube system (Krishtal and Pidoplichko, 1980; Fenwick et al., 1982), the concentration clamp technique (Akaike et al., 1986), the Y-tube system (Brett et al., 1986), and the liquid filament technique (Franke et al., 1987; Maconochie and Knight, 1989). These systems constitute valuable realizations for drug application in the range of 100 ms (U-tube), 1 ms (concentration clamp, Y-tube), and 0.2 ms (liquid filament). Faster systems using photoactivable compounds that are released in the μs time scale are also available (Niu et al., 1996; Matsubara et al., 1992), but they lack a precise control of the agonist concentration. Moreover, none of these methods allow rapid successive applications of different solutions on a same cell for studying dose-response relationships. In addition, successive application of drugs in sequences such as "control, agonist, agonist + drug, agonist, control," a protocol that is essential for the determination of agonist and allosteric modulator interactions (Bertrand et al., 1991; Valera et al., 1992; Buisson et al., 1996), is not feasible with such methods. To fill this gap, we have combined a multibarrel with a liquid filament device (Fig. 6-4).

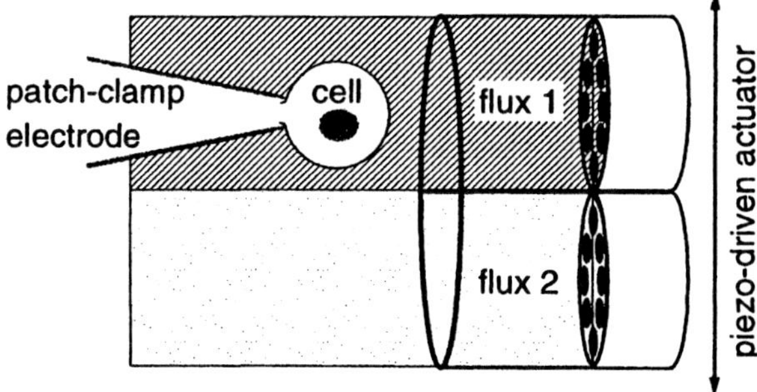

Figure 6.4. *Schematic drawing of the double barrel system used for drug application. Fast solution exchange is performed by switching from flux 1 to flux 2 in less than 100 μs. The mechanical displacement of the device (about up to 1 mm) is controlled by a piezo-quartz. The parallel fluxes are guided by a glass theta tube (two half circles separated by a thin septum) with a final diameter of about 200 μm. Each half of the theta tube is connected to a multibarrel that allows quick switching between any of the solutions.* Extracellular solution: *Most of the recordings were performed with a Hepes-buffered Artificial Cerebro-Spinal Fluid (Hepes-ACSF) of the following composition (in mM): 120 NaCl, 5 KCl, 2 $MgCl_2$, 2 $CaCl_2$, 25 glucose and 10 Hepes; pH was adjusted to 7.4 with NaOH.* Intracellular solution *(mM): 10 KCl, 120 KF, 5 NaCl, 2 $MgCl_2$, 10 Hepes, 10 BAPTA; pH 7.4 with KOH. To prevent the activation of endogenous muscarinic receptors, 0.5 or 1 μM atropine was added to the medium. In the experiments performed with voltage-ramp protocols, the potassium salts were replaced by CsF or Cs methane sulfonate in order to block the voltage-dependent potassium channels.* Data acquisition, recording, and analysis: *Currents were recorded with an Axopatch 200A or 200B amplifier using the whole cell or outside-out configurations of the patch-clamp technique (Hamill et al., 1981). To reduce the electrode capacitance in single-channel recordings, the electrode tips were coated with Sylgard 184 (Dow Corning). Data were filtered at 1 or 2 kHz, digitized at 2–5 kHz, and stored on a personal computer equipped with an analog-to-digital converter (ATMIO-16D, National Instrument, Austin, Texas). Data were analyzed on a MacIntosh Performa 5200 using the proprietary MacDATAC program.*

With this system it is now possible to extend the capacity of the liquid filament without affecting its time resolution. Solution exchange time was typically deduced on a living cell measurement by ionic substitution during a steady exposure to a low agonist concentration (Buisson et al., 1996).

Cell adhesion represents a key factor for the success of electrophysiological studies using fast application systems. This necessity arises from the fact that fast solution exchanges are obtained only with a rapid flux that easily detaches cells from the Petri dish. In order to increase the cell adhesion factor, dishes were coated with polyethyleneimine (PEI), which allows cells to strongly stick to the plastic. Briefly, 1 g PEI solution (SIGMA) is dissolved in 10 ml deionized water (stock solution) and stored at 4°C. For coating of the plastic dishes this solution is dissolved (1:10,000) in distilled water and aliquoted to 2 ml in each Petri dish. After 2 hours of incubation, the PEI solution is removed and dishes are washed three times with deionized water and dried in the laminar flow hood. Two to four days before the recording session, cells are plated at low density in freshly PEI-treated 35 mm Nunc dishes (about 500 cells per dish).

Endogenous nAChR Currents in HEK-293 Cells

After the establishment of the whole cell configuration, voltage-activated currents were observed in most of the cells investigated. Two main categories of currents can be distinguished: (1) outward potassium currents (100% of the cells) and in a fraction of about 10% of the cells (2) fast-activating/inactivating inward currents that display a strong rundown in the first minutes of recording. Typical voltage-dependent currents observed immediately after rupture of the cell membrane (whole cell configuration) are presented in Figure 6-5. The fast-activating/inactivating inward current resembles the tetrodotoxin-sensitive sodium currents of neurons but was not further characterized. Substitution of potassium ions by cesium in the pipette solution readily suppresses outward currents within a few minutes after establishment of the whole cell recording configuration, indicating that the outward current likely results from the activation of voltage-gated potassium channels. Application of 1 mM ACh on untransfected cells failed to evoke any detectable current (at -100 mV), even when atropine was not added to the extracellular medium. It was therefore concluded that HEK-293 cells do not express endogenous nAChRs that can be activated by ACh. This finding is also supported by results from radioligand binding and flux studies showing lack of specific [^{3}H]agonist binding or nicotine-stimulated $^{86}Rb^+$ efflux (Gopalakrishnan et al., 1995, 1996).

Macroscopic nAChR Currents

Since the patch-clamp technique implicates that cells are tested individually, it gives the opportunity to examine the heterogeneity of nAChR expression among the cells. Independently of the nAChR(s) subunit transfected ($\alpha7$, $\alpha4\beta2$, or $\alpha3\beta2$), we observed a very high level of expression of the different nAChR combinations with less than 5% of nonresponsive cells. In more than 50% of the cells recorded ($n > 400$), the maximal current amplitude elicited by a saturating concentration of ACh (1 mM) was above 1000 pA, with some cells reaching up to 6000 pA. Because receptors may preferentially be located on the portion of the cell membrane facing the Petri dish, this current amplitude may underestimate the total amount of nAChRs expressed by a given cell. Moreover, ligand accessibility to this membrane area may be restricted, a factor that can cause significant artifacts in the determination of receptor properties. This can, however, be minimized by "lifting" the cell from the bottom of the dish as demonstrated by Sather et al. (1992) in the case of NMDA receptors.

The K177 cell line has been continuously cultured over a period of 2 years (about 100 pas-

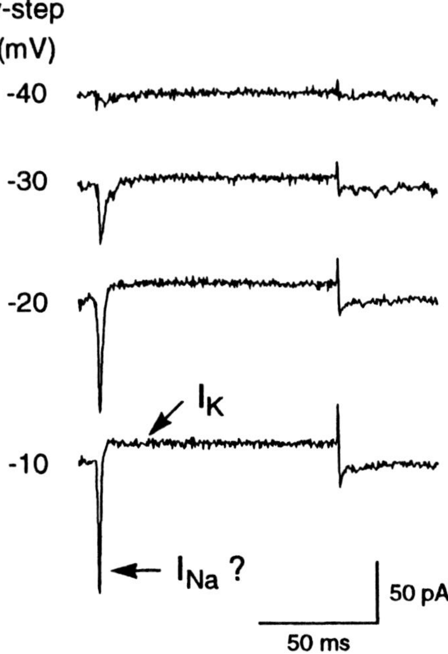

Figure 6.5. *Voltage-dependent currents of transfected HEK-293 cells. In this typical K177 cell (expressing the human α4β2 nAChR), the membrane potential was held at −100 mV and successive depolarizing steps (of 100 ms) were performed at −40, −30, −20, and −10 mV. The cell capacitance was electronically compensated (>90%) and the leak current was subtracted. Two different currents could be observed. The first one, which activates at about −40 mV, is a fast-activating/inactivating inward current similar to the tetrodotoxin-sensitive sodium current. The second, which activates at about −30 mV, is a noninactivating outward current that is readily suppressed in a few minutes when the patch electrode contains cesium ions instead of potassium ions.*

sages) and we have observed only a small decrease of responsive cells after about 75 passages. Since the degree of expression of the nAChRs is very high and stable for months, these cells transfected with the human α7, α4β2, and α3β2 subunits represent inestimable materials for the studies of these types of nAChRs.

In the whole cell recording configuration, human neuronal nAChRs display a strong rectification at positive potentials that preclude accurate determination of the reversal potential and measurement of the ionic selectivity of these channels. However, in the outside-out configuration (see below), the α4β2 nAChR channels do not rectify and display a linear current-voltage relationship (Buisson et al., 1996) that allows determination of the reversal potential of the ACh-evoked current (about −7 mV) that is in good accordance with a cationic selectivity.

To investigate the functional sensitivity of human nAChRs to different agonists, the simplest approach involves performing dose-response protocols. Typically, one cell is successively challenged with increasing concentrations of a given agonist as illustrated on the chart recording presented in Figure 6-6A. The dose-response relationship is determined by plotting the peak-current amplitude as a function of the agonist concentration (Fig. 6-6B). Experimental data can then be fitted by the empirical Hill equation, yielding an apparent affinity (EC_{50}) and a Hill coefficient (n_H) (Gopalakrishnan et al., 1995; Buisson et al., 1996).

Challenge of different cell lines with ACh revealed distinct apparent affinities (EC_{50}) for

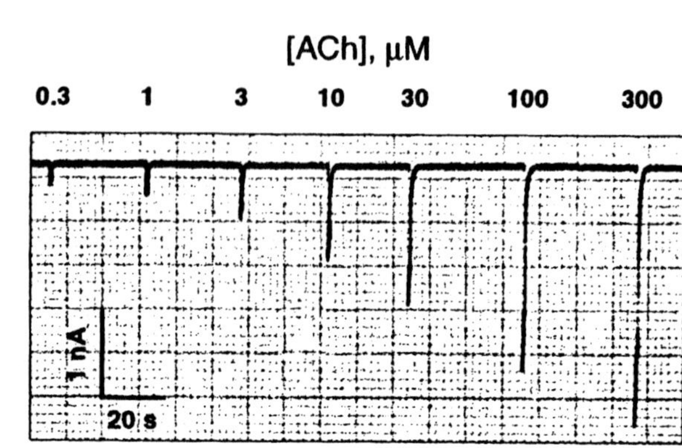

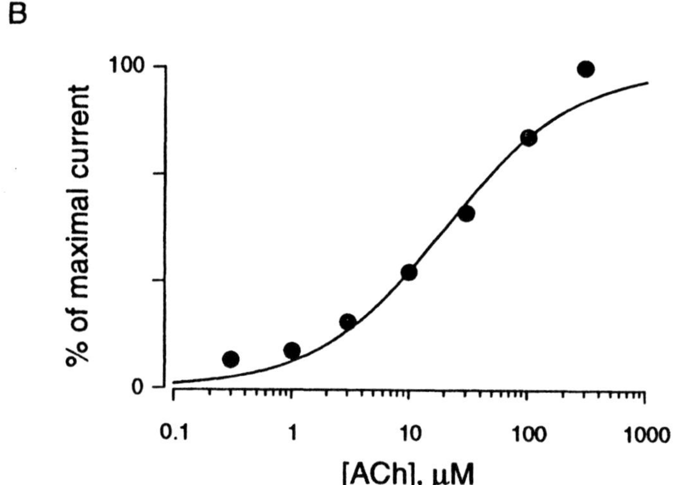

Figure 6.6. ACh sensitivity of the human α4β2 nAChR. (A) Chart paper recording of a K177 cell challenged with increasing concentrations of ACh (400 ms each, indicated on the top of the chart). The cell membrane was held at − 100 mV and the chart speed was 25 mm/min). (B) Dose-response curve measured from the same current traces stored in the computer. Continuous line corresponds to the empirical Hill equation with an EC_{50} of 10 μM and Hill coefficient of 1.

this ligand as well as characteristic time course of activation and desensitization (Fig. 6-7). With the piezo-driven liquid filament and a high-frequency sampling, the rising phases of the currents elicited by 1 mM ACh were measured in the ms range (Fig. 6-8A) but showed no significative differences between α7 and α4β2. However, the desensitization time course of the ACh-evoked current increases strongly from α4β2 to α3β2 and to α7 with respective time constants of 400, 70, and 20 ms at saturating ACh concentrations (Fig. 8B).

The patch-clamp technique allows the investigation of the voltage dependence of agonist and/or antagonists. These measurements are highly valuable for studying the voltage depen-

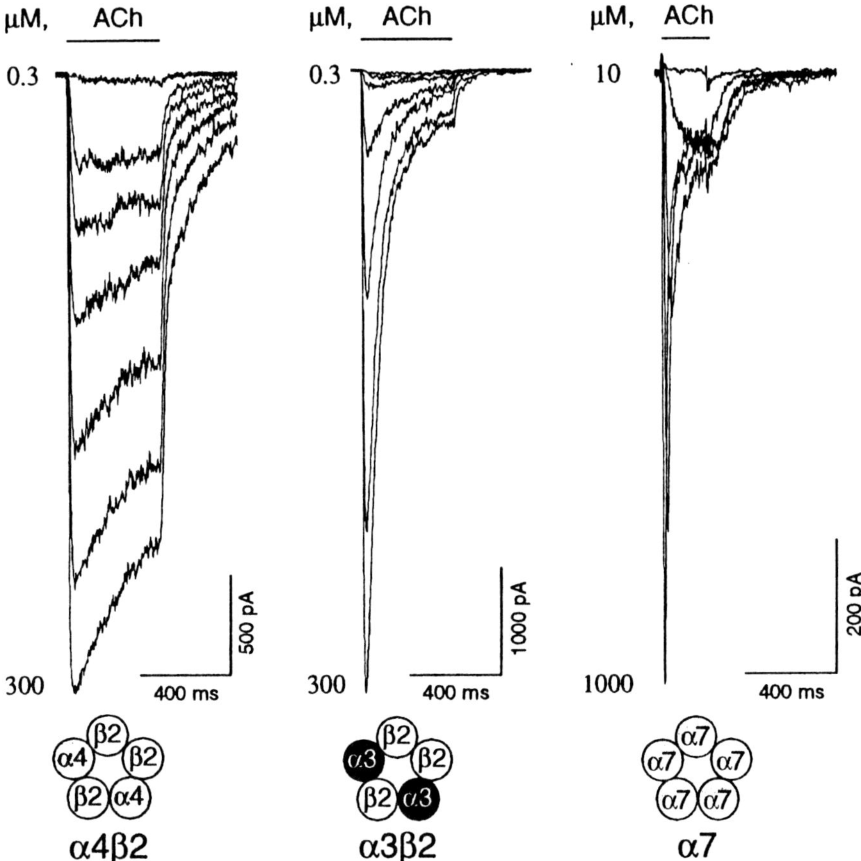

Figure 6.7. Responses of ACh at three nAChR subunit combinations. K177, K4-14, and K28 cells expressing respectively the α4β2, α3β2, and α7 nAChR subunits were challenged with several ACh concentrations (0.3, 1, 3, 10, 30, 100, and 300 μM for α4β2 and α3β2; 10, 30, 100, 300, and 1000 μM for α7). The ACh pulse duration is indicated by the horizontal bars (holding potential: −100 mV). The putative arrangement of the different subunits is represented under the current traces assuming a pentameric structure of the nAChRs.

dency of noncompetitive antagonists. Some of them, called open-channel blockers, can enter and block the channel of open and/or closed nAChRs by steric hindrance. Most of these compounds display at least one positive charge at a physiological pH that renders them sensitive to voltage (a charged molecule present in the electrical field being submitted to the electrical force $F = q.E$). When one of these compounds binds to its site within the nAChR pore, it senses a fraction of the transmembrane potential gradient. Thus, efficacy of its block is voltage dependent and is often reinforced at hyperpolarized membrane potentials. When the molecule presents a positive charge ($q > 0$), the driving force F follows the electrical field direction: from outside to inside of the cell at negative membrane potentials. For example, current–voltage relationships recorded with slow voltage ramps starting at positive values have revealed the volt-

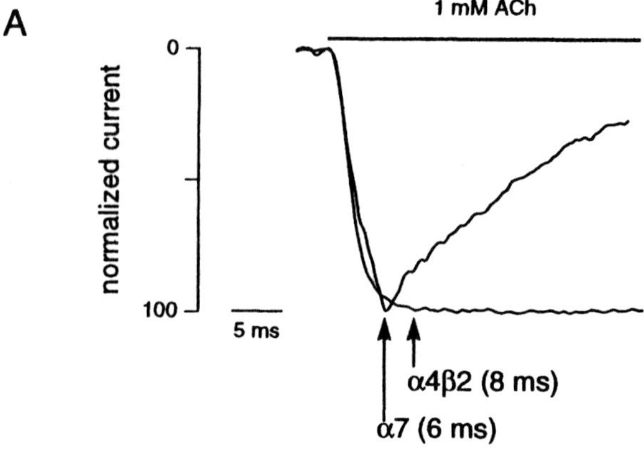

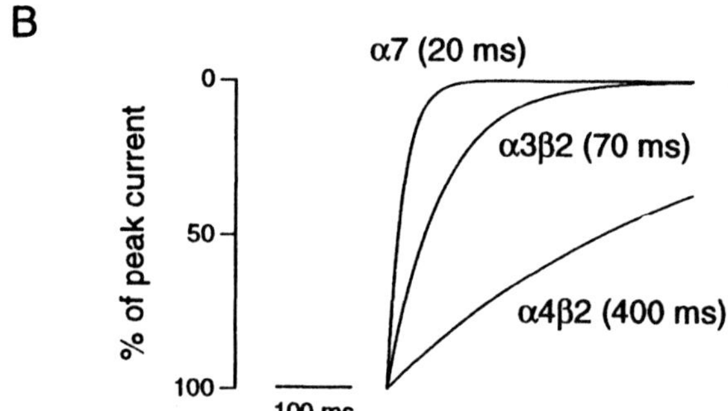

Figure 6.8. *Time course of activation and desensitization of the human nAChRs. (**A**) Comparison, on an expended time scale, of the rising phase of* α*7 and* α*4*β*2 ACh-evoked currents by a saturating concentration of ACh (1 mM). The time-to-peak for each subtype is indicated by the arrows. Currents were normalized at unity to their maximal values. (**B**) Desensitization kinetics of* α*7,* α*3*β*2, and* α*4*β*2 nAChRs. Starting from the peak current, the desensitizing phase of a current evoked by a saturating ACh pulse can be fitted by a single exponential function* $y = exp(-t/\tau)$*;* t *is the time and* τ *is the time constant of the desensitization process. The figure presents the theoretical exponential fits determined for* α*7 (1 mM ACh) and for* α*3*β*2 and* α*4*β*2 (300* μ*M ACh).*

age dependency of the amantadine block (Fig. 6-9A, B). This compound, originally characterized for its action on glutamate-operated channels (NMDA receptors), blocks the human α4β2 nAChR in the μM range (Buisson and Bertrand, 1998a).

Microscopic nAChR Currents

Protein activity at the single molecule level can be studied using the patch-clamp technique by recording the activity of one or a few channels in a small patch of the cell membrane (cell-attached, inside-out, and outside-out configurations; Hamill et al., 1981). It is of value to mention that unless altered by the investigation technique, summation of numerous elementary channel openings should lead to the "reconstruction" of the macroscopic events with charac-

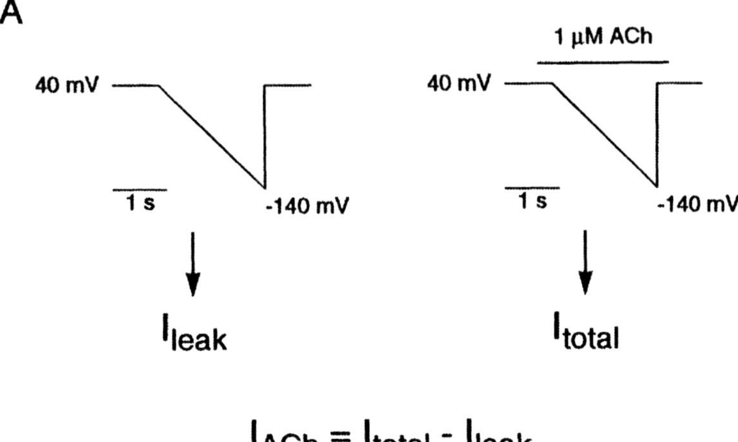

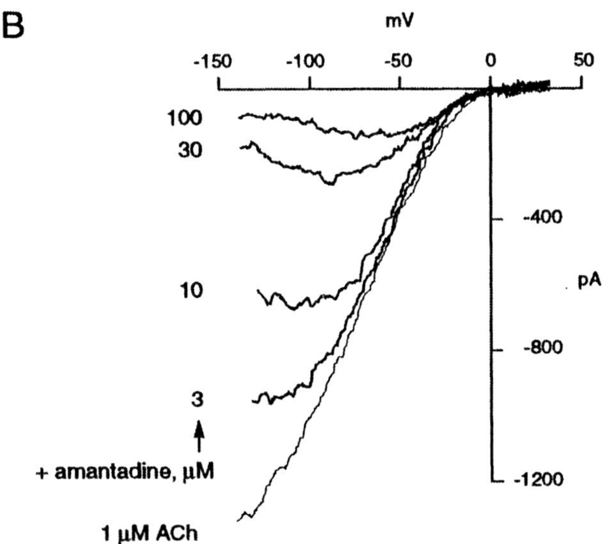

Figure 6.9. *Voltage-dependent blockade of α4β2 nAChRs by amantadine. (**A**) The voltage-ramp protocol was performed as follows: The membrane was held at 40 mV, and a first ramp (from 40 to −140 mV in 2 s) was applied in the standard saline medium (without agonist) for the determination of the leak current. Three seconds later, 1 µM ACh was delivered for 3 s, and the voltage ramp was applied 400 ms after the onset of delivery. I_{ACh} was computed as indicated. (**B**) Current–voltage relationship of a 1 µM ACh-evoked current in absence (thin line) or in presence (thick lines) of increasing concentrations of amantadine.*

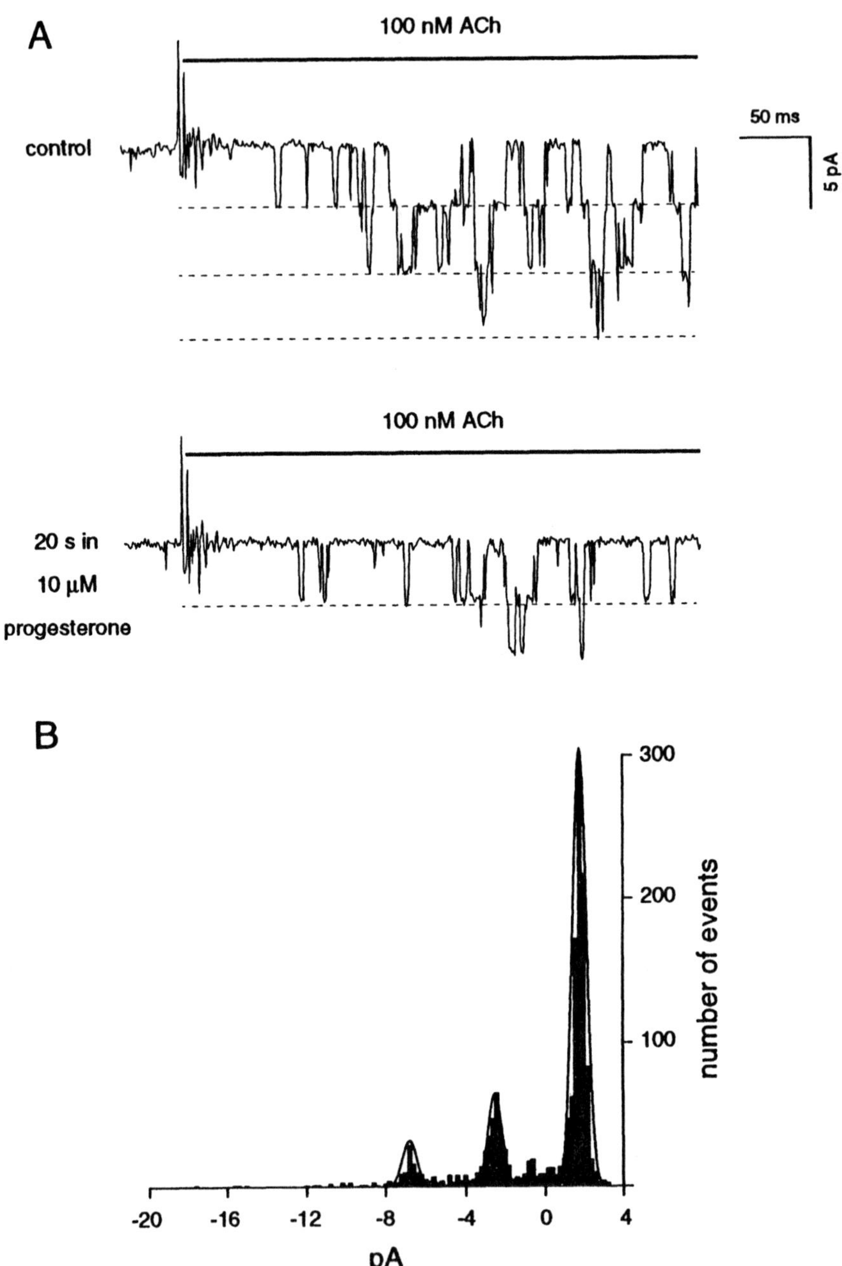

Figure 6.10. *Single-channel recording of the α4β2 nAChR. (**A**) Single-channel activities of α4β2 nAChRs were recorded in the outside-out configuration upon exposure to a low ACh concentration (indicated by the horizontal bar). Top trace was recorded in control. Lower-trace probability of opening is strongly reduced by exposure to progesterone (10 μM) but not current amplitude. (**B**) All-point amplitude histogram corresponding to the single-channel events measured in three traces recorded in control. Continuous line is the fit by the sum of three Gaussians yielding mean values of 1.8, −2.5, and −6.8 pA corresponding respectively to zero current and noise of the electrode, a single and dual level of opening. A single-channel amplitude of 4.3 pA allows the computation of an elementary conductance γ of 40 pS (γ = i/(V − Er), where V is the holding voltage (− 100 mV) and Er is the reversal potential (− 7 mV).*

teristics identical to those of whole cell currents. Since elementary transition of the nAChRs (opening, closing) occurs in the μs range, it is necessary to sample the currents at the highest frequency possible. The upper limit for time resolution is limited by the high-frequency noise of the electrode and amplifier (<10 kHz in the best conditions; Benndorf, 1995).

Single-channel conductance can readily be measured from the current traces recorded in these conditions (Fig. 6-10A). Moreover, computation of an all-point amplitude histogram can be used to determine the number of channels in the patch, their respective conductance, and possible flickering. When enough measurements can be obtained in a steady condition, the fraction of the time that the channel spends in a given conformation can be also deduced. However, the best estimation is obtained when only one channel is present in the patch, a condition that is often restrictive. The all-point histogram (Fig. 6-10B) derived from traces such as presented in Fig. 6-10A yields a single-channel conductance of 40 pS for the α4β2 nAChR (with 140 mM cesium methanesulfonate in the pipette).

Because neuronal nAChRs exhibit a fast rundown in the outside-out patches originating either from freshly dissociated neurons (Mulle et al., 1991) or cell lines (Buisson et al., 1996), kinetic analysis is very restricted. Rundown corresponds to the progressive decline of channel activity in the patch. A typical rundown time constant of about 50 seconds was measured for the α4β2 nAChR (Buisson et al., 1996). Although the mechanism underlying the rundown has not yet been identified, this may reflect internal phosphorylation/dephosphorylation of the receptor and/or breakdown of the protein cytoskeleton. When possible, recordings in the cell-attached configuration may allow for further insights about the receptor properties. However, in the case of LGCs, these recordings are rather limited because it is technically not possible to apply drugs in the appropriate time scale inside the patch pipette.

The effect of compounds such as allosteric modulators can be investigated at the single-channel level to further characterize their mechanism(s) of action. As an illustration, the effect of progesterone (Bertrand et al., 1991; Valera et al., 1992; Buisson and Bertrand, 1998b) on the α4β2 nAChR single channels is presented in Fig. 6-10A: Preapplication of 10 μM progesterone decreases the channel activity while leaving the elementary current amplitude unaffected. This readily indicates that progesterone must exert its action by altering the channel kinetics.

Calcium Modulation

In previous studies, Mulle et al. (1992) and Vernino et al. (1992) have demonstrated that extracellular calcium ions behave as positive allosteric modulators of some native and reconstituted neuronal nAChRs. High extracellular calcium decreases the single-channel amplitude but increases its mean open time (Mulle et al., 1992). The net effect is an augmentation of the macroscopic whole cell current. Furthermore, calcium binding domains involved in the mechanism of potentiation have been identified by site-directed mutagenesis in the homomeric chick α7 nAChR (Galzi et al., 1996).

With the human α4β2 nAChR, we have observed that a high extracellular calcium concentration (≥2 mM) inhibits the whole cell current as well as the single-channel main amplitude. But because of the high rundown, no statistical analysis of the mean open time could be achieved to further investigate this calcium inhibition. Nevertheless, the effects of calcium ions on the human α4β2 nAChR seem to be different from those previously described by Mulle et al. (1992) and Vernino et al. (1992).

Recent investigations indicate that the human α3β2 nAChR is potentiated by a high calcium concentration (Kuntzweiler et al., 1997). Thus, it follows that a high extracellular calcium concentration (≥2 mM) has two opposite effects depending on the nature of the α subunit that

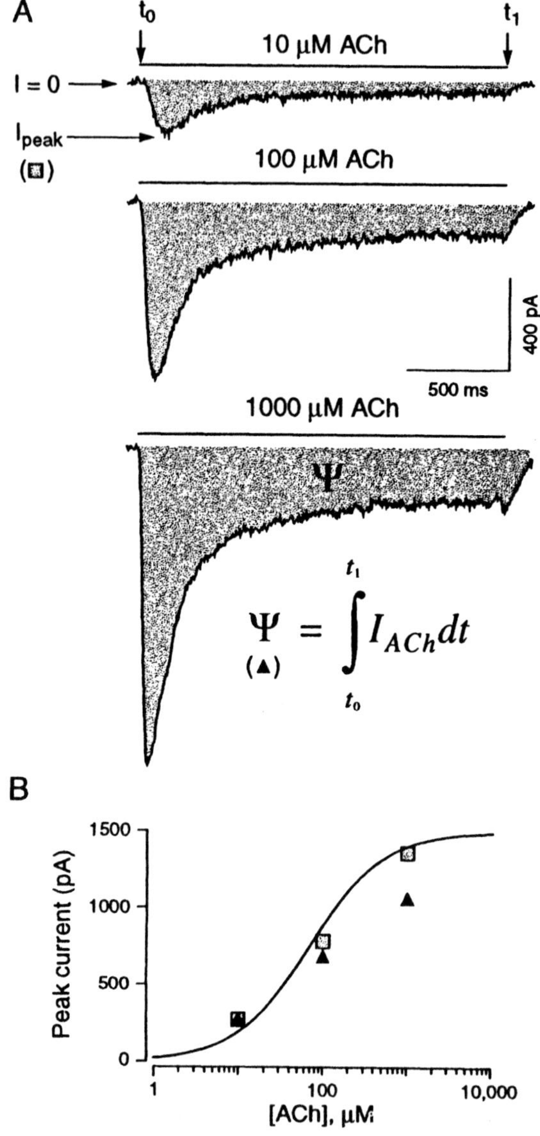

Figure 6.11. *Cumulative ACh-evoked current. (**A**) Traces evoked by a long ACh pulse in a K4-14 cell (α3β2 nAChR; holding potential, −100 mV). Responses evoked in the same cell by three ACh concentrations are represented. Pulses were separated by 20 s to allow recovery from desensitization. The cell was held at −100 mV throughout the experiment. Gray area can be computed as indicated in the formula. (**B**) Plot of the peak current amplitudes (I_{peak}) measured in (**A**) as a function of the ACh concentration (squares). Continuous line: ACh sensitivity of the human α3β2 nAChR (EC_{50} = 70 μM, n_H = 1.2; 13 cells; Kuntzweiler et al., 1997). Triangles correspond to the values of the cumulative current (ψ) integrated from 30 to 1750 ms and normalized to the peak current evoked by a 10 μM concentration of ACh (which induces a relatively "weak" desensitization). Note that discrepancy between peak and cumulative currents increases with the ACh concentration and desensitization.*

is coexpressed: a potentiation of the macroscopic current with α3 and an inhibition with α4. This illustrates that calcium ions can have differential effects according to the type of human nAChR subunit combinations.

Comparison of Rubidium Flux and Electrophysiology

Functional studies using both the rubidium efflux and the patch-clamp techniques were performed on the K177 cell line (Buisson et al., 1996; Gopalakrishnan et al., 1996). Comparison of these results reveals substantial differences for the ACh half activation (EC_{50} = 44 μM with the $^{86}Rb^+$ efflux and 3 μM with patch-clamp). However, this apparent discrepancy can be explained by the following considerations: Dose-response relationships obtained in patch-clamp are determined by plotting the peak of currents evoked by short ACh pulses (I_{peak}, Figs. 6-6 and 6-11) as a function of the agonist concentrations while in $^{86}Rb^+$ efflux it is the cumulative isotopic efflux. This method involves prolonged incubations in the presence of the agonist and the cumulative $^{86}Rb^+$ efflux corresponds, in a first approximation, to the integration of the ACh-evoked current over the period of agonist incubation (see equation in Fig. 6-11). Since the desensitization time course of the current increases with the agonist concentration (and could change from one agonist to the other), there is no proportional relationship between the peak current amplitude and the current integral. Thus, peak currents and cumulative effluxes cannot be compared based on this simple analysis. Moreover, in $^{86}Rb^+$ efflux experiments, cells are not clamped at a fixed potential, and following ACh stimulation, they must depolarize. This further alters the relationship between the total efflux and the current integral. Thus, results obtained in experiments performed with $^{86}Rb^+$ efflux or patch-clamp must be compared with caution.

STABLE CELL LINES AS MODEL SYSTEMS TO STUDY nAChR REGULATION

The availability of stably transfected cells has opened up possibilities for studying the regulation of various nAChR subtypes in an environment devoid of interference from other receptor types. Human α4β2 nAChRs expressed in K177 cells were found to be very sensitive to upregulation by prolonged (24 h) treatment with (−)-nicotine. The increase in receptor density was initiated by low concentrations of (−)-nicotine (100 nM) in a fairly rapid ($t_{0.5}$ = 4 h) and reversible manner (Fig. 6-12A). Treatment with other cholinergic channel modulators including (−)-cytisine, (±)-epibatidine and ABT-418 or by activators of protein kinase A- and C-dependent mechanisms also resulted in the upregulation of α4β2 nAChRs (Gopalakrishnan et al., 1997). In contrast, studies with the K28 cells stably expressing human α7 nAChRs revealed that relatively higher concentration of (−)-nicotine (about 1 mM) were required for upregulation of [^{125}I]α-bungarotoxin binding sites (Fig. 6-12B; Molinari et al., 1998). An understanding of the mechanisms underlying the differential modulation of various nAChR subtypes following chronic treatment with (−)-nicotine is important for the development of novel cholinergic channel therapeutics devoid of tolerance liabilities (Brioni et al., 1996).

CONCLUSIONS

Stable transfected cell lines represent the materials of choice for the physiological and pharmacological studies of the various human nAChRs. In particular, the patch-clamp recordings permit evaluation of properties at the whole cell level on a fast time scale.

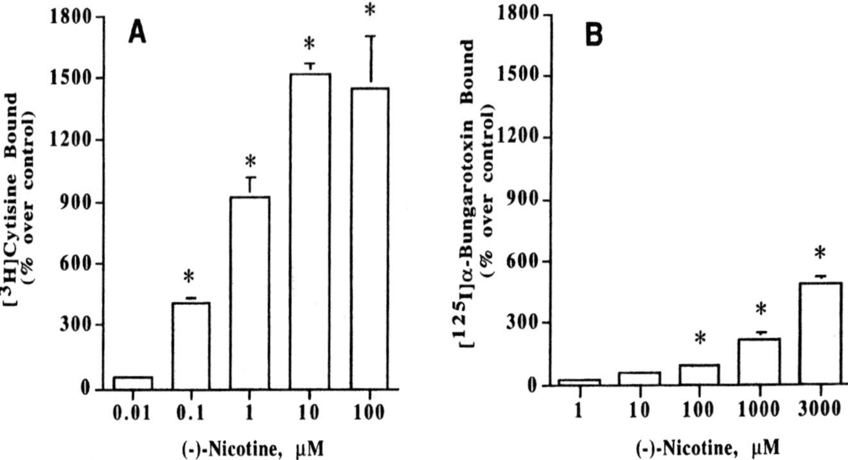

Figure 6.12. *Differential upregulation by (−)-nicotine of human nAChRs expressed in K177 (**A**) and K28 (**B**) cell lines. K177 cells stably expressing human α4β2 nAChRs or K28 cells stably expressing human α7 nAChRs were treated with varying concentrations of (−)-nicotine (indicated) and receptor levels were assessed by either [^{3}H]cytisine or [^{125}I]α-bungarotoxin binding. (*) represents values significantly different from untreated cells.*

The different nAChR subunits that have been investigated display characteristic pharmacological signatures (for more extensive data, see Gopalakrishnan et al., 1995, 1996; Buisson et al., 1996). Moreover, for a given human nAChR combination, substantial differences have been observed with the corresponding ones of other species investigated so far (McGehee and Role, 1995), thus emphasizing the necessity of reconstituted systems expressing human nAChRs for the design and evaluation of pharmacological compounds with potential clinically beneficial effects.

One should speculate that neuronal nAChRs that are expressed in a non-neuronal context (e.g., embryonic kidney cells) could present subtle differences with the native ones. At present, it is difficult to verify such a hypothesis. However, the α4β2 nAChRs in K177 cells are upregulated by chronic nicotine or other nicotinic ligands, indicating that in this system the nAChR protein contains the minimal information for its own regulation (Gopalakrishnan et al., 1997). Since the coexpression of the α5 subunit can profoundly modify the physiological as well as the pharmacological properties of the α4β2 combination (Ramirez-Latorre et al., 1996; Gillespie et al., 1997), cotransfection of α5 together with α4 and β2 represents a necessary step for the understanding of native properties of the human receptors.

An interesting alternative for the study of the influence of different subunits could be to realize a transient transfection of existing cell lines with the subunits that need to be investigated. For instance, genetic studies have revealed that the human nAChR α4 subunit is altered in autosomal dominant nocturnal frontal lobe epilepsy (ADNFLE; for a review, see Buisson et al., 1998). In separate families presenting the same clinical profile, two different mutations have been identified in the second transmembrane domain that lines the ionic pore of the nAChR. The first one is a missense mutation (Steinlein et al., 1995; Weiland et al., 1996) while the second one is a three-base pair insertion (Steinlein et al., 1997). Both mutations lead to altered physiological and pharmacological properties (Bertrand et al., 1997c). Transient or stable transfection of these mutated subunits would therefore help to investigate the physiological properties of these mutations at biochemical and single-channel levels and would therefore provide additional insights for linking genetic alterations with brain disorders.

ACKNOWLEDGMENTS

D. Bertrand is supported by the Swiss National Foundation (PNR-38) and by the Office Fédéral de l'Education et des Sciences.

REFERENCES

Akaike N, Inoue M, Krishtal OA (1986): "Concentration-clamp" study of γ-aminobutiric-acid-induced chloride current kinetics in frog sensory neurons. J. Physiol. (London) 379: 171–185.

Anand R, Lindstrom J (1990): Nucleotide sequence of the human nicotinic acetylcholine receptor β2 subunit gene. Nucleic Acid Res. 18: 4272.

Anand R, Conroy WG, Schoepfer R, Whiting P, Lindstrom J (1991): Neuronal nicotinic acetylcholine receptors expressed in *Xenopus* oocytes have a pentameric quaternary structure. J. Biol. Chem. 266: 11,192–11,198.

Benndorf K (1995): Low noise recording. In Sakmann B, Neher E, editors. Single channel recording. New York and London: Plenum Press, pp. 129–145.

Bertrand D, Changeux JP (1995): Nicotinic receptor: an allosteric protein specialized for intracellular communication. Seminar in Neuroscience 7: 75–90.

Bertrand D, Valera S, Bertrand S, Ballivet M, Rungger D (1991): Steroids inhibit nicotinic acetylcholine receptors. Neuroreport 2: 277–280.

Bertrand D, Buisson B, Krause RM, Hu HY, Bertrand S (1997a): Minireview: Electrophysiology: a method to investigate the functional properties of ligand-gated channels. J. Recept. Signal Transduct. Res. 17: 227–242.

Bertrand S, Devillers-Thiéry A, Palma E, Buisson B, Edelstein S, Corringer PJ, Changeux JP, Bertrand D (1997b): Paradoxical allosteric effects of competitive inhibitors on neuronal α7 nicotinic receptor mutants. Neuroreport 8: 3591–3596.

Bertrand S, Weiland O, Steinlein O, Bertrand D (1997c): Physiological effects of mutations of the α4 neuronal nicotinic acetylcholine receptor associated with epilepsy. Soc. Neurosci. 23: 389. Abstr. 156-14.

Brett RS, Dilger JP, Adams PR, Lancaster B (1986): A method for the rapid exchange of solutions bathing excised membrane patches. Biophys. J. 50: 987–992.

Brewer CB (1994): Cytomegalovirus plasmid vectors for permanent lines of polarized epithelial cells. In: Roth MG, editor. Methods in cell biology. San Diego: Academic Press, pp. 233–245.

Brioni JD, Morgan SJ, Oneill AB, Sykora TM, Postl SP, Pan JB, Sullivan JP, Americ SP (1996): "In vivo" profile of novel nicotinic ligands with CNS selectivity. Med. Chem. Res. 6: 487–510.

Buisson B, Bertrand D (1998a): Open-channel blockers at the human α4β2 neuronal nicotinic acetylcholine receptor. Mol. Pharmacol. 53: 555–563.

Buisson B, Bertrand D (1998b): Steroid modulation of the nicotinic acetylcholine receptor. In: Baulieu EE, Robel P, Schumacher M, editors. Neurosteroids: a new function in the nervous system. Contemporary Endocrinology. Totowa, NJ: Humana Press Inc.

Buisson B, Gopalakrishnan M, Arneric SP, Sullivan JP, Bertrand D (1996): Human α4β2 neuronal nicotinic acetylcholine receptor in HEK 293 cells: a patch-clamp study. J. Neurosci. 16: 7880–7891.

Buisson B, Curtis L, Bertrand D (1998): Neuronal nicotinic acetylcholine receptor and epilepsy. In: Genton P, Picard F, Marescaux C, Berkovic S editors. Genetics of focal epilepsies. London–Paris–Rome: John Libbey & Company Ltd.

Chini B, Raimond E, Elgoyhen AB, Moralli D, Balzaretti M, Heinemann S (1994): Molecular cloning and chromosomal localization of the human alpha 7-nicotinic receptor subunit gene (CHRNA7). Genomics 19: 379–381.

Colquhoun D, Sakmann B (1981): Fluctuations in the microsecond time range of the current through single acetylcholine receptor ion channels. Nature 294: 464–466.

Conroy WG, Vernallis AB, Berg DK (1992): The alpha 5 product assembles with multiple acetylcholine receptor subunits to form distinctive receptor subtypes in brain. Neuron 9: 679–691.

Cooper E, Couturier S, Ballivet M (1991): Pentameric structure and subunit stoichiometry of a neuronal nicotinic acetylcholine receptor. Nature 350: 235–238.

Corriveau RA, Berg DK (1993): Coexpression of multiple acetylcholine receptor genes in neurons: Quantification of transcripts during development. J. Neurosci. 13: 2662–2671.

Couturier S, Bertrand D, Matter JM, Hernandez MC, Bertrand S, Millar N, Valera S, Barkas T, Ballivet M (1990): A neuronal nicotinic acetylcholine receptor subunit (alpha 7) is developmentally regulated and forms a homo-oligomeric channel blocked by alpha-bungarotoxin. Neuron 5: 845–856.

Delbono O, Gopalakrishnan M, Renganathan M, Monteggia LM, Messi ML, Sullivan JP (1997): Activation of the recombinant human alpha(7) nicotinic acetylcholine receptor significantly raises intracellular free calcium. J. Pharmacol. Exp. Ther. 280: 428–438.

Elgoyhen AB, Johnson DS, Boulter J, Vetter DE, Heinemann S (1994): α9: An acetylcholine receptor with novel pharmacological properties expressed in rat cochlear hair cells. Cell 79: 705–715.

Felgner PL, Gadek TR, Holm M, Roman R, Chan HW, Wenz M, Northrop, JP, Ringold GM, Danielsen M (1987): Lipofection: a highly efficient, lipid-mediated DNA-transfection procedure. Proc. Natl. Acad. Sci. USA 84: 7413–7417.

Fenwick EM, Marty A, Neher E (1982): A patch-clamp study of bovine chromaffin cells and their sensitivity to acetylcholine. J. Physiol. (London) 331: 577–597.

Flores CM, Decamp RM, Kilo S, Rogers SW, Hargreaves KM (1996): Neuronal nicotinic receptor expression in sensory neurons of the rat trigeminal ganglion: demonstration of alpha 3 beta 4, a novel subtype in the mammalian nervous system. J. Neurosci. 16: 7892–7901.

Franke C, Hatt H, Dudel J (1987): Liquid filament switch for ultra-fast exchanges of solutions at excised patches of synaptic membranes of crayfish muscle. Neurosci. Lett. 77: 199–204.

Galzi JL, Changeux JP (1995): Neuronal nicotinic receptors: molecular organization and regulations. Neuropharmacology 34: 563–582.

Galzi JL, Bertrand S, Corringer PJ, Changeux JP, Bertrand D (1996): Identification of calcium binding sites that regulate potentiation of a neuronal nicotinic acetylcholine receptor. EMBO J. 15: 5824–5832.

Gillespie A, O'Neill Claeps B, Crona J, Jones J, Elliott K, Chavez-Noriega LE, Johnson EC, Stauderman K, Corey-Naeve, J (1997): Stable expression and characterization of the human α3β2α5 nicotinic acetylcholine receptor. Soc. Neurosci. 23: 57.6.

Gopalakrishnan M, Buisson B, Touma E, Giordano T, Campbell JE, Hu IC, Donnelly-Roberts D, Arneric SP, Bertrand D, Sullivan JP (1995): Stable expression and pharmacological properties of the human α7 nicotinic acetylcholine receptor. Eur. J. Pharmacol. Mol. Pharmacol. Sec. 290: 237–246.

Gopalakrishnan M, Molinari EJ, Sullivan JP (1997): Regulation of human alpha 4 beta 2 neuronal nicotinic acetylcholine receptors by cholinergic channel ligands and second messenger pathways. Mol. Pharmacol. 52: 524–534.

Gopalakrishnan M, Monteggia LM, Anderson DJ, Molinari EJ, Piattonikaplan M, Donnelly-Roberts D, Arneric SP, Sullivan JP (1996): Stable expression, pharmacologic properties and regulation of the human neuronal nicotinic acetylcholine alpha(4)beta(2) receptor. J. Pharmacol. Exp. Ther. 276: 289–297.

Hamill OP, Marty A, Neher E, Sakmann B, Sigworth FJ (1981): Improved patch-clamp techniques for high resolution current recording from cells and cell-free patches. Pflugers Archiv.—Eur. J. Physiol. 391: 85–100.

Heim R, Tsien RY (1996): Engineering green fluorescent protein for improved brightness, longer wavelengths and fluoresence energy transfer. Curr. Biol. 6: 178–182.

Horch H, Sargent PB (1996): Synaptic and extrasynaptic distribution of two distinct populations of nicotinic acetylcholine receptor clusters in the frog cardiac ganglion. J. Neurocytol. 25: 67–77.

Katz B, Miledi R (1973): The binding of acetylcholine to receptors and its removal from the synaptic clefts. J. Physiol. (London) 231: 549–574.

Krishtal OA, Pidoplichko VI (1980): A receptor for protons in the nerve cell membrane. Neuroscience 2: 2325–2327.

Kuntzweiler TA, Anderson DJ, Campbell JE, Manelli A, Buisson B, Bertrand D, Arneric SP, Donnelly-Roberts D (1997): Characterization of recombinant human $\alpha 2\beta 2$ and $\alpha 3\beta 2$ neuronal nicotinic acetylcholine receptors (nAChRs). Soc. Neurosci. Abstr. 23: 155.1.

Kuntzweiler, TA, Arneric, SP, Donnelly-Roberts D (1998): Rapid assessment of ligand actions at nicotinic acetylcholine receptors using calcium dynamics and FLIPR. Drug Dev. Res. In press.

Léna C, Changeux JP (1997): Role of Ca^{2+} ions in nicotinic facilitation of GABA release in mouse thalamus. J. Neurosci. 17: 576–585.

Lindstrom J (1996): Neuronal nicotinic acetylcholine receptors. Ion Channels 4: 377–450.

Lukas, RJ, Cullen MJ (1988): An isotopic ion efflux assay for the functional characterization of nicotinic acetylcholine receptors. Anal. Biochem. 175: 212–218.

Maconochie DJ, Knight DE (1989): A method for making solution changes in the sub-millisecond range at the tip of a patch pipette. Pflugers Archiv.—Eur. J. Physiol. 414: 589–596.

Matsubara, N, Billington AP, Hess GP (1992): How fast does an acetylcholine receptor channel open? Laser-pulse photolysis of an inactive precursor of carbamoylcholine in the microsecond time region. Biochemistry 31: 5507–5514.

McGehee DS, Role LW (1995): Physiological diversity of nicotinic acetylcholine receptors expressed by vertebrate neurons. Ann. Rev. Physiol. 57: 521–546.

Molinari EJ, Delbono O, Messi ML, Renganathan ML, Arneric SP, Sullivan JP, Gopalakrishnan M (1998): Upregulation of human $\alpha 7$ nicotinic receptors by chronic treatment with activator and antagonist ligands. Eur. J. Pharmacol. 374: 131–139.

Monteggia LM, Gopalakrishnan M, Touma E, Idler KB, Nash N, Arneric SP, Sullivan JP, Giordano T (1995): Cloning and transient expression of genes encoding the human alpha 4 and beta 2 neuronal nicotinic acetylcholine receptor (nAChR) subunits. Gene 155: 189–193.

Mulle C, Vidal C, Benoit P, Changeux JP (1991): Existence of different subtypes of nicotinic acetylcholine receptors in the rat habenulo-interpeduncular system. J. Neurosci. 11: 2588–2597.

Mulle C, Choquet D, Korn H, Changeux JP (1992): Calcium influx through nicotinic receptor in rat central neurons: its relevance to cellular regulation. Neuron 8: 135–143.

Niu L, Gee KR, Schaper K, Hess GP (1996): Synthesis and photochemical properties of a kainate precursor and activation of kainate and AMPA receptor channels on a microsecond time scale. Biochemistry 35: 2030–2036.

Peng X, Katz M, Gerzanich V, Anand R, Lindstrom J (1994): Human alpha 7 acetylcholine receptor: cloning of the alpha 7 subunit from the SH-SY5Y cell line and determination of pharmacological properties of native receptors and functional alpha 7 homomers expressed in *Xenopus* oocytes. Mol. Pharmacol. 45: 546–554.

Pereira EF, Alkondon M, Reinhardt S, Maelicke A, Peng X, Lindstrom J, Whiting P, Albuquerque EX (1994): Physostigmine and galanthamine: probes for a novel binding site on the $\alpha 4\beta 2$ subtype of neuronal nicotinic acetylcholine receptors stably expressed in fibroblast cells. J. Pharmacol. Exp. Ther. 270: 768–778.

Puchacz E, Buisson B, Bertrand D, Lukas RJ (1994): Functional expression of nicotinic acetylcholine receptors containing rat $\alpha 7$ subunits in human SH-SY5Y neuroblastoma cells. FEBS Lett. 354: 155–159.

Pugh PC, Corriveau RA, Conroy WG, Berg DK (1995): Novel subpopulation of neuronal acetylcholine receptors among those binding alpha-bungarotoxin. Mol. Pharmacol. 47: 717–725.

Quik M, Philie J, Choremis J (1997): Modulation of $\alpha 7$ nicotinic receptor-mediated calcium influx by nicotinic agonists. Mol. Pharmacol. 51: 499–506.

Ramirez-Latorre J, Yu CR, Qu X, Perin F, Karlin A, Role L (1996): Functional contributions of alpha 5 subunit to neuronal acetylcholine receptor channels. Nature 380: 347–351.

Role LW, Berg DK (1996): Nicotinic receptors in the development and modulation of CNS synapses. Neuron 16: 1077–1085.

Roth, MG (ed). (1994): Methods in Cell Biology, Vol. 43: Protein Expression in Animal Cells. Academic Press, San Diego.

Sacaan AI, Reid RT, Santori EM, Adams P, Correa LD, Mahaffy LS, Bleicher L, Cosford N, Stauderman KA, McDonald IA, Rao TS, Lloyd GK (1997): Pharmacological characterization of SIB-1765F: a novel cholinergic ion channel agonist. J. Pharmacol. Exp. Ther. 280: 373–383.

Sather W, Dieudonné S, McDonald JF, Ascher P (1992): Activation and desensitization of N-methyl-D-aspartate receptors in nucleated outside-out patches from mouse neurones. J. Physiol. (London) 450: 643–672.

Schroeder KS, Neagle B (1996): FLIPR: a new instrument for high throughput optical screening. Biomol. Screen. 1: 75–81.

Stauderman KA, Mahaffy LS, Chavez-Noriega L, Johnson E, Tran C, Elliott K, Corey-Naeve J (1995): Characterization of recombinant human neuronal nicotinic acetylcholine receptor subtypes $\alpha4\beta4$ and $\alpha2\beta4$ stably expressed in HEK-293 cells. Soc. Neurosci. Abstr. 21: 36.4.

Steinlein OK, Mulley JC, Propping P, Wallace RH, Phillips HA, Sutherland GR, Scheffer IE, Berkovic SF (1995): A missense mutation in the neuronal nicotinic acetylcholine receptor alpha 4 subunit is associated with autosomal dominant nocturnal frontal lobe epilepsy. Nature Genetics 11: 201–203.

Steinlein OK, Magnusson A, Stoodt J, Bertrand S, Weiland S, Berkovic SF, Nakken KO, Propping P, Bertrand D (1997): An insertion mutation of the CHRNA4 gene in a family with autosomal dominant nocturnal frontal lobe epilepsy. Hum. Mol. Genet. 6: 943–947.

Ullian EM, Sargent PB (1995): Pronounced cellular diversity and extrasynaptic location of nicotinic acetylcholine receptor subunit immunoreactivities in the chicken pretectum. J. Neurosci. 15: 7012–7023.

Valera S, Ballivet M, Bertrand D (1992): Progesterone modulates a neuronal nicotinic acetylcholine receptor. Proc. Natl. Acad. Sci. USA 89: 9949–9953.

Wang F, Nelson M, Lindstrom J (1997): Characterization of human $\alpha3$ AChRs stably expressed in tsa201 cells. Soc. Neurosci. Abstr. 23: 55.8.

Weiland S, Witzemann V, Villarroel A, Propping P, Steinlein O (1996): An amino acid exchange in the second transmembrane segment of a neuronal nicotinic receptor causes partial epilepsy by altering its desensitization kinetics. FEBS Lett. 398: 91–96.

Whiting P, Schoepfer R, Lindstrom J, Priestley T (1991): Structural and pharmacological characterization of the major brain nicotinic acetylcholine receptor subtypes stably expressed in mouse fibroblasts. Mol. Pharmacol. 40: 463–472.

Wonnacott S (1997): Presynaptic nicotinic ACh receptors. Trends Neurosci. 20: 92–98.

Xiao Y, Meyer EL, Thompson JM, Keller KJ (1997): Generation and characterization of a stably transfected cell line expressing rat $\alpha3\beta4$ neuronal nicotinic acetylcholine receptors. Soc. Neurosci. Abstr. 23: 55.9.

Part II

Physiological Roles of Nicotinic Receptors

7

Functional Anatomy of Nicotinic Acetylcholine Receptors in Mammalian Brain

Paul B. S. Clarke, Ph.D
Department of Pharmacology and Therapeutics
McGill University
Montréal, Québec, Canada

Two questions frame much of the current research on nicotine: Where are nicotinic acetylcholine receptors (nAChRs) located in mammalian brain and what do they do? These questions in turn raise several conceptual issues that form the focus of this chapter. For example, are all nAChRs ligand-gated ion channels? Are all brain nAChRs neuronal? Is there one nAChR subtype in the brain that can be regarded as predominant, by virtue of abundance, sensitivity to "smoking doses" of nicotine, or synaptic location? Do α4β2 nAChRs constitute the most prevalent subtype? Does systemic nicotine preferentially target somatodendritic or presynaptic nAChRs? This chapter does not provide a detailed cartography across brain areas; for this, there is no substitute for primary articles.

ARE ALL nAChRs LIGAND-GATED ION CHANNELS?

The classic nAChR is a ligand-gated ion channel, and the great majority of nicotinic responses detected in the brain have characteristics consistent with such a mechanism (Clarke, 1990).

Neuronal Nicotinic Receptors: Pharmacology and Therapeutic Opportunities, Edited by S. P. Arneric and J. D. Brioni
ISBN 0-471-24743-x, pages 127–139. Copyright © 1998 by Wiley-Liss, Inc.

However, several nicotinic responses have recently been described in brain and in other tissues that are suggestive of other mechanisms. In rat brain, the clearest evidence for a metabotropic nicotinic receptor comes from electrophysiological studies of the dorsolateral septal nucleus (Wong and Gallagher, 1989, 1991; Sorenson and Gallagher, 1993, 1996). In this brain area, nicotine produced a direct hyperpolarizing action mediated by increased membrane permeability to K^+. Unlike classic nicotinic responses, the effect was slow in onset and offset. It was not dependent on Na^+ influx through TTX-sensitive channels and was only somewhat reduced by removal of extracellular Ca^{++}. Instead, the response appeared completely dependent on intracellular Ca^{++}, consistent with a transduction mechanism involving second messengers (Wong and Gallagher, 1991; Sorenson and Gallagher, 1996).

Several characteristics indicate that this putative nAChR is unique. First, its function was not disrupted by chemical reduction, suggesting that it lacks the vicinal cysteine residues that are a hallmark of excitatory nAChRs (Sorenson and Gallagher, 1993). Second, the response was abolished by intracellular administration of GTPγS, indicating the possible involvement of G proteins (Sorenson and Gallagher, 1996). Third, the receptor has a unique pharmacology (Wong and Gallagher, 1991; Sorenson and Gallagher, 1996). At present, it is unclear whether the receptor in question is of the G protein-coupled superfamily or whether it contains an intrinsic K^+ channel that is modulated by G proteins. Its sluggish kinetics support the first suggestion, but the recognizably nicotinic pharmacology supports the latter more strongly.

ARE ALL BRAIN nAChRs NEURONAL?

In the periphery, nAChRs are situated not only on neurons and skeletal muscle but on a variety of other cell types, notably cochlear hair cells, keratinocytes, lymphocytes, erythrocytes, and spermatozoa. It is therefore surprising that few attempts have been made to determine whether brain nAChRs reside on glial cells, which outnumber neurons by approximately 10 to 1, and express a variety of other ligand-gated ion channels.

The evidence that brain nAChRs are exclusively neuronal is not overwhelming. Reports based on in situ hybridization either explicitly state that nAChR subunit mRNA is confined to neurons or omit mention of any glial component (Deneris et al., 1989; Wada et al., 1989, 1990; Dineley-Miller and Patrick, 1992; Seguela et al., 1993; Lobron et al., 1995; Le Novère et al., 1996). However, it is important to note that none of the above in situ hybridization studies employed a specific histological marker for glial cells. Instead, the neuronal localization rests on Nissl staining; with such a stain, it would be difficult to distinguish glia from small neurons unless visualized under high magnification.

It is relatively easy to test whether neurons in the brain possess functional nAChRs. Comparable assays for glia do not exist, and the nearest equivalent is to test glial cells in culture. This approach has provided evidence that glial cells can indeed express nAChRs (Hosli et al., 1988). In these cultures, nicotine (1 and 10 μM) induced a small hyperpolarization in most cells sampled. This action had a slow onset (15–55 s) and offset (1–3 min), and was blocked by mecamylamine, when tested in a small number of cells. More recently, nAChR-like immunohistochemical staining of these astrocytes has been reported (Hosli et al., 1994), but controls for nonspecific staining were not described.

Although several other attempts to identify glial nAChRs in primary culture have garnered completely negative results, several points must be considered. First, culture conditions can have a major impact on gene expression. Second, primary cultures are necessarily prepared from animals at an early stage of development. Hence, glia that are maintained in culture may not adequately represent glia in the living adult brain. Third, the electrophysiological characteristics attributed to glial nAChRs do not resemble those expected of a ligand-gated ion channel (Hosli et al., 1988). Thus, if glial nAChRs do exist in vivo, they might very well evade detection in assays designed to detect classic nAChRs.

DOES ONE SUBTYPE OF BRAIN nAChR PREDOMINATE BY VIRTUE OF ABUNDANCE?

Nicotinic Receptors Identified By ^{3}H-Nicotine and Other Agonist Radioligands

The main problem in assessing relative abundance of nAChRs is the paucity of nAChR subtype-selective probes. It is now 15 years since the basic distinction was made between nAChRs that bind agonists (e.g., ^{3}H-nicotine) with high affinity and those that bind ^{125}I-α-bungarotoxin. Subsequently, other ^{3}H-agonists were developed; of these, ^{3}H-cytisine (Pabreza et al., 1991), by virtue of its slow dissociation, proved particularly useful for immunoprecipitation experiments (Flores et al., 1992). All these agonists labeled the same nAChR population, comprising largely or exclusively nAChRs that contain α4 and β2 subunits (Whiting and Lindstrom, 1987, 1988; Flores et al., 1992).

The recent advent of ^{3}H-epibatidine marks a new departure, since this agonist ligand detects not only α4β2 receptors but certain other nAChR subtypes as well (Houghtling et al., 1995; Flores et al., 1996; Wang et al., 1996). The wider nAChR spectrum of binding that is obtained with ^{3}H-epibatidine compared to other ^{3}H-agonists is also reflected in a somewhat wider anatomical distribution in rat brain (Perry and Kellar, 1995).

The principal high-affinity ^{3}H-agonist binding site is associated with a single macroscopic affinity if assay contaminants are avoided. Experiments using brain and other tissue suggest that the labeled receptor is stabilized by the assay conditions in a desensitized state (Marks et al., 1996). Assuming that this interpretation is correct, the extent to which the binding equilibrium shifts to the high-affinity state is not known. Unless this shift is complete, determinations of B_{max} will underestimate true receptor density, making estimates of relative nAChR abundance unreliable.

Nicotinic Receptors Identified by ^{125}I-α-Bungarotoxin

Convergent evidence indicates that ^{125}I-α-bungarotoxin binds to nAChRs containing α7 subunits (Couturier et al., 1990; Schoepfer et al., 1990; Seguela et al., 1993; Dominguez Del Toro et al., 1994; Orr-Urtreger et al., 1997; see Clarke, 1992, for a review). It is not yet clear whether, in mammalian brain, ^{125}I-α-bungarotoxin identifies only a single nAChR subtype. An early report suggested that decamethonium might inhibit binding more effectively in some brain areas than others (Morley et al., 1977), but this finding has not been followed up. Saturation studies have typically revealed a single binding component of nanomolar affinity (Morley et al., 1979), but in such assays receptor heterogeneity could easily be missed, especially if one site is more prevalent or the affinities are similar. This is the case, for example, in embryonic chick brain, where the majority of α-bungarotoxin binding proteins (approximately 90%) correspond to α7 subunits, with an additional contribution from α8 subunits (Schoepfer et al., 1990). On present evidence, however, mammals do not appear to possess an α8 gene homologue, and as indicated above, α7 gene deletion completely abolishes ^{125}I-α-bungarotoxin binding in mice (Orr-Urtreger et al., 1997).

If mammalian α7-containing receptors are indeed heterogeneous, this could result from combination with other subunits, from alternative splicing, or from post-translational modifications. Only the first mechanism has been investigated, with conflicting conclusions. A number of groups have affinity-isolated α-bungarotoxin binding proteins from chick and rat brain, with subsequent separation of putative subunits by SDS-PAGE. Between one and four putative subunits have been reported (Clarke, 1992). In contrast, recent protein purification experiments, combined with immunological identification of subunits, have supported the notion that α7-containing nAChRs in mammalian brain are homo-oligomers (Chen and Patrick, 1997).

If other subunits do commonly assemble with α7, it is hard to see what these would be. Although the α7 subunit can certainly be coexpressed with other subunits in the same neuron (Rogers et al., 1992), in situ hybridization studies suggest no obvious "partner" subunit whose distribution correlates with that of α7. Moreover, a number of nAChR subunits have been shown not to form functional channels with α7 in oocytes (Couturier et al., 1990; Seguela et al., 1993). In addition, the β2 subunit, which appears to be the most widely expressed "structural" subunit (Wada et al., 1989; Hill et al., 1993), does not appear to associate with α7 subunits (Picciotto et al., 1995).

Nicotinic Receptors Identified by Other Antagonist Radioligands

In general, the most useful radioligands are often antagonists, since unlike agonists, they have less tendency to produce conformational changes that are associated with multiple affinity states. However, with the exception of ^{125}I-α-bungarotoxin, no nicotinic antagonist has been widely used for this purpose. The competitive antagonist ^{3}H-dihydro-β-erythroidine (^{3}H-DHBE) was shown to bind to nicotinic sites in rat brain membranes; however, the presence of a more abundant lower-affinity site and of high nondisplaceable binding precluded further analysis (Williams and Robinson, 1984).

The snake venom extract ^{125}I-neuronal bungarotoxin (^{125}I-κ-toxin, ^{125}I-toxin F) is an antagonist that possesses a unique binding profile in rat brain (Schulz et al., 1991). As in other tissues, two nicotinic sites have been distinguished. One site is also labeled by ^{125}I-α-bungarotoxin and presumably represents nAChR containing α7 subunits. Binding at the second site is inhibited by nicotinic agonists but not by α-bungarotoxin. These two putative populations of nAChR are physically segregated, as shown at the ultrastructural level in the periphery (Loring et al., 1988), and by low-resolution autoradiography in rat brain sections (Wang et al., 1991). The identity of the α-bungarotoxin-insensitive binding site is of particular interest. Three observations show that it does not correspond to the nAChR that represents the principal component of ^{3}H-agonist binding. First, the site is much less abundant. Second, its brain distribution is markedly different from that of ^{3}H-nicotine (Wang et al., 1991). Third, unlabeled nBTX does not inhibit ^{3}H-nicotine binding in the brain (Marks et al., 1993). Instead, there is a strong resemblance to the minority of ^{125}I-epibatidine binding sites that are not recognized by ^{3}H-nicotine or ^{3}H-cytisine and that have been proposed to contain α3 subunits (Perry and Kellar, 1995). In chick ciliary ganglion, ^{125}I-neuronal bungarotoxin labels a synaptic nAChR that contains α3, α5, β4, and possibly β2 subunits (Conroy and Berg, 1995; Ullian et al., 1997). Taken together, these findings indicate that ^{125}I-neuronal bungarotoxin identifies a minority population of sites in rat brain; by analogy with chick autonomic ganglion, these receptors may be synaptically located.

Currently available radioligands serve to recognize nAChRs but give little or no indication of their functional status. In contrast, drugs that bind inside the nAChR ion channel provide a means to study receptor function. The utility of such an approach has been demonstrated at peripheral nAChRs with ^{3}H-perhydrohistrionicotoxin, whose binding is greatly enhanced by application of agonists (Aronstam et al., 1981). With this application in mind, the nicotinic antagonist ^{3}H-mecamylamine was characterized in rat brain membranes (Banerjee et al., 1990).

However, nicotine inhibited ^{3}H-mecamylamine binding rather than increasing it, as would be expected if mecamylamine binding required that nAChR channel to be opened. In retrospect, this may not be too surprising; in the brain, blockade by mecamylamine does not appear to be use dependent and hence is unlikely to involve channel block (El-Bizri and Clarke, 1994).

Aside from ^{3}H-mecamylamine, no other putative nAChR channel blocker has been developed. One promising compound that might fulfill this role is chlorisondamine, a *bis*-quaternary compound that exerts an insurmountable blockade in the brain (Clarke and Reuben, 1996) via a use-dependent mechanism (El-Bizri and Clarke, 1994).

Nicotinic Receptors Identified by Immunohistochemistry

The distribution of nAChR-associated immunoreactivity has been surveyed across the entire rat brain in several studies. Deutch and colleagues (1987), employing a monoclonal antibody raised against Torpedo nAChR, observed heavy staining in several of the brain areas that are associated with non-$\alpha7$ nAChRs. However, the pattern of staining suggests that the antibody was not selective for any single known subunit, including recent additions such as $\alpha5$ and $\alpha6$. The $\alpha7$ subunit has also been localized in rat brain using monoclonal antibodies (Dominguez Del Toro et al., 1994). The distribution was extensive and appeared comparable to that produced by ^{125}I-α-bungarotoxin autoradiography.

Immunohistochemical mapping of rat and human brain nAChRs has been reported using a second monoclonal antibody (WF 6) raised directly from Torpedo nAChR (Schröder, 1992). However, this antibody is probably not selective for any particular α subunit. Swanson and coworkers (1987) employed a monoclonal antibody (mab 270) that is selective for $\beta2$ subunits (Whiting and Lindstrom, 1987; Forsayeth and Kobrin, 1997); in unfixed brain sections, immunoreactivity was widespread and highly reminiscent of ^{3}H-nicotine binding. Immunohistochemistry in *fixed* brain has been achieved using anti-$\beta2$ antisera (Hill et al., 1993); immunoreactivity was again widely distributed, but the regional pattern does not seem to correspond as closely to ^{3}H-nicotine autoradiography. This raises the important but largely ignored possibility that in unfixed tissue prepared for autoradiography a considerable fraction of nAChR antigenicity and binding sites are lost.

ARE $\alpha4\beta2$ nAChRs THE MOST PREVALENT SUBTYPE IN THE BRAIN?

Initial protein purification experiments using mab 270 indicated that $\alpha4$ and $\beta2$ subunits are strongly associated in mammalian brain (Whiting and Lindstrom, 1986, 1987, 1988). More specifically, it appeared that $\beta2$ subunit-containing nAChRs contained only one other type of subunit, subsequently identified as $\alpha4$ (Whiting et al., 1987; Nakayama et al., 1991). The strong association of $\alpha4$ and $\beta2$ was also consistent with anatomical mapping studies of rat brain using ^{3}H-agonist and ^{125}I-mab 270 (Clarke et al., 1985; Swanson et al., 1987; Happe et al., 1994). Immunoprecipitation experiments indicated that at least 85–90% of ^{3}H-nicotine binding proteins from whole rat brain contain $\alpha4$ (Nakayama et al., 1991) and $\beta2$ subunits (Whiting and Lindstrom, 1986). Given the possibility of regional heterogeneity, nine individual brain regions were subsequently analyzed, this time using ^{3}H-cytisine; in each region, virtually *all* binding sites appeared to represent $\alpha4\beta2$ nAChRs (Flores et al., 1992). A yet finer-grained analysis using autoradiography showed that genetic knockout of the $\beta2$ subunit abolishes ^{3}H-nicotine binding throughout the mouse brain (Picciotto et al., 1995). On the basis of these findings, not a few researchers have concluded that the most prevalent nAChR subtype in the brain is one comprising only $\alpha4$ and $\beta2$ subunits. However, several caveats are in order:

1. These results are seen through the prism of high-affinity ^{3}H-agonist binding, which may not detect the majority of nAChR subtypes.
2. Estimations of relative subunit abundance based on transcript assays are problematic. Importantly, mRNA levels, even if accurately known, are at best an approximate guide to relative protein abundance (Blumenthal et al., 1997; Forsayeth and Kobrin, 1997). In practical terms, comparing the transcript levels of different subunits is prone to error, because probes tend to differ in hybridization efficiency and specific activity. Only recently have these factors been taken into account (Le Novère et al., 1996). Some doubts about probe specificity have also been voiced (Dineley-Miller and Patrick, 1992; Le Novère, et al., 1996).
3. Estimating relative subunit abundance using immunological approaches is also problematic. Immunohistochemical staining can be highly sensitive to fixation (Deutch et al., 1987; Swanson et al., 1987) and is seldom quantitative. In addition, the possibility of immunoreactivity with a shared epitope, located on an unrelated protein, is seldom completely excluded. Such an eventuality has been ruled out, however, for the β2 subunit; not only do antisera raised against different nAChR domains produce the same staining pattern (Hill et al., 1993) but immunostaining is abolished by knockout of the β2 gene (Picciotto et al., 1995). A separate problem is that most nAChR-like immunoreactivity is commonly observed to be cytosolic (Schröder, 1992; Nakayama et al., 1995).
4. As-yet unidentified nAChR subunits may be expressed abundantly in the brain.
5. The existence of nAChRs comprising only two types of subunit (i.e., α4 and β2) would be a little surprising, given that all other native neuronal nAChRs characterized to date comprise more than two kinds of subunit (Conroy et al., 1992; Vernallis et al., 1993; Conroy and Berg, 1995; Wang et al., 1996), with the possible exception of nAChRs that bind α-bungarotoxin (Chen and Patrick, 1997). Furthermore, it is clear that in heterologous expression systems, α4 and β2 subunits can form functional nAChRs by combining with a third kind of subunit (Ramirez-Latorre et al., 1996; Fucile et al., 1997).
6. It is conceivable that the "α4β2 subtype" identified in published immunoprecipitation experiments may have obtained additional subunits that evaded detection, either through proteolysis or because they copurified with α4 or β2. In this connection, β2 subunits appear less susceptible to proteolysis than β3 and β4 subunits (Forsayeth and Kobrin, 1997).

If additional nAChR subunits coassemble with α4 and β2 in mammalian brain, which would they be? The participation of α7 subunit can be ruled out on anatomical and biochemical grounds (Clarke et al., 1985; Whiting and Lindstrom, 1987). The involvement of α3, α5, β3, and β4 subunits has been tested in two immunoprecipitation studies. In the first, antisera raised against these subunits failed to deplete ^{3}H-cytisine binding detectably in rat forebrain, suggesting that the nAChRs in question comprised α4β2 only (Flores et al., 1992). However, results from the second study suggest that α4 and β2 subunits can assemble with some of these other subunits in rat cerebellum, forming α4β2β3β4 nAChRs; these receptors would likely be labeled by ^{3}H-cytisine (Forsayeth and Kobrin, 1997). These findings are provocative, and it becomes important to establish whether the difference in outcomes reflects the different brain regions studied (i.e., forebrain vs. cerebellum).

Immunoprecipitation studies have been restricted either to whole brain or to large brain areas, whereas message for certain subunits is strongly expressed only in small nuclei. For example, in nigrostriatal and mesolimbic dopaminergic cells, α6 and β3 mRNA appears more abundant than α4 and β2 mRNA, and it would be interesting to know whether α6 and β3 subunits contribute to the abundant ^{3}H-nicotine binding that is associated with these neurons (Clarke and Pert, 1985).

Nicotinic Receptor Abundance: Conclusions

In summary, radioligand binding studies show that nAChR subtypes that are labeled with ^{3}H-agonists (nicotine, cytisine) or with ^{125}I-α-bungarotoxin are expressed in many brain nuclei. These populations differ markedly in their distributions, but their overall densities appear similar. Most or all nAChRs labeled with ^{3}H-nicotine and ^{3}H-cytisine contain α4 and β2 subunits, and most or all nAChRs identified by ^{125}I-α-bungarotoxin contain α7 subunits. At present, it is not clear whether either population represents a single subtype of receptor. Radiolabeled epibatidine recognizes not only α4β2-containing nAChRs but one or more less prevalent subtypes as well. Immunohistochemical mapping studies provide further evidence that nAChRs are widely distributed in mammalian brain but offer little information about the relative abundance of subtypes.

DOES ONE SUBTYPE OF BRAIN nAChR PREDOMINATE BY VIRTUE OF SENSITIVITY TO NICOTINE?

In the periphery, nAChRs at the neuromuscular junction are not appreciably activated in tobacco smokers, whereas ganglionic nAChRs are. In the brain, nAChR subtypes differ in their sensitivity to nicotine, but whether any are preferentially sensitive to "smoking doses" of the drug is not certain. Identifying such subtypes is complicated by a number of issues. The main problem relates to the kinetics of nicotine and of nAChR function. It is widely believed that the pharmacological actions of nicotine in cigarette smoke depend on the delivery of intermittent nicotine "boli" superimposed on a fluctuating baseline. However, at present, there is little hard evidence to suggest that individual cigarette puffs lead to discrete pharmacological events. In addition, animal studies suggest that certain brain nAChRs can desensitize or inactivate in response to continuous exposure to nicotine that occurs in smokers (Benwell and Balfour, 1992; Marks et al., 1994).

During smoking, nicotine levels in the blood and brain are subject to complex kinetics that are not adequately mimicked by any existing procedure of in vivo drug administration. Greater control over kinetics can be obtained in vitro. Here, acute sensitivity to nicotine can be assessed in isolated brain tissue by measuring electrophysiological or other changes in response to application of the drug. If a response is obtained at concentrations within the smoking range (perhaps 0.05–5 μM), then the nAChRs potentially play a role in smoking. Thus, many brain regions, and probably more than one nAChR subtype, represent possible targets (Marks et al., 1993; Clarke and Reuben, 1996; Albuquerque et al., 1997). Since nAChRs can be both activated and desensitized within this concentration range, this approach provides incomplete information. Alternatively, if a response requires higher concentrations of agonist, or desensitizes rapidly and virtually completely, then it is probably not involved in smoking. On both counts, evidence to date would tend to rule out α7-containing nAChRs (Albuquerque et al., 1997).

Recently, it has been suggested that "pre- rather than postsynaptic nAChRs mediate the broad psychophysical effects of nicotine in humans" (McGehee et al., 1995). However, this claim was based on a number of assumptions that are at best questionable. First, it was stated that even at synapses with presynaptic choline acetyltransferase expression and postsynaptic nAChRs, synaptic transmission is not mediated by ACh. This may be the case in certain brain nuclei (e.g., interpeduncular nucleus) but is almost certainly not true elsewhere (e.g., substantia nigra). Second, it was stated that the serum nicotine concentrations in a moderate smoker range from 10 to 100 nM. These values underestimate steady-state *venous* levels, and fail to take into account arteriovenous differences and puff-related fluctuations (Henningfield et al., 1993). Third, observations were restricted to embryonic cell cultures derived from a nonmammalian species (chick). Finally, the claim of greater sensitivity of pre- versus postsynaptic nAChRs was based on observations in only two neural systems. Postsynaptic (or more precisely, somatodendritic) sensitivity to nicotine has been qualified in very few studies using brain tissue, but in at least one case neuronal excitation has been observed at concentrations as low as 100 nM (Brioni et al., 1997).

CAN SYSTEMIC NICOTINE PREFERENTIALLY TARGET SOMATODENDRITIC nAChRs?

Other evidence suggests that following systemic administration of nicotine, the primary site of action may be somatodendritic. An early indication was provided by studies in which the actions of nicotine in the brain were mapped by quantifying changes in 2-deoxyglucose uptake following systemic drug administration (London et al., 1988). The resulting anatomical pattern resembled that of ^{3}H-nicotine binding, suggesting that most of the drug effects were probably local. The most affected sites tended to correspond to the location of somatodendritic nAChRs (e.g., thalamus, substantia nigra). In contrast, little if any effects were seen in several areas where presynaptic nAChRs are concentrated, such as the striatum (Clarke and Pert, 1985) and cerebral cortex (Lavine et al., 1997).

More direct evidence of the importance of somatodendritic nAChRs has been provided by studies of catecholamine release in the brain. In the mesolimbic dopamine system, nAChRs are located both somatodendritically and on terminals (Clarke and Pert, 1985). In isolated tissues, nicotine exerts excitatory actions on both somatodendritic and presynaptic receptors. In the whole animal, acute systemic nicotine increases dopamine (DA) outflow in the main terminal area (nucleus accumbens). Experiments employing local infusion of drugs suggest that this effect results exclusively from a somatodendritic site of action (Benwell et al., 1993; Nisell et al., 1994). Analogous findings have been obtained with respect to ascending noradrenergic neurons (Mitchell, 1993).

The relative contribution of somatodendritic versus presynaptic receptors in these responses may reflect several factors. For example, different receptor subtypes may be involved. Alternatively, somatodendritic nAChRs may be present in a higher density than presynaptic nAChRs. In addition, nAChRs on dopamine terminals are highly susceptible to desensitization (Grady et al., 1994); possibly, somatodendritic nAChRs are less so. Although brief application of nicotine can increase transmitter release through a direct presynaptic mechanism (Gray et al., 1996), it is not unlikely that more prolonged exposure would partially inactivate voltage-gated ion channels, thus inhibiting impulse-dependent release.

DOES ONE SUBTYPE OF BRAIN nAChR PREDOMINATE BY VIRTUE OF SYNAPTIC LOCATION?

nAChRs that are synaptically located might be expected to play an important role in cholinergic transmission and would also present targets for drug therapies aiming to reverse cholinergic hypofunction. However, evidence for nicotinic cholinergic transmission in the brain is fragmentary, as reviewed elsewhere (Clarke, 1993).

The traditional view that nAChRs subserve rapid, synaptic transmission in all nervous tissues needs reassessment, based on the following observations. In autonomic ganglia, two nAChR subtypes have been recognized: one predominantly synaptic, the other perisynaptic (Loring et al., 1988; Horch and Sargent, 1995). The perisynaptic receptors were initially thought to serve a primarily trophic function, but more recently a role in neurotransmission has been demonstrated (Zhang et al., 1996). This phenomenon was observed in chick embryonic tissue; whether it also occurs in adult mammalian ganglia remains to be determined. This finding suggests that cholinergic transmission can occur in part outside the synapse, as defined morphologically. Further evidence for this notion is provided by ultrastructural studies in rat striatum (Contant et al., 1996). In this brain area, the great majority of ACh varicosities were not found at synaptic junctions. If ACh indeed works in a paracrine fashion, this might provide a function for nAChRs that are located on striatal dopaminergic nerve terminals, since in this structure axoaxonic contacts are rare.

An alternative possibility is that in the brain, newly assembled nAChRs are not positioned with any precision on the cell membrane. Several observations lend support to this view. First, in the periphery, nAChRs have been detected on axons (Armett and Ritchie, 1961), and pharmacological evidence has also been presented for the existence of "preterminal" receptors in the brain (Léna et al., 1993). Second, nAChRs possessing a predominantly extrasynaptic location have been detected immunohistochemically in rat (Hill et al., 1993) and chick brain (Ullian and Sargent, 1995). Third, nAChRs have been found at the somatodendritic level and at the level of terminals in several neural pathways, notably the thalamocortical (Prusky et al., 1987; Lavine et al., 1997), nigrostriatal and mesolimbic (Clarke and Pert, 1985), retinofugal (Swanson et al., 1987), and habenulo-interpeduncular projections (Clarke et al., 1986).

CONCLUSIONS

It appears that not all brain nAChRs are ligand-gated ion channels. Some nAChRs may be located on glia. On present evidence, it is unlikely that one nAChR subtype predominates in the brain, either by virtue of abundance, sensitivity to "smoking doses" of nicotine, or synaptic location. The relative importance of somatodendritic versus presynaptic nAChR is still largely undecided.

REFERENCES

Albuquerque EX, Alkondon M, Pereira EFR, Castro NG, Schrattenholz A, Barbosa CTF, Bonfante-Cabarcas R, Aracava Y, Eisenberg HM, Maelicke A (1997): Properties of neuronal nicotinic acetylcholine receptors: pharmacological characterization and modulation of synaptic function. J. Pharmacol. Exp. Ther. 280: 1117–1136.

Armett CJ, Ritchie JM (1961): The action of acetylcholine and some related substances on conduction in mammalian non-myelinated nerve fibres. J. Physiol. (London) 155: 372–384.

Aronstam RS, Eldefrawi AT, Pessah IN, Daly JW, Albuquerque EX, Eldefrawi ME (1981): Regulation of [3H]perhydrohistrionicotoxin binding to *Torpedo ocellata* electroplax by effectors of the acetylcholine receptor. J. Biol. Chem. 256: 2843–2850.

Banerjee S, Punzi JS, Kreilick K, Abood LG (1990): [^{3}H]Mecamylamine binding to rat brain membranes. Studies with mecamylamine and nicotine analogues. Biochem. Pharmacol. 40: 2105–2110.

Benwell MEM, Balfour DJK (1992): The effects of acute and repeated nicotine treatment on nucleus accumbens dopamine and locomotor activity. Br. J. Pharmacol. 105: 849–856.

Benwell MEM, Balfour DJK, Lucchi HM (1993): Influence of tetrodotoxin and calcium on changes in extracellular dopamine levels evoked by systemic nicotine. Psychopharmacology (Berlin) 112: 467–474.

Blumenthal EM, Conroy WG, Romano SJ, Kassner PD, Berg DK (1997): Detection of functional nicotinic receptors blocked by bungarotoxin on PC12 cells and dependence of their expression on post-translational events. J. Neurosci. 17: 6094–6104.

Brioni JD, Kim DJB, O'Neill A, Brodie MS, Decker MW, Arneric SP (1997): ABT-089 [2-methyl-3-(2-(S)-pyrrolidinylmethoxy) pyridine dihydrochloride]: discriminative stimulus properties and electrophysiological actions. Drug Dev. Res. 40: 259–266.

Chen DN, Patrick JW (1997): The α-bungarotoxin-binding nicotinic acetylcholine receptor from rat brain contains only the α7 subunit. J. Biol. Chem. 272: 24,024–24,029.

Clarke PBS (1990): The central pharmacology of nicotine: electrophysiological approaches. In: Wonnacott S, Russell MAH, Stolerman IP, editors. Nicotine psychopharmacology: molecular, cellular, and behavioural aspects. Oxford: Oxford University Press, pp. 158–193.

Clarke PBS (1992): The fall and rise of neuronal alpha-bungarotoxin binding proteins. Trends Pharmacol. Sci. 11: 407–413.

Clarke PBS (1993): Nicotinic receptors in mammalian brain: localization and relation to cholinergic function. Prog. Brain Res. 98: 77–83.

Clarke PBS, Pert A (1985): Autoradiographic evidence for nicotine receptors on nigrostriatal and mesolimbic dopaminergic neurons. Brain Res. 348: 355–358.

Clarke PBS, Reuben M (1996): Release of [^{3}H]-noradrenaline from rat hippocampal synaptosomes by nicotine: mediation by different nicotinic receptor subtypes from striatal [^{3}H]-dopamine release. Br. J. Pharmacol. 117: 595–606.

Clarke PBS, Schwartz RD, Paul SM, Pert CB, Pert A (1985): Nicotinic binding in rat brain: autoradiographic comparison of ^{3}H-acetylcholine, ^{3}H-nicotine, and ^{125}I-alpha-bungarotoxin. J. Neurosci. 5: 1307–1315.

Clarke PBS, Hamill GS, Nadi NS, Jacobowitz, DM, Pert A (1986): ^{3}H-nicotine- and ^{125}I-alpha-bungarotoxin-labeled nicotinic receptors in the interpeduncular nucleus of rats. II. Effects of habenular deafferentation. J. Comp. Neurol. 251: 407–413.

Conroy WG, Berg DK (1995): Neurons can maintain multiple classes of nicotinic acetylcholine receptors distinguished by different subunit compositions. J. Biol. Chem. 270: 4424–4431.

Conroy WG, Vernallis AB, Berg DK (1992): The α5 gene product assembles with multiple acetylcholine receptor subunits to form distinctive receptor subtypes in brain. Neuron 9: 679–691.

Contant C, Umbriaco D, Garcia S, Watkins KC, Descarries L (1996): Ultrastructural characterization of the acetylcholine innervation in adult rat neostriatum. Neuroscience 71: 937–947.

Couturier S, Bertrand D, Matter JM, Hernandez MC, Bertrand S, Millar N, Valera S, Barkas T, Ballivet M (1990): A neuronal nicotinic acetylcholine receptor subunit (alpha 7) is developmentally regulated and forms a homo-oligomeric channel blocked by alpha-BTX. Neuron 5: 847–856.

Deneris ES, Boulter J, Swanson LW, Patrick J, Heinemann S (1989): Beta 3: a new member of nicotinic acetylcholine receptor gene family is expressed in brain. J. Biol. Chem. 264: 6268–6272.

Deutch AY, Holliday J, Roth RH, Chun LL, Hawrot E (1987): Immunohistochemical localization of a neuronal nicotinic acetylcholine receptor in mammalian brain. Proc. Natl. Acad. Sci. USA 84: 8697–8701.

Dineley-Miller K, Patrick J (1992): Gene transcripts for the nicotinic acetylcholine receptor subunit, beta4, are distributed in multiple areas of the rat central nervous system. Brain Res. Mol. Brain Res. 16: 339–344.

Dominguez Del Toro E, Juiz JM, Peng X, Lindstrom J, Criado M (1994): Immunocytochemical localization of the α7 subunit of the nicotinic acetylcholine receptor in the rat central nervous system. J. Comp. Neurol. 349: 325–342.

El-Bizri H, Clarke PBS (1994): Blockade of nicotinic receptor-mediated release of dopamine from striatal synaptosomes by chlorisondamine and other nicotinic antagonists administered in vitro. Br. J. Pharmacol. 111: 406–413.

Flores CM, Rogers SW, Pabreza LA, Wolfe BB, Kellar KJ (1992): A subtype of nicotinic cholinergic receptor in rat brain is composed of alpha4 and beta2 subunits and is up-regulated by chronic nicotine treatment. Mol. Pharmacol. 41: 31–37.

Flores CM, DeCamp RM, Kilo S, Rogers SW, Hargreaves KM (1996): Neuronal nicotinic receptor expression in sensory neurons of the rat trigeminal ganglion: demonstration of $\alpha3\beta4$, a novel subtype in the mammalian nervous system. J. Neurosci. 16: 7892–7901.

Forsayeth JR, Kobrin E (1997): Formation of oligomers containing the $\beta3$ and $\beta4$ subunits of the rat nicotinic receptor. J. Neurosci. 17: 1531–1538.

Fucile S, Barabino B, Palma E, Grassi F, Limatola C, Mileo AM, Alemà S, Ballivet M, Eusebi F (1997): α_5 subunit forms functional $\alpha_3\beta_4\alpha_5$ nAChRs in transfected human cells. NeuroReport 8: 2433–2436.

Grady SR, Marks MJ, Collins AC (1994): Desensitization of nicotine-stimulated [^{3}H]dopamine release from mouse striatal synaptosomes. J. Neurochem. 62: 1390–1398.

Gray R, Rajan AS, Radcliffe KA, Yakehiro M, Dani JA (1996): Hippocampal synaptic transmission enhanced by low concentrations of nicotine. Nature 383: 713–716.

Happe HK, Peters JL, Bergman DA, Murrin LC (1994): Localization of nicotinic cholinergic receptors in rat brain: Autoradiographic studies with [^{3}H]cytisine. Neuroscience 62: 929–944.

Henningfield JE, Stapleton JM, Benowitz NL, Grayson RF, London ED (1993): Higher levels of nicotine in arterial than in venous blood after cigarette smoking. Drug Alcohol Depend. 33: 23–29.

Hill JA, Zoli M, Bourgeois J-P, Changeux J-P (1993): Immunocytochemical localization of a neuronal nicotinic receptor: the beta2-subunit. J. Neurosci. 13: 1551–1568.

Horch HLW, Sargent PB (1995): Perisynaptic surface distribution of multiple classes of nicotinic acetylcholine receptors on neurons in the chicken ciliary ganglion. J. Neurosci. 15: 7778–7795.

Hosli L, Hosli E, Della Briotta G, Quadri L, Heuss L (1988): Action of acetylcholine, muscarine, nicotine and antagonists on the membrane potential of astrocytes in cultured rat brainstem and spinal cord. Neurosci Lett. 92: 165–170.

Hosli L, Hosli E, Winter T, Stauffer S (1994): Coexistence of cholinergic and somatostatin receptors on astrocytes of rat CNS. NeuroReport 5: 1469–1472.

Houghtling RA, Dávila-García MI, Kellar KJ (1995): Characterization of ($\pm$)-[^{3}H]Epibatidine binding to nicotinic cholinergic receptors in rat and human brain. Mol. Pharmacol. 48: 280–287.

Lavine N, Reuben M, Clarke PBS (1997): A population of nicotinic receptors is associated with thalamocortical afferents in the adult rat: laminal and areal analysis. J. Comp. Neurol. 380: 175–190.

Léna C, Changeux J-P, Mulle C (1993): Evidence for "preterminal" nicotinic receptors on GABAergic axons in the rat interpeduncular nucleus. J. Neurosci. 13: 2680–2688.

Le Novère N, Zoli M, Changeux JP (1996): Neuronal nicotinic receptor $\alpha6$ subunit mRNA is selectively concentrated in catecholaminergic nuclei of the rat brain. Eur. J. Neurosci. 8: 2428–2439.

Lobron C, Wevers A, Dämgen K, Jeske A, Rontal D, Birtsch C, Heinemann S, Reinhardt S, Maelicke A, Schröder H (1995): Cellular distribution in the rat telencephalon of mRNAs encoding for the $\alpha3$ and $\alpha4$ subunits of the nicotinic acetylcholine receptor. Mol. Brain Res. 30: 70–76.

London ED, Connolly RJ, Szikszay M, Wamsley JK, Dam M (1988): Effects of nicotine on local cerebral glucose utilization in the rat. J. Neurosci. 8: 3920–3928.

Loring RH, Sah DW, Landis SC, Zigmond RE (1988): The ultrastructural distribution of putative nicotinic receptors on cultured neurons from the rat superior cervical ganglion. Neuroscience 24: 1071–1080.

Marks MJ, Farnham DA, Grady SR, Collins AC (1993): Nicotinic receptor function determined by stimulation of rubidium efflux from mouse brain synaptosomes. J. Pharmacol. Exp. Ther. 264: 542–552.

Marks MJ, Grady SR, Yang J-M, Lippiello PM, Collins AC (1994): Desensitization of nicotine-stimulated $^{86}Rb^+$ efflux from mouse brain synaptosomes. J. Neurochem. 63: 2125–2135.

Marks MJ, Robinson SF, Collins AC (1996): Nicotinic agonists differ in activation and desensitization of $^{86}Rb^+$ efflux from mouse thalamic synaptosomes. J. Pharmacol. Exp. Ther. 277: 1383–1396.

McGehee DS, Heath MJS, Gelber S, Devay P, Role LW (1995): Nicotine enhancement of fast excitatory synaptic transmission in CNS by presynaptic receptors. Science 269: 1692–1696.

Mitchell SN (1993): Role of the locus coeruleus in the noradrenergic response to a systemic administration of nicotine. Neuropharmacology 32: 937–949.

Morley BJ, Lorden JF, Brown GB, Kemp GE, Bradley RJ (1977): Regional distribution of nicotinic acetylcholine receptors in rat brain. Brain Res. 134: 161–166.

Morley BJ, Kemp GE, Salvaterra P (1979): Alpha-bungarotoxin binding sites in the CNS. Life Sci. 24: 859–872.

Nakayama H, Nakashima T, Kurogochi Y (1991): Alpha 4 is a major acetylcholine binding subunit of cholinergic ligand affinity-purified nicotinic acetylcholine receptor from rat brains. Neurosci. Lett. 121: 122–124.

Nakayama H, Shioda S, Okuda H, Nakashima T, Nakai Y (1995): Immunocytochemical localization of nicotinic acetylcholine receptor in rat cerebral cortex. Mol. Brain Res. 32: 321–328.

Nisell M, Nomikos GG, Svensson TH (1994): Systemic nicotine-induced dopamine release in the rat nucleus accumbens is regulated by nicotinic receptors in the ventral tegmental area. Synapse 16: 36–44.

Orr-Urtreger A, Göldner FM, Saeki M, Lorenzo I, Goldberg L, De Biasi M, Dani JA, Patrick JW, Beaudet AL (1997): Mice deficient in the $\alpha7$ neuronal nicotinic acetylcholine receptor lack α-bungarotoxin binding sites and hippocampal fast nicotinic currents. J. Neurosci. 17: 9165–9171.

Pabreza LA, Dhawan S, Kellar KJ (1991): [^{3}H]cytisine binding to nicotinic cholinergic receptors in brain. Mol. Pharmacol. 39: 9–12.

Perry DC, Kellar KJ (1995): [^{3}H]epibatidine labels nicotinic receptors in rat brain: an autoradiographic study. J. Pharmacol. Exp. Ther. 275: 1030–1034.

Picciotto MR, Zoli M, Léna C, Bessis A, Lallemand Y, LeNovère N, Vincent P, Pich EM, Brûlet P, Changeux J-P (1995): Abnormal avoidance learning in mice lacking functional high-affinity nicotine receptor in the brain. Nature 374: 65–67.

Prusky GT, Shaw C, Cynader MS (1987): Nicotine receptors are located on lateral geniculate nucleus terminals in cat visual cortex. Brain Res. 412: 131–138.

Ramirez-Latorre J, Yu CR, Qu X, Perin F, Karlin A, and Role L (1996): Functional contributions of $\alpha5$ subunit to neuronal acetylcholine receptor channels. Nature 380: 347–351.

Rogers SW, Mandelzys A, Deneris ES, Cooper E, Heinemann S (1992): The expression of nicotinic acetylcholine receptors by PC12 cells treated with NGF. J. Neurosci. 12: 4611–4623.

Schoepfer R, Conroy WG, Whiting P, Gore M, Lindstrom J (1990): Brain alpha-bungarotoxin binding protein cDNAs and MAbs reveal subtypes of this branch of the ligand-gated ion channel gene superfamily. Neuron 5: 35–48.

Schröder H (1992): Immunohistochemistry of cholinergic receptors. Anat. Embryol. 186: 407–429.

Schulz DW, Loring RH, Aizenman E, Zigmond RE (1991): Autoradiographic localization of putative nicotinic receptors in the rat brain using ^{125}I-neuronal bungarotoxin. J. Neurosci. 11: 287–297.

Seguela P, Wadiche J, Dineley-Miller K, Dani JA, Patrick JW (1993): Molecular cloning, functional properties, and distribution of rat brain α_7: a nicotinic cation channel highly permeable to calcium. J. Neurosci. 13: 596–604.

Sorenson EM, Gallagher JP (1993): The reducing agent dithiothreitol (DTT) does not abolish the inhibitory nicotinic response recorded from rat dorsolateral septal neurons. Neurosci. Lett. 152: 137–140.

Sorenson EM, Gallagher JP (1996): The membrane hyperpolarization of rat dorsolateral septal nucleus neurons is mediated by a novel nicotinic receptor. J. Pharmacol. Exp. Ther. 277: 1733–1743.

Swanson LW, Simmons DM, Whiting PJ, Lindstrom J (1987): Immunohistochemical localization of neuronal nicotinic receptors in the rodent central nervous system. J. Neurosci. 7: 3334–3342.

Ullian EM, Sargent PB (1995): Pronounced cellular diversity and extrasynaptic location of nicotinic acetylcholine receptor subunit immunoreactivities in the chicken pretectum. J. Neurosci. 15: 7012–7023.

Ullian EM, McIntosh JM, Sargent PB (1997): Rapid synaptic transmission in the avian ciliary ganglion is mediated by two distinct classes of nicotinic receptors. J. Neurosci. 17: 7210–7219.

Vernallis AB, Conroy WG, Berg DK (1993): Neurons assemble acetylcholine receptors with as many as three kinds of subunits while maintaining subunit segregation among receptor subtypes. Neuron 10: 451–464.

Wada E, Wada K, Boulter J, Deneris E, Heinemann S, Patrick J, Swanson LW (1989): Distribution of alpha 2, alpha 3, alpha 4, and beta 2 neuronal nicotinic receptor subunit mRNAs in the central nervous system: a hybridization histochemical study in the rat. J. Comp. Neurol. 284: 314–335.

Wada E, McKinnon D, Heinemann S, Patrick J, Swanson LW (1990): The distribution of mRNA encoded by a new member of the neuronal nicotinic acetylcholine receptor gene family (alpha 5) in the rat central nervous system. Brain Res. 526: 45–53.

Wang F, Gerzanich V, Wells GB, Anand R, Peng X, Keyser K, Lindstrom J (1996): Assembly of human neuronal nicotinic receptor α5 subunits with α3, β2, and β4 subunits. J. Biol. Chem. 271: 17,656–17,665.

Wang YT, Neuman RS, Bieger D (1991): Somatostatin inhibits nicotinic cholinoceptor mediated-excitation in rat ambigual motoneurons in vitro. Neurosci. Lett. 123: 236–239.

Whiting P, Lindstrom J (1986): Pharmacological properties of immuno-isolated neuronal nicotinic receptors. J. Neurosci. 6: 3061–3069.

Whiting P, Lindstrom J (1987): Purification and characterization of a nicotinic acetylcholine receptor from rat brain. Proc. Natl. Acad. Sci. USA 84: 595–599.

Whiting PJ, Lindstrom JM (1988): Characterization of bovine and human neuronal nicotinic acetylcholine receptors using monoclonal antibodies. J. Neurosci. 8: 3395–3404.

Whiting P, Esch F, Shimasaki S, Lindstrom J (1987): Neuronal nicotinic acetylcholine receptor beta-subunit is coded for by the cDNA clone alpha 4. FEBS Lett. 219: 459–463.

Williams M, Robinson JL (1984): Binding of the nicotinic cholinergic antagonist, dihydro-beta-erythroidine, to rat brain tissue. J. Neurosci. 4: 2906–2911.

Wong LA, Gallagher JP (1989): A direct nicotinic receptor-mediated inhibition recorded intracellularly in vitro. Nature 341: 439–442.

Wong LA, Gallagher JP (1991): Pharmacology of nicotinic receptor-mediated inhibition in rat dorsolateral septal neurones. J. Physiol. 436: 325–346.

Zhang ZW, Coggan JS, Berg DK (1996): Synaptic currents generated by neuronal acetylcholine receptors sensitive to α-bungarotoxin. Neuron 17: 1231–1240.

8

Nicotinic Receptor Modulation of Neurotransmitter Release

Sergio Kaiser Ph.D and Susan Wonnacott Ph.D

Department of Biology and Biochemistry
University of Bath
Bath, United Kingdom

INTRODUCTION

Background

Until 1961, the accepted model of synaptic transmission between neurons was that a chemical neurotransmitter released from a nerve terminal in response to an action potential crossed the synapse to interact with its postsynaptic receptor on the cell soma or dendrites of the receptive neuron. This view was modified when Koelle (1961) proposed that acetylcholine (ACh) facilitates its own release from sympathetic ganglia, in addition to its postsynaptic action. Since then, the term *presynaptic receptor* has been used to define those receptors present on nerve terminals that, in the presence of their cognate neurotransmitter or neuromodulator, either facilitate or inhibit Ca^{2+}-dependent transmitter release. The presynaptic cholinergic receptors first recognized by Koelle represent the class of *autoreceptors*, presynaptic receptors respon-

Neuronal Nicotinic Receptors: Pharmacology and Therapeutic Opportunities, Edited by S. P. Arneric and J. D. Brioni
ISBN 0-471-24743-x, pages 141–159. Copyright © 1998 by Wiley-Liss, Inc.

sive to the substance released from the same terminals. Their physiological role is easily recognized as positive or negative feedback regulation, whereby they may fine-tune the magnitude of the chemical signal that originates from the nerve ending.

The term *heteroreceptors* is used to describe those presynaptic receptors that respond to a substance not released from the nerve terminal on which they reside. Heteroreceptors may be postsynaptic to another input forming an axo-axonic synapse, although in most cases this is not proven. Alternatively, these receptors may be activated by substances released by adjacent or more distant terminals acting in a paracrine manner. Heteroreceptors provide a mechanism for chemical crosstalk between neurons.

The majority of presynaptic receptors were recognized as being metabotropic. Their coupling via G-proteins to cell signaling pathways offers a tangible route for mediating their modulatory effects. These receptors can either facilitate (e.g., β adrenoceptors) or attenuate (e.g., α2 adrenoceptors) evoked release. The concept of presynaptic ionotropic receptors was viewed with skepticism by pharmacologists: Why would one need a fast-acting ion channel receptor in this modulatory role? It is now evident that ligand-gated cation channels, including nicotinic ACh receptors (nAChRs; Wonnacott, 1997) and ionotropic glutamate receptors (McGehee et al., 1996; see section on "Receptor interplay"), are located on presynaptic terminals, where they can elicit Ca^{2+}-dependent transmitter release. This chapter focuses on the properties, mechanisms, and implications of presynaptic nAChRs in the mammalian brain.

nAChRs: Subcellular Localization

Because nAChRs mediate neuromuscular and ganglionic transmission, an analogous postsynaptic role in fast excitatory transmission in the central nervous system (CNS) was expected. But the difficulties in demonstrating such a function, together with the high relative permeability to Ca^{2+} shown by neuronal nAChRs, has recently suggested a more modulatory role for these receptors in the brain (Role and Berg, 1996). The well-established ability of nicotine to elicit transmitter release by acting on presynaptic nAChRs is consistent with such a modulatory activity. Indeed, it has been suggested that presynaptic modulation might be the predominant role of nAChRs in the CNS. Nevertheless, nAChRs have been found on somatodendritic regions of neurons (e.g., Pidoplichko et al., 1997) in areas where there is evidence for local cholinergic inputs, although whether the nAChRs are postsynaptic at cholinergic synapses remains uncertain. Nevertheless, the modulation of transmitter release by nAChRs on nerve terminals is more evident than the relatively low abundance of neuronal nAChRs might predict, reinforcing the proposition that this is a significant function of nAChRs in the brain.

The nature of neuronal nAChRs as ligand-gated cation channels has implications for their mechanism of presynaptic modulation. In contrast to metabotropic presynaptic receptors, presynaptic nAChRs have their major effect under resting or hyperpolarized conditions, or at low levels of stimulation. This is possible because the cation flux through the receptor channel can itself trigger transmitter release (the mechanisms are considered below), but the characteristic inward rectification shown by neuronal nAChRs will decrease this cation influx if the membrane becomes depolarized, for example, in response to an action potential invading the nerve terminal.

Where exactly on axon terminals might nAChRs reside? The tetrodotoxin (TTX) insensitivity of nicotinic agonist-elicited responses has been taken to define presynaptic nAChRs in integrated preparations, such as brain slices and neuronal cultures. TTX sensitivity of nicotinic responses in the absence of nerve conduction has been interpreted in favor of a "preterminal" location—i.e., on the axon rather than the terminal bouton (Léna et al., 1993). However, synaptosome preparations, in which nerve terminals should constitute the only functional elements, have given controversial results with various degrees of TTX sensitivity of nicotine-stimulated transmitter release (see Table 8-1). Our recent data suggest that at least two populations of presynaptic nAChRs (TTX sensitive and TTX insensitive) participate in nicotine-stimulated dopamine release from rat striatal synaptosomes (Marshall et al., 1996; see Figure 8-1b). The nAChRs that elicit TTX-insensitive release may be located close to the exocytotic machinery such that their activation produces sufficient Na^+ and/or Ca^{2+} influx (see below) to trigger release directly. It is probably most useful to consider TTX-sensitive release as arising from nAChRs located at some distance from the synapse (see Figure 8-1c3) but not necessarily on the axon or preterminal regions. The TTX sensitivity of the response to activation of these nAChRs may depend on various factors (such as the resting potential of the nerve terminals), which may contribute to the reports of variable TTX sensitivity.

Although nAChRs may have been viewed as simple ligand-gated ion channels unsuited for a subtle modulatory role, it is evident that their responses may be influenced by their location with respect to the synapse and by its activity. In addition, neuronal nAChRs constitute a potentially huge family, whose members differ in their properties, including agonist sensitivity, conductance, propensity to desensitize, and relative permeability to Ca^{2+}. These factors conspire to produce a class of presynaptic receptors exquisitely tuned for enhancing transmitter release with the most effect under hyperpolarizing or resting conditions.

CURRENT KNOWLEDGE

Methods for Studying Presynaptic Receptor Functions in the Brain

There are many questions concerning presynaptic nAChRs, from their subunit composition and mechanism of action and regulation to their physiological significance. The ability to address such questions is constrained by methodological considerations. The virtues and limitations of different approaches are briefly considered. All of these methods assume that transmitter release is the principal consequence of nAChR activation.

Synaptosome Preparations. Perfused synaptosomes have been extensively used to demonstrate the nicotinic stimulation of dopamine release from striatal (Rapier et al., 1990; Grady et al., 1992; El-Bizri and Clarke, 1994) and frontal cortex synaptosomes (Whiteaker et al., 1995), noradrenaline (Clarke and Reuben, 1996), GABA (Wonnacott et al., 1988), and ACh release (Wilkie et al., 1996) from hippocampal synaptosomes (see Table 8-1). Superfusion of synaptosomes as a monolayer allows the study of nerve terminals in isolation (Raiteri et al., 1974): Nicotinic responses can be confidently ascribed to presynaptic nAChRs on terminals releasing the transmitter measured. However, the lack of physiological context (e.g., other inputs; feedback regulation) makes it difficult to ascertain what the contribution of such presynaptic receptors would be in more holistic preparations. Transmitter pools are usually radiolabeled by uptake of a tritiated precursor (e.g., Wilkie et al., 1996) or tritiated transmitter (e.g., Clarke and Reuben, 1996) during a preincubation period to enhance the sensitivity of detection; the labeled pool will diminish over time. The anatomical resolution is limited to rather gross brain areas.The temporal resolution of this technique is usually seconds to minutes, although a fast superfusion system for measuring transmitter release on a subsecond timescale has been pioneered by Turner, Dunlap, and colleagues (e.g., Turner and Dunlap, 1995). This system has facilitated the demonstration of a nicotinic enhancement of [^{3}H]glutamate release from hippocampal synaptosomes (T.J. Turner, L. Colquohoun, and K. Dunlap, personal communication).

TABLE 8.1. Selected Examples of Neurotransmitters Modulated by Presynaptic nAChRs in the Brain

Neurotransmitter	Brain Area (Species)	Technique (Preparation)	Pharmacology[a] (Subunits)[b]	Reference
ACh	Hippocampus (adult rat)	Superfusion (synaptosomes)	• DHβE, mecamylamine and pempidine sensitive 1 uM MLA insensitive ($\alpha4\beta2$?)	Wilkie et al. (1996)
	Hippocampus and frontal cortex (adult rat)	Superfusion (slices)	• DHβE and dTC sensitive α-Bgt insensitive TTX insensitive	Araujo et al. (1988)
	Cortex (adult rat)	Superfusion (synaptosomes and slices)	• Mecamylamine sensitive	Marchi & Raiteri (1996)
	Frontal cortex (adult rat)	Superfusion (synaptosomes)	• Neuronal-Bgt sensitive α-Bgt insensitive ($\alpha3\beta2$?)	Ochoa & O'Shea (1994)
	Cerebellum (rat)	Static release assay (slices)	• DHβE, dTC and neuronal-Bgt sensitive α-Bgt insensitive TTX insensitive	Lapchak et al. (1989)
Dopamine	Striatum (adult rat)	Superfusion (synaptosomes and slices)	• Partially α-CTx-MII sensitive Mecamylamine sensitive ($\alpha3\beta2$)	Kaiser et al. (1998)
		Superfusion (slices)	• Mecamylamine and DHβE sensitive dTC insensitive TTX insensitive	Sacaan et al. (1995)
		Superfusion (synaptosomes)	• Partially α-CTx-MII sensitive α-CTx-MI and α-CTx-ImI insensitive ($\alpha3\beta2$)	Kulak et al. (1997)
		Superfusion (synaptosomes)	• chlorisondamine sensitive	El-Bizri & Clarke (1994)
		Superfusion (synaptosomes)	• DHβE, mecamylamine, pempidine, and neosurugatoxin sensitive α-Bgt insensitive	Rapier et al. (1990)
	Striatum (adult mouse)	Superfusion (synaptosomes)	• Neuronal-Bgt sensitive α-Bgt insensitive ($\alpha3$?) ($\alpha3\beta2$)	Grady et al. (1992)
Noradrenaline	Hippocampus (adult rat)	Superfusion (slices)	• Neuronal-Bgt sensitive α-Bgt and α-CTx-ImI insensitive TTX sensitive	Sershen et al. (1997)

	Hippocampus and thalamus (adult rat)	Superfusion (slices)	• dTC and mecamylamine sensitive DHβE insensitive TTX sensitive	Sacaan et al. (1995, 1996)
	Hippocampus (adult rat)	Superfusion (synaptosomes)	• α-CTx-MII insentitive	Kulak et al. (1997)
		Superfusion (synaptosomes)	• Chlorisondamine sensitive TTX insensitive (α3, β4 ?)	Clarke & Reuben (1996)
L-glutamate	Hippocampus (postnatal rat)	Whole cell patch clamp (CA3 field slices, cultured neurones)	• MLA (low nM) and α-Bgt sensitive TTX insensitive (α7)	Gray et al. (1996)
	Olfactory bulb (fetal rat)	Whole cell patch clamp (primary cell culture)	• MLA (low nM) sensitive TTX insensitive (α7)	Alkondon et al. (1996)
	Co-cultures of medial habenula nucleus and IPN explants (embryonic chick)	Whole cell patch clamp (primary co-culture, antisense)	• α-Bgt sensitive TTX insensitive (Nonhomomeric, α7-containing)	McGehee et al. (1995)
	Ventral lateral geniculate nucleus (embryonic chick)	Whole cell patch clamp (slices)	• DHβE sensitive MLA insensitive	Guo et al. (1997)
GABA	Hippocampus (rat)	Whole cell patch clamp (postnatal hippocampus CA1 field slices)	• Interneurons: Choline sensitive (α7-containing?) DHβE-sensitive (α4β2 ?) Neurons: MLA and α-Bgt insensitive TTX sensitive	Alkondon et al. (1997)
		Superfusion (synaptosomes)	• DHβE, PCP and MK801 sensitive α-Bgt insensitive	Wonnacott et al. (1988)
	Thalamus (β2-knockout mouse)	Patch clamp (slices)	• (β2)	Léna & Changeux (1997)
	Medial septum (rat)	In vivo extracellular single-unit recordings	• DHβE sensitive	Yang et al. (1996)
	Lateral spiriform nucleus (embryonic chick)	Whole cell patch clamp (slices)	• Neuronal- and α-Bgt insensitive TTX sensitive	McMahon et al. (1994a)
	Ventral lateral geniculate nucleus (embryonic chick)	Whole cell patch clamp (slices)	• TTX sensitive	McMahon et al. (1994b)

continued

TABLE 8.1. Selected Examples of Neurotransmitters Modulated by Presynaptic nAChRs in the Brain

Neurotransmitter	Brain Area (Species)	Technique (Preparation)	Pharmacology[a] (Subunits)[b]	Reference
GABA	Interpeduncular nucleus (adult rat)	Whole cell patch clamp (slices)	• Hexamethonium, DHβE, and mecamylamine sensitive TTX sensitive	Léna et al. (1993)
	Globus pallidus (adult rat)	Superfusion (slices)	• Pempidine sensitive	Kayadjanian et al. (1994)
5HT	Hippocampus (rat)	Superfusion (slices)	• Mecamylamine sensitive	Lendvai et al. (1996)

[a]Abbreviations:

Bgt	bungarotoxin	dTC	d-tubocurarine	PCP	phencyclidine
CTx	conotoxin	MK-801	dixocilpine	TTX	tetrodotoxin
DHβE	dihydro-β-erythroidine	MLA	methyllycaconitine		

[b]Subunits implicated by or deduced from pharmacological profile.

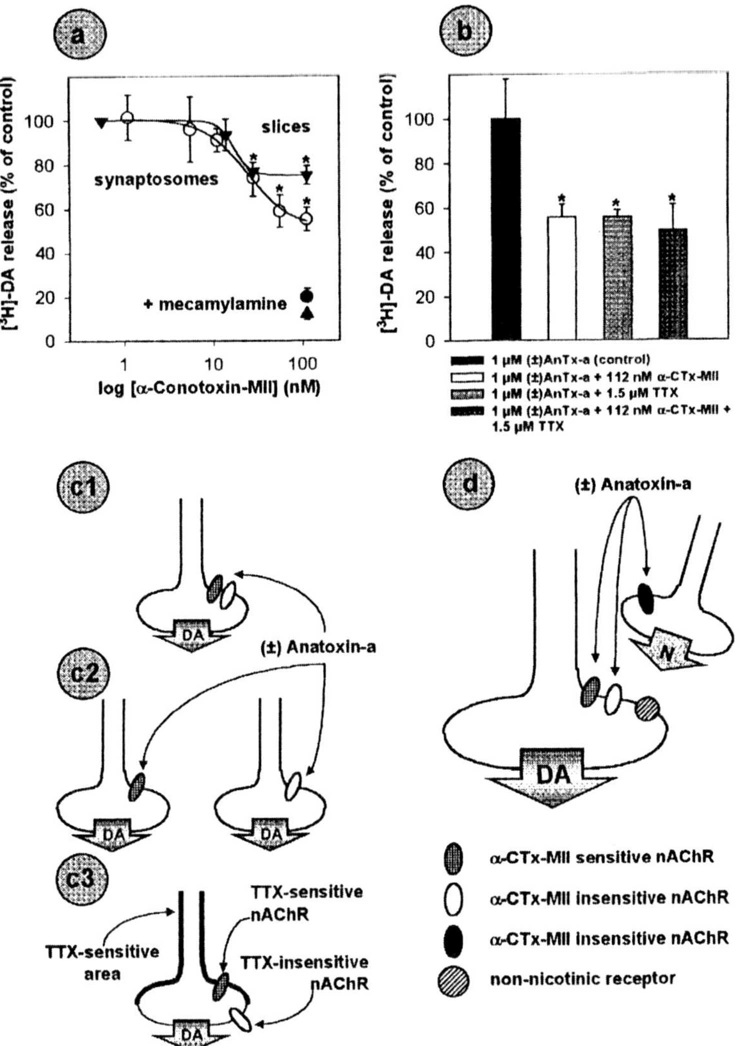

Figure 8.1. *(**a**) Dose-dependent inhibition by αCTx-MII of [³H]dopamine release evoked by 1 μM (±)-anatoxin-a from rat striatal synaptosomes (○) and slices (▼). Maximum inhibition was 44.9 ± 5.5% (n = 8) and 25.0 ± 4.1% (n = 8) respectively in the presence of 112 nM α-CTx-MII (*p < 0.05 significantly different from control without antagonist). Mecamylamine (10 μM) inhibited control responses by 80.2 ± 3.8% and 88.0 ± 2.6% in synaptosomes (●) and slices (▲) respectively. (**b**) α-CTx-MII-sensitive nAChRs on striatal synaptosomes are TTX sensitive. α-CTx-MII (112 nM) and TTX (1.5 μM) inhibited control responses by 44.4 ± 5.7% and 44.4 ± 3.0% respectively. Both applied together blocked the control response by 50.4 ± 11.4% (n = 3; *p < 0.05 significantly different from control). (**c1**) to (**d**) Schematic representation of nicotinic modulation dopamine release based on (**c**) synaptosome studies and (**d**) slice experiments. (±)-Anatoxin-a can directly stimulate dopamine release, acting at presynaptic nAChRs. α3β2-like nAChRs and α-CTx-MII-insensitive nAChRs may be present on the same terminals (**c1**) or segregated on different terminals (**c2**). Segregation of nAChR subtypes to different domains may coincide with TTX sensitivity (**c3**). In more intact preparations, represented by slices (**d**), (±)-anatoxin-a may indirectly enhance dopamine release via presynaptic nAChRs located on nondopaminergic terminals. The indirectly released neurotransmitter(s) (N) acts at its receptors on the dopaminergic terminal.*

Slices. Superfused slice, prism, or mince preparations are technically easier to study and are employed in a similar manner to synaptosome preparations, with comparable temporal, chemical, and anatomical resolutions. The nicotinic stimulation of the release of dopamine from striatum (Giorguieff-Chesselet et al., 1979; Sacaan et al., 1995), noradrenaline (Sershen et al., 1997), and ACh (Araujo et al., 1988) from hippocampus has been described with similar properties to those seen in synaptosome preparations. One difference is the generally greater sensitivity to TTX in slice preparations (Marshall et al., 1996), a reflection of the more integrated tissue. Therefore, the nicotinic stimulation of transmitter release from a slice preparation may not reflect a direct presynaptic action but could arise from an indirect effect via interneurons or terminals releasing another transmitter that in turn influences the release of the transmitter measured (see Figure 8-1d).

Microdialysis. In vivo microdialysis is a useful technique for examining drug effects in the brain of a living animal. Application of nicotine, or nicotinic antagonist, via the dialysis probe that samples neurotransmitters from a selected brain area can provide evidence for local effects in the vicinity of the terminals releasing the transmitter in question (Quirion et al., 1994; Marshall et al., 1997). The same limitations as discussed above for in vitro slice preparations apply to thė interpretation of such data with respect to presynaptic receptors. However, the technique allows comparison of local drug effects with responses to drug applications in the area containing the cell bodies of the neurons in question. For example, in the mesolimbic pathway dopamine release in the nucleus accumbens can be provoked by application of nicotine into either the ventral tegmentum (VTA; cell bodies) or accumbens. However, the response to systemic nicotine is only blocked by administration of mecamylamine into the VTA and not by mecamylamine applied to the accumbens (Nisell et al., 1994a,b). This implies that activation of dopamine neurons via the cell body, with firing of action potentials, precludes or masks any contribution from presynaptic nAChRs at the dopamine terminals. Thus this approach is important for exploring the physiological relevance of presynaptic receptors. The anatomical resolution is high, but it is difficult to estimate drug concentrations reaching the receptors. Following local application, a concentration gradient across a few millimeters from the probe will be established, while many factors determine the drug concentration reaching central synapses following systemic drug application. Thus, only qualitative comparisons with data from in vitro preparations are possible. The technique also suffers from low temporal resolution: Sample collection times of at least 15 min are required for detectable amounts of transmitter. In vivo voltametry offers high temporal as well as anatomical resolution, and was recently exploited to study the effects of systemic nicotine on dopamine release from the shell and core of the nucleus accumbens (Nisell et al., 1997).

Electrophysiology. The highest temporal and anatomical resolution can be achieved by electrophysiological recordings of receptor activity at the level of single neurons or even single channels. However, with a few exceptions (e.g., Coggan et al., 1997), direct recordings from nerve terminals are not feasible because of their tiny dimensions. Instead, information on presynaptic nAChRs has been obtained by recording from postsynaptic cells, where the postsynaptic response is used as a detection method for transmitter release. Modulation of transmitter release is seen as an increase in the frequency of spontaneous miniature postsynaptic currents. TTX can be used to demonstrate that modulation is a local effect. Coupled with imaging techniques to reveal changes in intracellular Ca^{2+} in nerve terminals presumed to be responsive to nicotinic agonists, this approach has provided some elegant analyses of putative presynaptic nAChRs (e.g., McGehee et al., 1995; Gray et al., 1996; Alkondon et al., 1996; Léna and Changeux, 1997).

Subtypes of Presynaptic nAChRs

The extensive heterogeneity of nAChR subunits presents a major challenge to define the receptor subtypes present on various nerve terminals. In the absence of definitive subtype-selective antagonists (of which there are very few) or agonists (which are even more rare), attempts to determine the subunit composition of presynaptic nAChRs have used other strategies.

Pharmacological comparisons. The overall pharmacology of a presynaptic response can be compared with the pharmacological profiles of heterologously expressed nAChRs. In most cases the latter comprise pairwise combinations of subunits, whereas evidence is accumulating that native nAChRs may be more complex (Ramirez-Latorre et al., 1996). Such a comparison led Clarke and Reuben (1996) to propose that the presynaptic nAChRs mediating noradrenaline release from hippocampal synaptosomes most closely resemble α3β4 nAChRs expressed in mammalian cells.

In situ hybridization. Examination of the nicotinic subunits expressed, according to data from in situ hybridization experiments, requires analysis of the anatomical region containing the cell bodies of the neurons in question. Every region expresses mRNAs for a plethora of subunits (Wada et al., 1989, see Wonnacott, 1997). Thus, this approach cannot provide definitive answers, but it can tentatively exclude certain subunits that are not expressed and can support subunit combinations suggested on the basis of other evidence. The initial suggestion (Goldberg et al., 1986) that a particular subunit combination might always be targeted to nerve terminals is no longer tenable, as evidence for presynaptic nAChRs differing in pharmacological properties and subunit composition emerges (see Table 8-1).

Antisense and transgenic strategies. In principle, the antisense approach to knock out candidate subunits by targeting their mRNA to prevent translation is an attractive one. In practice, this approach has enjoyed success in cell culture systems, in the laboratory of Lorna Role. The technique has provided compelling evidence for the involvement of α7 subunits in the modulation of glutamate and ACh release from medial habenular and sympathetic neurons, respectively (McGehee et al., 1995). Interestingly, in this study the antisense ablation of the α7 subunit abolished α-bungarotoxin (α-Bgt) sensitivity of presynaptic responses, but presynaptic modulation still occurred, suggestive of plasticity in the assembly of subunits to create nAChRs. Reported transgenic approaches to studying nAChRs have so far been limited to knocking out the β2 subunit (Picciotto et al., 1995). Whereas nicotinic agonists acting at presynaptic nAChRs elicit GABA release from mouse thalamus in wild-type animals, mice lacking the β2 subunit gene were deficient in this response (Léna and Changeux, 1997).

Tentative subunit assignments. In no case has it been possible to fully define the subunit composition of a presynaptic nAChR. Table 8-1 attempts to summarize the current knowledge of some of the better characterized examples. The nicotinic facilitation of glutamate release from habenular, olfactory bulb and hippocampal neurons involves a presynaptic nAChR that contains the α7 subunit, based on sensitivity to α-Bgt and methyllycaconitine (MLA) (McGehee et al., 1995; Alkondon et al., 1996; Gray et al., 1996) and antisense knockout experiments (McGehee et al., 1995). nAChRs comprised of α4 and β2 subunits have been proposed as candidates for the autoreceptors present on cholinergic terminals in the rat hippocampus (Wilkie et al., 1996), based on comparative pharmacology and desensitization profiles with rat α4β2 nAChRs expressed in *Xenopus* oocytes. This view is consistent with the loss of [^{3}H]nicotine binding sites (which can be equated with α4β2 nAChRs) when the septohippocampal projection degenerates in Alzheimer's disease. A loss of [^{3}H]nicotine binding sites was also encountered when the nigrostriatal tract was lesioned in rats (Clarke and Pert, 1985), favoring presynaptic nAChRs composed of α4 and β2 subunits on dopamine terminals in the striatum. However, sensitivity of nicotine-evoked [^{3}H]dopamine release from striatal synaptosomes to neuronal bungarotoxin has been interpreted in favor of α3-type nAChRs (Schulz and Zigmond, 1989; Grady et al., 1992).

Recently, a novel *Conus* toxin, α-conotoxin-MII, was shown to selectively inhibit nAChRs formed from α3 and β2 subunits in *Xenopus* oocytes (Cartier et al., 1996). Examination of the ability of this toxin to inhibit the presynaptic nicotinic stimulation of transmitter release has been very revealing (Kulak et al., 1997; Kaiser et al., 1998). Low nanomolar concentrations of α-conotoxin-MII dose-dependently inhibited [^{3}H]dopamine release from striatal synaptosomes, elicited by nicotine (Kulak et al., 1997) or (±)-anatoxin-a (Kaiser et al., 1998; see Fig. 8-1a). The IC_{50} value for this inhibition is close to that for blockade of α3β2 nAChRs in *Xenopus* oocytes. However, the toxin is only capable of inhibiting 40–60% of the evoked release, compared with antagonism by the nonselective nicotinic antagonist mecamylamine of greater than 80%. These data strongly imply heterogeneity of presynaptic nAChRs on striatal nerve terminals. Thus, a proportion containing α3 and β2 subunits (possibly in association with other, additional subunits) has sensitivity to nanomolar concentrations of α-conotoxin-MII, and a proportion lacking one or both of these subunits is insensitive to α-conotoxin-MII. Does a single dopamine terminal express two types of nAChRs or are they segregated to different populations of terminals (Fig. 8-1c)? We do not know at present. However, we noted that the degree of inhibition by α-conotoxin-MII resembles the partial inhibition of nicotine-evoked [^{3}H]dopamine release achieved by TTX (Marshall et al., 1996; see Fig. 8-1b). Maximally effective concentrations of α-conotoxin-MII and TTX were not additive in their degree of inhibition (Fig. 8-1b). This provokes the speculation that α-conotoxin-MII-sensitive nAChRs may occur at some distance from the release sites, requiring the intervention of voltage-operated Na^+ channels in order to elicit release, whereas the α-conotoxin-MII-insensitive nAChRs may reside close to the release sites (Fig. 8-1c3). Kulak et al. (1997) showed that nicotine-evoked noradrenaline release from hippocampal synaptosomes is not blocked at all by α-conotoxin-MII, excluding a nAChR with an α3β2 interface from these nerve terminals. This supports the conclusion by Clarke and Reuben (1996) that presynaptic nAChRs mediating noradrenaline release are pharmacologically and structurally distinct from those involved in the regulation of dopamine release in the striatum.

PHYSIOLOGICAL CONSIDERATIONS

From the brief account above and the examples cited in Table 8-1 it is clear that a variety of nAChR subtypes can elicit neurotransmitter release. However, the rules governing (1) subunit assembly from the array of subunit transcripts that may be present, (2) targeting of particular nAChR subtypes to nerve terminals, and (3) segregation of nAChR subtypes to different terminal domains (with respect to proximity to the synapse or to voltage-operated channels) are largely unknown at present. The contribution of presynaptic nAChRs to synaptic transmission will be determined by the concentration and time course of agonist interaction, in turn determined by the source of ACh (synaptic or paracrine) or alternative agonist. nAChR subtypes with different agonist sensitivities and rates of desensitization will show different profiles of activation. Responses will also be shaped by the state of the terminal with respect to depolarization from action potentials and other local inputs (e.g., via other presynaptic receptors).

In this section we first discuss how nAChRs function to elicit transmitter release at the level of the nerve terminal, then consider the local integration of neuronal elements, and finally speculate on the contribution of presynaptic nAChRs to synaptic transmission in a broader physiological context.

Mechanisms Mediating nAChR-evoked Transmitter Release

Ion Dependence and the Involvement of Voltage-operated Ion Channels. As already discussed, the TTX sensitivity of neurotransmitter release elicited by putative presynaptic nAChRs varies between preparations and labs (see Table 8-1). However, it would seem that in some cases, at least, both TTX-sensitive and -insensitive mechanisms are involved, implicating voltage-operated Na^+ channels in a portion of the response. In both cases, transmitter release is Ca^{2+} dependent. The presynaptic locus of neuronal nAChRs could take advantage of their high relative permeability to Ca^{2+} by the direct coupling of nAChR activation to exocytosis. However, neurochemical studies have generally shown that nAChR-evoked transmitter release is Na^+ dependent and significantly reduced by Cd^{2+} and/or blockers of N-type voltage-operated Ca^{2+} channels (VOCC; Soliakov and Wonnacott, 1996; Sershen et al., 1997). We suggest that in the case of TTX-insensitive nicotinic stimulation of transmitter release, Na^+ influx through the nicotinic channel produces sufficient local depolarization to activate N-type channels. These channels colocalize with syntaxin (Sheng et al., 1994) and presynaptic nAChRs may reside in close proximity at the active zone. The notion that different nAChR subtypes may be differentially distributed at nerve terminals (see the section on "Subtypes of Presynaptic nAChRs"; Figure 1c3) introduces additional functional consequences.

Electrophysiological analyses of presynaptic nAChR activity have provided evidence that Ca^{2+} flux through the receptor channel may contribute to transmitter release. Nicotine increased the frequency of postsynaptic currents in a hippocampal slice preparation in the presence of TTX and Cd^{2+} (Gray et al., 1996), and an increase in Fura-2 fluorescence in presynaptic terminals occurred under the same conditions. The nAChR examined in this study had the characteristics of an $\alpha 7$-type nAChR: The exceedingly high relative permeability to Ca^{2+} of this subtype may be sufficient for triggering release. Similar techniques have found that at 50% of chick medial habenular interpeduncular nucleus synapses in culture, nicotine-induced glutamate release was dependent upon VOCC. At the remaining 50% of synapses, Ca^{2+} entry through the nAChR channel was sufficient (Girod et al., 1997). $\alpha 7$ is also a probable contributor to these nAChRs (McGehee et al., 1995). Similarly in mouse thalamus where the $\beta 2$ subunit contributes to the presynaptic nAChR, nicotine-evoked GABA release (monitored by electrophysiological recording from the postsynaptic cell) was blocked by Cd^{2+} and Ni^{2+} at some synapses but not others (Léna and Changeux, 1997). Thus, rather like the contribution of TTX-sensitive voltage-operated Na^+ channels, VOCC may be activated to boost nicotinic responses in some instances but not others, depending on their relative localizations and regulated status.

Direct Versus Indirect Nicotinic Stimulation of Transmitter Release: Receptor Interplay. We heve briefly considered how presynaptic nAChRs interface with voltage-operated channels to elicit transmitter release. But nAChR and other ion channels will not be the only intrinsic signaling proteins present. Other presynaptic receptors may engage in crosstalk that will influence the outcome of nicotinic stimulation. This area is largely unexplored but can be illustrated by reference to the nicotinic modulation of dopamine-glutamate interactions in rat striatum: The established circuitry in the striatum (Di Chiara et al., 1994) provides a physiological context for such interactions to occur.

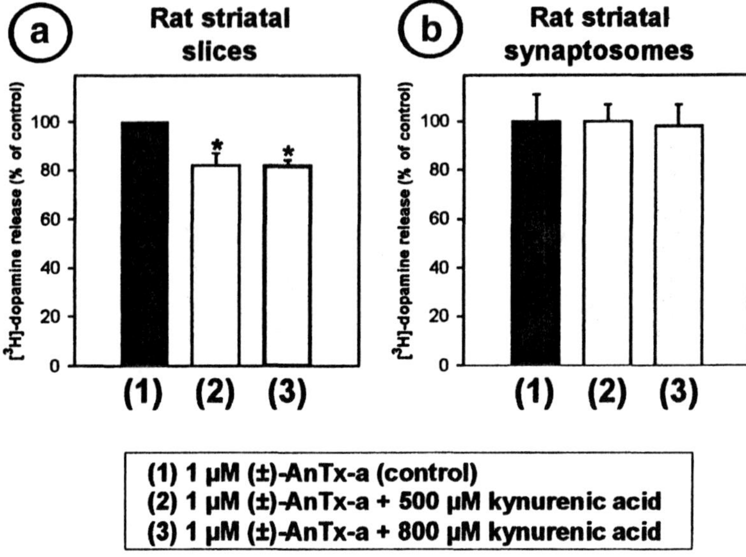

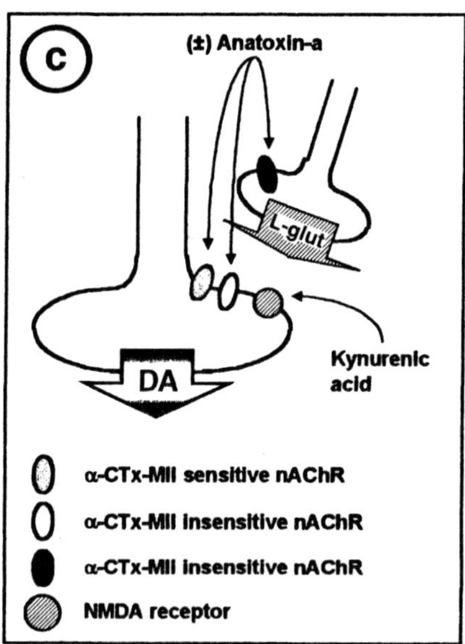

Figure 8.2 *In rat striatum (±)-anatoxin-a evokes dopamine release by acting at nAChRs present on dopaminergic and nondopaminergic terminals. In striatal slices (**a**) kynurenic acid, a broad spectrum, ionotropic glutamate receptor antagonist, significantly inhibited (*p < 0.05) control responses by 18.0% (n = 3). The same concentrations of kynurenic acid tested on striatal synaptosomes (**b**) produced no inhibition of evoked dopamine release, indicating that the antagonist has no nonspecific effects on nAChRs. These results are interpreted in a model (**c**) showing that (±)-anatoxin-a evokes dopamine release directly, by acting on dopaminergic terminal nAChRs, and indirectly releasing L-glutamate from striatal glutamatergic terminals.*

As long ago as 1979, glutamate was reported to stimulate DA release from striatal slices in vitro (Roberts and Anderson, 1979), interpreted as a reflection of presynaptic glutamate receptors on dopaminergic terminals. This notion has been reinforced in subsequent studies, with evidence for the presence of both AMPA and NMDA receptors on dopamine terminals (e.g., Wang et al., 1991; Desce et al., 1992). It was speculated that activation of AMPA receptors could relieve the Mg^{2+} block of the NMDA receptors. Similarly, activation of presynaptic nAChRs on dopamine terminals has been inferred to relieve the Mg^{2+} block of the NMDA receptors, ACh or nicotine having a synergistic effect with NMDA itself (Cheramy et al., 1996; Kaiser and Wonnacott, unpublished observations). Using in vivo microdialysis, local application of NMDA antagonists to the terminal regions of the mesolimbic (Museo and Pert, 1994) and nigrostriatal (Toth et al., 1993) pathways markedly decreased the ability of locally applied nicotine to elicit DA release. Furthermore, locally applied nicotine has been shown to increase extracellular levels of excitatory amino acids, predominantly glutamate (Toth et al., 1993), and this effect was mecamylamine sensitive. Recently, the potential importance of presynaptic nAChRs in modulating glutamate release was highlighted in electrophysiological studies (McGehee et al., 1995; Gray et al., 1996).

We have used rat striatal slice and synaptosome preparations to explore the relationship between nAChR, ionotropic glutamate receptors, and dopamine release. In slices, the broad spectrum, ionotropic glutamate receptor antagonist kynurenic acid partially blocked dopamine release evoked by the potent nicotinic agonist (±)-anatoxin-a (Fig. 8-2a), whereas kynurenic acid had no effect on (±)-anatoxin-a-evoked dopamine release from striatal synaptosomes (Fig. 8-2b). This is consistent with the presence of presynaptic nAChRs on glutamate terminals; when activated by agonists these nAChRs elicit the release of glutamate, which in turn stimulates the release of dopamine via presynaptic ionotropic glutamate receptors on dopamine terminals (Fig. 8-2c). This conclusion is consistent with recent electrophysiological recordings from striatum in situ (Garcia-Muñoz et al., 1996). A similar interaction between glutamate and dopamine is suggested in the frontal cortex (Jedema and Moghaddam, 1996).

The presence of inwardly rectifying nAChRs together with voltage-dependent NMDA receptors offers the prospect of the direct modulation of transmitter release by ACh at hyperpolarized, resting, and slightly depolarized membrane potentials, and indirect modulation via glutamate acting at NMDA receptors under depolarizing conditions.

Modulation Versus Mediation of Transmitter Release

Stimulation of presynaptic nAChRs results in increased transmitter release. Is this their role in vivo and, if so, what are the consequences for synaptic transmission? The conditions under which nAChRs can provoke transmitter release will be limited by the characteristics of the nAChR subtype, with respect to inward rectification and desensitization. Nicotinic agonists are less likely to be effective when terminals are already depolarized by invading action potentials. But nicotinic stimulation that precedes an action potential could enhance its effect by increasing the probability of transmitter release (Gray et al., 1996). It has been proposed that feedback activation of nicotinic autoreceptors on motor nerve endings prepares the terminals for sustained release by mobilizing a reserve pool of synaptic vesicles (Bowman et al., 1988). This raises the prospect that presynaptic nAChRs may adjust the sensitivity or sustainability of evoked release by triggering second messenger cascades through the transient influx of Ca^{2+} via the nicotinic channel. The measurement of transmitter release as an index of presynaptic nAChR function may not reflect the physiological impact of nAChR activation on nerve terminal function and regulation.

Activation Versus Desensitization

The properties of desensitization and inward rectification displayed by nAChRs could constitute a protective mechanism to avoid "chemical resonance," whereby enhancement of transmitter release could escalate, especially in an autoreceptor feedback loop. The balance of activation and desensitization will shape the profile of nicotinic responses. The spatial relationship between presynaptic nAChRs and sites of ACh release will be of importance in this context: Local delivery of ACh through an axo-axonic synapse will provide millimolar concentrations of agonist for a brief duration, before it is quickly hydrolyzed by acetylcholinesterase. Alternatively, paracrine delivery of ACh will yield much lower concentrations over a longer time frame. The recent observation that millimolar concentrations of choline can activate α7-type nAChRs is intriguing (Papke et al., 1996; Alkondon et al., 1997). Such high concentrations would only be approached in the cholinergic synapse, but a concentration of 10 μM, similar to that found in the extracellular fluid at rest, can partially desensitize the receptor.

In tobacco use, there will be a balance between activation and desensitization of nAChRs that may be tailored by tobacco users through manipulation of their basal nicotine levels and the frequency and magnitude of nicotine boli. Those using tobacco for stress relief will aim to achieve desensitization through constant intake (e.g., chain smoking) while those seeking stimulation will maintain lower basal levels between intermittent boli (Russell, 1990). Low nanomolar concentrations of nicotine have been shown to inactivate nAChRs present on striatal and thalamic synaptosomes without prior activation (Grady et al., 1994; Rowell and Hillebrand, 1994), so smokers may readily render some subtypes of presynaptic nAChRs nonfunctional. This cannot necessarily be equated with a general depression of CNS activity, which will depend on whether the transmitter modulated by nAChR is excitatory or inhibitory.

CONCLUSIONS: UNANSWERED QUESTIONS AND FUTURE PROSPECTS

In this chapter we have considered the evidence for presynaptic nAChRs in the mammalian CNS and discussed the possible mechanisms and physiological relevance of these receptors. It is evident from Table 8-1 that presynaptic nAChRs are widespread in the brain and are associated with many different transmitters. The pharmacological characteristics of nAChRs in different preparations vary, consistent with several different nAChR subtypes fulfilling a presynaptic role. Despite this wealth of documentation, we are still far from understanding their contribution to normal synaptic transmission and brain function. Some of the questions that have already been alluded to in the text are:

- What is the subunit composition of presynaptic nAChRs associated with particular transmitters? Incomplete pharmacological profiles and, more importantly, the dearth of definitive subtype-selective tools have prevented the identification of subtypes. Elucidation of the full complement of constituent subunits is even more challenging, and this task has not been accomplished for any particular nAChR yet. Antisense strategies (McGehee et al., 1995) or immunoisolation of nAChRs in synaptosome preparations are plausible approaches.
- Where on nerve terminals are nAChRs localized with respect to (1) release sites and (2) cholinergic innervation that provides the endogenous agonist for activation? Does a single nerve terminal have more than one subtype of nAChR and, if so, are they spatially segregated? These are crucial questions that can only be addressed by microscopic techniques such as immunogold labeling of nAChR subunits and colabeling of transmitter markers for visualization at the electron microscope level. This approach has proved very

illuminating in the case of $GABA_A$ and glutamate receptors (e.g., Nusser et al., 1996). In addition to establishing the proximity to release sites, electron microscopy can also demonstrate the colocalization of different receptor subunits. However, microscopy does not necessarily reflect functional receptors.

- What is the purpose of presynaptic nAChRs? At the neuronal level, electrophysiological studies are perhaps the best means of assessing receptor function in a physiological context (e.g., Gray et al., 1996). However, such experiments are technically demanding and measurements are indirect because terminals are too small to impale. Recourse to cultured neurons for this purpose imposes the caviat that the culture system may provide a developmental snapshot, the cultured neurons generally being of perinatal origin. Biochemical studies are also needed to investigate the downstream events that follow nAChR activation. It will be important to elucidate changes produced in the nerve terminal (e.g., phosphorylation events, vesicle recruitment), in addition to the release of transmitter. To evaluate the contribution of presynaptic nAChRs on a broader scale—i.e., in terms of brain function—in vivo antisense or transgenic knockout strategies could be deployed, but neither of these approaches is selective for presynaptic nAChRs: Subunit deletion will affect nAChRs on soma and dendrites as well as terminals. If it emerges that certain subtypes are only localized to terminals, the deletion approaches would have considerable merit, although interpretation of results can be confounded by compensatory or plastic adaptations (McGehee et al., 1995).
- Are presynaptic nAChRs valid therapeutic targets? Given our poverty of understanding of the significance of these receptors in normal brain function, this question is difficult to answer. Putative presynaptic nAChRs are lost in Alzheimer's and Parkinson's diseases and offer a rational site of drug action to boost residual activity, especially in the earlier stages of degeneration. Indeed, presynaptic nAChRs may become more effective under suboptimal conditions (Marchi et al., 1996). Moreover, the modulatory actions of presynaptic nAChRs make them attractive therapeutic targets, as modulation is preferable to direct manipulation of on/off signals.

Clearly we have a long way to go before fully comprehending the significance of the presynaptic location of diverse subtypes of this family of fast-acting ligand-gated cation channels. On the journey it is likely that several more preconceptions of synaptic functioning will be overturned.

ACKNOWLEDGMENTS

Work in the authors' laboratory is supported financially by the Biotechnology and Biological Sciences Research Council (BBSRC)Medical Research Council (MRC), Wellcome Trust, and British American Tobacco (BAT) Co. Ltd.

REFERENCES

Alkondon M, Rocha ES, Maelicke A, Alburquerque EX (1996): Diversity of nicotinic acetylcholine receptors in rat brain. V. α-Bungarotoxin-sensitive nicotinic receptors in olfactory bulb neurons and presynaptic modulation of glutamate release. J. Pharmacol. Exp. Ther. 278: 1460–1471.

Alkondon M, Pereira EFR, Barbosa CTF, Alburquerque EX (1997): Neuronal nicotinic acetylcholine activation modulates GABA release from CA1 neurons of rat hippocampal slices. J. Pharmacol. Exp. Ther. 283: 1396–1411.

Araujo DM, Lapchak PA, Robitaille Y, Gauthier S, Quirion R (1988): Differential alteration of various cholinergic markers in cortical and subcortical regions of human brain in Alzheimer's disease. J. Neurochem. 50: 1914–1923.

Bowman WC, Marshall IG, Gibb AJ, Harborne AJ (1988): Feedback-control of transmitter release at the neuromuscular-junction. Trends Pharmacol. Sci. 9: 16–20.

Cartier GE, Yoshikami D, Gray WR, Luo S, Olivera BM, McIntosh JM (1996): A new α-conotoxin which targets α3β2 nicotinic acetylcholine receptors. J. Biol. Chem. 271: 7522–7528.

Chéramy A, Godeheu G, L'Hirondel M, Glowinsky J (1996): Cooperative contributions of cholinergic and NMDA receptors in the presynaptic control of dopamine release from synaptosomes of the rat striatum. J. Pharmacol. Exp. Ther. 276: 616–625.

Clarke PBS, Reuben M (1996): Release of [^{3}H]noradrenaline from rat hippocampal synaptosomes by nicotine: mediation by different receptor subtypes from striatal [^{3}H]dopamine release. Br. J. Pharmacol. 117: 595–606.

Coggan JS, Paysan J, Conroy WG, Berg DK (1997): Direct recording of nicotinic responses in presynaptic nerve terminals. J. Neurosci. 17: 5798–5806.

Desce JM, Godeheu G, Galli T, Artaud F, Chéramy A, Glowinsky J (1992): L-glutamate-evoked release of dopamine from synaptosomes of the rat striatum: Involvement of AMPA and NMDA receptors. Neuroscience 47: 333–339.

Di Chiara G, Morelli M, Consolo S (1994): Modulatory functions of neurotransmitters in the striatum: ACh/dopamine/NMDA interactions. Trends Neurosci. 17: 228–233.

El-Bizri H, Clarke PBS (1994): Blockade of nicotinic receptor-mediated release of dopamine from striatal synaptosomes by chlorisondamine and other nicotinic antagonists administered in vitro. Br. J. Pharmacol. 111: 406–413.

Garcia-Munoz M, Patino P, Young SJ, Groves PM (1996): Effects of nicotine on dopaminergic nigrostriatal axons require stimulation of presynaptic glutamatergic receptors. J. Pharmacol. Exp. Ther. 277: 1685–1693.

Giorguieff-Chesselet MF, Kernel ML, Wandscheer D, Glowinsky J (1979): Regulation of dopamine release by presynaptic receptors in rat striatal slices: effect of nicotine in a low concentration. Life Sci. 25: 1257–1262.

Girod R, McGehee DS, Role L (1997): Routes of nicotinic receptor induced increase of presynaptic calcium and effects of chronic treatment with nicotine. Soc. Neurosci. Abstr. 265: 5.

Grady SR, Marks M, Wonnacott S, Collins AC (1992): Characterisation of nicotinic receptor mediated [^{3}H]dopamine release from synaptosomes prepared from mouse striatum. J. Neurochem. 59: 848–856.

Grady SR, Marks MJ, Collins AC (1994): Desensitization of nicotine-stimulated [^{3}H]dopamine release from mouse striatal synaptosomes. J. Neurochem. 62: 1390–1398.

Gray R, Rajan AR, Radcliffe KA, Yakehiro M, Dani J (1996): Hippocampal synaptic transmission enhanced by low concentrations of nicotine. Nature 383: 713–716.

Guo JZ, Tredway TL, Chiappinelli VA (1998): Glutamate and GABA release are enhanced by different subtypes of presynaptic nicotinic receptors in the lateral geniculate nucleus. J. Neurosci. 18: 1963–1969.

Jedema HP, Moghaddam B (1996): Characterization of excitatory amino-acid modulation of dopamine release in the prefrontal cortex of conscious rats. J. Neurochem. 66: 1448–1453.

Kaiser SA, Soliakov L, Harvey SC, Luetje CW, Wonnacott S (1998): Differential inhibition by α-conotoxin-MII of the nicotinic stimulation of [^{3}H]dopamine release from rat striatal synaptosomes and slices. J. Neurochem. 70: 1069–1076.

Kayadjanian N, Retaux S, Menetrey A, Besson MJ (1994): Stimulation by nicotine of the spontaneous release of [^{3}H]gamma-aminobutyric-acid in the substantia-nigra and in the globus-pallidus of the rat. Brain Res. 649: 129–135.

Koelle GB (1961): A proposed dual neurohumoral role of acetylcholine: its functions at the pre- and postsynaptic sites. Nature 190: 208–211.

Kulak JM, Nguyen TA, Olivera BM, McIntosh JM (1997): α-Conotoxin-MII blocks nicotine-stimulated dopamine release in rat striatal synaptosomes. J. Neurosci. 17: 5263–5270.

Harvey SC, McIntosh JM, Cartier GE, Maddox FN, Luetje CW (1997): Determinants of specificity for α-Conotoxin MII on α3β2 neuronal nicotinic receptors. Mol. Pharmacol. 51: 336–342.

Lapchak PA, Araujo DM, Quirion R, Collier B (1989): Presynaptic cholinergic mechanisms in the rat cerebellum: evidence for nicotinic, but not muscarinic autoreceptors. J. Neurochem. 53: 1843–1851.

Léna C, Changeux J-P (1997): Role of Ca^{2+} ions in nicotinic facilitation of GABA release in mouse thalamus. J. Neurosci. 17: 576–585.

Léna C, Changeux J-P, Mulle C (1993): Evidence for "preterminal" nicotinic receptors on GABAergic axons in the rat interpeduncular nucleus. J. Neurosci. 13: 2680–2688.

Lendvai B, Sershen H, Lajtha A, Santha E, Baranyi M, Vizi ES (1996): Differential mechanisms involved in the effect of nicotinic agonists DMPP and lobeline to release [^{3}H]5-HT from rat hippocampal slices. Neuropharmacology 35: 1769–1777.

Lipiello PM, Bencherif M, Caldwell WS, Arrington SR, Fowler KW, Lovette ME, Reeves LK (1996): Metanicotine: a nicotinic agonist with central nervous system selectivity—In vitro and in vivo characterisation. Drug Develop. Res. 38: 169–176.

Marchi M, Raiteri M (1996): Nicotinic autoreceptors mediating enhancement of acetylcholine release become operative in conditions of "impaired" cholinergic presynaptic function. J. Neurochem. 67: 1979–1984.

Marshall DL, Soliakov L, Redfern PH, Wonnacott S (1996): Tetrodotoxin sensitivity of nicotine-evoked dopamine release from rat striatum. Neuropharmacology 35: 1531–1536.

Marshall DL, Redfern P, Wonnacott S (1997): Presynaptic nicotinic modulation of dopamine release in the three ascending pathways studied by in vivo microdialysis: comparison of naive and chronic nicotine-treated rats. J. Neurochem. 68: 1511–1519.

McGehee DS, Role LW (1996): Presynaptic ionotropic receptors. Curr. Opin. Neurobiol. 6: 342–349.

McGehee DS, Heath MJS, Gelber S, Devay P, Role L (1995): Nicotine enhancement of fast excitatory synaptic transmission in CNS by presynaptic receptors. Science 269: 1692–1696.

McMahon LL, Yoon KW, Chiappinelli VA (1994a): Nicotinic receptor activation facilitates GABAergic neurotransmission in the avian lateral spiriform nucleus. Neuroscience 59: 689–698.

McMahon LL, Yoon KW, Chiappinelli VA (1994b): Electrophysiological evidence for presynaptic nicotinic receptors in the avian ventral lateral geniculate nucleus. J. Neurophysiol. 71: 826–829.

Nisell M, Nomikos GG, Svensson TH (1994a): Systemic nicotine induced dopamine release in the rat nucleus accumbens is regulated by nicotinic receptors in the ventral tegmental area. Synapse 16: 36–44.

Nisell M, Nomikos GG, Svensson TH (1994b): Infusion of nicotine in the ventral tegmental area or the nucleus accumbens of the rat differentially affects accumbal dopamine release. Pharmacol. Toxicol. 75: 348–352.

Nisell M, Marcus M, Nomikos GG, Svensson TH (1997): Differential effects of acute and chronic nicotine on dopamine output in the core and shell of the rat nucleus accumbens. J. Neural Transmission 104: 1–10.

Nusser Z, Sieghart W, Benke D, Fritschy JM, Somogyi P (1996): Differential synaptic localization of 2 major gamma-aminobutyric-acid type-A receptor-alpha subunits on hippocampal pyramidal cells. Proc. Natl. Acad. Sci. USA 93: 11,939–11,944.

Ochoa ELM, O'Shea SM (1994): Concomitant protein phosphorylation and endogenous acetylcholine release induced by nicotine: dependency on neuronal nicotinic receptors and desensitization. Cell. Mol. Neurobiol. 14: 315–340.

Papke RL, Bencherif M, Lipiello P (1996): An evaluation of neuronal nicotinic acetylcholine receptor activation by quaternary nitrogen compounds indicates that choline is selective for the α7 subtype. Neurosci. Lett. 213: 201–204.

Picciotto MR, Zoll M, Léna C, Bessis A, Llallemand Y, LeNovere N, Vincent P, Pich EM, Brulet P,

Changeux J-P (1995): Abnormal avoidance-learning in mice lacking functional high-affinity nicotine receptor in the brain. Nature 374: 65–67.

Pidoplichko VI, DeBiasi M, Williams JT, Dani J (1997): Nicotine activates and desensitizes midbrain dopamine neurons. Nature 390: 401–404.

Quirion R, Richard J, Wilson A (1994): Muscarinic and nicotinic modulation of cortical acetylcholine release monitored by in vivo microdialysis in freely moving rats. Synapse 17: 92–100.

Raiteri M, Angelini F, Levi G (1974): A simple apparatus for studying the release of neurotransmitters from synaptosomes. Eur. J. Pharmacol. 25: 411–414.

Ramirez-Latorre J, Yu CR, Qu F, Perin F, Karlin A, Role L (1996): Functional contributions of α5 subunit to neuronal acetylcholine receptor channels. Nature 380: 347–351.

Rapier C, Lunt GG, Wonnacott S (1990): Nicotinic modulation of [^{3}H]dopamine from striatal synaptosomes: pharmacological characterisation. J. Neurochem. 54: 937–945.

Roberts PJ, Anderson SD (1979): Stimulatory effect of L-glutamate and related amino acids on [^{3}H]dopamine release from rat striatum: an in vitro model for glutamate actions. J. Neurochem. 32: 1539–1545.

Role LW, Berg DK (1996): Nicotinic receptors in the development and modulation of CNS synapses. Neuron 16: 1077–1085.

Rowell PP, Hillebrand JA (1994): Characterization of nicotine-induced desensitization of evoked dopamine release from rat striatal synaptosomes. J. Neurochem. 63: 561–569.

Rusell MAH (1990): Nicotine intake and its control over smoking. In: Wonnacott S, Rusell MAH, Stolerman IP, editors. Nicotine psychopharmacology—molecular, cellular and behavioural aspects. Oxford: Oxford University Press, pp. 374–418.

Sacaan AI, Dunlop JL, Lloyd GK (1995): Pharmacological characterization of neuronal acetylcholine gated ion channel receptor-mediated hippocampal norepinephrine and striatal dopamine release from rat brain slices. J. Pharmacol. Exp. Ther. 274: 224–230.

Sacaan AI, Menzaghi F, Dunlop JL, Correa LD, Whelan KT, Lloyd GK (1996): Epibatidine: a nicotinic acetylcholine receptor agonist releases monoaminergic neurotransmitters: in vitro and in vivo evidence in rats. J. Pharmacol. Exp. Ther. 276: 509–515.

Schulz DW, Zigmond RE (1989): Neuronal bungarotoxin blocks the nicotinic stimulation of endogenous dopamine release from rat striatum. Neurosci. Lett. 98: 310–316.

Sershen H, Balla A, Lajtha A, Vizi ES (1997): Characterization of nicotinic receptors involved in the release of noradrenaline from the hippocampus. Neuroscience 77: 121–130.

Sheng Z, Rettig J, Takahashi M, Catterall WA (1994): Identification of a syntaxin-binding site on N-type calcium channels. Neuron 13: 1303–1313.

Soliakov L, Wonnacott S (1996): Voltage sensitive Ca^{2+} channels involved in nicotinic receptor-mediated [^{3}H]dopamine release from rat striatal synaptosomes. J. Neurochem. 67: 163–170.

Toth E, Vizi ES, Lajtha A (1993): Effect of nicotine on levels of extracellular amino acids in regions of the rat-brain in-vivo. Neuropharmacology 34: 1535–1541.

Turner TJ, Dunlap K (1995): Prolonged time course of glutamate release from nerve terminals: relationship between stimulus duration and the secretory event. J. Neurochem. 64: 2022–2033.

Wada E, Wada K, Boulter J, Deneris E, Heinemann S, Patrick J, Swanson LW (1989): Distribution of alpha2, alpha3, alpha4 and beta2 neuronal nicotinic subunit mRNA in the central nervous system: a hybridization histochemical study in the rat. J. Comp. Neurol. 284: 314–335.

Wang JK (1991): Presynaptic glutamate receptors modulate dopamine release from striatal synaptosomes. J. Neurochem. 57: 819–822.

Whiteaker P, Garcha HS, Wonnacott S, Stolerman IP (1995): Locomotor activation and dopamine release produced by nicotine and isoarecolone in rats. Br. J. Pharmacol. 116: 2097–2105.

Wilkie GI, Hutson P, Sullivan JP, Wonnacott S (1996): Pharmacological characterization of a nicotinic autoreceptor in rat hippocampal synaptosomes. Neurochem. Res. 21: 1141–1148.

Wonnacott S (1997): Presynaptic nicotinic ACh receptors. Trends Neurosci. 20: 92–98.

Wonnacott S, Fryer L, Irons H, Lunt GG (1988): Presynaptic sites of nicotine's action in the brain. In: Rand MJ, Thurau K, editors. The pharmacology of nicotine. Oxford–Washington, DC: IRL Press, pp. 342–343.

Yang XH, Criswell HE, Breese GR (1996): Nicotine-induced inhibition in medial septum involves activation of presynaptic nicotinic cholinergic receptors on gamma-aminobutyric acid-containing neurons. J. Pharmacol. Exp. Ther. 276: 482–489.

9

Functional Diversity of Nicotinic Acetylcholine Receptors in the Mammalian Central Nervous System: Physiological Relevance

E.F.R. Pereira, Ph.D., M. Alkondon, Ph.D., and Edson X. Albuquerque, M.D., Ph.D.

Department of Pharmacology and Experimental Therapeutics
University of Maryland School of Medicine
Baltimore, Maryland

A. Maelicke, Ph.D.

Institute of Physiological Chemistry
Johannes-Gutenberg University School of Medicine
Mainz, Germany

Edson X. Albuquerque, M.D., Ph.D.

Department of Basic and Clinical Pharmacology
Institute of Biomedical Sciences
Center of Health Sciences
Federal University of Rio de Janeiro
Rio de Janeiro, Brazil

Neuronal Nicotinic Receptors: Pharmacology and Therapeutic Opportunities, Edited by S. P. Arneric and J. D. Brioni
ISBN 0-471-24743-x, pages 161–186. Copyright © 1998 by Wiley-Liss, Inc.

Although nicotine has been recognized for more than a century as a psychopharmacological drug, nicotinic acetylcholine receptors (nAChRs) in the central nervous system (CNS) have only recently become accessible to in-depth biochemical and functional analysis. On the basis of initial binding studies with [^{3}H]nicotine and [^{125}I]α-BGT, two populations of neuronal nAChRs were postulated: one that could bind [^{3}H]nicotine with high affinity, and the other that could bind α-BGT with high affinity (e.g., Pauly et al., 1989). However, molecular genetic studies of the past decade revealed that there may be a number of different subtypes of nAChRs in the CNS.

At present, eight neuronal nAChR agonist binding α subunits (α2–α9) and three neuronal nAChR "structural" β subunits (β2–β4) have been identified in various species (reviewed in Galzi and Changeux, 1995; Lindstrom, 1995). The neuronal nAChR α subunits have two adjacent cysteines (cys) that are homologous to cys192 and cys193 present on the muscle nAChR α subunit, whereas the neuronal nAChR β subunits lack the adjacent residues of cys. Although the β subunits are referred to as structural subunits, several lines of evidence indicate that both α and β subunits contribute to the pharmacological profile of the neuronal nAChRs (Luetje and Patrick, 1991). Based on ectopic expression studies in *Xenopus* oocytes and other expression systems, functional nAChRs can be homopentamers composed of α7, α8, or α9 subunits, and heteropentamers formed by combinations of α and β subunits (reviewed in Lindstrom, 1995; McGehee and Role, 1995). Although limited by the fact that subtype-specific agonists/antagonist have become available for only some nAChRs, electrophysiological studies confirmed the existence of more than just two populations of nAChRs in neurons of the mammalian brain.

This review is focused on (1) data that are presently available on the pharmacological and kinetic properties and on the possible physiological functions of nAChR subtypes that are naturally expressed by neurons of the mammalian CNS; (2) recent findings about regulation of nAChR function by a variety of ligands that act via allosteric sites, with particular emphasis on excitatory allosteric effectors referred to as noncompetitive agonists; and (3) the agonistic effect of choline on some nAChR subtypes, which suggests an important modulatory role in the CNS of this metabolic precursor/product of ACh. An ever-increasing body of evidence indicates that neuronal nAChRs are not just receptors for ACh; rather, neuronal nAChRs are regulated signal transducers capable of sensing and regulating the activity state of other neurotransmitter systems.

NATURALLY OCCURRING SUBTYPES OF NICOTINE RECEPTORS IN NEURONS OF THE CENTRAL NERVOUS SYSTEM

Most of the information regarding the subunits that can form functional nAChR subtypes has been derived from ectopic expression studies using frog oocytes and mammalian tumor cell lines (for a review, see Gotti et al., 1997). Although these studies have revealed the pharmacological and kinetic properties of homopentameric nAChRs formed by α7, α8, or α9 subunits, and heteromeric nAChRs formed by combinations of α and β subunits (Luetje and Patrick, 1991; Whiting et al., 1991; Anand et al., 1993; Bertrand et al., 1993; Séguéla, 1993; Stetzer et al., 1996; Ramirez-Latorre et al., 1996; Gotti et al., 1997), they could not prove whether the same nAChR subtypes are naturally expressed in vivo. In particular, it remained open whether the specific membrane environment, post-translational modifications, of associated proteins in CNS neurons could modify the pharmacological and kinetic properties of nAChR subtypes. As summarized below, there indeed exist considerable differences in properties of nAChRs expressed in a native cellular environment as compared to ectopic expression systems.

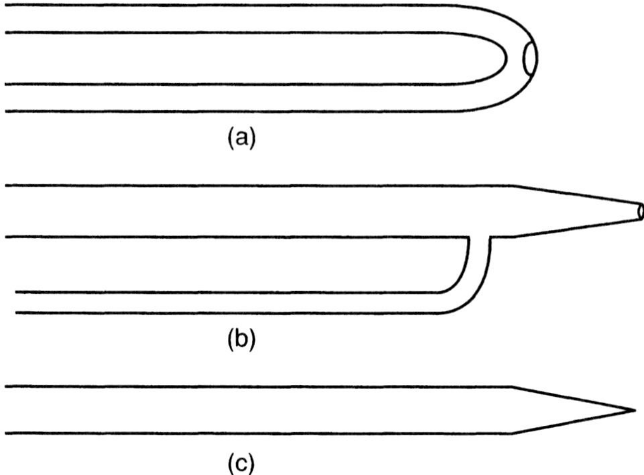

Figure 9.1. Schematic diagram showing the shapes of different drug delivery systems. **(A)** Regular U-tube, which is made from glass tubing (i.d. = 400 μm) and has a pore with a 200–300 μm diameter. Agonist is applied to the neurons by closing the outlet for a specified period of time. **(B)** Modified U-tube fabricated from a glass tubing (i.d. 800 μm) pulled to achieve a tip diameter of 70–90 μm. Another bent glass tubing (i.d. = 400 μm) is connected to the large tube about 6–8 mm from the tip. Agonist is applied to the neurons by closing the outlet for a specified period of time. **(C)** A patch pipette pulled to achieve a tip diameter of 0.4–0.7 μm. Agonist is applied to the neurons by pressure ejection.

Experimental Requirements for Monitoring Functional Nicotinic Receptors

The rather fast kinetics of activation and inactivation of some subtypes of neuronal nAChRs has made it experimentally quite challenging to monitor the activity of neuronal nAChRs in their natural environment. There should be no leakage of agonist onto the neurons and agonists should be quickly removed from the surroundings of the neurons so that receptor desensitization by continuous exposure to agonists does not take place. This challenge has been further aggravated by the fact that the surface of neurons in their natural environment could not be easily accessed. Thus, the use of cultured neurons and the development of the U-tube (Fig. 9-1) (Alkondon and Albuquerque, 1991) as a system that guarantees agonists to be rapidly applied to and removed from the vicinity of neurons have been pivotal for the identification and characterization of nicotinic responses recorded from neurons taken from various areas of the brain (for a review, see Albuquerque et al., 1997). In fact, on the basis of the pharmacological and kinetic properties of nicotinic responses recorded from CNS neurons, we have been able to establish a set of criteria, or fingerprints, that help in identifying the nAChR subtype that subserves a given nicotinic response.

A posteriori, with the establishment of a technique in which individual neurons with dendritic trees ranging from 200 to 400 μm in length can be acutely dissociated from different brain areas without the aid of enzymes, it became possible to study the nicotinic responses from neurons that developed in their natural environment in vivo (Barbosa et al., 1996; Albuquerque et al., 1997). Finally, with the development of new systems of drug application, it was possible to extend our study to the characterization of nicotinic responses of neurons in slices and to study the distribution of functional nAChRs on the surface of CNS neurons (Alkondon et al., 1996a; Albuquerque et al., 1997a). A modified glass U-tube (see Fig. 9-1), which bears a pore of 70–90 μm diameter and consists of a straight tube of 800-μm internal diameter, assembled together with the aid of glue to a 400-μm internal diameter tube, has been used to deliver agonist to neurons in slices; the agonist-containing solutions are fed by gravity. A gentle suction is also applied to prevent leak of agonist through the pore. On the other hand, a straight tube bearing a pore of 0.4–0.7-μm diameter (see Fig. 9-1), through which agonist-containing solutions can be delivered to the neurons by 15–50-ms pulses of 10–20 pSi, has been essential for focal application of agonists to well-defined areas on the neuronal surface. In both techniques, the agonists are removed from the vicinity of the neuronal regions by the continuous background perfusion of the neurons with agonist-free solutions.

Studies of the nicotinic responses of neurons in slices and of the distribution of functional nAChRs on the neuronal surface depended not only on the improvement of the agonist-delivery systems but also on the improvement of the optical systems to visualize the neurons and on the use of a computerized system of micromanipulators that could allow for the recording pipettes and the drug-delivery systems to be positioned on well-defined areas on the surface of neurons. Infrared-assisted videomicroscopy was essential for the visualization of individual neurons in brain slices and for in-depth morphological analysis of the surface of neurons in culture, and with the coupling of infrared-assisted videomicroscopy to fluorescence microscopy it became possible to visualize and study spine-rich dendritic regions (Alkondon et al., 1996a, 1997a).

Pharmacological and Kinetic Properties of Functional Nicotinic Receptors Present in Hippocampal Neurons

Understanding synaptic transmission and synaptic integration in the hippocampus is one of the most intriguing challenges in neurobiology, because of the involvement of this brain area in learning and memory. The hippocampus, one of the major components of the limbic system, contains two interlocked C-shaped layers of cortex: the pyramidal cells of the Ammon's horn (CA1–CA3 fields) and the granular cells of the dentate gyrus. Also, three major pathways can be identified in the hippocampus: (1) the perforant pathway, which runs from the entorhinal cortex to the granule cells in the hilus of the dentate gyrus; (2) the mossy fiber pathway, which is made of axons of the granular cells of the dentate gyrus and runs to the pyramidal neurons in the CA3 region of the hippocampus; and (3) the Schaffer collaterals, which run from pyramidal neurons in the CA3 field to pyramidal cells in the CA1 region. The major cholinergic input to the hippocampus arises from the septum, and it is by the end of the third postnatal week of development that this input is fully developed in the rat. Because the majority of the synapses in the hippocampus (and throughout the CNS) terminate on dendrites, and more specifically onto dendritic spines, it has long been assumed that dendrites can integrate a multitude of synaptic inputs to provide a neuronal output. As of today, however, the dendritic properties that provide this integrative function and the nature of the integration are still poorly understood. Therefore, our studies have been focused on characterizing the nicotinic responses of hippocampal neurons and on determining the physiological relevance of neuronal nAChRs to the ongoing synaptic activity of hippocampal neurons.

Characterization of Nicotinic Whole Cell Currents in Cultured Hippocampal Neurons: Pharmacological Evidence of Distinct Subtypes. Functional nAChRs present in cultured hippocampal neurons from fetal rats have been characterized by means of the whole cell patch-clamp technique, using a number of structurally divergent agonist and highly selective neurotoxins.

In the presence of the Na^+ channel blocker tetrodotoxin (TTX), hippocampal neurons in

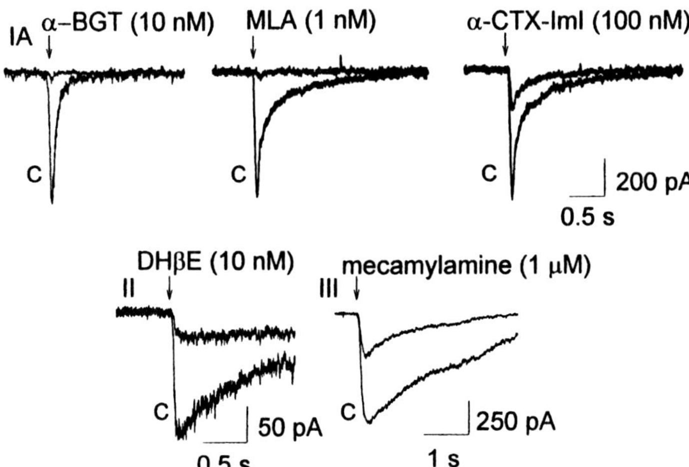

Figure 9.2. Family of nicotinic whole cell currents elicited in cultured hippocampal neurons and their sensitivity to selective pharmacological agents. Whole cell currents evoked by ACh (1–3 mM) at −50 mV under control conditions (C) or after exposing the neurons to a given antagonist for 10–15 min.

culture respond to nicotinic agonists with at least one of three types of nicotinic whole cell currents, namely type IA, type II, and type III, which are distinguished from each other on the basis of their kinetic and pharmacological properties (Fig. 9-2) (Alkondon and Albuquerque, 1993). Type IA currents, by far the predominant response of hippocampal neurons to nicotinic agonists, are fast-desensitizing currents that have an exquisitely high sensitivity to blockade by the snake toxin α-bungarotoxin (α-BGT), the plant alkaloid methyllycaconitine (MLA), and the *Conus* peptide α-conotoxin-ImI (α-CTx-ImI, see Fig.9-2) (Alkondon et al., 1992; Alkondon and Albuquerque, 1993; Pereira et al., 1996).

In contrast to type IA currents, type II and III currents, which desensitize very slowly, can be recorded from a small population of hippocampal neurons in culture (Alkondon and Albuquerque, 1993, 1995). Approximately 10% of cultured hippocampal neurons respond to nicotinic agonists with type II currents, whereas no more than 2% of the neurons in culture respond to the agonists with type III currents. Type II and III currents are differentiated from each other on the basis of their sensitivity to nicotinic antagonists. Activation of type II currents is inhibited by dihydro-β-erythroidine (DHβE, 10–100 nM), and that of type III currents is inhibited by mecamylamine (1 μM; see Fig. 9-2).

A comparison of the kinetic and pharmacological properties of the nicotinic currents evoked in hippocampal neurons to those of currents elicited in oocytes ectopically expressing distinct subtypes of functional nAChRs led to the suggestions that an α7-bearing nAChR subserves type IA currents, an α4β2 nAChR subserves type II currents, and an α3β4 nAChR subserves type III currents. These suggestions were supported not only by the finding of mRNAs coding for α7, α4, and β2 subunits in hippocampal neurons but also by the proportion of hippocampal neurons that bind with high affinity either [^{125}I]α-BGT (a probe to label α-BGT-sensitive neuronal nAChRs) or [^{3}H]nicotine (a probe to label high affinity, presumably α4β2, neuronal nAChRs; Alkondon et al., 1994; Barrantes et al., 1995). Evidence that a single neuron can express more than one subtype of functional nAChR was also obtained when a hybrid current that had two components, a fast and a slowly decaying component, was recorded from approximately 10% of the neurons in culture. The fast-decaying component, being sensitive to blockade by α-BGT and MLA, is likely to be subserved by the α7-bearing nAChR, whereas the slowly decaying component, being sensitive to blockade by DHβE, is likely to be subserved by an α4β2 nAChR (Alkondon and Albuquerque, 1993; Alkondon et al., 1994).

The nAChRs subserving each of the whole cell currents recorded from hippocampal neurons can also be identified on the basis of the apparent potency and efficacy of various agonists in activating the different types of currents (Alkondon and Albuquerque, 1995). The rank order of potency of various agonists, as determined either from their EC_{50} values or from their threshold concentration to evoke the currents, is (−)epibatidine > (+)-anatoxin-a (AnTX) > (+)epibatidine > (−)nicotine > dimethyl phenyl piperazinium (DMPP) > cytisine > ACh > choline for type IA currents, and (+)epibatidine > (−)epibatidine > cytisine > AnTX > (−)nicotine > ACh > DMPP for type II currents. Although these agonists produce nearly identical maximal type IA responses, the size of the maximal type II responses varies with the agonist. The finding that cytisine is one of the least efficacious agonists in inducing α4β2 nAChR-mediated type II currents is in agreement with the report that cytisine, in addition to acting as an agonist at α4β2 nAChRs, can act as an open-channel blocker of these receptor channels (Papke and Heinemann, 1994).

Choline as a Specific Agonist of Naturally Occurring α7 Nicotinic Receptors. At the neuromuscular junction, the actions of ACh are terminated by the enzyme acetylcholinesterase (AChE), which hydrolyzes ACh to choline and acetate. Both products are unable to activate postsynaptic muscle nAChRs to any significant extent, and choline appears to increase the onset rate of desensitization of muscle nAChRs (delCastillo and Katz, 1957). In the mammalian CNS, it is not clear whether AChE is linked to brain nicotinic cholinergic transmission in the same fashion as it is to the neuromuscular transmission. In view of earlier binding studies that identified CNS nAChRs as possible targets for choline (Costa and Murphy, 1984; Morley et al., 1977; Coutcher et al. 1992), we hypothesized that choline could act as an agonist at some neuronal nAChRs. This being the case, the evolutionarily oldest nAChRs—i.e., homomers made up of the α7, α8, or α9 subunits (LeNovere and Changeux, 1995)—could be putative nicotinic targets for choline in the mammalian CNS. In fact, recent studies have shown that choline can activate the homomeric α7 nAChR ectopically expressed in *Xenopus* oocytes (Mandelyzs et al., 1995; Papke et al., 1996), and investigating the ability of choline to interact with the various native nAChR subtypes present on neurons of the rat hippocampus, olfactory bulb, and thalamus, as well as on PC12 cells, we have demonstrated that choline acts as a selective agonist at native α7-containing nAChRs (Alkondon et al., 1997b).

In hippocampal, thalamic, and olfactory bulb neurons, either ACh or choline can activate the fast-desensitizing, α-bungarotoxin-sensitive type IA currents, and in olfactory bulb neurons either compound can increase glutamate release by activating α7-containing nAChRs present on glutamatergic terminals (Fig. 9-3; Alkondon et al., 1997b). The magnitude of the maximal response of ACh was the same as that of the response evoked by ACh, whereas the EC_{50} values for choline and ACh in activating native α7-containing nAChRs were 1.6 mM and 130 μM, respectively. A 10-min exposure of the hippocampal neurons to micromolar concentrations of choline inhibited the activation of α7-containing nAChRs (Alkondon et al., 1997b). The IC_{50} for choline was approximately 37 μM (Fig. 9-4). The inhibitory effect of choline on the α7-containing nAChRs, being noncompetitive with respect to the agonist, is most likely due to its ability to desensitize these receptors. The desensitizing action of nicotinic agonists has been reported for agonists other than choline. Choline is also capable of regulating the number of functional α7-containing nAChRs in hippocampal neurons, given that prolonged (10 days) exposure of these neurons to a low concentration of choline (~ 32 μM) decreases their nicotinic sensitivity (Alkondon et al., 1997).

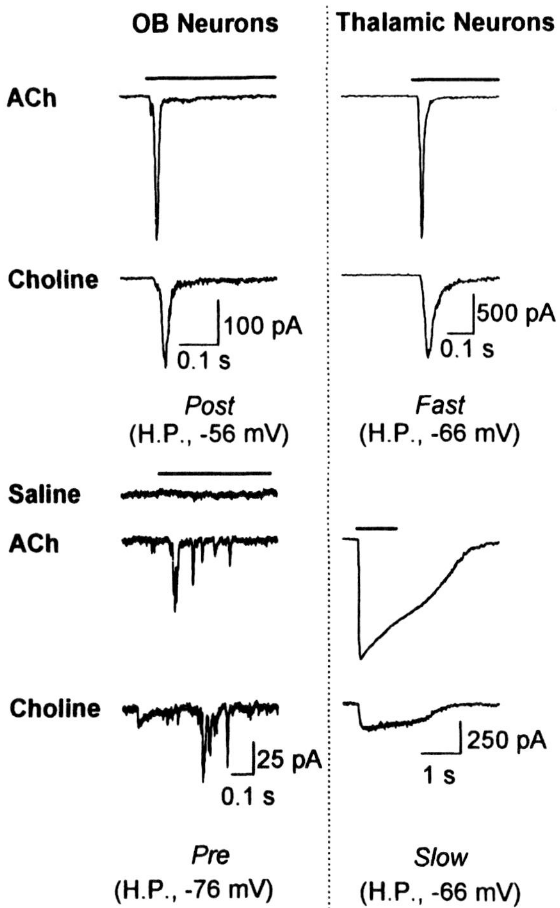

Figure 9.3. *Family of nicotinic currents elicited by ACh and choline in neurons cultured from rat olfactory bulb (OB) and thalamus. Top panel of traces represent type IA currents elicited by both agonists in the neurons of two CNS regions. Bottom left panel represents glutamate-mediated currents evoked by the activation of presynaptic* (Pre) *nAChRs. Bottom right panel represents slowly decaying nicotinic currents (resembling type III currents) evoked by the two agonists in a thalamic neuron.*

Our studies have also indicated that choline is unable to activate α4β2 nAChRs (Alkondon et al., 1997b) and is much less effective than ACh in activating the α3β4 nAChRs that are likely to subserve the slowly decaying currents recorded from some thalamic neurons (see Fig. 9-3). These findings, taken together, indicate that choline can effectively alter nicotinic functions mediated by α7-containing nAChRs in the mammalian CNS.

The Diverse Functional Properties of the Different Neuronal nAChR Subtypes Present in Hippocampal Neurons in Culture. The nAChRs present in hippocampal neurons differ one from the other not only with respect to their kinetics of inactivation and their sensitivity to a variety of agonists and antagonists but also with respect to functional properties such as single-channel kinetics, rectification, ion permeability, and modulation by Ca^{2+}.

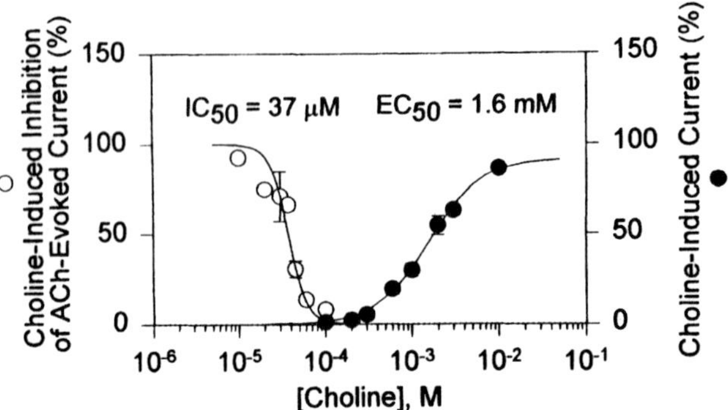

Figure 9.4. *Hill-plot analysis of the inhibition of ACh-evoked type IA currents by choline and of choline-evoked type IA currents in cultured hippocampal neurons.*

Type IA currents—i.e., those subserved by α7-bearing nAChRs—in addition to desensitizing quickly in the continuous presence of nicotinic agonists, show an intense rundown that can be prevented to a great extent by intracellular ATP-regenerating compounds and an inward rectification that is dependent on the levels of intracellular Mg^{2+} (Alkondon and Albuquerque, 1993; Alkondon et al., 1994). The α7-nAChR channels are also unique because of their brief open time (~ 100 μs at − 80 mV), high conductance (~ 73 pS), and high Ca^{2+} permeability (approximately 60% that of the NMDA channels; Castro and Albuquerque, 1993, 1995).

The α7 nAChRs in hippocampal neurons are also sensitive to changes in extracellular concentrations of Ca^{2+} ($[Ca^{2+}]_o$) (Bonfante-Cabarcas et al., 1996). Extracellular Ca^{2+} modulates the affinity of the α7 nAChRs for ACh, the cooperativity between ACh-binding sites as well as the inward rectification, the decay phase, and rundown of α7-nAChR-mediated type IA currents. Upon increasing the $[Ca^{2+}]_o$ from 10 μM to 1 mM, the apparent affinity of the α7 nAChRs for ACh increases and the cooperativity between ACh-binding sites decreases (Bonfante-Cabarcas et al., 1996). Several lines of evidence indicate that the effects of Ca^{2+} on the interaction of ACh with the α7 nAChRs in the hippocampus are mediated by the interactions of Ca^{2+} with specific sites on the receptor rather than through nonspecific Ca^{2+} sites on the membrane or through changes of surface potential (Bonfante-Cabarcas et al., 1996). Not only do the effects of Ca^{2+} on the receptor function follow sigmoid functions, which is expected for effects mediated by specific binding sites, but also calculated surface potential varies by less than 10% for different values of surface charge density. It is likely that binding of Ca^{2+} to the α7 nAChR in the hippocampus controls the concerted transformation of the subunits to yield various open-channel states (Bonfante-Cabarcas et al., 1996). This concept is in agreement with the multiple-channel conductance states reported for the α7 nAChRs in hippocampal neurons (Castro and Albuquerque, 1993; Pereira et al., 1993).

The decay phase of type IA currents evoked by saturating concentrations of ACh (≥1 mM) is accounted for by the desensitization of the α7 nAChRs (Castro and Albuquerque, 1993) and is accelerated by increasing the $[Ca^{2+}]_o$ from 2 mM to 10 mM. In the presence of 2 mM $[Ca^{2+}]_o$, ACh (1 mM)-evoked type IA currents have a decay-time constant of about 20 ms, whereas in the presence of 10 mM $[Ca^{2+}]$, the currents have a decay-time constant of about 10 ms (Castro and Albuquerque, 1995; Bonfante-Cabarcas et al., 1996). Thus, extracellular Ca^{2+} plays an important role in the rate of desensitization of the α-BGT-sensitive α7 nAChRs.

Changes in $[Ca^{2+}]_o$ also affect the inward rectification of type IA currents, which is dependent on the intracellular concentrations of Mg^{2+}. When recording type IA currents using a F^--containing internal solution, no inward rectification is observed. However, when a nominally Mg^{2+}-free, malate-based internal solution is used and the extracellular solution contains 2 mM Ca^{2+}, the rectification of type IA currents is confined to a short range of membrane potentials (0–30 mV). Upon raising the intracellular concentration of Mg^{2+} to 10 mM, the inward rectification persists up to 50 mV, and if concomitantly the $[Ca2^+]_0$ is lowered to 0.3 mM or less, the rectification persists up to 70 mV. Thus, the inward rectification of type IA currents, which is maximal in the presence of low extracellular Ca^{2+} concentrations and added intracellular Mg^{2+}, can be reversed when the extracellular Ca^{2+} concentration is increased to levels considered to be within the physiological range, indicating that depending on the ongoing synaptic activity and levels of Ca^{2+} surrounding the α7 nAChR, the receptor activity at positive potentials can range from negligible to considerably high levels.

In contrast to the α7-bearing nAChRs, the α4β2 nAChRs have a long-lasting open time, a low conductance, and do not desensitize as much in the continuous presence of nicotinic agonists (Castro and Albuquerque, 1993). Also, type II currents, in contrast to type IA currents, show an inward rectification that is independent of intracellular Mg^{2+}, and do not run down with time (Alkondon and Albuquerque, 1993; Alkondon et al., 1994).

Characteristics of the Nictonic Responses Recorded from Neurons in the CA1 Field of Hippocampal Slices. A substantial amount of information regarding the physiological roles of some ligand-gated and voltage-gated ion channels has been gathered from studies on brain slices. Although in-depth analysis of the pharmacological and kinetic properties of such channels are made difficult due to the barriers for drug access to the cells in the slices, the use of this preparation has the advantage of allowing for the identification of receptor function within its natural environment. Thus, we have investigated the functions mediated by nAChRs expressed on neurons visualized in the CA1 field of hippocampal slices (Alkondon et al., 1997a).

Using infrared-assisted videomicroscopy, neurons in the CA1 region of hippocampal slices can be morphologically identified as pyramidal neurons in the stratum pyramidale and interneurons in the stratum pyramidalé and in the stratum radiatum (Spruston et al., 1995; Alkondon et al., 1997a). The interneurons have nonpyramidal-shaped soma and thin dendrites (Fig. 9-5). Fluorescence microscopy of Lucifer yellow-filled neurons not only confirms the morphological features of pyramidal and nonpyramidal cells but also enables identification of spiny dendritic regions (see figs. 1 and 2 in Alkondon et al., 1997a).

In the absence of TTX, application of ACh and other nicotinic agonists to CA1 neurons results in the elicitation of at least one of four types of nicotinic responses: (1) slowly decaying whole-cell currents and (2) GABA-mediated postsynaptic currents (PSCs) that can be recorded from pyramidal neurons and interneurons, (3) fast-decaying nicotinic currents and (4) fast current transients that can be recorded only from interneurons (Fig. 9-6). All the responses are nicotinic in nature because they are sensitive to blockade by the nAChR antagonist d-tubocurarine, are insensitve to blockade by the muscarinic acetylcholine receptor antagonist atropine, and can be evoked by nAChR-selective agonists (Alkondon et al. 1997a).

The use of a number of nicotinic agonists and antagonists that have been found to be useful, to a certain extent, in discriminating the nAChR subtypes present in CNS neurons provided some clues regarding the nAChR subtype subserving the nicotinic responses under study. In pyramidal neurons, the nicotinic agonists (+)epibatidine, dimethyl phenyl piperazinium (DMPP), and (+)anatoxin-a (AnTX), unlike ACh, induced responses in which the PSCs were more prominent than the slow inward currents, suggesting that the nAChR subtype

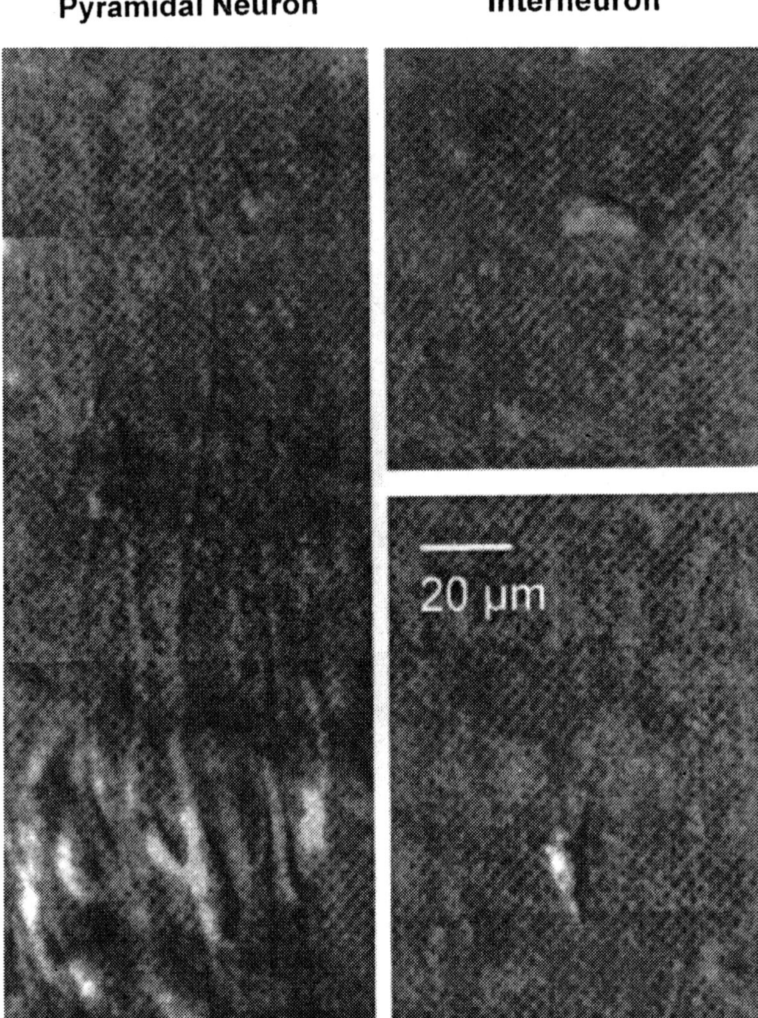

Figure 9.5. Infrared images of pyramidal neurons and interneurons in a hippocampal slice obtained from a 16-day-old rat.

subserving the slowly decaying currents is different from that mediating that agonist-evoked PSCs. In addition, choline, an α7-specific nAChR agonist (Alkondon et al., 1997b; Papke et al., 1996), acted as a weak agonist when applied to pyramidal neurons that responded to ACh with slowly decaying currents accompanied by PSCs.The slowly decaying currents and the PSCs recorded from the pyramidal neurons were also insensitive to blockade by the α7-nAChR-specific antagonist MLA (Alkondon et al., 1997a). Additional clues regarding the nAChR subtype that subserves the nicotinic responses in the pyramidal neurons were obtained from the finding that cytisine and (−)nicotine evoked slowly decaying currents and PSCs that

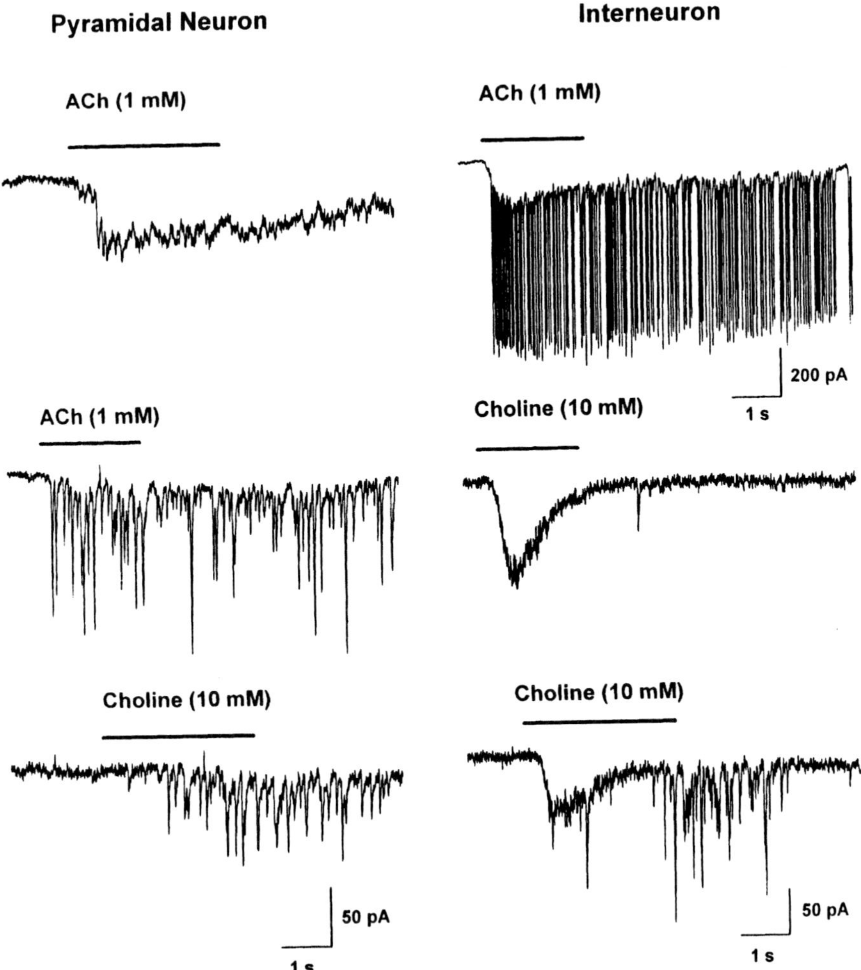

Figure 9.6. Family of nicotinic responses recorded from pyramidal neurons and interneurons in hippocampal slices. In pyramidal neurons, ACh evoked either slowly decaying currents accompanied by PSCs (top trace in left panel) or mainly PSCs (middle trace in left panel). Choline also induces PSCs (bottom trace in left panel). In interneurons, ACh induces slowly decaying currents mostly associated with fast current transients (top trace in right panel). Choline induces fast-decaying nicotinic current (resembling type IA current) either alone (middle trace in right panel) or in association with PSCs (bottom trace in right panel). Holding potential = −60 mV.

had approximately the same magnitude as those evoked by ACh. Considering that cytisine is a very poor agonist at $\alpha 4\beta 2$ nAChRs, it is unlikely that such receptors account for these nicotinic responses in pyramidal neurons.

In the majority of interneurons sampled, choline induced inward currents that decayed to the baseline values before termination of the agonist pulse and that were sometimes accompanied by PSCs. The kinetics of inactivation of choline-evoked whole cell currents resembled that of $\alpha 7$ nAChR-mediated currents recorded from hippocampal neurons in culture (Alkondon et al., 1997b). A few CA1 interneurons did not respond to choline and others showed only PSCs in response to choline or a slowly decaying inward current accompanied by fast current transients that resembled action potentials. Recent reports from other laboratories have also confirmed the presence of functional α7 nAChRs in CA1 interneurons (Jones and Yakel, 1997; Frazier et al., 1998).

The responses evoked by choline in the interneurons were strikingly different from those induced by ACh and by the nicotinic agonists cytisine and nicotine. In neurons that responded to choline with fast-decaying currents, PSCs, or fast-decaying currents accompanied by PSCs, ACh, cytisine, or nicotine always evoked a slowly decaying current. This finding could be explained by the ability of such nonspecific agonists to activate both rapidly and slowly desensitizing nAChRs, and by the fact that, depending on its magnitude, a slowly decaying current can mask a rapidly decaying current. Thus, in agreement with the results obtained from cultured hippocampal neurons, a single neuron in hippocampal slices can express more than one subtype of functional nAChR, one of which may be composed of the α7 subunit, as indicated by the ability of choline to evoke fast-decaying currents that resemble nAChR-mediated type IA currents. The fact that the responses evoked by cytisine in the interneurons had smaller amplitude than those evoked by ACh suggests that an α4β2 nAChR is being activated. This contention is supported by the finding that the slowly decaying currents evoked by application of nicotinic agonists to the interneurons were sensitive to blockade by the α4β2 nAChR-specific antagonist DHβE (Alkondon et al., 1997a).

Of major interest was the finding that in the CA1 field of the hippocampus, TTX (200 nM) inhibited the ACh-evoked, GABA-mediated PSCs, indicating that depolarization resulting from activation of nAChRs in presynaptic neurons leads to the generation of action potentials that propagate along the neuronal axon and invade the presynaptic terminal causing the release of GABA (Albuquerque et al., 1997; Alkondon et al., 1997a). Such "preterminal" nAChRs have also been found to be present in neurons of rat interpeduncular nucleus (Léna et al., 1993) and avian lateral spiriform nucleus (McMahon et al., 1994), whereby their activation leads to release of GABA.

Based on the results obtained from the experiments in hippocampal slices, it is reasonable to propose that activation of nAChRs present in various interneurons results in the release of GABA at synapses made onto other interneurons and onto pyramidal neurons, and that GABA, in turn, by activating $GABA_A$ receptors, leads to the activation of PSCs. Accordingly, in a hypothetical situation of a neuronal network composed of the series: interneuron ⇒ interneuron ⇒ pyramidal neuron, nAChR activation in the second interneuron should be very effective in increasing the inhibitory tonus to the pyramidal neuron, thereby overriding a possible functional disinhibition of the pyramidal neuron (Tóth et al., 1997). This disinhibition could occur in such a network only if GABA release from the first interneuron in the circuitry were sufficient to silence the interneuron that synapses onto the pyramidal neuron. Thus, the frequency and timing of cholinergic stimuli arriving at the various CA1 interneurons may be essential in determining the magnitude of the output signals of the CA1 pyramidal neurons.

Developmental Regulation of the Expression of Functional α7-containing nAChRs in Hippocampal Neurons In Vivo and In Vitro. Studies of the nicotinic responses of hippocampal neurons cultured for up to 40 days and of neurons acutely dissociated without the aid of enzymes from the hippocampus of rats at different stages of development provided valuable information regarding the developmental regulation of expression of functional α7-bearing nAChRs in vitro and in vivo (Alkondon and Albuquerque, 1991; Ishihara et al., 1995; Barbosa et al., 1996). The nicotinic sensitivity of cultured hippocampal neurons increases as the neurons age in culture, as seen by a progressive increase in the amplitude of AnTX-evoked currents along with the number of days that the neurons are in culture (Alkondon and Albuquerque, 1991). These findings are supported by the report that the number of α-BGT binding sites increases along with the number of days that hippocampal neurons remain in culture (Barrantes et al., 1995).

As in cultured neurons, the average peak amplitude of ACh-induced currents increases with the maturation of neurons enzymatically dissociated from the rat hippocampus (Ishihara et al., 1995). A major difference between cultured and enzymatically dissociated neurons is that lower current amplitudes evoked by ACh are generally recorded in the latter. Such an observation led to the suggestion that nAChR-bearing dendrites may be lost during the process of acute dissociation.

The nicotinic currents recorded from neurons mechanically dissociated from the CA1 area of the hippocampus of rats of various ages show the same age-related changes in the amplitude as those observed in currents recorded from cultured neurons and from enzymatically dissociated neurons. However, the nicotinic currents recorded from the mechanically dissociated neurons have much larger amplitudes (50–800 pA) than the currents recorded from the enzymatically dissociated hippocampal neurons. These findings indicated that the expression of functional nAChRs in hippocampal neurons increases with neuronal maturation either in vivo or in vitro and supported the hypothesis that functional nAChRs are present on the dendrites of the hippocampal neurons. This hypothesis was confirmed when a system of computerized micromanipulators and infrared-assisted videomicroscopy was used to map the distribution of functional nAChRs on the surface of hippocampal neurons (Alkondon et al., 1996a).

SUBCELLULAR DISTRIBUTION OF NICOTINIC RECEPTORS IN HIPPOCAMPAL NEURONS

As already mentioned, our studies with hippocampal neurons in culture and in slices showed that more than one nAChR subtype can be expressed in a single neuron. This prompted the question of whether individual nAChR subtypes are present on different regions of the neuronal surface and serve specific cellular functions. Thus, we have addressed the distribution of nAChRs on the surface of hippocampal neurons by recording whole-cell currents evoked by focal application of ACh from the tip of a micropipette to well-defined areas on the surface of hippocampal neurons (Alkondon et al., 1996a). The neuronal area covered by ACh was determined on the basis of the estimates obtained from the spread of phenol red solution released from the tip of the drug-delivery pipette while using a 20-pSi and 25-msec pulse. When the pipette was positioned about 2 μm from the dendrite, assuming a trapezoidal shape for the agonist spread, the length of the dendrite segment in the immediate vicinity of the pipette (labeled b) was estimated to be about 3μm. With increasing width of the dendrite (labeled w) the length of dendritic segment on the opposite side of the dendrite (labeled B) would be equal to $3 + (w \times 0.6)$ μm. Using infrared images of the dendrites, the values for w can be determined, and then B can be calculated. Once the value of b, B, and w are known, the area in μm^2 can calculated from the formula, Area $= w(B + b)/2$.

Application of ACh to the dendritic extensions resulted in the activation of type IA currents whose amplitudes were smaller than those of currents evoked by application of ACh to the cell body (Fig. 9-7). Plots of the current density (estimated as [current amplitude recorded from the

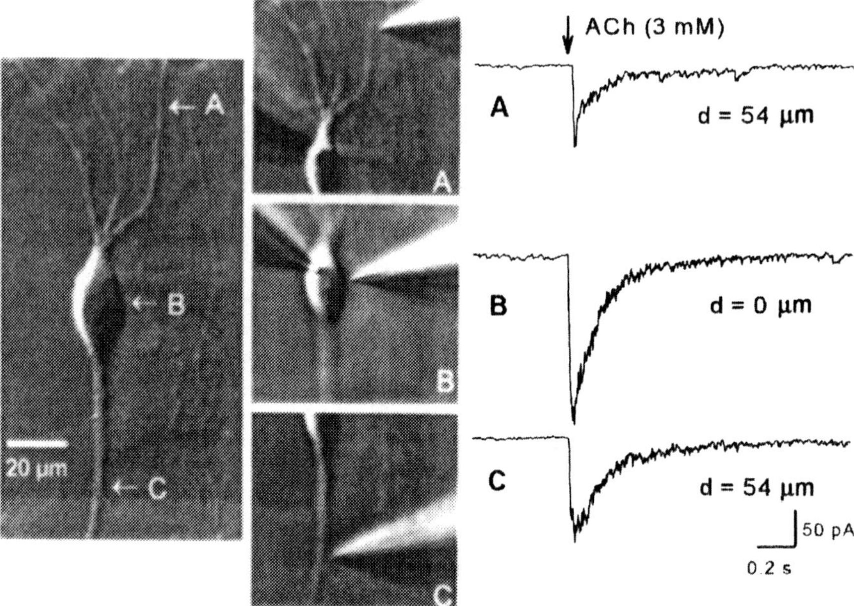

Figure 9.7. Mapping type IA currents in cultured hippocampal neurons. Infrared image of part of a hippocampal neuron and the position of the agonist delivery pipette are shown. Type IA currents elicited by focal application (25 msec, 20 psi) of ACh to different locations are shown at the right. Holding potential = −50 mV.

soma]/[membrane area exposed to the agonist, i. e., pA/μm^2]) against the distance from the soma at which the agonist was applied revealed that the density of type IA currents is substantially higher on the apical and basal dendrites of pyramidal neurons and on the dendrites of bipolar neurons than on the soma of these neurons. Considering that the same single-channel conductance accounts for type IA currents evoked at the soma or at the dendrites, it is clear that the density of α7-bearing nAChRs is higher on the dendrites than on the cell body of the neurons. The same analysis of type II currents evoked by application of ACh to various areas of the surface of hippocampal neurons and recorded from the cell body provided evidence that α4β2 nAChRs are also at higher density on the dendrites than on the soma of these neurons (Alkondon et al., 1996a). Analysis of type IA and type II currents generated at more remote dendritic areas (up to 85 μm from the center of the soma) indicated that the current density in dendritic areas increases with the distance from the center of the soma.

The issue of nAChR distribution on the neuronal surface becomes critical provided that several lines of evidence support the concept that segregation of Ca^{2+}-permeable channels on the neuronal surface is vital for integration and processing of a synaptic input to the neurons. It has been suggested that NMDA receptors, being at high density on distal dendritic regions, may serve a direct role in the induction of LTP at active synapses, whereas L-type Ca^{2+} channels, being at high density on the cell body and proximal dendritic areas, may mediate intracellular regulatory events in the cell body in response to the same synaptic inputs that lead to LTP at the distal dendritic areas of hippocampal neurons (Westenbroek et al., 1990; Regehr and Tank, 1990).

INVOLVEMENT OF NICOTINIC RECEPTORS IN HIGHER BRAIN FUNCTIONS: NICOTINIC RECEPTORS AS SENSORS OF SYNAPTIC ACTIVITY

Analysis of the behavioral effects induced by a number of nicotinic agonists and antagonists have led to the suggestion that neuronal nAChRs in the CNS may be involved in modulating pain, convulsions, learning, and memory. However, direct evidence of mediation of synaptic transmission by neuronal nAChRs in the mammalian CNS is scarce. One rate example, however, is the cholinergic transmission between neurons of the zona intermedialis reticularis and a subset of neurons within the nucleus ambiguus, which appears to be mediated by an $\alpha4\beta2$ nAChR (Zhang et al., 1993). The difficulties in detecting nicotinic postsynaptic currents, either spontaneous or evoked by stimulation of presynaptic elements, could be attributed to cholinergic projections and to nAChR-containing targets being scattered throughout the CNS (Woolf, 1991), and thus being difficult to address in situ. Also, given that we and others have shown that neuronal nAChRs, particularly those made up of $\alpha7$ subunits, are susceptible to modulation by the levels of extracellular Ca^{2+} (Vermino et al., 1992; Amador and Dani, 1995; Bonfante-Cabarcas et al., 1996), it is possible that during intense synaptic activity, changes in the extracellular concentrations of Ca^{2+} overshadow the nicotinic synaptic transmission. It should be emphasized that although α-BGT binding sites have been visualized in postsynaptic areas on hippocampal neurons (Hunt and Schmidt, 1978), only recently have evoked α-BGT-sensitive nicotinic synaptic currents been recorded from hippocampal neurons (Alkondon et al., 1998).

Many of the biological functions of neuronal nAChRs identified to date can be associated with the modulation of intracellular Ca^{2+} levels, e.g., modulation of neurite outgrowth (Pugh and Berg, 1994), control of synthesis and/or release of neurotrophins, regulation of early gene (e.g., c-fos) transcript levels (Greenberg et al., 1986), modulation of release of various neurotransmitters including dopamine, glutamate, and GABA (Wonnacott et al. 1988; Rocha et al., 1995; McGehee et al., 1995; Alkondon et al., 1996b, 1997a; Gray et al., 1996; Guo et al, 1998), and activation of second messengers (MacNicol and Schulman, 1992; Vijayaraghavan et al., 1995). Elevation of the intracellular levels of Ca^{2+} could be due to activation of voltage-gated Ca^{2+} channels triggered by membrane depolarization consequent to activation of the nAChRs, or due to Ca^{2+} permeation through the nAChR channels. In fact, the α-BGT-sensitive $\alpha7$-containing nAChRs in the CNS, similar to the homomeric $\alpha7$ nAChRs expressed in *Xenopus* oocytes, are highly permeable to Ca^{2+} (Bertrand et al., 1993, Sands et al., 1993, Séguéla et al., 1993; Castro and Albuquerque, 1995). Based on our permeation studies, Ca^{2+} entry into hippocampal neurons through α-BGT-sensitive nAChRs is approximately 60% of that through NMDA receptors. However, considering that NMDA receptor channels have longer open time and slower kinetics of inactivation than $\alpha7$ nAChR channels in hippocampal neurons (Castro and Albuquerque, 1993; Nelson and Albuquerque, 1995), the Ca^{2+} influx through the NMDA receptor should last longer than that through the $\alpha7$ nAChR. In addition, these receptors could have nonoverlapping roles if the amount of Ca^{2+} entering the neurons via the NMDA receptors differs from that entering the neurons via the $\alpha7$ nAChRs, so that distinct Ca^{2+} signaling pathways are affected depending on the receptor being activated.

It should be noticed that LTP is, so far, the best studied model for memory acquisition. It is induced rapidly by small bursts of activity and it is characterized by increased efficacy of neurotransmission. In addition, some forms of LTP, in particular, NMDA-dependent LTP, are consistent with Hebbian learning mechanisms (Hebb, 1949)—i.e., the conjunction of presynaptic and postsynaptic activation that may underlie associative learning. In NMDA-dependent LTP, the NMDA receptor channel only opens in response to presynaptically released glutamate if the postsynaptic cell is sufficiently depolarized. The calcium ions that enter the NMDA receptor channel could then activate intracellular kinases, which by modifying postsynaptic receptors can increase the postsynaptic sensitivity to glutamate and/or cause the release of retrograde factors that act on the presynaptic transmitter release machinery.

Mounting evidence indicates that neuronal nAChRs are intimately involved in learning and memory. The level of expression of these receptors is substantially reduced in the brain of patients with Alzheimer's disease—a neuropathological condition characterized by progressive memory impairment —(Schröder et al., 1989). Also, the incidence of Alzheimer's disease is approximately 60% lower in smokers than in nonsmokers (vanDujin et al., 1995), and nicotinic agonists can improve learning and memory in experimental animals (Decker et al., 1995). Therefore, it is tempting to speculate on ways by which nicotinic receptors could participate in LTP generation.

Considering that the predominant functional nAChR in hippocampal neurons is an α7-containing receptor and that cognition is mainly processed in the hippocampus, it is very likely that the activity of this receptor plays a significant role in learning and memory. By activating presynaptic α7-containing nAChRs, nicotinic agonists can control the release of glutamate onto neurons that coexpress NMDA and AMPA receptors. Then activation of AMPA receptors by glutamate could result in sufficient membrane depolarization to reduce the blockade of

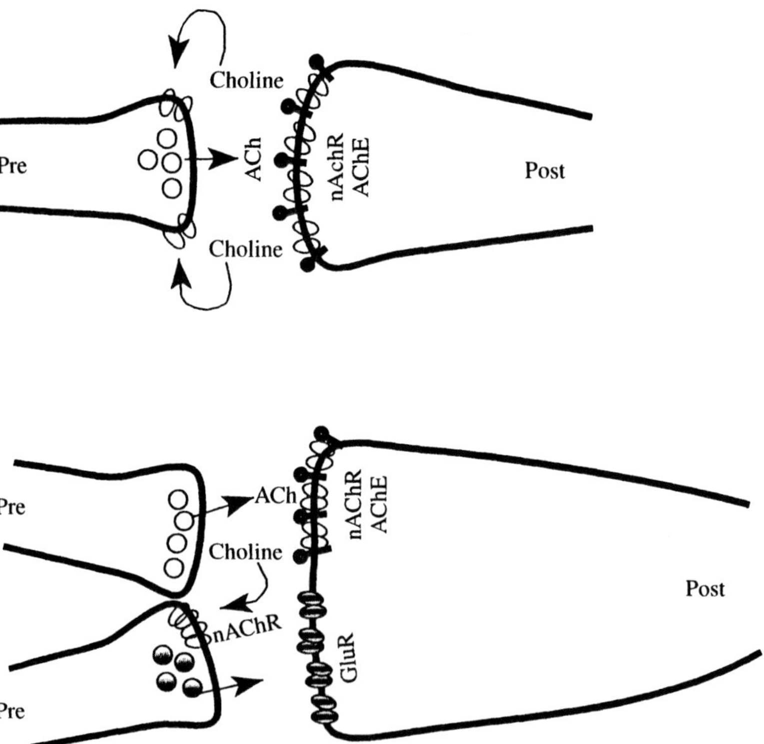

Figure 9.8. Postulated model for the actions of choline at central synapses. Choline serves as a retrograde messenger in increasing ACh release (scheme at the top) or glutamate release (scheme at the bottom) by acting on the α7 nAChRs at the respective nerve terminals.

NMDA receptors by Mg^{2+} physiologically present in the extracellular milieu. This cascade of events triggered by activation of presynaptic α7 nAChRs can lead to full activation of the NMDA receptors, thereby contributing to the NMDA-dependent form of LTP. On the other hand, activation of postsynaptic nAChRs by ACh could provide the level of postsynaptic depolarization that is a prerequisite of NMDA receptor channel gating by glutamate. Alternatively, activation of postsynaptic α7-containing nAChRs could result in a significant increase of the local cytosolic Ca^{2+} concentration, which is assumed to play a central role in the determination of the physiological response of CNS neurons (Teyler et al. 1994).

Choline may also be critical for cognition. In this regard, it is important to consider that the concentration of choline in cholinergic synapses can build up to considerably high levels, given that the uptake of choline into presynaptic terminals and into glial cells is substantially slower than the hydrolysis of ACh by AChE (see review by Wecker, 1990, for details). Thus, if a glutamatergic synapse made of a presynaptic terminal-bearing presynaptic α7-containing nAChRs is nearby a cholinergic synapse, it is feasible to postulate that during synaptic activation, choline generated from ACh hydrolysis could diffuse away and reach the presynaptic glutamatergic terminal (Fig. 9-8). Then activation by choline of the presynaptic α7 nAChR could lead to the release of glutamate, and under conditions of repeated synaptic activity, such a situation in which choline serves as a retrograde transmitter may be essential for LTP. Alternatively, diffusion of choline into neighboring synapses containing extrasynaptic α7 nAChRs could result in activation of these receptors and cause localized depolarization. In this case, because activation of postsynaptic α7 nAChRs could also cause a significant elevation of the cytosolic concentration of Ca^{2+} (Castro and Albuquerque, 1995), choline action could induce a cascade of intracellular events, including modification of postsynaptic receptors by phosphorylation and release of messengers such as arachidonic acid, NO, and CO. Finally, should choline accumlate near synapses, it would desensitize the α7 nAChR and would control the amount of signal transmitted via activation of these recptors by ACh.

REGULATORY CONTROL OF NICOTINIC RECEPTORS BY ALLOSTERIC LIGANDS

The activity of many ligand-gated ion channels is subject to modulation by ligands other than the natural agonist. Various typical examples are noted. For instance, NMDA receptor activity has been shown to be positively modulated by glycine (Johnson and Ascher, 1987; Scaton, 1993), and the $GABA_A$ receptor activity is known to be enhanced by benzodiazepines and steroids (McDonald and Twyman, 1992). In this section we provide an overview of the knowledge about compounds that can positively modulate the activity of neuronal nAChRs and are referred to as noncompetitive agonists.

Studies from our laboratories have provided evidence that some anticholinesterases, particularly the carbamate physostigmine, activate the muscle nAChR in frog single muscle fibers and in *Torpedo* vesicles, and that this effect is unrelated to blockade of AChE (Shaw et al., 1985; Albuquerque et al., 1988; Kuhlmann et al., 1991; Okonjo et al., 1991). A posteriori, the nicotinic agonist action of physostigmine was shown to be insensitive to blockade by competitive nicotine antagonists, being sensitive only to inhibition by the nAChR-specific monoclonal antibody FK1 (Okonjo et al., 1991). It was then suggested that physostigmine activates the nAChR channel by binding to a site distinct from that for ACh. By means of photoaffinity labeling of the *Torpedo* nAChR with [^{3}H]physostigmine it was demonstrated that physostigmine binds to a region on the nAChR α subunits that includes and/or surrounds the amino acid lys-125. The fact that the epitope for the antibody FK1 is located on the amino acid sequence 118 to 142 of the nAChR α subunits, and that this antibody antagonizes the agonist action of physostigmine without affecting that of ACh, supported the concept that physostigmine activates the muscle-type nAChR by binding to the region including and/or surrounding the amino acid lys-125 on the nAChR α subunit.

The agonist effect of physostigmine is not confined to the muscle nAChR (Pereira et al., 1993, 1994; Storch et al., 1995), and many compounds including the anticholinesterase galanthamine, the muscle relaxant benzoquinonium, and the opioid codeine, all of which are structurally related to physostigmine, were also found to activate nAChR channels via the physostigmine binding site (Pereira et al., 1993a,b, 1994; Storch et al., 1995). These compounds, referred to as *noncompetitive agonists,* do not evoke nicotinic macroscopic currents; instead, they potentiate the nAChR activity induced by classical nicotinic agonists in different preparations (Fig. 9-9).

The fact that the region of the nAChR α subunits that bears the binding site for noncompetitive agonists has unique characteristics that resemble those of the ACh-binding site on acetylcholinesterase may explain why many cholinesterase inhibitors can interact with the region including and/or surrounding the residue lys-125 of the nAChR subunits (for a review, see Albuquerque et al., 1997). Because some studies have indicated that indolamines, including the neurotransmitter 5-hydroxytryptamine (5-HT), can interact with the cholinesterases found in the placques of patients with Alzheimer's disease (Wright et al., 1993), 5-HT was tested for its ability to modulate ACh-evoked responses. In PC12 cells, 5-HT was shown to mimic the potentiating action of galanthamine on ACh-evoked currents (Schrattenholz et al., 1996). This result and the previous finding that the opioid codeine can interact with the physostigmine binding site on various nAChR subtypes led to the hypothesis that 5-HT and, by inference, en-

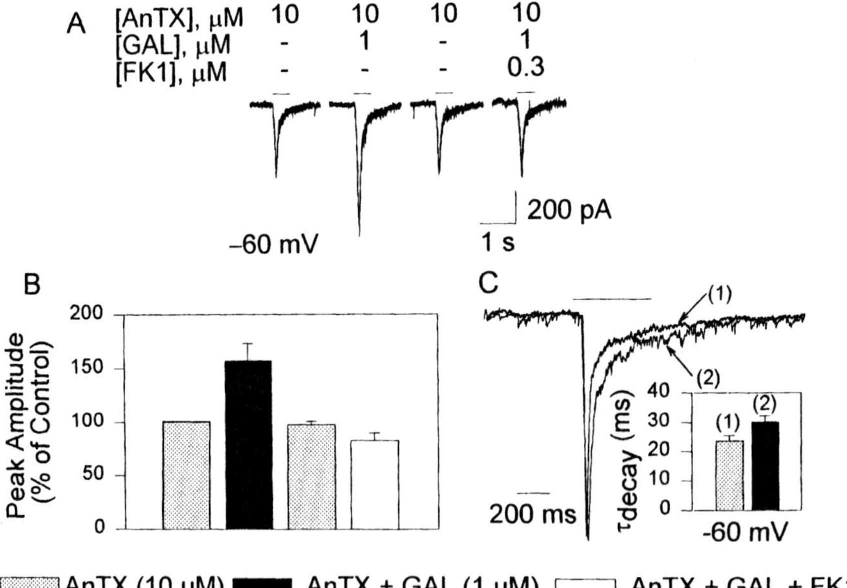

Figure 9.9. *Characterization of the potentiating effect of galanthamine on type IA currents recorded from hippocampal neurons in culture. (**A** and **B**) Galanthamine-induced increase in the peak amplitude of ACh-evoked type IA currents. (**C**) The decay phase of type IA currents is prolonged in the presence of galanthamine, indicating that galanthamine reduces the rate of desensitization of the α7 nAChR that mediates these nicotinic responses.*

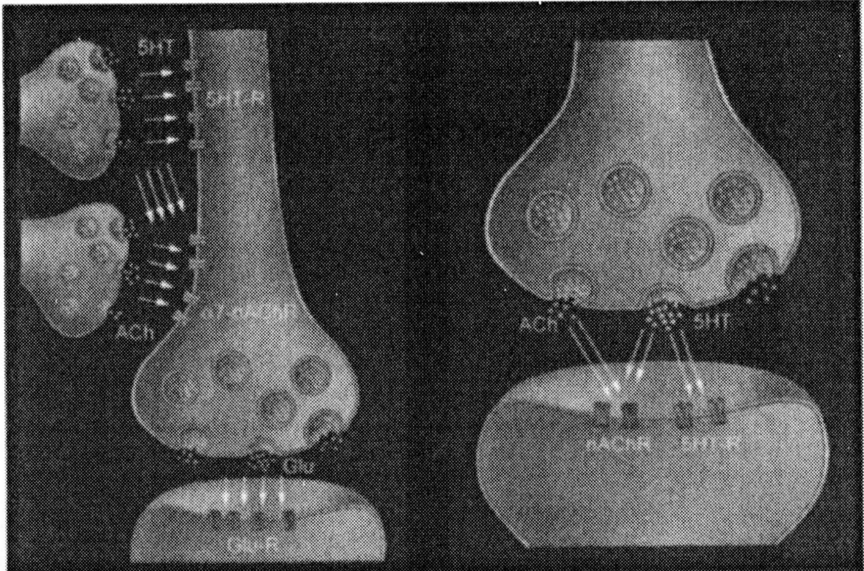

Figure 9.10. Hypothetical schemes illustrating modulation of nicotinic cholinergic function by 5-HT.

dogenous opiates could act as endogenous allosteric modulators of the nAChR function by binding to this novel nAChR site.

Endogenous modulators of the nAChR activity could be stored together with ACh in cholinergic terminals. In this respect, it is worthy to mention that enkephalines are present in splanchnic cholinergic terminals that synapse on adrenal medullary chromaffin cells (Guidotti et al., 1983), and that there may be areas in the CNS where enkephalines or other endogenous opiates are costored with ACh, because it is common to find the simultaneous presence of different neurotransmitters and small peptides in a particular nerve terminal (Guidotti et al., 1983). Thus, it is likely that such endogenous opiates co-released with ACh positively modulate the nAChR activity by acting through the physostigmine binding site. Alternatively, endogenous modulators of the nAChR activity could have a paracrine action. In this regard, it is tempting to speculate that in areas where tryptaminergic and cholinergic synapses are colocalized, 5-HT released from its terminal could diffuse away and control the activation of the nicotinic channels in nearby cholinergic synapses by binding to the physostigmine site on the nAChRs (Schrattenholz et al., 1996). The finding that submicromolar concentrations of 5-HT are enough to potentiate nicotinic responses is in agreement with the concept that diffusing 5-HT could modulate the nAChR activity. Indeed, if 5-HT can act as a neurotransmitter at serotoninergic synapses and as a modulator of the nAChR activity at cholinergic synapses, a cross-talk between the neurotransmitter systems could take place whenever cholinergic and serotoninergic terminals are close to each other so that the cholinergic function would be maximal when activation of serotoninergic terminals is triggered. Under this condition, synchronous activation of nicotinic cholinergic synapses and serotoninergic synapses that are in close proximity would result in maximal responsiveness of the nAChRs to ACh released from the cholinergic terminal.

Figure 9-10 illustrates some putative modulatory functions of 5-HT in light of the present findings. Let us consider the axo-dendritic synapse shown on the left. According to this scheme, the presynaptic glutamatergic terminal receives synaptic inputs from cholinergic terminals and from serotoninergic terminals. In fact, mounting evidence indicates that both ACh and 5-HT can control the release of glutamate in various CNS areas (Maura et al., 1991; Shupliakov et al., 1995; McGehee et al., 1995; Rocha et al., 1995; Alkondon et al., 1996). Whereas ACh increases the release of glutamate, 5-HT decreases it. Thus, if there are glutamatergic terminals in the CNS bearing both nicotinic and 5-HT receptors, and receiving serotoninergic as well as cholinergic innervation, it is very likely that the amount of glutamate released upon depolarization of the glutamatergic terminal would be modulated by the ongoing activity in the cholinergic and serotoninergic terminals. If, for instance, there is coincident activation of both cholinergic and serotoninergic terminals, 5-HT diffusing away from the serotoninergic terminal could increase the nAChR activity in the cholinergic terminal, it is possible that the effect of the cholinergic input onto the glutamatergic terminal would predominate, so the final output would be an increase in glutamatergic transmission.

There is also some evidence that more than one neurotransmitter can be released from a single synaptic terminal. As shown in the right panel of Figure 9-10, if ACh and 5-HT are coreleased from a single terminal impinging onto a postsynaptic element that bears both nAChR and 5-HT receptors, it is possible that 5-HT has two functions: It acts as a neurotransmitter on its own by activaing postsynaptic 5-HT receptors and it acts as a positive allosteric modulator of the nicotinic channel activity by binding to the physostigmine site on the nAChR.

CONCLUDING REMARKS

Very recent studies have been critical to unveil the possible functions of neuronal nAChRs in the CNS and particularly in the hippocampus. The finding that nicotinic agonists increase the release of GABA from CA1 neurons (Alkondon et al., 1997a) has provided the basis for a better understanding of physiological functions and physiopathological conditions in which neuronal nAChRs appear to be involved.

The Θ oscillatory activity that is induced by the perisomatic $GABA_A$-receptor-mediated synaptic events in GABAergic interneurons and can be brought about in the hippocampus by cholinergic neurons has been shown to be very effective in synchronizing pyramidal cell activity (Cobb et al., 1995; Fox, 1989; Soltesz and Deschênes, 1993). Since the Θ rhythm is frequently recorded from the hippocampus of animals given learning tasks (Winson, 1978; Otto et al., 1991), it is plausible to suggest that nAChR-mediated modulation of GABA release onto CA1 pyramidal neurons is a mechanism by which nicotinic agonists may improve learning and memory.

The finding that activation of a neuronal nAChR on CA1 hippocampal neurons modulates the release of GABA may also provide the basis for the understanding of the mechanism by which an intact forebrain cholinergic innervation is essential to suppress kindling epileptogenesis in the hippocampus (Kokaia et al., 1996). Selective immunolesioning of basal forebrain cholinergic neurons with the immunotoxin 192IgG-saporin results in a marked facilitation of the initial stages of seizure development in hippocampal kindling (Kokaia et al., 1996), which appears to be directly associated with a reduction of GABAergic inhibition in the CA1 field of the hippocampus (Lothman et al., 1993). Because activation of muscarinic receptors (Pitler and Alger, 1992) and nAChRs increases the release of GABA from CA1 interneurons, it is conceivable that facilitation of the development of seizures due to the destruction of the major cholinergic input to the hippocampus is a result of the reduced activation of both muscarinic and nicotinic receptors in this brain area. In addition, our finding that the nAChR controlling GABA release from CA1 hippocampal neurons is insensitive to MLA sheds new light on the mechanism underlying the ability of nicotine to induce seizures in experimental animals, given that this effect of nicotine, which is believed to initiate in the hippocampus (Brown, 1967; Stumpf and Gogolak, 1967), cannot be prevented by MLA (Gasior et al., 1996). It is possible that nicotine-induced desensitization of the MLA-insensitive nAChRs present on CA1 hippocampal neurons accounts for its ability to induce seizures in the experimental animals.

A recent study has also implicated α7 nAChRs in schizophrenia by providing linkage data supporting the involvement of the α7-nAChR gene in the pathophysiological aspect of the illness (Freedman et al., 1997). The authors of that study attributed the attentional deficit seen in schizophrenics to a decrease in the inhibitory control to the hippocampal pyramidal neurons. Our present results provide direct evidence for the existence of a mechanism by which nAChRs may control the inhibitory tonus in the hippocampus.

Finally, considering that choline can control the function and number of α7 nAChRs, caution should be taken when using anticholinesterase agents in the treatment of pathological conditions such as Alzheimer's disease. Blockade of AChE would lead to a reduction of the amount of choline generated from ACh hydrolysis in the synaptic cleft, and if an equilibrium between the concentration of choline and ACh is necessary in order to maintain the normal function mediated by α7 nAChRs, disruption of this equilibrium could result in impairment of such function. In this regard, one could consider the use of noncompetitive agonists as the means by which the nAChR activity could be increased. Such compounds, if specific for a given nAChR subtype, could be very helpful as therapeutic agents to treat pathological conditions in which the nAChR activity is known to be impaired.

REFERENCES

Albuquerque EX, Aracava Y, Cintra WM, Brossi A, Schonenberger B, Deshpande SS (1988): Structure-activity relationship of reversible cholinesterase inhibitors: activation, channel blockade and stereospecificity of the nicotinic acetylcholine receptor-ion channel complex. Brazilian J. Med. Res. 21: 1173–1196.

Albuquerque EX, Alkondon M, Pereira EFR, Castro NG, Schrattenholz A, Barbosa CTF, Bonfante-Cabarcas R, Aracava Y, Eisenberg HM, Maelicke A (1997): Properties of neuronal nicotinic acetylcholine receptors: pharmacological characterization and modulation of synaptic function. J. Pharmacol. Exp. Ther. 280: 1117–1136.

Alkondon M, Albuquerque EX (1991): Initial characterization of the nicotinic acetylcholine receptors in rat hippocampal neurons. J. Receptor Res. 11: 1001–1022.

Alkondon M, Albuquerque EX (1993): Diversity of nicotinic acetylcholine receptors in rat hippocampal neurons. I. Pharmacological and Functional evidence for distinct structural subtypes. J. Pharmacol. Exp. Ther. 265: 1455–1473.

Alkondon M, Albuquerque EX (1995): Diversity of nicotinic acetylcholine receptors in rat hippocampal neurons. III. Agonist actions of the novel alkaloid epibatidine and analysis of type II current. J. Pharmacol. Exp. Ther. 274: 771–782.

Alkondon M, Pereira EFR, Wonnacott, S, Albuquerque EX (1992): Blockade of nicotinic currents in hippocampal neurons defines methyllycaconitine as a potent and specific receptor antagonist. Mol. Pharmacol. 41: 802–808.

Alkondon M, Reinhardt S, Lobron C, Hermsen B, Maelicke A, Albuquerque EX (1994): Diversity of nicotinic acetylcholine receptors in rat hippocampal neurons. II. The rundown and inward rectification of agonist-elicited whole-cell currents and identification of receptor subunits by in situ hybridization. J. Pharmacol. Exp. Ther. 271: 494–506.

Alkondon M, Pereira EFR, Albuquerque EX (1996a): Mapping the location of functional nicotinic and γ-aminobutyric $acid_A$ receptors on hippocampal neurons. J. Pharmacol. Exp. Ther. 279: 1491–1506.

Alkondon M, Pereira EFR, Albuquerque EX (1998) α-Bungarotoxin- and methyllycaconitine-sensitive nicotinic receptors mediate fast synaptic transmission in interneurons of rat hippocampal slices. Brain Res. *in press.*

Alkondon M, Pereira EFR, Barbosa CTF, Albuquerque EX (1997a): Neuronal nicotinic acetylcholine receptor activation modulates γ-aminobutyric acid release from CA1 neurons of rat hippocampal slices. J. Pharmacol. Exp. Ther. 283: 1396–1411.

Alkondon M, Pereira EFR, Cortes WS, Maelicke A, Albuquerque, EX (1997b): Choline is a selective agonist of α7 nicotinic acetylcholine receptors. Eur. J. Neurosci. 9: 2734–2742.

Alkondon M, Rocha ES, Maelicke A, Albuquerque EX (1996b): Diversity of nicotinic acetylcholine receptors in rat brain. V. α-Bungarotoxin-sensitive nicotinic receptors in olfactory bulb neurons and presynaptic modulation of glutamate release. J. Pharmacol. Exp. Ther. 278: 1460–1471.

Amador M, Dani JA (1995): Mechanism of modulation of nicotinic acetylcholine receptors that can influence synaptic transmission. J. Neurosci. 15: 4525–4532.

Anand R, Peng X, Lindstrom J (1993): Homomeric and native α7 acetylcholine receptors exhibit remarkably similar but non-identical pharmacological properties, suggesting that the native receptor is a heteromeric protein complex. FEBS Lett. 327: 241–246.

Barbosa CTF, Alkondon M, Aracava Y, Maelicke A, Albuquerque EX (1996): Ligand-gated ion channels in acutely dissociated rat hippocampal neurons with long dendrites. Neurosci. Lett. 210: 177–180.

Barrantes GE, Rogers AT, Lindstrom J, Wonnacott S (1995): α-Bungarotoxin binding sites in rat hippocampal and cortical cultures: initial characterization, colocalization with α7 subunits and upregulation by chronic nicotine treatment. Brain Res. 672: 228–236.

Bertrand D, Galzi JL, Devillers-Thiery A, Bertrand S, Changeux JP (1993): Mutations at two distinct sites within the channel domain M2 alter calcium permeability of neuronal α7 nicotinic receptor. Proc. Natl. Acad. Sci. USA 90: 6971–6975.

Bonfante-Cabarcas R, Swanson KL, Alkondon M, Albuquerque EX (1996): Diversity of nicotinic acetylcholine receptors in rat hippocampal neurons. IV. Regulation by external Ca^{++} of α-bungarotoxin-sensitive receptor function and of rectification induced by internal Mg^{++}. J. Parmacol. Exp. Ther. 277: 432–444.

Brown BB (1967): Relationship between evoked response changes and behavior following small doses of nicotine. Ann. NY Acad. Sci. 142: 190–200.

Castro NG, Albuquerque EX (1993): Brief-lifetime, fast-inactivating ion channels account for the α-bungarotoxin-sensitive nicotinic response in hippocampal neurons. Neurosci. Lett. 164: 137–140.

Castro NG, Albuquerque EX (1995): α-Bungarotoxin-sensitive hippocampal nicotinic receptor channel has a high calcium permeability. Biophys. J. 68: 516–524.

Cobb SR, Buhl EH, Halasy K, Paulsen O, Somogyi P (1995): Synchronization of neuronal activity in hippocampus by individual GABAergic interneurons. Nature 378: 75–78.

Costa LG, Murphy SD (1984): Interaction of choline with nicotinic and muscarinic cholinergic receptors in the rat brain in vitro. Clinical Exp. Pharmacol. Physiol. 11: 649–654.

Coutcher JB, Cawley G, Wecker L (1992): Dietary choline supplementation increases the density of nicotine binding sites in rat brain. J. Pharmacol. Exp. Ther. 262: 1128–1132.

Decker MW. Brioni JD, Bannon AW, Arneric SP (1995): Diversity of neuronal nicotinic acetylcholine receptors: lessons from behavior and implications for CNS therapeutics. Life Sci. 56: 545–570.

delCastillo J, Katz B (1957): Interaction at the end-plate receptors between different choline derivatives. Proc. Roy. Soc. London Ser. B 146: 369–381.

Fox SE (1989): Membrane potential and impedance changes in hippocampal pyramidal cells during theta rhythm. Exp. Brain Res. 77: 283–294.

Freedman R, Coon H, Myles-Worsley M, Orr-Urtreger A, Olincy A, Davis A, Polymeropoulos M, Holik J, Hopkins J, Hoff M, Rosenthal J, Waldo MC, Reimherr F, Wender P, Yaw J, Young DA, Breese CR, Adams C, Patterson D, Adler LE, Kruglyak L, Leonard S, Byerley W (1997): Linkage of a neurophysiological deficit in schizophrenia to a chromosome 15 locus. Proc. Natl. Acad. Sci. USA 94: 587–592.

Galzi J-L, Changeux J-P (1995): Neuronal nicotinic receptors: Molecular organization and regulations. Neurpharmacology 34: 563–582.

Gasior M, Goldberg SR, Shoiab M (1996): Central nicotinic receptors mediate nicotine-induced seizures in rats. Abstr. Neurosci. Soc. 22: 2093.

Gotti C, Fornasari D, Clementi F (1997): Human neuronal nicotinic receptors. Prog. Neurobiol. 53: 199–237.

Gray R, Rajan AS, Radicliffe K, Yakehiro M, Dani J (1996): Hippocampal synaptic transmission enhanced by low concentrations of nicotine. Nature 383: 713–716.

Greenberg E, Ziff EB, Green LA (1986): Stimulation of neuronal acetylcholine receptors induces rapid gene transcription. Science 234: 80–83.

Guidotti A, Saiani L, Wise BC, Costa E (1983): Cotransmitters: pharmacological implications. J. Neural Transm. 18: 213–225.

Guo J-Z, Tredway TL, Chiappinelli VA (in press): Glutamate and GABA release are enhanced by different subtypes of presynaptic nicotinic receptors in the lateral geniculate nucleus. J. Neurosci. 18: 1963–1969.

Hebb DO (1949): The organization of behaviour. New York: Wiley.

Hunt SP, Schmidt J (1978): The electron microscopic autoradiographic localization of alpha-bungarotoxin binding sites within the central nervous system of the rat. Brain Res. 142: 152–159.

Ishihara K, Alkondon M, Montes JG, Albuquerque EX (1995): Nicotinic response in acutely dissociated rat hippocampal neurons and the selective blockade of fast-desensitizing nicotinic currents by lead. J. Pharmacol. Exp. Ther. 273: 1471–1482.

Johnston JW, Ascher P (1987): Glycine potentiates the NMDA response in cultured mouse brain neurons. Nature 325: 529–531.

Kokaia M, Ferencz I, Leanza G, Elmér E, Metsis M, Kokaia Z, Wiley RG, Lindval O (1996): Immunolesioning of basal forebrain cholinergic neurons facilitates hippocampal kindling and perturbs neurotrophin messenger RNA regulation. Neuroscience 70: 313–327.

Kuhlmann J, Okonjo KO, Maelicke A (1991): Desensitization is a property of the cholinergic binding region of the nicotinic acetylcholine receptor, not of the receptor-integral ion channel. FEBS Lett. 279: 216–218.

Léna C, Changeux J-P, Mulle C (1993): Evidence for "preterminal" nicotinic receptors on GABAergic axons in the rat interpeduncular nucleus. J. Neurosci. 13: 2680–2688.

LeNovere N, Changeux J-P (1995): Molecular evolution of the nicotinic acetylcholine receptor family: an example of multigene family in excitable cells. J. Mol. Evol. 40: 155–172.

Lindstrom J (1995): Nicotinic acetylcholine receptors. In: Alan R. North, editor. Ligand- and voltage-gated ion channels. Boca Raton, FL: CRC Press, pp. 153–175.

Lothman EW, Stringer JL, Bertram EH (1993): Rapidly recurring seizures and status epilepticus: ictal density as a factor in epileptogenesis. In: Schwartzkroin PA, editor. Epilepsy: models, mechanisms, and concepts. Cambridge: Cambridge University Press, pp. 323–355.

Luetje CW, Patrick J (1991): Both α- and β- subunits contribute to the agonist sensitivity of neuronal nicotinic acetylcholine receptors. J. Neurosci. 11: 837–845.

MacNicol M, Schulman H (1992): Multiple Ca^{2+} signaling pathways converge on CaM kinase in PC12 cells. FEBS Lett. 304: 237–240.

Mandelzys A, De Koninck P, Cooper E (1995): Agonist and toxin sensitivities of ACh-evoked currents on neurons expressing multiple nicotinic ACh receptor subunits. J. Neurophysiol. 74: 1212–1221.

Maura G, Carbone R, Guido M, Pestarino M, Raiteri M (1991): 5-HT2 presynaptic receptors mediate inhibition of glutamate release from cerebellar mossy fibre terminals. Eur. J. Pharmacol. 202: 185–190.

McDonald RL, Twyman RE (1992): Kinetic properties and regulation of GABA receptor channels. Ion Channels 3: 315–343.

McGehee DS, Role LW (1995): Physiological diversity of nicotinic acetylcholine receptors expressed by vertebrate neurons. Annu. Rev. Physiol. 57: 521–546.

McGehee DS, Heath MJS, Gelber S, Devay P, Role LW (1995): Nicotine enhancement of fast excitatory synaptic transmission in CNS by presynaptic receptors. Science 269: 1692–1696.

McMahon LL, Yoon KW, Chiappinelli VA (1994): Nicotinic receptor activation facilitates GABAergic neurotransmission in the avian lateral spiriform nucleus. Neuroscience 59: 689–698.

Morley BJ, Robinson GR, Brown GB, Kemp GE, Bradley RJ (1977): Effects of dietary choline on nicotinic acetylcholine receptors in brain. Nature 266: 848–850.

Nelson ME, Albuquerque EX (1994): 9-Aminoacridines act at a site different from that for Mg^{2+} in blockade of N-methyl-D-aspartate receptor channel. Mol. Pharmacol. 46: 151–160.

Okonjo KO, Kuhlmann J, Maelicke A (1991): A second pathway for the activation of the *Torpedo* acetylcholine receptor. Eur. J. Biochem. 200: 671–677.

Otto T, Eichenbaum H, Wiener SI, Wible SG (1991): Learning-related patterns of CA1 spike trains parallel stimulation parameters optimal for inducing hippocampal long-term potentiation. Hippocampus 1: 181–192.

Papke RL, Heinemann SF (1994): Partial agonist properties of cytisine on neuronal nicotinic receptors containing the β2 subunit. Mol. Pharmacol. 45: 142–149.

Papke RL, Bencherif M, Lippiello P (1996): An evaluation of neuronal nicotinic acetylcholine receptor activation by quaternary nitrogen compounds indicates that choline is selective for the α7 subtype. Neurosci. Lett. 213: 201–204.

Pauly JR, Stitzel JA, Marks MJ, Collins AC (1989): An autoradiographic analysis of cholinergic receptors in mouse brain. Brain Res. Bull. 22: 453–459.

Pereira EFR, Alkondon M, Tano T, Castro NG, Froes-Ferrao MM, Rosental R, Aronstam RS, Albuquerque EX (1993a): A novel agonist binding site on nicotinic acetylcholine receptors. J. Recep. Res. 13: 413–436.

Pereira EFR, Reinhardt-Maelicke S, Schrattenholz A, Maelicke A, Albuquerque EX (1993b): Identification and functional characterization of a new agonist site on nicotinic acetylcholine receptors of cultured hippocampal neurons. J. Pharmacol. Exp. Ther. 265: 1474–1491.

Pereira EFR, Alkondon M, Reinhardt S, Maelicke A, Peng X, Lindstrom J, Whiting P, Albuquerque EX (1994): Physostigmine and galanthamine: probes for a novel binding site on the α4β2 subtype of neuronal nicotinic acetylcholine receptors stably expressed in fibroblast cells. J. Pharmacol. Exp. Ther. 270: 768–778.

Pereira EFR, Alkondon M, McIntosh JM, Albuquerque EX (1996): α-conotoxin-ImI: a competitive antagonist at α-bungarotoxin-sensitive neuronal nicotinic receptors in hippocampal neurons. J. Pharmacol. Exp. Ther. 278: 1471–1483.

Pidoplichko VI, DeBiasi M, Williams JT, Dani JA (1997): Nicotine activates and desensitizes midbrain doapmine neurons. Nature 390: 401–404.

Pitler TA, Alger BE (1992): Cholinergic excitation of GABAergic interneurons in the rat hippocampal slice. J. Physiol. (London) 450: 127–142.

Pugh CP, Berg DK (1994): Neuronal acetylcholine receptors that bind α-bungarotoxin mediate neurite retraction in a calcium-dependent manner. J. Neurosci. 14: 889–896.

Ramirez-Latorre J, Yu CR, Perin F, Karlin A, Role L (1996): Functional contributions of α5 subunit to neuronal acetylcholine receptor channels. Nature 380: 347–351.

Regehr WG, Tank DW (1990): Postsynaptic NMDA receptor-mediated calcium accumulation in hippocampal CA1 pyramidal cell dendrites. Nature 345: 807–810.

Rocha ES, Alkondon M, Burt D, Albuquerque EX (1995): A presynaptic modulatory role for the α-bungarotoxin-sensitive nicotinic acetylcholine receptors (nAChRs) in rat olfactory bulb neurons. Soc. Neurosci. Abstr. 21: 1331.

Sands SB, Costa ACS, Patrick J (1993): Barium permeability of neuronal nicotinic acetylcholine receptor channels in PC12 cells. Brain Res. 560: 38–42.

Scatton B (1993): The NMDA receptor complex. Fund. Clin. Pharmacol. 7: 389–400.

Schrattenholz A, Pereira EFR, Roth U, Weber KH, Albuquerque EX, Maelicke A (1996): Agonist responses of neuronal nicotinic acetylcholine receptors are potentiated by a novel class of allosterically acting ligands. Mol. Pharmacol. 49: 1–6.

Schroder H, Giacobini E, Struble RG, Zilles K, Maelicke A (1991): Nicotinic cholinoceptive neurons of the frontal cortex are reduced in Alzheimer's disease. Neurobiol. Aging 12: 259–262.

Séguéla P, Wadiche J, Dineley-Miller K, Dani JA, Patrick JW (1993): Molecular cloning, functional properties, and distribution of rat brain α7: a nicotinic cation channel highly permeable to calcium. J. Neurosci. 13: 596–604.

Shaw KP, Aracava Y, Akaike A, Daly JW, Rickett DL, Albuquerque EX (1985): The reversible cholinesterase inhibitor physostigmine has channel-blocking and agonist effects on the acetylcholine receptor-ion channel complex. Mol. Pharmacol. 28: 527–538.

Shupliakov O, Pieribone VA, Gad H, Brodin L (1995): Synaptic vesicle depletion in reticulospinal axons is reduced by 5-hydroxytryptamine: direct evidence for presynaptic modulation of glutamatergic transmission. Eur. J. Neurosci. 7: 1111–1116.

Soltesz I, Deschênes M (1993): Low- and high-frequency membrane potential oscillations during theta activity in CA1 and CA3 pyramidal neurons of the rat hippocampus under ketamine-xylazine anesthesia. J. Neurophysiol. 70: 97–116.

Spruston N, Jonas P, Sakmann B (1995): Dendritic glutamate receptor channels in rat hippocampal CA3 and Ca1 pyramidal neurons. J. Physiol. (London) 482: 325–352.

Stetzer E, Ebbinghaus U, Storch A, Poteur L, Schrattenholz A, Kramer G, Methfessel C, Maelicke A (1996): Stable expression in HEK-293 cells of the rat α3/β4 subtype of neuronal nicotinic acetylcholine receptor. FEBS Lett. 397: 39–44.

Storch A, Schrattenholz A, Cooper JC, Abdel Ghani EM, Gutbrod O, Weber KH, Reinhardt S, Lobron C, Hermsen B, Soskic V, Pereira EFR, Albuquerque EX, Methfessel C, Maelicke A (1995): Physostigmine, galanthamine and codeine act as noncompetitive nicotinic agonists on clonal rat pheochromocytoma cells. Eur. J. Parmacol. 290: 207–219.

Stumpf C, Gogolak G (1967): Actions of nicotine upon the limbic system. Ann. NY Acad. Sci. 142: 143–158.

Teyler TJ, Cavus I, Coussens C, Discenna P, Grover L, Lee YP, Little Z (1994): Multideterminant role of calcium in hippocampal synaptic plasticity. Hippocampus 4: 623–634.

Tóth K, Freund TF, Miles R (1997): Disinhibition of rat hippocampal pyramidal cells by GABAergic afferents from the septum. J. Physiol. (London) 500: 463–474.

vanDujin CM, Havekees LM, VanBroeckhoven C, deKnijff P, Hoffman A (1995): Apolipoprotein E genotype and association between smoking and early onset Alzheimer's disease. Neurobiol. Aging 112: 259–262.

Vernino S, Amador M, Luetje CW, Patrick J, Dani JA (1992): Calcium modulation and high calcium permeability of neuronal nicotinic acetylcholine receptors. Neuron 8: 127–134.

Vijayaraghavan, S, Huang B, Blumenthal EM, Berg DK (1995): Arachidonic acid as a possible negative feedback inhibitor of nicotinic acetylcholine receptors on neurons. J. Neurosci. 15: 3679–3687.

Wecker L. (1990): Choline utilization by central cholinergic neurons. In Wurtman RJ, Wurtman JJ, editors, *Nutrition and the Brain,* vol. 8. New York: Raven Press, pp. 147–162.

Westenbroek RE, Ahlijanian MK, Catterall WA (1990): Clustering of L-type Ca^{2+} channels at the base of major dendrites in hippocampal pyramidal neurons. Nature 347: 281–284.

Whiting P, Schoepfer R, Lindstrom J, Priestley T (1991): Structural and pharmacological characterization of the major brain nicotinic acetylcholine receptor subtype stably expressed in mouse fibroblasts. Mol. Pharmacol. 40: 463–472.

Winson J (1978): Loss of hippocampal theta rhythm results in spatial memory deficit in the rat. Science 201: 160–163.

Wonnacott S, Irons J, Lunt GG, Rapier CM, Albuquerque EX (1988): α-Bungarotoxin and presynaptic nicotinic receptors: Functional studies. In: Clementi F et al., editors. Nicotinic acetylcholine receptors in the nervous system. Berlin: Springer-Verlag, pp. 41–60.

Woolf NJ (1991): Cholinergic systems in mammalian brain and spinal cord. Prog. Neurobiol. 37: 475–524.

Wright CI, Geula C, Mesulam MM (1993): Protease inhibitors and indolamines selectively inhibit cholinesterases in the histopathologic structures of Alzheimer's disease. Ann. NY Acad. Sci. 695: 65–68.

Zhang M, Wang YT, Vyas DM, Neuman RS, Bieger D (1993): Nicotinic cholinoceptor-mediated excitatory postsynaptic potentials in rat nucleus ambiguus. Exp. Brain Res. 96: 83–88.

10

Diverse Functions of Neuronal Nicotinic Receptors Determined by Subcellular Localization

Darwin K. Berg Ph.D, William G. Conroy Ph.D, and Paul D. Kassner Ph.D

Department of Biology
University of California
San Diego

Broad interest in nicotinic acetylcholine receptors (nAChRs) in the nervous system has been driven, in part, by the biomedical relevance of the receptors and, in part, by the challenge they pose for understanding nicotinic cholinergic signaling in neural networks. Though nAChRs are widely expressed in the central and peripheral nervous systems, their levels are substantially below those of several other ligand-gated ion channels such as $GABA_A$ or glutamate receptors, and until recently, little was known about their actual functions. New information now suggests a diversity of roles including presynaptic and postsynaptic actions as well as potential instructive effects during development.

The most abundant nicotinic receptor in the vertebrate nervous system appears to be a species containing the nAChR α7 gene product (Couturier et al., 1990; Schoepfer et al., 1990; Conroy and Berg, 1998). When the α7 gene product is expressed heterologously in *Xenopus* oocytes, it produces a functional nAChR that is cation selective, relatively permeable to calcium, and blocked by α-bungarotoxin (α-Bgt; Couturier et al., 1990; Bertrand et al., 1993; Seguela et al., 1993). Though the α7 gene product can be found coassembled with α8 subunits

Neuronal Nicotinic Receptors: Pharmacology and Therapeutic Opportunities, Edited by S. P. Arneric and J. D. Brioni
ISBN 0-471-24743-x, pages 187–196. Copyright © 1998 by Wiley-Liss, Inc.

in chick to produce heteromeric nAChRs (Schoepfer et al., 1990; Keyser et al., 1993), α7 subunits are also thought likely to form homopentameric receptors both in heterologous expression systems and in vivo (Schoepfer et al., 1990; Anand et al., 1993; Gopalakrishnan et al., 1995; Quik et al., 1996; Chen and Patrick, 1997). The functional roles of native nAChRs containing the α7 gene product (α7 nAChRs) and how they depend on subcellular localization make up the focus of this chapter. The biochemical, biophysical, and regulatory properties of the receptors are addressed in other chapters.

PRESYNAPTIC ACTIONS OF α7 nAChRs

Increasing evidence indicates that nAChRs act at presynaptic sites to modulate neurotransmitter release (for review, see Wonnacott, 1997). The first demonstration that α7 nAChRs are likely to be among them was provided by Role and her colleagues. They used whole cell patch-clamp recording to show that bath-applied nicotine increased the frequency of spontaneous synaptic currents in dissociated interpeduncular neurons when co-cultured with medial habenula nucleus explants (McGehee et al., 1995). Since the endogenous transmitter responsible for the synaptic events was glutamate, and since the frequency but not the amplitude of the spontaneous events was affected by nicotine, the site of action was judged to be presynaptic. Moreover, nicotine increased the amplitude of stimulus-evoked synaptic currents, consistent with presynaptic nAChRs modulating synaptic transmission. A similar phenomenon was found at cholinergic synapses between spinal cord explants and sympathetic neurons in culture where bath-applied nicotine increased the frequency of spontaneous synaptic currents in a calcium-dependent, tetrodotoxin (TTX)-insensitive manner. Evidence that the presynaptic receptors were α7 nAChRs was provided by showing that the nicotinic effect was blocked by α-Bgt and that α7 antisense oligonucleotides eliminated the α-Bgt-sensitive nicotine-induced enhancement.

Similar conclusions were reached by John Dani and his colleagues examining the effects of nicotine on hippocampal neurons (Gray et al., 1996). Nicotine applied either to dissociated neurons in culture or to hippocampal slices increased the frequency but not the amplitude of spontaneous synaptic currents (Fig. 10-1A). The effect was calcium dependent, TTX insensitive, and blocked by α-Bgt. Fura-2 fluorescence imaging of calcium further showed that puffer-applied nicotine increased the calcium content of mossy-fiber presynaptic terminals in hippocampal slices. The increases were comparable in amplitude to those caused by action potentials invading the terminal and activating voltage-gated calcium channels, and, again, were blocked by α-Bgt.

Recently, whole cell patch-clamp recording has been used to measure directly nicotine-induced currents in presynaptic terminals in situ (Coggan et al., 1997). This was made possible by the large preganglionic terminals present on chick ciliary ganglion neurons at a stage when the chemical synapses they form are not yet complemented by electrical synapses. Bath-applied nicotine induced long-lasting inward currents in the calyces that were TTX resistant and blocked by α-Bgt (Fig. 10-1B). In the absence of TTX, the currents were sufficient to elicit action potentials in the preganglionic terminals, which in turn evoked synaptic currents in postganglionic neurons. Biochemical analysis of preganglionic nerve extracts with subunit-specific monoclonal antibodies confirmed the presence of α7 nAChRs.

The results show that α7 nAChRs are present on presynaptic terminals and can influence transmitter release. Their high relative permeability to calcium and subcellular positioning may enable even low numbers of such receptors to have a major impact on synaptic transmission. Challenges for the future include determining how and when the receptors are activated in vivo and how they shape the signaling capacity of a network.

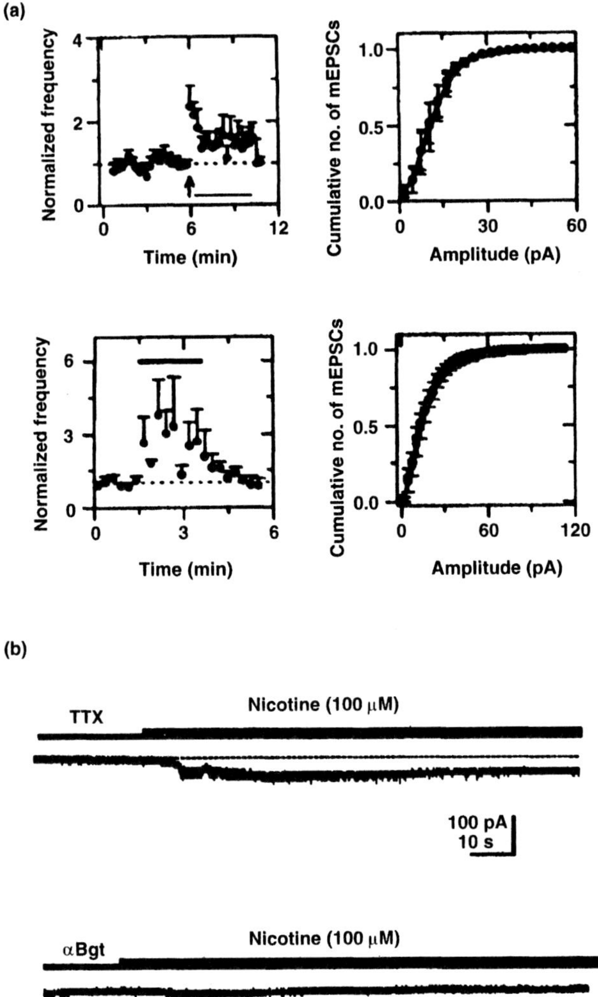

*Figure 10.1. Presynaptic α7 nAChRs detected electrophysiologically in situ. **(A)** Nicotine-induced increase in frequency of spontaneous synaptic currents in CA3 neurons of rat hippocampal slices. Left: Frequency of events before, during, and after 5-second applications of 20 μM nicotine in the presence of TTX and cadmium (normalized to 1 before nicotine). Right: Cumulative amplitude distribution (normalized to 1) for the events before (open circles) and during (filled circles) nicotine application (from Gray et al., 1996). **(B)** Nicotine-induced α-Bgt-sensitive currents in a preganglionic calyx. Whole cell patch-clamp recordings were made from preganglionic calyces in whole ganglia while bath applying 100 μM nicotine in the presence of TTX (upper trace) or αBgt (lower trace) (from Coggan et al., 1997).*

POSTSYNAPTIC ACTIONS OF α7 nAChRs

Neuronal nAChRs have rarely been demonstrated to function as postsynaptic receptors except in the case of transmission through autonomic ganglia. In these instances, however, the postsynaptic receptors are thought not to be α7 nAChRs because they are not blocked by α-Bgt.

(a)

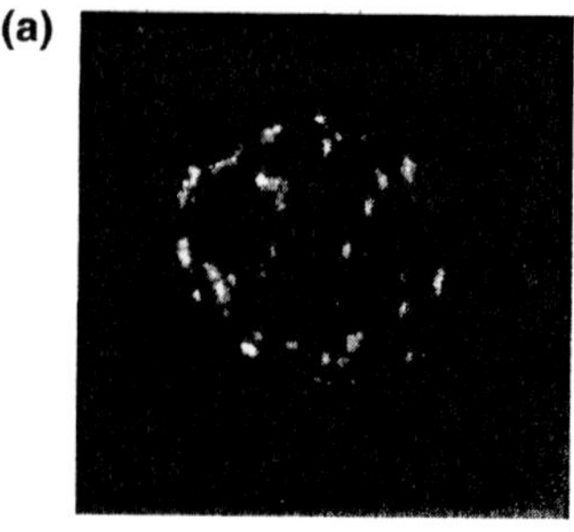

(b)

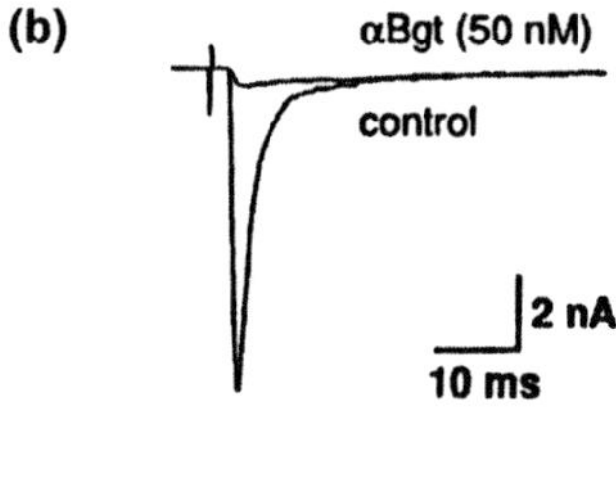

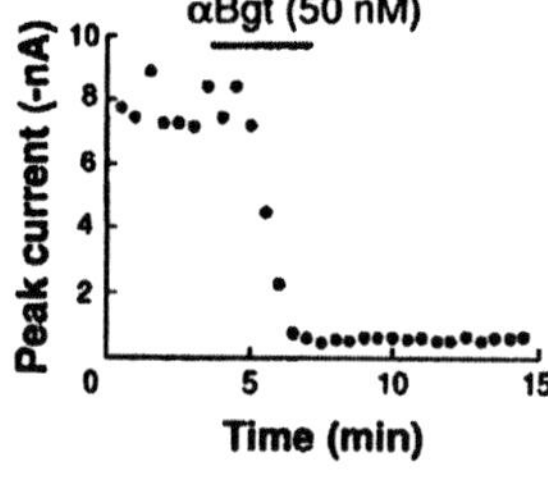

Figure 10.2. Synaptic responses in ciliary ganglion neurons attributed to perisynaptic α7 nAChRs. (A) Perisynaptic clusters of α7 nAChRs labeled with an anti-α7 monoclonal antibody, followed by a fluorescently labeled secondary antibody, and visualized with confocal laser microscopy (from Wilson Horch and Sargent, 1995). (B) α-Bgt-sensitive synaptic currents evoked in a ciliary ganglion neuron in situ and attributed to α7 nAChRs. Upper: Superimposed current traces showing evoked responses before (control) and after (α-Bgt) a 3-minute incubation with 50 nM α-Bgt. Lower: Time course of α-Bgt blockade in which synaptic currents were evoked every 30 seconds in the presence and absence of α-Bgt. The horizontal bar indicates the period when α-Bgt was present (from Zhang et al., 1996).

Autonomic neurons almost invariably have α7 nAChRs, but their functional significance has remained a mystery.

One of the richest sources of α7 nAChRs is the chick ciliary ganglion (Chiappinelli and Giacobini, 1978), where electron microscopic analysis indicates the receptors are excluded from postsynaptic regions demarcated by traditional membrane thickenings (Jacob and Berg, 1983). Immunofluorescence analysis with confocal laser microscopy shows that α7 nAChRs are concentrated in perisynaptic clusters on the neurons (Wilson Horch and Sargent, 1995). The clusters are adjacent to, but not overlapping with, presumed sites of transmitter release identified immunochemically (Fig. 10-2A). Other classes of neuronal nAChRs have been identified in postsynaptic membrane on the neurons (Jacob et al., 1984; Loring and Zigmond, 1987) and have previously been shown capable of mediating synaptic transmission through the ganglion (Chiappinelli et al., 1981).

Surprisingly, α7 nAChRs appear capable of generating large amounts of synaptic current even when confined to perisynaptic locations on neurons. Thus, whole cell patch-clamp recording from neurons in intact ciliary ganglia while stimulating the preganglionic nerve root with a suction electrode reveals a biphasic synaptic current (Zhang et al., 1996; Ullian et al., 1997). The large, rapid component, which accounts for most of the peak current and about half of the total charge entering the cell during the response, resembles that previously attributed to α7 nAChRs on dissociated ciliary ganglion neurons (Zhang et al., 1994). The response rapidly desensitizes and is blocked by low concentrations of α-Bgt applied to the ganglion (Fig. 10-2B). The α-Bgt-sensitive nAChRs are positioned sufficiently close to points of transmitter release such that they also contribute importantly to spontaneously occurring miniature synaptic currents. Estimates of the diffusion time required for a transmitter to reach the receptors is sufficient to account for activation of perisynaptic α7 nAChRs during the evoked response.

The results show for the first time that α7 nAChRs can contribute importantly to synaptic currents in neurons. They also indicate that the functional domain of the postsynaptic membrane may encompass regions not previously thought to participate. However, the results leave unanswered the question of what functions the receptors provide if they are not necessary for synaptic transmission. One possibility is that the receptors support high-frequency transmission through the ganglion either because of their rapid effects on the membrane potential (Zhang et al., 1996) or because of their ability to elevate intracellular calcium levels and thereby activate calcium-dependent currents that may influence the firing capabilities of the neuron (Dryer et al., 1991). Alternatively, calcium influx through α7 nAChRs may regulate calcium-dependent events in the cells such as activation of second messenger cascades (Vijayaraghavan et al., 1995). Another unanswered question is whether α7 nAChRs on cell bodies differ in subunit composition or post-translational modifications from presynaptic α7 nAChRs because preliminary indications are that the two classes of receptors may differ in their affinities for α-Bgt and their rates of desensitization (Alkondon and Albuquerque, 1993; McGehee et al., 1995; Zhang et al., 1996; Coggan et al., 1997).

DEVELOPMENTAL ROLES OF α7 nAChRs

The early appearance during embryogenesis of neuronal nAChRs and the enzyme required to synthesize ACh suggest that nicotinic cholinergic transmission may play an important formative role during development (for review, see Role and Berg, 1996). Studies in cell culture have suggested effects of α7 nAChRs on pathfinding and target selection. Thus, nicotinic agonists can alter neurite outgrowth from the rat pheochromocytoma cell line PC12 and α-Bgt blocks the effects (Chan and Quik, 1993). Similar findings have been obtained with dissociated ciliary ganglion neurons in culture where application of (−)-nicotine or ACh to the growing neurite induces retraction and the effect is largely blocked by α-Bgt (Pugh and Berg, 1994). Both cell types express the α7 gene product and assemble it into α-Bgt binding nAChRs (Rogers et al., 1992; Vernallis et al., 1993).

Recently, presynaptic nAChRs have been studied on motoneurons and inferred to be of the α7 type. Mixed cultures of *Xenopus* spinal cord and muscle cells were grown to obtain neuron-myotube synapses (Fu and Liu, 1997). Bath application of (−)-nicotine caused a large increase in the abilities of ATP and glutamate to enhance the frequency of spontaneously occurring synaptic currents in the myotubes. The nicotinic effect was completely abolished by low concentrations of α-Bgt as expected for α7 nAChRs. Moreover, whole cell patch-clamp recordings directly demonstrated the presence of nicotinic-induced currents in the nerve terminals but not in the neuronal somata in response to local application of nicotine.

The results indicate that α7 nAChRs can be preferentially concentrated in the vicinity of the growth cone and neurite terminal. The appearance of the receptors on motoneuron nerve terminals may reflect a precocious expression anticipating modulatory functions at the mature neuromuscular junction. Equally likely, however, is the possibility that α7 nAChRs on the growing terminal act to provide information about growth cone location and to participate in the subsequent steps of synapse formation. The finding that α7 transcripts and protein can also be found in other tissues such as skeletal muscle, tendon, and periosteum during early development suggests that the gene product may have diverse roles in embryogenesis (Corriveau et al., 1995; Romano et al., 1997).

ASSEMBLY AND MEMBRANE TARGETING OF α7 nAChRs

The fact that α7 nAChRs can display multiple functions depending on their subcellular localization directs attention to the mechanisms responsible for the targeting of receptors to specific sites. Best understood is the vertebrate neuromuscular junction where the 43 kD protein rapsyn clusters muscle nAChRs at the endplate (Gautam et al., 1995). Gephyrin appears to play a similar role for glycine receptors on neurons (Kirsch et al., 1993) while PSD-95 protein clusters NMDA receptors on neurons (Kornau et al., 1995).

Little is known at present about the assembly, stabilization, and membrane clustering of most neuronal nAChRs, but several lines of evidence suggest that special mechanisms may be involved for α7 nAChRs. Cell lines stably transfected with an α7 cDNA can produce α7 nAChRs that are indistinguishable from native receptors, but the cells appear to be far less efficient at doing so than are neurons (Cooper and Millar, 1997; Kassner and Berg, 1997). In addition, neurons convert a portion of the α7 nAChRs they express on the cell surface to a relatively stable subpopulation with a long metabolic half-life and a resistance to extraction with nonionic detergents (Kassner and Berg, 1997). Stably transfected non-neuronal cells appear unable to do so. Another example is provided by a PC12 cell variant that has recently been shown to be post-translationally blocked in the ability to express functional α7 nAChRs (Blumenthal et al., 1997). Other PC12 cell variants can express the receptors, even though they contain fewer α7 transcripts. Lastly, it has been suggested that cyclophilin is necessary as an isomerase or chaperone for the heterologous expression of functional α7 nAChRs in *Xenopus* oocytes (Helekar et al., 1994; Helekar and Patrick, 1997). A requirement for cyclophilin does not appear to explain the ability of neurons to express high levels of the receptors (Cooper and Millar, 1997; Kassner and Berg, 1997).

One component that has recently emerged as a candidate for organizing α7 nAChRs on the neuron surface is rapsyn. Rapsyn transcripts have been identified in various brain regions (Yang et al., 1997) and in chick ciliary ganglion neurons that express high levels of α7 nAChRs (Burns et al., 1997). Cotransfection of QT6 cells with rapsyn and α7 cDNA constructs produces α7 nAChR clusters that codistribute with rapsyn protein (Fig. 10-3). If a green fluorescent protein cDNA construct is substituted for rapsyn in the cotransfections as a negative control, only a diffuse distribution of α7 protein is detected in the cells. In addition, detergent extraction with Triton-X100 indicates that rapsyn converts nearly half of the α7 nAChRs on the cell surface to an extraction-resistant form and dramatically extends the metabolic half-life of such receptors (Kassner et al., 1998). The effect is specific because rapsyn appears to have no effect on the distribution of numerous other membrane proteins that have been examined, nor does it affect the detergent solubility of a number of other endogenously expressed or co-transfected proteins tested.

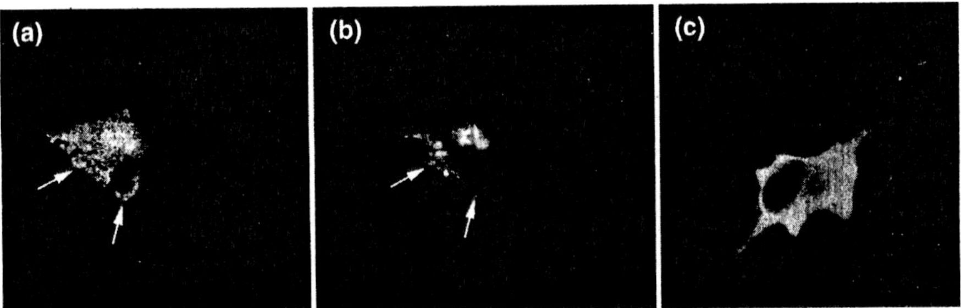

Figure 10.3. *Rapsyn-induced clusters of α7 nAChRs on cotransfected cells. QT6 cells were cotransfected with an α7 and either a rapsyn* ***(A, B)*** *or a green fluorescent protein* ***(C)*** *cDNA construct. After 48 hours, the transfected cells were fixed, permeabilized, and stained either with an anti-α7* ***(A,C)*** *or antirapsyn* ***(B)*** *antibody, followed by Cy3- and FITC-labeled secondary antibodies, and examined by confocal laser microscopy. Arrows indicate examples of codistributed α7 and rapsyn protein.*

The results suggest that rapsyn may induce clustering of α7 nAChRs in neurons. In fact, rapsyn may recognize several types of receptors belonging to the family of ligand-gated ion channels thought to have four transmembrane domains such as nAChRs and $GABA_A$ receptors (Gautam et al.,1995; Yang et al., 1997; Kassner et al., 1998). Neurons also appear to express several variants of rapsyn, including one that lacks exon 2 (Burns et al., 1997), which contains the leucine zipper and may be crucial for receptor clustering (Phillips et al., 1991). It is too early to guess how many distinct factors are responsible for the assembly, stabilization, and clustering of α7 nAChRs at specific sites on the neuron surface, but judging from the distinctive locations of neuronal nAChRs, cellular machinery must clearly be able to distinguish not only among different kinds of receptors but also among different neuronal nAChR subtypes.

SUMMARY AND FUTURE DIRECTIONS

Presynaptic α7 nAChRs can influence transmitter release and modulate synaptic transmission. Perisynaptic α7 nAChRs are likely to generate substantial synaptic currents. Because α7 nAChRs can quickly elevate intracellular calcium levels and because the receptors appear early and in a variety of places, they are positioned to influence many events during development as well. Major challenges for the future include determining the conditions that activate the receptors in vivo and the consequences of receptor activation for neural development and network function. A specific challenge will be identifying the machinery responsible for targeting α7 nAChRs to specific locations on the cell surface. A related issue is whether α7 nAChRs differ in subunit composition or post-translational modifications in such a way as to influence either their final locations or their kinetic and regulatory properties. Answers to these questions will provide insights into the physiological significance of nicotinic cholinergic signaling in the nervous system and may also suggest intervention strategies of biomedical relevance.

ACKNOWLEDGMENTS

We thank Dr. John Dani (Baylor University, Houston) and Dr. Peter Sargent (University of California, San Francisco) for allowing us to reprint figures from their manuscripts. Grant support was provided by NIH grants NS12601 and NS35469, and the Tobacco-Related Disease Research Program grant 6RT-0050.

REFERENCES

Alkondon M, Albuquerque EX (1993): Diversity of nicotinic acetylcholine receptors in rat hippocampal neurons. I. Pharmacological and functional evidence for distinct structural subtypes. J. Pharm. Exper. Ther. 265: 1455–1473.

Anand R, Peng X, Lindstrom J (1993): Homomeric and native α7 acetylcholine receptors exhibit remarkably similar but non-identical pharmacological properties, suggesting that the native receptor is a heteromeric protein complex. FEBS Lett. 327: 241–246.

Bertrand D, Galzi JL, Devillers-Thiery A, Bertran T, Ballivet M (1990): Mutations at two distinct sites within the channel domain M2 alter calcium permeability of neuronal α7 nicotinic receptor. Proc. Natl. Acad. Sci. USA 90: 6971–6975.

Blumenthal EM, Conroy WG, Romano SJ, Kassner PD, Berg DK (1997): Detection of functional nicotinic receptors blocked by α-bungarotoxin on PC12 cells and dependence of their expression on post-translational events. J. Neurosci. 17: 6094–6104.

Burns AL, Benson D, Howard MJ, and Margiotta JF (1997): Chick ciliary ganglion neurons contain transcripts coding for acetylcholine receptor-associated protein at synapses (rapsyn). J. Neurosci. 17: 5016–5026.

Chan J, Quik M (1993): A role for the nicotinic α-bungarotoxin receptor in neurite outgrowth in PC12 cells. Neuroscience 56: 441–451.

Chen D, Patrick JW (1997): The α-bungarotoxin-binding nicotinic acetylcholine receptor from rat brain contains only the α7 subunit. J. Biol. Chem. 272: 24,024–24,029.

Chiappinelli VA, Giacobini E (1978): Time course of appearance of α-bungarotoxin binding sites during development of chick ciliary ganglion and iris. Neurochem. Res. 3: 465–478.

Chiappinelli VA, Cohen JB, Zigmond RE (1981): The effects of α- and β-neurotoxins from the venoms of various snakes on transmission in autonomic ganglia. Brain Res. 211: 107–126.

Coggan JS, Paysan J, Conroy WG, Berg DK (1997): Direct recording of nicotinic responses in presynaptic nerve terminals. J. Neurosci. 17: 5798–5806.

Conroy WG, Berg DK (1998): Nicotinic receptor subtypes in the developing chick brain: appearance of a species containing the α4, β2, and α5 gene products. Molec. Pharmaco. 53: 392–401.

Cooper ST, Millar NS (1997): Host cell-specific folding and assembly of the neuronal nicotinic acetylcholine receptor α7 subunit. J. Neurochem. 68: 2140–2151.

Corriveau RA, Romano SJ, Conroy WG, Oliva L, Berg DK (1995): Expression of neuronal acetylcholine receptor genes in vertebrate skeletal muscle during development. J. Neurosci. 15: 1372–1383.

Couturier S, Bertrand D, Matter J-M, Hernandez M-C, Bertrand S, Millar N, Valera S, Barkas T, Ballivet M (1990): A neuronal nicotinic acetylcholine receptor subunit (α7) is developmentally regulated and forms a homo-oligomeric channel blocked by α-Btx. Neuron 5: 847–856.

Dryer SE, Dourado MM, Wisgirda ME (1991): Characteristics of multiple Ca^{2+}-activated K^+ channels in acutely dissociated chick ciliary ganglion neurones. J. Physiol. (London) 443: 601–627.

Fu W-M, Liu J-J (1997): Regulation of acetyocholine release by presynaptic nicotinic receptors at developing neuromuscular synapses. Molec. Pharmacol. 51: 390–398.

Gautam M, Noakes PG, Mudd J, Nichol M, Chu GC, Sanes JR, Merlie JP (1995): Failure of postsynaptic specialization to develop at neuromuscular junctions of rapsyn deficient mice. Nature 377: 232–236.

Gopalakrishnan M, Buisson B, Touma E, Giordano T, Campbell JE, Hu IC, Donnelly-Roberts D, Arneric SP, Bertrand D, Sullivan JP (1995): Stable expression and pharmacological properties of the human α7 nicotinic acetylcholine receptor. Eur. J. Pharmacol. 290: 237–246.

Gray R, Rajan AS, Radcliffe KA, Yakehiro M, Dani JA (1996): Hippocampal synaptic transmission enhanced by low concentrations of nicotine. Nature 383: 713–716.

Helekar SA, Char D, Neff S, Patrick J (1994): Prolyl isomerase requirement for the expression of functional homo-oligomeric ligand-gated ion channels. Neuron 12: 179–189.

Helekar SA, Patrick J (1997): Peptidyl prolyl cis-trans isomerase activity of cyclophilin A in functional homo-oligomeric receptor expression. Proc. Natl. Acad. Sci. USA 94: 5432–5437.

Jacob MH, Berg DK (1983): The ultrastructural localization of α-bungarotoxin binding sites in relation to synapses on chick ciliary ganglion neurons. J.Neurosci. 3: 260–271.

Jacob MH, Berg DK, Lindstrom JM (1984): Shared antigenic determinant between the *Electrophorus* acetylcholine receptor and a synaptic component on chicken ciliary ganglion neurons. Proc. Natl. Acad. Sci USA 81: 3223–3227.

Kassner PD, Berg DK (1997): Differences in the fate of neuronal acetylcholine receptor protein expressed in neurons and stably transfected cells. J. Neurobiol. 33: 968–982.

Kassner PD, Conroy WG, Berg DK (1998): Clustering and stabilization of neuronal nicotinic acetylcholine receptors by rapsyn. Molec. Cell. Neurosci. 10: 258–270.

Keyser KT, Britto LRG, Schoepfer R, Whiting P, Cooper J, Conroy W, Brozozowska-Prechtl A, Karten HJ, Lindstrom J (1993): Three subtypes of α-bungarotoxin-sensitive nicotinic acetylcholine receptors are expressed in chick retina. J. Neurosci. 13: 442–454.

Kirsch J, Wolters I, Triller A, Betz H (1993): Gephyrin antisense oligonucleotides prevent glycine receptor clustering in spinal neurons. Nature 366: 745–748.

Kornau HC, Schenker LT, Kennedy MB, Seeburg PH (1995): Domain interaction between NMDA receptor subunits and the postsynaptic density protein PSD-95. Science 269: 1737–1740.

Loring RH, Zigmond RE (1987): Ultrastructural distribution of ^{125}I-toxin F binding sites on chick ciliary neurons: synaptic localization of a toxin that blocks ganglionic nicotinic receptors. J. Neurosci. 7: 2153–2162.

McGehee D, Heath M, Gelber S, Role LW (1995): Nicotine enhancement of fast excitatory synaptic transmission in CNS by presynaptic receptors. Science 269: 1692–1697.

Phillips WD, Maimone M, Merlie JP (1991): Mutagenesis of the 43-kD postsynaptic protein defines domains involved in plasma membrane targeting and nAChR clustering. J. Cell Biol. 115: 1713–1723.

Pugh PC, Berg DK (1994): Neuronal acetylcholine receptors that bind α-bungarotoxin mediate neurite retraction in a calcium-dependent manner. J. Neurosci. 14: 889–896.

Quik M, Choremis J, Komourian J, Lukas RJ, Puchacz E (1996): Similarity between rat brain nicotinic α-bungarotoxin receptors and stably expressed α-bungarotoxin binding sites. J. Neurochem. 67: 145–154.

Rogers SW, Madelzys A, Deneris ES, Cooper E, Heinemann S (1992): The expression of nicotinic acetylcholine receptors by PC12 cells treated with NGF. J. Neurosci. 12: 4611–4623.

Role LW, Berg DK (1996): Nicotinic receptors in the development and modulation of CNS synapses. Neuron 16: 1077–1085.

Romano SJ, Corriveau RA, Schwarz R, Berg DK (1997): Expression of the nicotinic receptor α7 gene in tendon and periosteum during early development. J. Neurochem. 68: 640–648.

Schoepfer R, Conroy WG, Whiting P, Gore M, Lindstrom J (1990): Brain α-bungarotoxin binding protein cDNAs and mAbs reveal subtypes of this branch of the ligand-gated ion channel gene superfamily. Neuron 5: 35–48.

Seguela P, Wadiche J, Dineley-Miller K, Dani JA, Patrick JW (1993): Molecular cloning, functional properties, and distribution of rat brain α7: a nicotinic cation channel highly permeable to calcium. J. Neurosci. 13: 596–604.

Ullian EM, McIntosh JM, Sargent PB (1997): Rapid synaptic transmission in the avian ciliary ganglion is mediated by two distinct classes of nicotinic receptors. J. Neurosci. 17: 7210–7219.

Vernallis AB, Conroy WG, Berg DK (1993): Neurons assemble acetylcholine receptors with as many as three kinds of subunits while maintaining subunit segregation among receptor subtypes. Neuron 10: 451–464.

Vijayaraghavan S, Huang B, Blumenthal EM, Berg DK (1995): Arachidonic acid as a possible negative feedback inhibitor of nicotinic acetylcholine receptors on neurons. J. Neurosci. 15: 3679–3687.

Wilson Horch HL., Sargent PB (1995): Perisynaptic surface distribution of multiple classes of nicotinic acetylcholine receptors on neurons in the chicken ciliary ganglion. J. Neurosci. 15: 7778–7795.

Wonnacott S (1997): Presynaptic nicotinic nACh receptors. TINS 20: 92–98.

Yang S-H, Armson PF, Cha J, Phillips WD (1997): Clustering of $GABA_A$ receptors by rapsyn/43kD protein in vitro. Molec. Cell. Neurosci. 8: 430–438.

Zhang Z-W, Vijayaraghavan S, Berg DK (1994): Neuronal acetylcholine receptors that bind α-bungarotoxin with high affinity function as ligand-gated ion channels. Neuron 12: 167–177.

Zhang Z-W, Coggan JS, Berg DK (1996): Synaptic currents generated by neuronal acetylcholine receptors sensitive to α-bungarotoxin. Neuron 17: 1231–1240.

11

Immunosuppressive and Anti-Inflammatory Properties of Nicotine

Mohan Sopori, Ph.D.
Pathophysiology Division
Lovelace Respiratory Research Institute
Albuquerque, New Mexico

Cigarette smoke is a major cause of mortality and morbidity among cigarette smokers and, in the United States alone, it is estimated to cause over 400,000 premature deaths (Peto et al., 1992; US Department of Health and Human Services, 1994). Smoking significantly increases the risks of heart disease, several cancers, and acute/chronic respiratory tract infections (reviewed in Holt and Keast, 1977; Johnson et al., 1990; Sopori et al., 1994). It has been speculated that this increased susceptibility to infections and cancer may result from adverse effects of cigarette smoke on the immune system (Holt and Keast, 1977). Chronic inhalation of cigarette smoke has been shown to affect a wide range of immunologic functions in humans and animals. In experimental animals, chronic exposure to cigarette smoke effects both humoral and cell-mediated immune responses (reviewed in Johnson et al., 1990; Sopori et al., 1994). In humans, smoking increases leukocytosis, reduces antibody titer to influenza, and impairs granulocyte chemotaxis (reviewed in Sopori et al., 1994). Although debatable, recent reports suggest that tobacco smoking may be a risk factor in faster development of AIDS (acquired immunodeficiency syndrome) and *Pneumocytis carinii* infections in HIV-1 seropositive individuals, and higher frequency of transmission of HIV-1 infections from smoking mothers to their offspring (reviewed in Sopori and Kozak, 1998).

While the health risks of tobacco smoke are well established, there is increasing evidence

Neuronal Nicotinic Receptors: Pharmacology and Therapeutic Opportunities, Edited by S. P. Arneric and J. D. Brioni
ISBN 0-471-24743-x, pages 197–209. Copyright © 1998 by Wiley-Liss, Inc.

suggesting that cigarette smokers have a lower incidence of several autoimmune/inflammatory-type diseases (reviewed in Baron, 1996). Therefore, some component(s) of tobacco smoke may modulate inflammatory responses. Also, there is increasing evidence showing that cigarette smoke may prevent or ameliorate Parkinson's and Alzheimer's diseases (reviewed in Aisen, 1997; Perry et al., 1997); however, the mechanism(s) of these effects is unclear.

Tobacco smoke is a complex mixture of several thousand chemicals, many of which have toxic and/or carcinogenic activity (Hoffmann et al., 1979). One of the major pharmacological components of tobacco is nicotine (Cryer et al., 1976; Seyler et al., 1986; Van Loon et al., 1987), and in this chapter I review the recent evidence suggesting that nicotine may also be a major contributor to the immunosuppressive and anti-inflammatory properties of tobacco smoke.

EFFECTS OF NICOTINE ON THE IMMUNE SYSTEM

Nicotine May Affect the Cigarette Smoke–Induced Immunosuppression

Rodents (rats and mice), when exposed chronically to cigarette smoke, exhibit inhibition of the primary antibody response to both T-dependent and T-independent antigens (Sopori et al., 1989; Chang et al., 1990; Goud et al., 1992). Moreover, cigarette smoke, containing higher levels of tar and nicotine, induces immunological changes faster than cigarette smoke containing lower levels of these components (Holt et al., 1976). This observation suggests that tar and/or nicotine may represent the immunosuppressive component(s) of cigarette smoke. We have reported that chronic exposure of rats to the vapor phase of cigarette smoke does not significantly affect their antibody response, indicating that the immunosuppressive properties of cigarette smoke are associated primarily with the particulate phase of cigarette smoke (Sopori et al., 1993). This phase contains thousands of different compounds including most of the nicotine (Hoffmann et al., 1979; Dube and Green, 1982). Because nicotine is one of the biologically very active compounds, it was possible that nicotine caused or contributed to the immunosuppressive properties of tobacco smoke.

To determine the role of nicotine in the tobacco smoke–induced immunosuppression, rats were exposed to saline (control) or -(−)nicotine (hereafter referred to as nicotine) via the subcutaneously implanted miniosmotic pumps. These pumps delivered nicotine at a constant rate

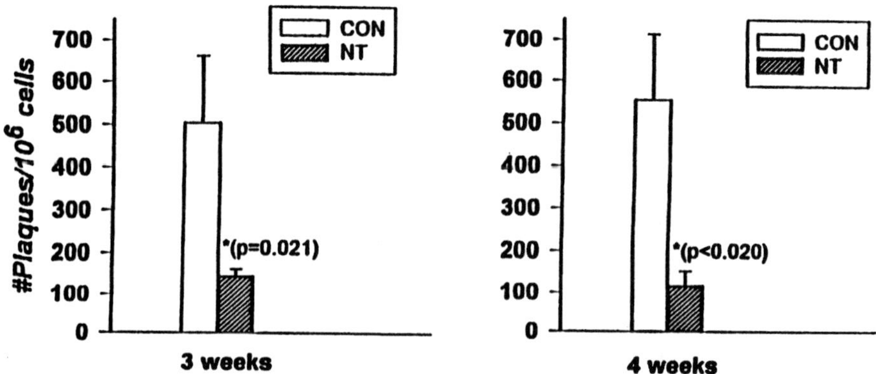

Figure 11.1. Chronic exposure of rats to nicotine inhibits anti-SRBC AFC response of spleen cells. LEW rats (six to eight in each group) were implanted with (−)-nicotine- or saline-containing miniosmotic pumps for 3 (left) or 4 weeks (right). Animals were immunized with SRBC 4 days prior to assay (Geng et al., 1995).

(1 mg/kg body weight /24 h) for 4 weeks; the serum nicotine level of these rats was comparable to that of humans smoking approximately one pack of cigarettes/day. Under these conditions, the antibody-forming cell (AFC) response of spleen cells was significantly reduced in animals treated with nicotine for 3 to 4 weeks (Fig. 11-1). However, as with chronic cigarette smoking (Sopori et al., 1989; Savage et al., 1991), this inhibition of the antibody response is not accompanied with significant changes in leukocyte subpopulations (Geng et al., 1995). From the above observations we conclude that chronic exposure to nicotine causes immunosuppression.

Nicotine Impairs Antigen-mediated Signaling in T Lymphocytes

We previously observed that chronic exposure of rats to cigarette smoke inhibits the antigen-induced elevation of intracellular calcium concentration $[Ca^{2+}]_i$, in T cells (Sopori et al., 1993), and T cells from chronically nicotine-treated rats also exhibit a similar defect in the Ca^{2+} response (Fig. 11-2), suggesting that both cigarette smoke and nicotine affect the antigen-mediated signaling in T lymphocytes (Geng et al., 1996). To understand the mechanism by which nicotine inhibits the antigen-dependent Ca^{2+} response in T cells, we examined the major steps proximal to Ca^{2+} mobilization in the T-cell antigen receptor (TCR)-induced signaling pathway. Signaling through antigen receptors can lead to profound biological responses, including activation, tolerance, and/or differentiation depending on the nature of the stimulus and the differentiation state of the lymphocyte (Chan et al., 1994). For the most part, experimentally, activation of T cells by an antigen is achieved by ligation of TCR with anti-CD3 or anti-αβTCR antibodies, resulting in augmented protein tyrosine kinase (PTK) activity (Weiss and Littman, 1994; Robey and Allison, 1995). A consequence of increased PTK activity in T cells is the activation of phospholipase C-γ1 (PLC-γ1) through tyrosine phosphorylation (Nishibi et al., 1990). Activated (i.e., tyrosine phosphorylated) PLC-γ1 catalyzes the hydrolysis of phosphatidylinositol 4,5-bisphosphate to inositol 1,4,5-trisphosphate (IP3) and diacylglycerol, and IP3 acts as a second messenger for mobilizing Ca^{2+} from intracellular Ca^{2+} stores leading to elevated $[Ca^{2+}]_i$ (Berridge, 1993; Clapham, 1995).

Interestingly, T cells from chronic nicotine-treated animals exhibited significantly higher basal levels of IP3 (Geng et al., 1996). The latter may result from a higher background activity of PTK (Fig 11-3A) and increased tyrosine phosphorylation of PLC-γ1 (Fig. 11-3B) in T

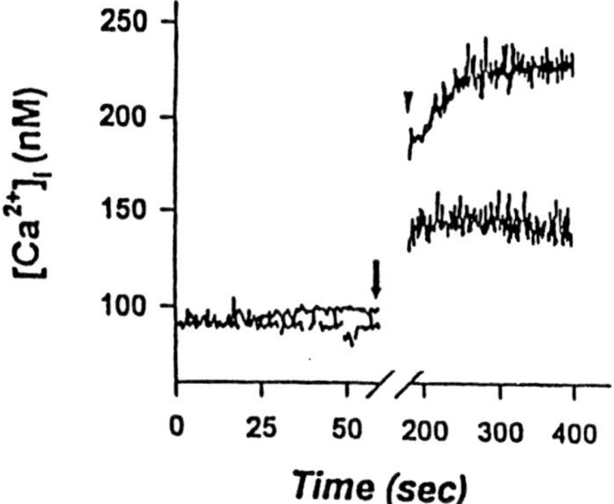

Figure 11.2 Chronic exposure to nicotine inhibits T-cell antigen receptor-mediated rise in $(Ca^{2+})_i$ Spleen cells from control and (−)-nicotine-treated (4 weeks) LEW rats were loaded with acetoxymethyl ester of indo-1 under previously described conditions (Geng et al., 1995). At the indicated time (↓), cells were stimulated with anti-CD3 and analyzed by spectrofluorometry.

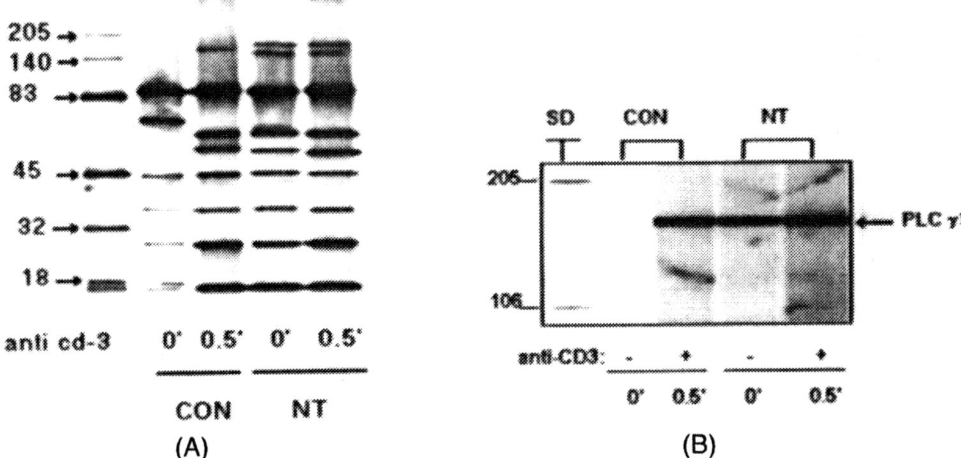

Figure 11.3. *Chronic nicotine treatment causes constitutive activation of protein tyrosine kinase activity* ***(A)*** *and tyrosine phosphorylation (activation) of PLC-γ1* ***(B)*** *in T cells. Spleen cells from control and (−)-nicotine-treated (4 weeks) rats were incubated with (+) and without (−) anti-CD3 antibody for 0.5 min. Cell lysates were prepared and analyzed by Western blots (Geng et al., 1996).*

cells frem nicotine-treated animals. However, unlike control T cells, the PTK and the PLC-γ1 activities of nicotine-treated cells are refractory to up-regulation by anti-CD3 (fig. 11-3A and B). Increased background PTK and IP3 levels are also seen in antigen-specific T-cell clones that become unresponsive (anergic) to antigen-specific activation (Gajewski et al., 1994; Sloan-Lancaster et al., 1994). Therefore, it is possible that chronic nicotine exposure also induces a similar anergic state in T cells.

The observation that, in the absence of anti-CD3, the background $[Ca^{2+}]_i$ in nicotine-treated cells is not different from control cells (Geng et al., 1996), suggests that the higher IP3 level in nicotine-treated T cells does not translate into mobilization of intracellular IP3-sensitive Ca^{2+} stores. Therefore, these stores in nicotine-treated T cells are either insensitive to elevated IP3 levels or do not contain releasable Ca^{2+} in the Ca^{2+} stores. Our recent results (Fig. 11-4) indicate that T cells from nicotine-treated animals release substantially less Ca^{2+} when treated with thapsigargin, an inhibitor of Na-K ATPases, releasing Ca^{2+} by irreversibly blocking the filling of IP3-sensitive endoplasmic Ca^{2+} stores (Thastrup et al., 1990). Thus, nicotine appears to deplete IP3-sensitive Ca^{2+} stores in T cells, and recent evidence suggests that these stores are critical for transport of molecules across the nuclear pore complex (Greber and Gerace, 1995) and the movement of G1 into the S phase of the cell cycle (Takuwa et al., 1995). In support of this possibility, we have demonstrated that, following stimulation with the T-cell mitogen, concanavalin A, spleen cells from nicotine-treated animals fail to enter normally into the S phase and are arrested in the G0/G1 phase of the cell cycle (Geng et al, 1995). Thus, chronic nicotine treatment depletes IP3-sensitive intracellular Ca^{2+} stores leading to the G0/G1 arrest and nonresponsiveness of T cells to antigen receptor–mediated stimulation.

Anti-inflammatory Properties of Nicotine

Long-term smoking causes chronic pulmonary inflammation accompanied by accumulation of "activated" macrophages in alveoli and respiratory bronchioli (reviewed in Sopori et al., 1994).

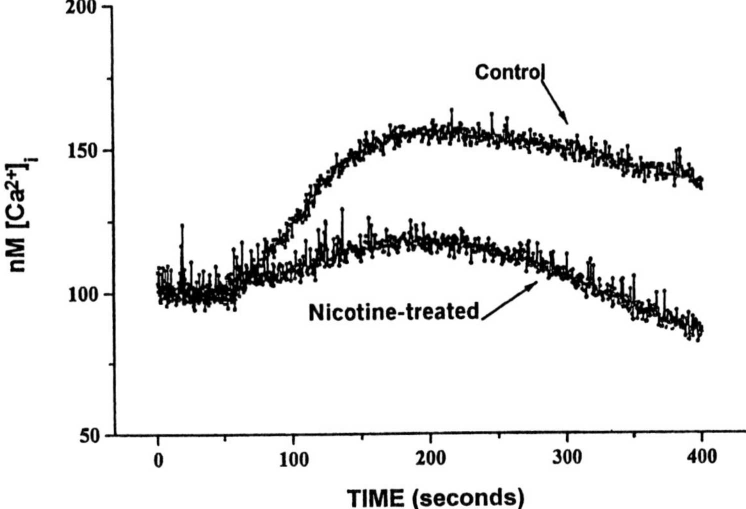

Figure 11.4. *Chronic nicotine depletes intracellular IP3-sensitive calcium stores of T cells. T cells from control and (−)-nicotine-treated (4 weeks) rats were labeled with indo-1, suspended in calcium-free medium, and treated with 50 nM thapsigargin (Razani-Boroujerdi et al., 1994) to release IP3-sensitive Ca^{2+} stores. Changes in $[Ca^{2+}]_i$ were analyzed by spectrofluorometry.*

However, the ability of alveolar macrophages (AMs) from cigarette smokers to secrete proinflammatory cytokines (IL-1, IL-6, and TNF-α) is significantly reduced (Brown et al., 1989; McCrea et al., 1994; Sauty et al., 1994). Moreover, as discussed before, cigarette smokers tend to have a lower frequency of non-lung-inflammatory diseases (Sopori et al., 1994). Therefore, in general, cigarette smoke tends to be anti-inflammatory; however, in the lung, constant inhalation and deposition of the particulate matter may favor inflammation in this organ. To examine the question of whether nicotine is anti-inflammatory, we have used the following two animal models of inflammation.

Turpentine Abscess-induced Febrile Response. It is a well-established fact that infections, regardless of their origin, stimulate various parameters of the inflammatory response including a rise in the body temperature (Kluger, 1991). Sterile tissue damages (burns, surgical traumas, ischemic necrosis) also trigger an inflammatory response (Kushner, 1982), and an abscess caused by subcutaneous injection of turpentine is a widely used model for localized inflammation (Kozak et al., 1997). To evaluate whether nicotine suppresses inflammation in this model, mice were implanted intra-abdominally with biotelemeters to record constantly the deep-body temperature (Sopori et al., 1997). One week after biotelemeter implantation, animals were implanted subcutaneously with saline (control) or nicotine-containing miniosmotic pumps; two weeks later animals were challenged with steam-distilled turpentine or pyrogen-free saline into the hind limb. Fig. 11-5 shows that prior to turpentine injection there is no significant difference in the body temperature between control and nicotine-treated animals; however, a few hours after turpentine injection, body temperatures of control animals started to rise and a difference of approximately 2°C before and after turpentine treatment is observed. On the other hand, body temperatures of nicotine-treated animals does not significantly differ between before and after turpentine injection (Fig.11-5). These results suggest that nicotine attenuates the febrile component of the inflammatory response.

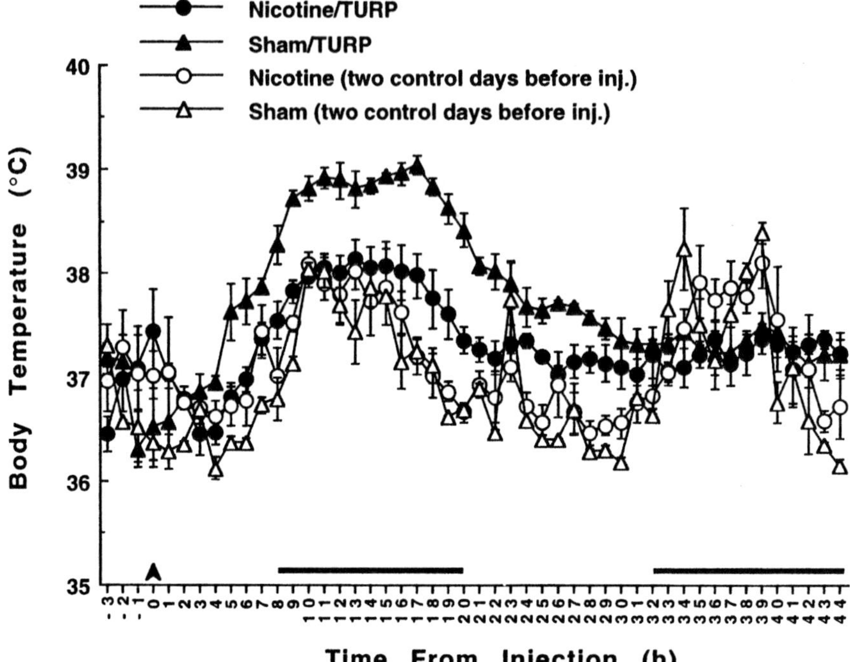

Figure 11.5. Nicotine inhibits the rise in body temperature in response to turpentine-induced sterile abscess. SWR mice were implanted with biotelemeters to record deep-body temperature, and saline-(control) or (−)-nicotine-containing miniosmotic pumps. Two weeks after the implantation, mice were injected with 50 μL Turpentine (Turp), and body temperature was recorded for 48 h (Sopori et al., 1997). Open symbols represent the recorded body temperature for 48 h before turpentine injection. Animals were kept in 12 h dark: light cycle, and horizontal dark lines represent the dark (active) period.

Influenza Virus-Induced Pneumonitis. Infection of mice with lethal doses of influenza A virus causes death within a few days, resulting primarily from an acute severe inflammatory response and respiratory distress (Pepper and Van Campen, 1995). To determine whether nicotine would attenuate morbidity/mortality of influenza-infected mice by suppressing the lung inflammation, mice were treated with saline-or nicotine-containing miniosmotic pumps for 2 weeks before lethal infection of influenza virus. Results presented in Fig. 11-6 show that, whereas all animals in the control group died by day 11, about 50% of the nicotine-treated mice were still alive at that time. This attenuated mortality was not the result of reduced viral titer, because nicotine-treated animals actually had a 3- to 5-fold higher viral burden in the lung than control animals (Sopori et al., 1997). These results strongly suggest that nicotine moderates some parameters of inflammation (Sopori and Kozak, 1998).

Neuroimmune Effects of Nicotine

There is no unequivocal evidence for the presence of classical nicotine acetylcholine receptors on lymphocytes (Richman and Arnason, 1979; Mihoviloc and Roses, 1993). We have observed

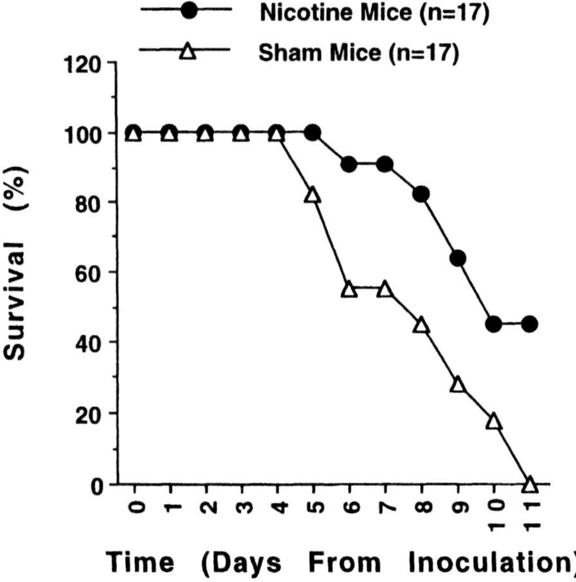

Figure 11.6. Nicotine prolongs survival of SWR mice after a lethal influenza virus infection. Mice were implanted with (−)-nicotine-containing miniosmotic pumps and after 2 weeks were inoculated intranasally with influenza A virus (Sopori et al., 1997).

that relatively high concentrations of nicotine are required to stimulate PTK activity, IP3 synthesis, and Ca^{2+} response in lymphocytes, but physiological concentration has no discernible effect on these cells (Sopori et al., 1997). While the possibility for the presence of low-affinity nicotinic receptors on lymphocytes exists, it is hard to gauge the role of such receptors in mediating immunosuppression under physiological conditions.

Moreover, because nicotine-induced immunosuppression takes at least 2–3 weeks to manifest (Geng et al., 1995), it is likely that the major immunologic effects of nicotine in vivo do not result from its direct interaction with lymphocytes but indirectly through its effects on the central nervous system (CNS). Therefore, it is possible that concentrations of nicotine, which are too low to affect the immune system through peripheral administration, may suppress the immune system when given into the CNS. To ascertain this, cannulae were implanted intracerebroventricularly (icv) into the lateral ventricles of rat brains and connected to miniosmotic pumps delivering either artificial cerebrospinal fluid (aCSF) (sham control) or relatively small quantities of nicotine (28 μg/day/kg body weight) dissolved in aCSF (Sopori et al., 1997). After a 2-week treatment, compared to control animals,the anti-sheep red blood cell (SRBC) AFC response of nicotine-treated rats was dramatically reduced (Fig. 11-7), indicating a strong immunosuppressive response. Plasma concentrations of nicotine in these animals were below the assay detection limits and subcutaneous administration of this amount of nicotine (28 μg/kg body weight/day) for 3 weeks had no detectable effect on the AFC response (Fig. 11-7). It is likely, therefore, that in vivo the primary effect of nicotine on the immune system is mediated through its effects on the brain.

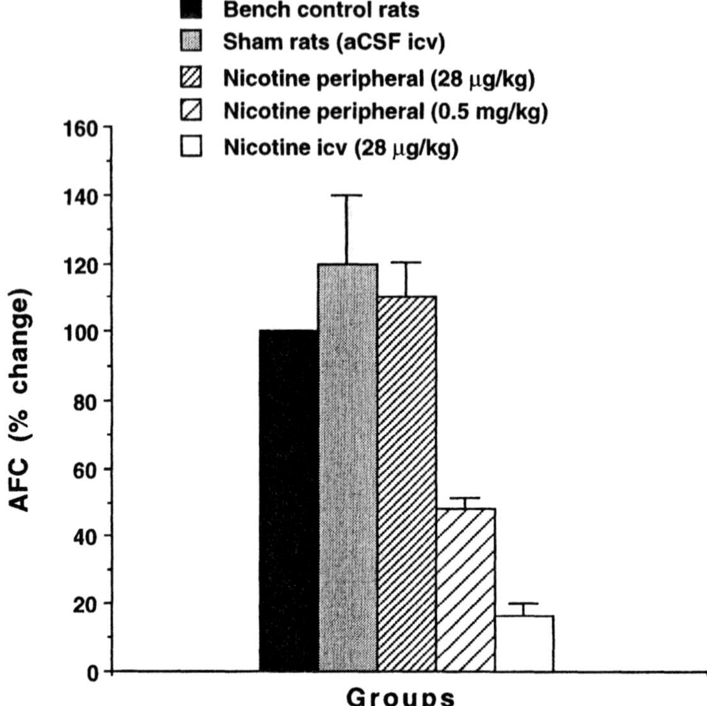

Figure 11.7. Intracerebroventricular (icv) administration of (−)-nicotine (28 μg/day/kg body weight) inhibits the AFC response to SRBC. LEW rats were implanted subcutaneously with nicotine-containing pumps delivering either 28 μg or 0.5 mg of (−)-nicotine/day/kg body weight) or aCSF (sham) was delivered icv via a cannula attached to a miniosmotic pump. Fourteen days after the implantation, animals were injected with SRBC and sacrificed 4 days later. Each group contained four to seven animals.

DISCUSSION AND CONCLUSIONS

Chronic exposure to tobacco smoke adversely affects the immune system in humans and experimental animals. Cigarette smoke is a mixture of thousands of chemicals and recent evidence suggests that nicotine is one of the major immunosuppressive components of cigarette smoke (reviewed in Sopori and Kozak, 1998), affecting the function of both T and B cells. In T cells, nicotine induces functional unresponsiveness (anergy) (Geng et al., 1995, 1996; Sopori et al., 1997), and this anergy may arise through its effects on the antigen-mediated signal transduction pathway leading to "partial activation" and arrest of T cells in the G0/G1 phase of the cell cycle (Geng et al., 1995, 1996). Recent results from our laboratory suggest that T cells from chronically nicotine-treated animals have depleted IP3-sensitive intracellular Ca^{2+} stores (Sopori et al., 1997). These Ca^{2+} pools are critical in the active transport of molecules between the cytoplasm and the nucleus (Greber and Gerace, 1995), and their depletion in nicotine-treated cells may impair the cytoplasm–nucleus communication, leading to the loss of T-cell function.

The manner in which nicotine interacts with lymphocytes is not clear at the present time. While, in vitro, lymphocytes may respond to relatively high concentrations of nicotine (Sopori et al., 1997), so far, the experiments do not support the presence of high-affinity nicotinic acetylcholine receptors on peripheral lymphocytes. It is possible that immunosuppressive effects of nicotine do not require direct interaction of nicotine with lymphocytes but involve interaction with nonimmune cells. In fact, concentrations of nicotine that are too low to affect the immune response when given subcutaneously are highly immunosuppressive when given directly into the brain. Therefore, it is likely that under physiological conditions effects of nicotine on the immune system are mediated through the CNS.

An intimate relationship exists between the neuroendocrine and immune systems during development, maturation, and aging processes, and substantial evidence supports the existence of a complex, bidirectional link between the central nervous system and the immune system. These systems communicate with each other through shared signal molecules such as cytokines, hormones, and neurotransmitters acting on receptors common to both systems (Blalock, 1994). Nicotine is a classical sympathoadrenal stimulant (Braubar, 1995) and stimulates the hypothalamus–pituitary–adrenal (HPA) axis, causing secretion of ACTH and glucocorticoids (Seyler et al., 1986). However, our recent results (Sopori et al., in preparation), using adrenalectomized animals, do not support a major role for the HPA axis in the immunosuppression caused by chronic nicotine treatment. Nicotine has also been shown to cause a temporary acute inhibition of T-cell proliferation (Dr. P. Puttfarcken, personal communication), and it is likely that these immunosuppressive effects are mediated via the activation of the HPA axis. Therefore, other neuroimmune pathway(s) may be operative during chronic nicotine exposure, which affects the immune system and remains to be identified. There are two, not necessarily mutually exclusive, possibilities: (1) Catecholamines have been shown to inhibit the immune response (Fuchs and Sanders, 1994), and nicotine increases the plasma concentration of both norepinephrine and epinephrine (Cryer et al., 1976; Seyler et al., 1986; Van Loon et al., 1987). (2) Neurohormones/peptides such as substance P, prolactin, and opioids are known to modulate the immune system (Cavagnaro et al., 1988). However, at present, there is no strong evidence to include or exclude any of these candidates.

It is becoming clear that cigarette smoking limits the risk of several inflammatory-type diseases, and smokers have lower incidence of ulcerative colitis, sarcoidosis, pigeon breeder's disease, farmer's lung, environmental allergies, endometriosis, uterine fibroids, and acne (Mills et al., 1993; Sopori et al., 1994; Baron, 1996). We believe most of these beneficial effects are associated with the anti-inflammatory property of nicotine, which we have demonstrated in two models of inflammation (i.e., turpentine-induced sterile abscess and influenza virus–induced pulmonary inflammation). Moreover, recent reports suggest an inverse relationship between the risk of Alzheimer's and Parkinson's diseases and cigarette smoking (reviewed in Birtwistle and Hall, 1996; Newhouse et al., 1997). While many explanations have been offered for the ameliorating effects of nicotine in these disorders (Arneric et al., 1995; James and Nordberg, 1995; Salomon et al., 1996), it is possible that the anti-inflammatory property of nicotine is, at least partially, responsible for limiting the risk of these diseases. This premise is based on recent observations that many neurological diseases including Alzheimer's and Parkinson's are associated with the presence of inflammatory cells and/or other components of inflammation in the central nervous system (Pereira et al., 1996; Barger and Harmon, 1997; Du Yan et al., 1997; Ishizuka et al., 1997).

Nicotine/cigarette smoke may affect inflammation through modulation of the proinflammatory cytokines IL-1, IL-6, and TNF-α, and compared to nonsmokers, alveolar macrophages from smokers secrete significantly lower levels of these cytokines (Brown et al., 1989; Yamaguchi et al., 1989; Soliman et al., 1992; McCrea et al., 1994; Sauty et al., 1994). Nicotine treatment is also associated with the inhibition of IL-1 and TNF-α production from colonic mucosa (Van Dijk et al., 1995) and could explain the lower incidence of ulcerative colitis in smokers. These proinflammatory cytokines are also known to play an important role in clear-

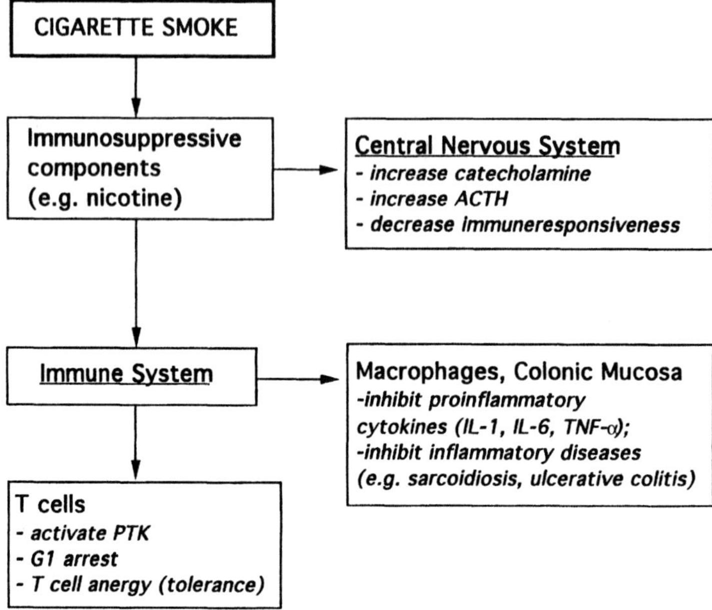

Figure 11.8. *An oversimplified model by which cigarette smoke/nicotine could affect the immune system.*

ing infections from the body (Balkwill, 1993; Marrack and Kappler, 1994). These observations may account for the higher susceptibility of smokers to infections but a lower frequency of some autoimmune diseases. Some potential interactions between nicotine and the immune system are shown schematically in Fig. 11-8. Understanding the molecular mechanisms of these interations may prove useful in the development of treatments for many important human diseases.

ACKNOWLEDGMENTS

This work was supported in part by grants from the National Institute of Drug Abuse (DA04208 and DA05662) and a donation from Abbot Laboratories, Abbott Park, IL.

REFERENCES

Aisen PS (1997): Inflammation and Alzheimer's disease: mechanisms and therapeutic strategies. Gerontology 43: 143–149.

Arneric SP, Sullivan JP, Decker MW, Brioni JD, Bannon AW, Briggs CA, Donnelly-Roberts D, Marsh KC, Kyncl J, Williams M, Buccafusco, JJ (1995): Potential treatment of Alzheimer's disease using cholinergic channel activators (ChCAs) with cognitive enhancement, anxiolytic-like, and cytoprotective properties. Alzheimer Dis. Assoc. Disord. 9: 50– 61.

Balkwill F (1993): Cytokines in health and disease. Immunol. Today 14: 149–150.

Barger WB, Harmon AD (1997): Microglial activation by Alzheimer amyloid precursor protein and modulation by apolipoprotein E. Nature 388: 878–881.

Baron JA (1996): Beneficial effects of nicotine and cigarette smoking: the real, the possible, and the spurious. Br. Med. Bull. 52: 58–73.

Berridge MJ (1993): Inositol trisphosphate and calcium signaling. Nature 361: 315–325.

Birtwistle J, Hall K (1996): Does nicotine have beneficial effects in the treatment of certain diseases? Br. J. Nurs. 5: 1195–1202.

Blalock JE (1994): The immune system: our sixth sense. Immunologist 2: 8–15.

Braubar N (1995): Direct effects of nicotine on the brain: evidence for chemical addiction. Arch. Environ. Health 50: 263–266.

Brown GP, Iwamoto GK, Monick MM, Hunninghake GW (1989): Cigarette smoking decreases interleukin-1 release by human alveolar macrophages. Am. J. Physiol. 256: C260–C264.

Cavagnaro J, Waterhouse JAW, Lewis RM (1988): Neuroendocrine-immune interactions: immunoregulatory signals mediated by neurohumoral agents. In Cruse JM, Lewis RE, editors. The year in immunology 1986–1987, vol. 3. Basal: Karger, pp. 228–246.

Chan A, Desai D, Weiss A (1994): The role of tyrosine kinases and protein tyrosine phosphatases in the antigen receptor signal transduction. Annu. Rev. Immunol. 12: 555–592.

Chang JCC, Distler SG, Kaplan AM (1990): Tobacco smoke suppresses T cell but not antigen-presenting cells in the lung-associated lymph nodes. Toxicol. Appl. Pharmacol. 102: 514–523.

Clapham DE (1995): Calcium signaling. Cell 80: 258–268.

Cryer PE, Hymond MW, Santiago JV, Shah SD (1976): Norepinephrine and epinephrine release and adrenergic mediation of smoking-associated hemodynamic and metabolic events. N. Engl. J. Med. 295: 573–577.

Dube MF, Green CR (1982): Methods of collection of smoke for analytical purposes. Recent Adv. Tobacco Sci. 8: 42–102.

Du Yan S, Zhu H, Fu J, Yan SF, Roher A, Tourtellotte WW, Rajavashisth T, Chen X, Godman GC, Stern D, Schmidt AM (1997): Amyloid-beta peptide receptor for advanced glycation end product interaction elicits neuronal expression of macrophage-colony stimulating factor: a proinflammatory pathway in Alzheimer's disease. Proc. Natl. Acad. Sci. USA 94: 5296–5301.

Fuchs BA, Sanders VM (1994): The role of brain-immune interaction in immunotoxicology. Crit. Rev. Toxicol. 24: 151–176.

Gajewski TF, Qian D, Fields P, Fitch FW (1994): Anergic T-lymphocyte clones have altered inositol phosphate, calcium, and tyrosine kinase signaling pathways. Proc. Natl. Acad. Sci. USA 91: 38–42.

Geng Y, Savage SM, Johnson LJ, Seagrave J, Sopori ML (1995): Effects of nicotine on the immune response. I. Chronic exposure to nicotine impairs antigen receptor-mediated signal transduction in lymphocytes. Toxicol. Appl. Pharmacol. 135: 268– 278.

Geng Y, Savage SM, Razani-Boroujerdi S, Sopori ML (1996): Effects of nicotine on the immune response. II. Chronic nicotine treatment induces T cell anergy. J. Immunol. 156: 2384–2390.

Goud SN, Kaplan AM, Subbarao B (1992): Effects of cigarette smoke on the antibody response to thymic independent antigens from different lymphoid tissues of mice. Arch. Toxicol. 66: 164–169.

Greber U, Gerace L (1995): Depletion of calcium from the lumen of endoplasmic reticulum reversibly inhibits passive diffusion and signal-mediated transport into the nucleus. J. Cell Biol. 128: 5–14.

Hoffmann D, Rivenson A, Hecht SS, Hilfrich I, Kobayashi N, Wynder EL (1979): Model studies in tobacco carcinogenesis with Syrian golden hamster. Prog. Exp. Tumor Res. 24: 370–390.

Holt PG, Keast D (1977): Environmentally induced changes in immunological function: acute and chronic effects of inhalation of tobacco smoke and other atmospheric contaminants in man and experimental animals. Bacteriol. Rev. 41: 205–216.

Holt PG, Chalmer J, Roberts LM, Papadimitriou JM, Thomas WR, Keast D (1976): Low-tar high-tar cigarettes: comparison of effects in mice. Arch. Environ. Health 31: 258–265.

Ishizuka K, Kimura T, Igata-yi R, Katsuragi S, Takamatsu J, Miyakawa T (1997): Identification of monocyte chemoattractant protein-1 in senile plaque and reactive microglia of Alzheimer's disease. Psychiatry Clin. Neurosci. 51: 135–138.

James JR, Nordberg A (1995): Genetic and environmental aspects of the role of nicotinic receptors in neurodegenerative disorders: emphasis on Alzheimer's disease and Parkinson's disease. Behav. Genet. 25: 149–159.

Johnson JD, Hauchens DP, Kluwe WM, Craig DK, Fisher GL (1990): Effects of mainstream and environmental tobacco smoke on the immune system in animals and humans: a review. Crit. Rev. Toxicol. 20: 369–395.

Kluger MJ (1991): Fever: role of pyrogens and cryogens. Physiol. Rev. 71: 93–127.

Kozak W, Poli V, Soszynski D, Conn CA, Leon LR, Kluger MJ (1997): Sickness behavior in mice deficient in interleukin-6 during turpentine and influenza pneumonitis. Am. J. Physiol. R621–R630.

Kushner I (1982): The phenomenon of the acute phase response. Ann. NY Acad. Sci. 389: 39–48.

Marrack P, Kappler J (1994): Subversion of immune system by pathogens. Cell 76: 323–332.

McCrea KA, EnSor JE, Nall K, Bleecker ER, Hasday JD (1994): Altered cytokine regulation in the lungs of cigarette smokers. Am. J. Respir. Crit. Care Med. 150: 696–703.

Mihovilovic M, Roses AD (1993): Expression of α-3, α-5, and β-4 neuronal acetylcholine receptor subunit transcripts in normal and myasthenia gravis thymus. J. Immunol. 151: 6517–6524.

Mills CM, Peters TJ, Finlay AY (1993): Does smoking influence acne? Clin. Exp. Dermatol. 18: 100–101.

Newhouse PA, Potter A, Levine ED (1997): Nicotinic system involvement in Alzheimer's and Parkinson's diseases. Implication for therapeutics. Drugs Aging 11: 206–228.

Nishibe S, Wahl M, Hernandez-Sotomayor SMT, Tonks NK, Rhee SG, Carpenter G (1990): The phosphorylation of PLC-γ1 results in increase of its catalytic activity. Science 250: 1253–1256.

Peper RL, Van Campen H (1995): Tumor necrosis factor as a mediator of inflammation in influenza A viral pneumonia. Microbial Pathogen 19: 175–183.

Pereira HA, Kumar P, Grammas P (1996): Expression of CAP37, a novel inflammatory mediator, in Alzheimer's disease. Neurobiol. Aging 17: 753–759.

Perry VH, Anthony DC, Bolton SJ, Brown HC (1997): The blood-brain barrier and the inflammatory response. Mol. Med. Today 3: 335–341.

Peto R, Lopez A, Boreham J, Thun M, Heath C Jr. (1992): Mortality from tobacco in developed countries: indirect estimation from national vital statistics. Lancet 339: 1268–1278.

Razani-Boroujerdi S, Partridge LD, Sopori ML. (1994): Intracellular calcium signaling induced by thapsigargin in excitable and inexcitable cells. Cell Calcium 16: 467–474.

Richman D, Amason BG (1979): Nicotinic acetylcholine receptor: evidence for functionally distinct receptor on human lymphocytes. Proc. Natl. Acad. Sci. USA 76: 4632–4635.

Robey E, Allison JP (1995): T cell activation: integration of signals from the antigen receptor and constimulatory molecules. Immunol. Today 16: 306–310.

Salomon AR, Marcinowski KG, Friedland RP, Zagorski MG (1996): Nicotine inhibits amyloid formation by the beta-peptide. Biochemistry 35: 13,568–13,578.

Sauty A, Mauel J, Philippeaus M-M, Leuenberger P (1994): Cytostatic activity of alveolar macrophages from smokers and nonsmokers. Role of interleukin-1β, interleukin-6, and tumor necrosis factor-α. Am. J. Respir. Cell Mol. Biol. 11: 631–637.

Savage SM, Donaldson LA, Cherian S, Chilukuri R, White VA, Sopori ML (1991): Effects of cigarette smoke on the immune response. II. Chronic exposure to cigarette smoke inhibits surface immunoglobulin-mediated responses in B cells. Toxicol. Appl. Pharmacol. 111: 523–529.

Seyler LE Jr., Pomerleau OF, Fertig JB, Hunt D, Parker K (1986): Pituitary hormone response to cigarette smoking. Pharmacol. Biochem. Behav. 24: 159–162.

Sloan-Lancaster J, Shaw AS, Rothbard JB, Allen PM (1994): Partial T cell signaling, altered phospho-ξ and lack of ZAP-70 recruitment in APL-induced T cell anergy. Cell 79: 913–922.

Soliman DM, Twigg HL III (1992): Cigarette smoke decreases bioactive interleukin-6 secretion by alveolar macrophages. Am. J. Physiol. 263: L471–L478.

Sopori ML, Kozak W (1998): Immunomodulatory effects of cigarette smoke. J. Neuroimmunol. 148–156.

Sopori ML, Cherian S, Chilukuri R, Shopp GM (1989): Cigarette smoke causes inhibition of the immune response to intratracheally administered antigens. Toxicol. Appl. Pharmacol. 97: 489–499.

Sopori ML, Savage SM, Christner RF, Geng Y, Donaldson LA (1993): Cigarette smoke and the immune response: mechanism of nicotine-induced immunosuppression. Adv. Biosci. 86: 663–672.

Sopori ML, Goud NS, Kaplan AM (1994): Effects of tobacco smoke on the immune system. In Dean JH, Luster MI, Munson AE, Kimber I, editors. Immunotoxicology and immunopharmacology. New York: Raven Press, pp. 413–434.

Sopori ML, Kozak W, Savage SM, Geng Y, Soszynski D, Kluger MJ, Perryman EK, Snow GE (1998): Effect of nicotine on the immune system: possible regulation of immune responses by central and peripheral mechanisms. Psychoneuroendocrinology 23: 189–204.

Takuwa N, Zhou W, Kumada M, Takuwa Y (1995): Involvement of intact inositol-1,4,5-trisphosphate-sensitive Ca^{2+} stores in cell cycle progression at the G1/S boundary in serum-stimulated human fibroblasts. FEBS Letters 360: 173–176.

Thastrup O, Cullen PJ, Droback BK, Hanley MR, Dawson AP (1990): Thapsigargin, a tumor promoter discharges intracellular Ca^{2+} stores by specific inhibition of the endoplasmic reticulum Ca^{2+}-ATPases. Proc. Natl. Acad. Sci. USA 87: 2466–2470.

U.S. Department of Health and Human Services (1994): CDC Morbidity and Mortality Weekly Report 43: 469–472.

Van Dijk JP, Madretsma GS, Keuskamp ZJ, Zijlstra FJ (1995): Nicotine inhibits cytokine synthesis by mouse colonic mucosa. Eur. J. Pharmacol. 278: R11–R12.

Van Loon GR, Kiritsy-Roy JA, Brown LV, Bobbitt FA (1987): Nicotinic regulation of sympathoadrenal catecholamine secretion: cross tolerance to stress. In Martin WR, Van Loon GR, Iwamato ET, Davis L, editors. Tobacco smoking and nicotine: a neurobiological approach. New York: Plenum Press, pp. 263–276.

Weiss A, Littman DR (1994): Signal transduction by lymphocyte antigen receptors. Cell 76: 263–274.

Yamaguchi E, Okazaki N, Itoh A, Abe S, Kawakami Y, Okuyama H (1989): Interleukin-1 production by alveolar macrophages decreased in smokers. Am. Rev. Respir. Dis. 140: 397–402.

Part III

Pharmacokinetics of Nicotine-like Alkaloids

12

Pharmacokinetics and Metabolism of Nicotine and Related Alkaloids

Neal L. Benowitz, MD, and Peyton Jacob III Ph.D.
Clinical Pharmacology Unit of the Medical Service
San Francisco General Hospital Medical Center and the Departments of Medicine, Psychiatry, and Biopharmaceutical Sciences
University of California, San Francisco

This chapter reviews current knowledge about the pharmacokinetics and metabolism of nicotine, some naturally occurring tobacco alkaloids, and nicotine analogs that are under development as potential therapeutic agents. The focus is on studies in humans, but animal data are mentioned when relevant to the interpretation of human data.

NICOTINE AND OTHER ALKALOIDS IN TOBACCO PRODUCTS

Nicotine (Fig. 12-1) is the principal tobacco alkaloid occurring to the extent of about 1.5% by weight in commercial cigarette tobacco and comprising about 95% of the total alkaloid content (Benowitz et al., 1983a; Schmeltz and Hoffmann, 1977). The nicotine in tobacco is largely, if not entirely, the levorotatory (S)-isomer. Tobacco smoke has been reported to contain a small amount (up to 10%) of the (R)-isomer, presumably resulting from racemization occurring during combustion (Pool et al., 1985; Klus and Kuhn, 1977). Interestingly, nornicotine in tobacco is almost racemic, occurring as a 40:60 mixture of R/S isomers.

In most tobacco strains, nornicotine and anatabine are the most abundant of the minor al-

Neuronal Nicotinic Receptors: Pharmacology and Therapeutic Opportunities, Edited by S. P. Arneric and J. D. Brioni
ISBN 0-471-24743-x, pages 211–234. Copyright © 1998 by Wiley-Liss, Inc.

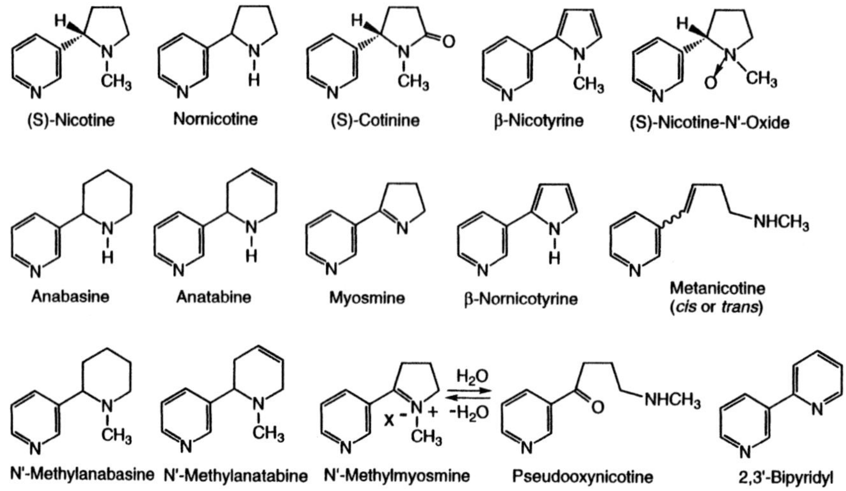

Figure 12.1. *Structures of tobacco alkaloids.*

kaloids, followed by anabasine (Fig. 12-1) (Schmeltz and Hoffmann, 1977). Small amounts of the N′-methyl derivatives of anabasine and anatabine are found in tobacco and tobacco smoke. Several of the minor alkaloids are thought to arise by bacterial action or oxidation during tobacco processing rather than by biosynthetic processes in the living plant (Leete, 1983). These include myosmine, N′-methylmyosmine, cotinine, nicotyrine, nornicotyrine, nicotine N′-oxide, 2,3′-bipyridyl, and metanicotine (Fig. 12-1). N-nitroso derivatives of tobacco alkaloids arise by the action of nitrous acid on nicotine, nornicotine, anabasine, and anatabine. These nitroso compounds are important because some are carcinogenic (Hecht and Hoffmann, 1989). Their metabolism has been reviewed extensively elsewhere and will not be discussed in this chapter.

Of the minor alkaloids that have been studied, only nornicotine, metanicotine, and anabasine have been shown to have significant pharmacologic activity (Clark et al., 1965). Qualitatively, their actions are similar to those of nicotine but are generally less potent, the relative potency depending upon the test system. Anabasine administered orally or sublingually has been reported to aid smoking cessation and to have cardiovascular effects (Nasirov et al., 1978). To our knowledge, no studies of the pharmacologic effects of any of the other minor alkaloids in humans have been reported.

ABSORPTION OF NICOTINE

Nicotine is distilled from burning tobacco and is carried proximally on tar droplets that are inhaled. Absorption of nicotine across biological membranes depends on pH. Nicotine is a weak base with a pKa of 8.0. In its ionized state, such as in acidic environments, nicotine does not rapidly cross membranes. The pH of smoke from flue-cured tobaccos, found in most cigarettes, is acidic (pH 5.5). At this pH, nicotine is primarily ionized. As a consequence, there is little buccal absorption of nicotine from cigarette smoke, even when it is held in the mouth (Gori et al., 1986). Smoke from air-cured tobaccos, the predominant tobacco used in pipes, cigars, and

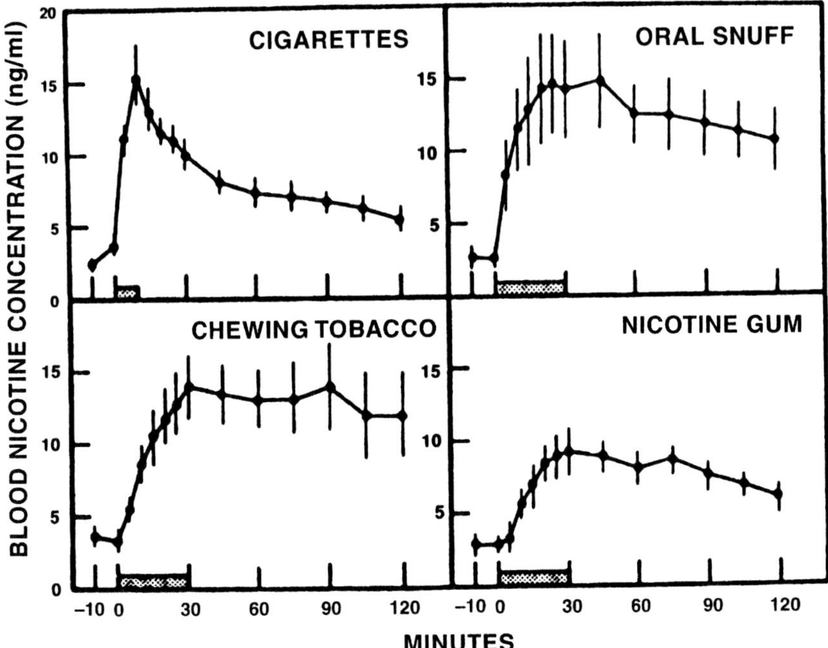

Figure 12.2. Blood nicotine concentrations during and after cigarette smoking for 9 minutes, oral snuff (2.5 grams), chewing tobacco (average 7.9 grams), and nicotine gum (two 2 mg pieces). Average values for 10 subjects (± SEM). (From Benowitz et al., 1988.)

European cigarettes, is less acidic (pH 6.5 or higher), and considerable nicotine is un-ionized. Smoke from these products is well absorbed through the mouth (Armitage et al., 1978).

When tobacco smoke reaches the small airways and alveoli of the lung, the nicotine is rapidly absorbed independent of pH. Blood concentrations of nicotine rise quickly during and peak at the completion of cigarette smoking (Fig. 12-2). The rapid absorption of nicotine from cigarette smoke through the lungs, presumably because of the huge surface area of the alveoli and small airways, and dissolution of nicotine in the fluid of pH 7.4 in the human lung facilitates transfer across membranes.

Chewing tobacco, snuff, and nicotine gum are buffered to alkaline pH to facilitate absorption of nicotine through mucous membranes. Concentrations of nicotine in the blood rise gradually with the use of smokeless tobacco and plateau at about 30 minutes, with levels persisting and declining only slowly over 2 hours or more (Fig. 12-2) (Benowitz et al., 1988).

Absorption of nicotine from nicotine gum is gradual with evidence of continued release through the entire period of chewing (Benowitz et al., 1988). Blood levels of nicotine are considerably lower, even when chewing 4 mg gum doses, than with the use of tobacco. The absolute dose of nicotine absorbed systemically from nicotine gum is much less than the nicotine content of the gum, in part because considerable nicotine is swallowed, with subsequent first-pass metabolism.

Nicotine is poorly absorbed from the stomach because it is protonated in the acidic gastric fluid, but it is well absorbed in the small intestine, which has a more alkaline pH and a large surface area. Following the administration of nicotine capsules, peak concentrations are

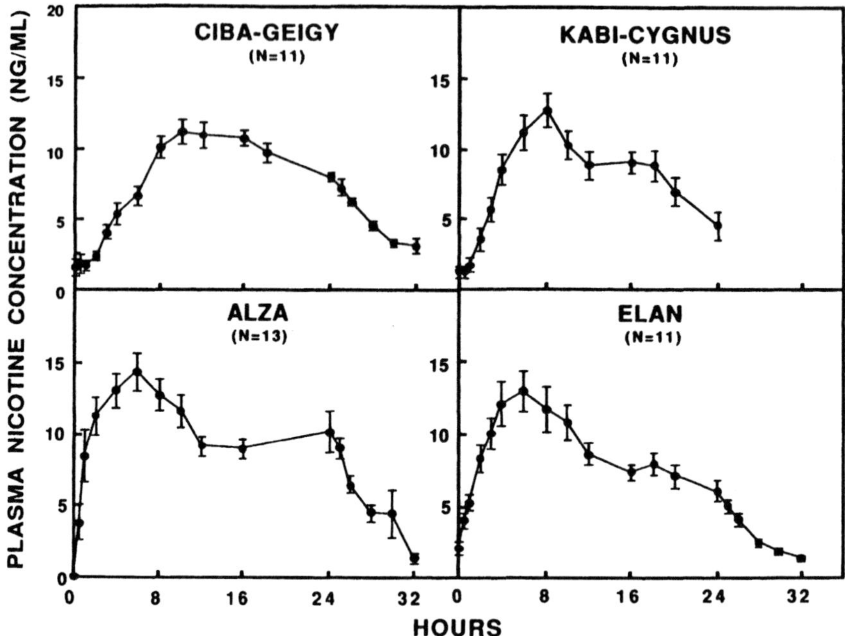

Figure 12.3. Plasma nicotine concentration time curves during and after application of four different transdermal nicotine delivery systems. The Ciba Geigy Habitrol, the Alza Nicoderm, and the Elan Prostep patches were worn for 24 hours, while the Kabi–Cygnus Nicotrol was worn for 16 hours. (Adapted from Benowitz, 1993.)

reached in just over 1 hour (Benowitz et al., 1991). The oral bioavailability of nicotine is about 45%. Bioavailability is incomplete because of first-pass metabolism.

Nicotine base is well absorbed through the skin, which is the basis for transdermal delivery technology (Benowitz, 1995). Currently in the U.S. four different nicotine transdermal systems are marketed. All are multilayer patches. The rate of release of nicotine into the skin is controlled by permeability of the skin, rate of diffusion through a polymer matrix, and/or rate of passage through a membrane in the various patches.

Plasma nicotine concentrations vary among the different transdermal systems (Fig. 12-3) (Benowitz, 1995). In all cases, there is an initial lag time of about 1 hour before nicotine appears in the bloodstream, and there is continued absorption (about 10% of the total dose) after the patch is removed due to residual nicotine in the skin.

Cotinine, the major metabolite of nicotine, is much more polar than nicotine, is metabolized more slowly, and undergoes little, if any, first-pass metabolism after oral dosing (Benowitz et al., 1983b).

DISTRIBUTION OF NICOTINE IN BODY TISSUES

This section describes the distribution of nicotine to various body organs but does not discuss distribution to specific regions of the brain or to specific nicotinic neuronal binding sites. Brain and neuronal distribution of nicotine and related alkaloids is described elsewhere in this volume (refer to Chapter 13).

After absorption, nicotine enters the bloodstream where, at pH 7.4, it is about 69% ionized and 31% un-ionized. Binding to plasma proteins is less than 5% (Benowitz et al., 1982a). The drug is distributed extensively to body tissues with steady-state volume of distribution averaging 2.6 × body weight. Steady-state nicotine tissue distribution has been studied in rats and rabbits (Benowitz et al., 1990). Liver, lungs, and brain have high affinity, and adipose relatively low affinity for nicotine.

The time course of nicotine in the brain and in other body organs, and resultant pharmacologic effects, are highly dependent on the route and rate of dosing. Smoking a cigarette delivers nicotine rapidly to the pulmonary venous circulation, from which it moves quickly to the left ventricle of the heart and to the systemic arterial circulation and the brain. The lag time between a puff of a cigarette and nicotine reaching the brain is 10–15 seconds. Nicotine concentrations in arterial blood after a puff of cigarette can be quite high, reaching or exceeding 100 ng/ml, many-fold higher than levels measured in venous blood (Henningfield et al., 1993; Gourlay and Benowitz, 1997). The rapid rate of delivery of nicotine by smoking (or intravenous injection, which presents similar distribution kinetics) results in high levels of nicotine in the central nervous system with little time for development of tolerance. The result is a more intense pharmacologic action. The short time interval between puffing and nicotine entering the brain also allows the smoker to titrate the dose of nicotine to a desired pharmacologic effect, further reinforcing drug self-administration and facilitating the development of addiction.

In contrast, slow delivery of nicotine, such as by transdermal systems, results in little, if any, arterial-venous disequilibrium. The resultant brain levels of nicotine are much lower than after smoking, and the gradual rise in levels of nicotine in the central nervous system allows for the development of considerable tolerance to pharmacologic effects. Thus, the intensity of central nervous system effects is much less and the addiction liability with the use of transdermal nicotine is virtually nil (Henningfield and Keenan, 1993). Routes of dosing that are associated with intermediate rates of delivery, such as chewing nicotine gum or taking nicotine orally, are expected to result in an intermediate intensity of effects and intermediate addiction liability. These same considerations regarding rate of delivery and pharmacologic effects are expected to apply to nicotine-related compounds.

METABOLISM OF NICOTINE

Nicotine is extensively metabolized, primarily in the liver (Fig. 12-4), but also to a small extent by other tissues, including the lung, kidney, and brain (Gorrod and Jenner, 1975; Vahakangas and Pelkonen, 1993; Jacob et al., 1997).

Quantitatively, the most important metabolite of nicotine in most mammalian species is the lactam derivative cotinine (Fig. 12-5). In humans, about 70–80% of nicotine is converted to cotinine (Benowitz and Jacob, 1994). This transformation involves two steps. The first is mediated by a cytochrome P450 system to produce nicotine-$\Delta 1',5'$-iminium ion (Murphy, 1973). The second step is catalyzed by a cytoplasmic aldehyde oxidase (Brandange and Lindblom, 1979; Gorrod and Hibberd, 1982). Nicotine iminium ion has received considerable interest since it is an alkylating agent and, as such, could play a role in the pharmacology of nicotine or carcinogenicity of tobacco (Jacob et al., 1997; Gorrod and Jenner, 1975; Hibberd and Gorrod, 1981; Shigenaga et al., 1988).

Nicotine N′-oxide is another primary metabolite (Fig. 12-5) of nicotine, although only about 4% of nicotine absorbed by smokers is metabolized via this route (Benowitz et al., 1994). The conversion of nicotine to nicotine N′-oxide involves a flavoprotein enzyme system, flavin mono-oxygenase (FMO), which results in formation of both possible diasteriomers, the 1′-(R)-

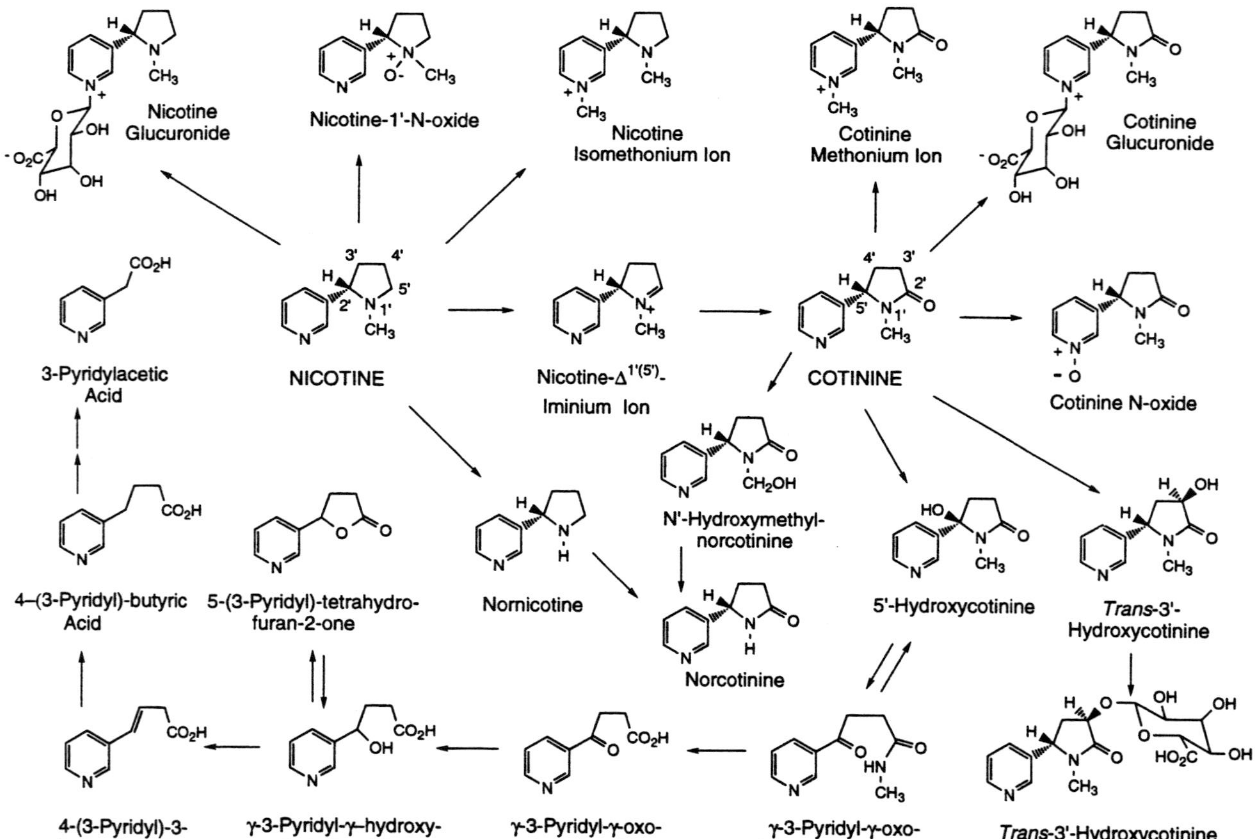

Figure 12.4. *Nicotine metabolic pathways.*

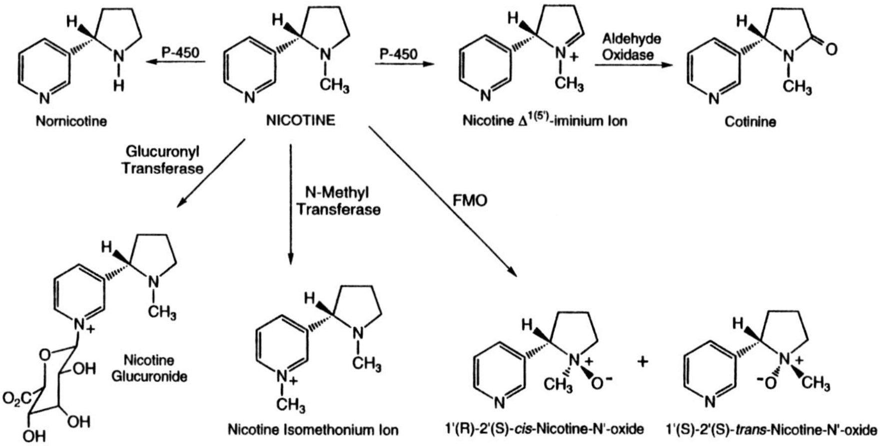

Figure 12.5. *Primary routes of nicotine metabolism.*

2′-(S)-*cis* and 1′-(S)-2′-(S)-*trans* isomers. In humans, this pathway is highly selective for the *trans* isomer (Cashman et al., 1992). It appears that nicotine N′-oxide is not further metabolized to any significant extent, except by reduction back to nicotine, which may lead to recycling of nicotine in the body (Dajani et al., 1975). A study by Beckett et al. indicated that reduction of nicotine N′-oxide to nicotine in humans is mediated by bacterial action in the large intestine (Beckett et al., 1970). These investigators found that nicotine N′-oxide administered intravenously was excreted largely, if not entirely, in the urine unchanged, whereas administration rectally as an enema resulted in extensive conversion to nicotine and cotinine, which appeared in the urine. Oral administration of nicotine N-oxide resulted in small but significant urinary excretion of nicotine and cotinine.

In addition to oxidation of the pyrrolidine ring, nicotine is metabolized by two nonoxidative pathways, methylation of the pyridine nitrogen giving nicotine isomethonium ion and glucuronidation (Fig. 12-5). The methylation pathway was first reported by McKennis, who found it in dogs dosed with (S)-nicotine as well as in human smokers (McKennis et al., 1962a). Recently, Crooks and co-workers have carried out detailed studies of the N-methylation pathway using animal models and liver homogenates (Nwosu and Crooks, 1988; Crooks and Godin, 1988). They found that S-adenosyl-methionine is the source of the methyl group in a reaction catalyzed by a heterocycle N-methyl transferase. The enzyme from different species exhibited remarkably different specificities for the (R)- and (S)-isomers of nicotine (Nwosu and Crooks, 1988; Crooks and Godin, 1988). In the guinea pig, only (R)-nicotine was found to be a substrate. Human liver cytosol was capable of methylating both enantiomers, but the (R)-isomer was methylated more rapidly than the (S)-isomer. Rat liver homogenates were incapable of methylating either enantiomer. In light of the reported pharmacologic activity of nicotine isomethonium ion (Dwoskin et al., 1992), further studies of this metabolite are warranted. It was recently reported that nicotine is metabolized to an N-quaternary glucuronide in humans (Curvall et al., 1991; Benowitz et al., 1994).

Although about 80% of nicotine is metabolized via the cotinine pathway in humans, only 10–20% of the nicotine absorbed by smokers appears in the urine as unchanged cotinine (Benowitz et al., 1994). A number of cotinine metabolites have been structurally characterized (Fig. 12-4). Indeed, it appears that most of the reported urinary metabolites of nicotine are derived from cotinine. Five primary metabolites of cotinine have been reported in humans: 3′-hydroxycotinine (Bowman and McKennis, 1962; Dagne and Castagnoli, 1972b); 5′-hydroxycotinine (Dagne and Castagnoli, 1974; Nguyen et al., 1981), which exists in tantomeric equilibrium with the open-chain derivative γ-(3-pyridyl)-γ-oxo-N-methylbutyramide (McKennis et al., 1962b, 1964b); cotinine N-oxide (Dagne and Castagnoli, 1972a; Shulgin et al., 1987); cotinine methonium ion (McKennis et al., 1962a); and cotinine glucuronide (Curvall et al., 1991; Benowitz et al., 1994). The conversion of cotinine to 3′-hydroxycotinine in humans is highly stereoselective for the *trans*-isomer (Jacob et al., 1990). This metabolite is also excreted as the glucuronide conjugate (Fig. 12-4) (Benowitz et al., 1994; Curvall et al., 1991).

As with nicotine N-oxide, cotinine N-oxide can be reduced back to the parent amine in vivo (Yi et al., 1977). Norcotinine has been detected in smokers' urine (McKennis et al., 1962c; Byrd et al., 1995), but it is not clear whether it is produced by demethylation of cotinine or oxidative metabolism of nornicotine absorbed from tobacco. Animal studies have demonstrated the existence of both of these pathways. Wada et al. (1961) and Papadopolous (1964) reported that norcotinine is a metabolite of nornicotine, and Harke et al. (1974) detected norcotinine in urine of pigs following cotinine administration.

Cotinine is a lactam, and it is reasonable to expect that the open-chain form γ-(3-pyridyl)-γ-methylaminobutyric acid, an amino acid, might be an intermediate in cotinine formation or might be formed via hydrolysis of cotinine. This open-chain derivative has indeed been reported to occur in urine of smokers (Bowman et al., 1959), but based on in vitro studies with liver homogenates it appears that cotinine is formed directly by oxidation of the nicotine Δ1′(5′) iminium ion (Brandange and Lindblom, 1979; Gorrod and Hibberd, 1982) rather than via the amino acid. The ketoamide γ-(3-pyridyl)-γ-oxo-N-methyl-butyramide derived from 5′-hydroxycotinine is presumably the precursor of a number of nicotine metabolites that result from degradation of the pyrrolidine ring. These include the ketoacid γ-(3-pyridyl)-γ-oxobutyric acid, its reduction product γ-(3-pyridyl)-γ-hydroxybutyric acid, a hydroxyacid that is in equilibrium with the lactone 5-(3-pyridyl)-tetrahydrofuran-2-one, and 3-pyridylacetic acid, the so-called terminal metabolite of nicotine. Both the ketoacid γ-(3-pyridyl)-γ-oxobutyric acid and 3-pyridylacetic acid have been characterized in human urine following oral administration of cotinine (Schwartz and McKennis, 1963; McKennis et al., 1964a). It is speculated (McKennis et al., 1964a) that 3-pyridylacetic acid is formed via dehydration of the hydroxyacid γ-(3-pyridyl)-γ-hydroxybutyric acid to give 4-(3-pyridyl)-3-butenoic acid, reduction to 4-(3-pyridyl)-butyric acid, β-oxidation, and cleavage to 3-pyridylacetic acid. This metabolic scheme is analogous to the catabolism of fatty acids, although there is no experimental evidence for the intermediacy of 4-(3-pyridyl)-3-butenoic acid or of 4-(3-pyridyl)-butyric acid.

Oxidative N-demethylation is frequently an important pathway in the metabolism of xenobiotics, but this route is in most species a minor pathway in the metabolism of nicotine. Conversion of nicotine to nornicotine in humans has been demonstrated. We found that small amounts (<1% of the dose) of deuterium-labeled nornicotine are excreted in the urine of smokers administered deuterium-labeled nicotine (Jacob and Benowitz, 1991). Metabolic formation of nornicotine from nicotine has also been reported by Neurath et al. (1991). Formation of an iminium ion as an intermediate in the demethylation of nicotine was reported by Castagnoli and co-workers, who characterized N′-cyanomethylnornicotine in extracts obtained following incubation of rabbit liver microsomes with nicotine and sodium cyanide (Nguyen et al., 1979). This observation implies the intermediacy of N′-methyleneiminium ion, which is captured by cyanide ion to form the stable cyano adduct. In the absence of cyanide, the iminium ion would be expected to hydrolyze to nornicotine and formaldehyde.

METABOLISM OF THE MINOR ALKALOIDS

Compared to nicotine, relatively little is known about the metabolism of the minor tobacco alkaloids. To our knowledge, the only published study involving administration of minor alkaloids to humans was reported by Beckett et al. (1972). This study involved determining urinary excretion of nicotine, nornicotine, anabasine, N-methyl-anabasine, β-nicotyrine, β-nornicotyrine, and myosmine following oral administration of 2 mg in two subjects. Excretion of nicotine, nornicotine, methylanabasine, and anabasine was dependent on urinary pH, with several-fold higher amounts excreted under acidic (pH 4.8) as compared to fluctuating urinary pH. With acidic urinary pH, the percentages of unchanged alkaloids recovered in urine ranged from about 15% for nicotine and methylanabasine to 70% for nornicotine and anabasine. Myosmine, β-nicotyrine, and β-nornicotyrine could not be detected in urine for the 24 h period following administration. No biotransformation products of any of the alkaloids other than nicotine were reported.

The metabolism of nornicotine in dogs following a slow intravenous infusion was reported by Wada et al. (1961). Both unchanged nornicotine and norcotinine (desmethylcotinine) (Fig. 12-6) were recovered from urine extracts, but no quantitative data were obtained. Metabolism of nornicotine by rabbits was reported by Papadopoulos (1964). Unchanged nornicotine, norcotinine, and an unidentified metabolite were detected by thin-layer chromatographic analysis following an intravenous injection or after incubation with liver homogenates.

The metabolism of nornicotine by rabbit liver microsomal preparations was studied by Nguyen (1976). Using specifically deuterated analogs of nornicotine, they were able to identify two isomeric pyrrolines, myosmine and 2′-(3-pyridyl)-Δ1′(5′)-pyrroline, as metabolites (Fig. 12-6). No norcotinine was detected in experiments with microsomal preparations, but GC-MS analysis of extracts from incubation of nornicotine with the 10,000 × g supernate fraction of rabbit liver did reveal the presence of norcotinine. Presumably, a soluble enzyme present in the 10,000 g supernate fraction is required for the conversion of 2′-(3-pyridyl)-Δ1′ (5′)-pyrroline to norcotinine. The analogous transformation of nicotine-Δ1′ (5′)-iminium ion to

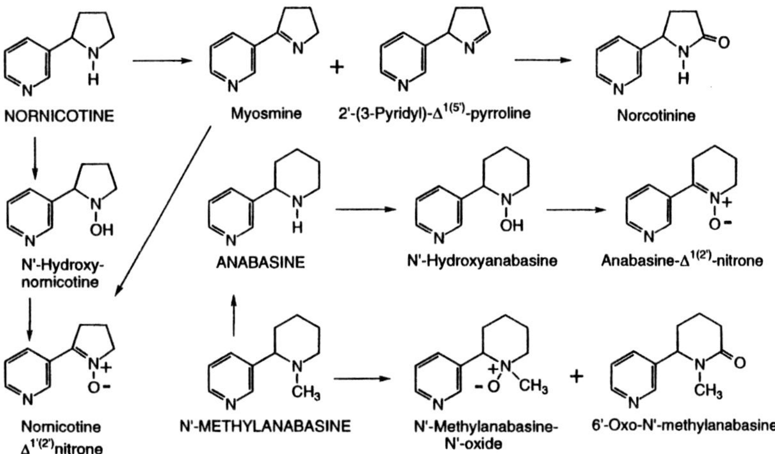

Figure 12.6. Metabolism of nornicotine, anabasine, and N′-methylanabasine.

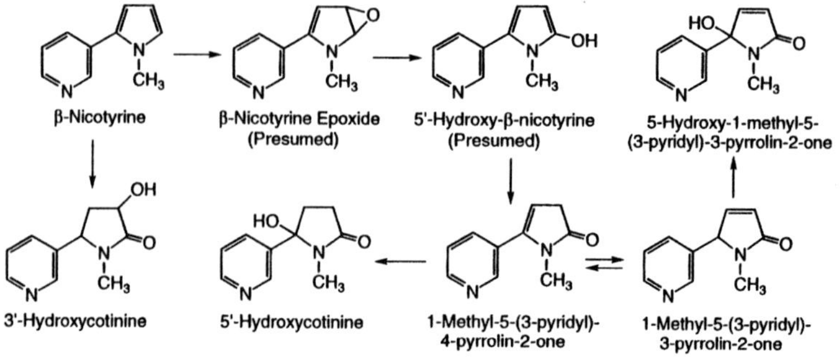

Figure 12.7. *Metabolism of* β*-nicotyrine.*

cotinine is catalyzed by a soluble aldehyde oxidase (Brandange and Lindblom, 1979; Gorrod and Hibberd, 1982). Nornicotine is also converted in vitro to a nitrone, presumably via N-hydroxynornicotine (Aislaitner et al., 1992).

In vitro metabolism of the minor alkaloid N-methylanabasine with liver homogenates was reported by Jenner and Gorrod (1973). Unchanged methylanabasine and the diasteriomeric 1′-N-oxides (Fig. 12-6) were measured in 10,000 × g hepatic supernate preparations from hamsters, mice, rats, and guinea pigs. Beckett and Sheikh found, in addition to the diasteriomeric N-oxides, products of N-demethylation. N′-hydroxy-anabasine and anabasine Δ1′,2′-nitrone (Fig. 12-6) are formed in vitro with rat, rabbit, and guinea pig liver homogenates (Beckett and Sheikh, 1973). The same two metabolites were formed during incubations of anabasine with lung tissue homogenates. Metabolic conversion of N′-methylanabasine to the cotinine analog 6′-oxo-N′-methylanabasine (Fig. 12-6) in vitro has been demonstrated recently (Gorrod and Aislaitner, 1994). Extensive metabolism of β-nicotyrine both in vitro (Shigenaga et al., 1989) and in vivo (Liu et al., 1993) has been reported (Fig. 12-7).

In reactions catalyzed by rabbit liver and lung microsomes, β-nornicotyrine is converted to a mixture of two pyrrolinones (Fig. 12-7), presumably via an epoxide intermediate (Shigenaga et al., 1989). Since epoxides are electrophilic compounds capable of alkylating biomacromolecules, this pathway could have toxicological significance. It is interesting to note that nicotyrine inhibits the metabolism of nicotine in mice, resulting in higher nicotine tissue levels, without increasing its toxicity (Stalhandski and Slanina, 1982). In vivo in rabbits (Liu et al., 1993), the major metabolite of β-nicotyrine was found to be 3′-hydroxycotinine along with small amounts of 5′-hydroxycotinine and 5-hydroxy-1-methyl-5-(3-pyridyl)-3-pyrrolin-2-one (Fig. 12-7). The pathway for the remarkable transformation of β-nicotyrine to 3′-hydroxycotinine remains to be elucidated.

LIVER ENZYMES RESPONSIBLE FOR NICOTINE AND COTININE METABOLISM

In vitro metabolism studies suggest that CYP2A6 is the enzyme that is primarily responsible for the oxidation of nicotine and cotinine. The evidence for the role of CYP2A6 includes human liver microsome studies in which the activities of nicotine oxidation to cotinine and cotinine oxidation to 3′-hydroxycotinine are highly correlated with coumarin 7-hydroxylase activity (known to be mediated by CYP2A6) and correlated with immunochemically determined CYP2A6 levels (Cashman et al., 1992; Nakajima et al., 1996a,b; Berkman et al., 1995). Nicotine oxidation to cotinine has also been shown to be inhibited by coincubation with coumarin (a competitive inhibitor of CYP2A6) and by the administration of rabbit anti-rat CYP2A1 antibody, an antibody that blocked cotinine 3′-hydroxylase activity and coumarin 7-hydroxylase activity to the same extent (Nakajima et al., 1996b). Finally, microsomes of B-lymphoblastoid cells with C-DNA expressed CYP2A6 have been shown to have high activity for metabolizing nicotine and cotinine, although other C-DNA expressed enzymes, including CYP2D6, CYP2C9, and CYP2E1, have also been reported to have some nicotine metabolizing activity (Nakajima et al., 1996b; McCracken et al., 1992; Flammang et al., 1992). CYP2B6 has been shown to have considerable activity in metabolizing nicotine as well, but this enzyme is not constitutively present in human liver (Nakajima et al., 1996b; McCracken et al., 1992; Flammang et al., 1992). Of note is that human liver specimens exhibit marked individual variability in levels of CYP2A6 mRNA and coumarin 7-hydroxylase activity, consistent with the known wide variability in the rate of nicotine metabolism in people (Nakajima et al., 1996b).

In support of the role of CYP2A6 in nicotine metabolism, we have recently found that a woman who had unusually slow metabolism of nicotine and markedly reduced conversion of nicotine to cotinine (Benowitz et al., 1995) was homozygous for the v1 variant of the CYP2A6 gene (unpublished data).

To ascertain the importance of 2D6 in nicotine and cotinine metabolism, we performed a nested case control study in which 11 individuals who were phenotyped as CYP2D6-poor metabolizers (using dextromethorphan as a probe) were compared to 33 control subjects who were extensive metabolizers (Benowitz et al., 1996). There were no differences in nicotine or cotinine metabolism in poor versus extensive metabolizers.

As described previously, nicotine and cotinine undergo phase II metabolic reactions via N-glucuronidation and hydroxycotinine via O-glucuronidation. The specific glucuronidation enzymes involved have not been identified. We have, however, observed a high degree of correlation between the extent of conjugation of nicotine and cotinine within individuals, and no correlation between the extent of nicotine or cotinine versus 3′-hydroxycotinine cojugation (Benowitz et al., 1994). This suggests that the same enzyme is responsible for nicotine and cotinine conjugation, and a different enzyme is responsible for 3′-hydroxycotinine conjugation.

QUANTITATIVE ASPECTS OF NICOTINE METABOLISM

Quantitative aspects of the pattern of nicotine metabolism have been fairly well worked out in people (Fig. 12-8). About 90% of nicotine metabolites can be accounted for in the urine (Benowitz et al., 1994). Based on studies with simultaneous infusion of labeled nicotine and cotinine, it has been determined that 70–80% of nicotine is converted to cotinine (Benowitz and Jacob, 1994). About 4% of nicotine is excreted as nicotine-1′-N-oxide and 4% as nicotine glucuronide. Cotinine is excreted unchanged in the urine to a small degree (10–15%). The remainder is converted to metabolites, primarily *trans*-3′-hydroxycotinine, cotinine glucuronide, and *trans*-3′-hydroxycotinine glucuronide.

The rate of metabolism of nicotine can be determined by measuring blood levels after administration of a known dose of nicotine. We have studied cigarette smokers and nonsmokers given intravenous infusions of nicotine for 30 to 60 minutes (Benowitz and Jacob, 1993). Total and renal clearance were computed directly, and the nonrenal or metabolic clearance was computed as the difference between the total and renal clearance. Total clearance of nicotine averaged about 1200 ml/min. Nonrenal clearance represents about 70% of liver blood flow. Assuming most nicotine is metabolized by the liver, this means about 70% of the drug is extracted from the blood in each pass through the liver.

The metabolism of cotinine is much slower than that of nicotine. Cotinine clearance aver-

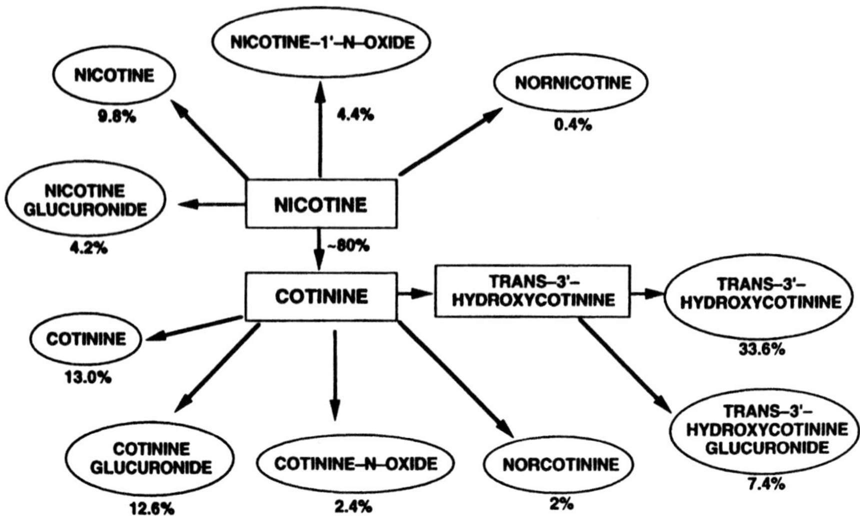

Figure 12.8. Quantitative scheme of nicotine metabolism, based on average excretion of metabolites as percent of systemic dose during transdermal nicotine application. (Reprinted from Benowitz et al., 1994.)

ages about 40 ml/min (Benowitz and Jacob, 1994). As it is slowly metabolized, the rate of elimination of cotinine is predicted not to be substantially influenced by changes in liver blood flow.

FACTORS INFLUENCING THE RATE OF NICOTINE METABOLISM

There is considerable interindividual variability in the rate of elimination of nicotine and cotinine (Benowitz et al., 1982a). A number of factors that might explain interindividual variability have been studied.

An implication of the high degree of hepatic extraction is that clearance of nicotine should be dependent upon liver blood flow. Thus, physiological events, such as meals, posture, exercise, or other drugs that perturb hepatic blood flow, are predicted to affect the rate of nicotine metabolism. We have found that meals consumed during a steady-state infusion of nicotine result in consistent decline in nicotine concentrations, the maximal effect seen 30 minutes after the end of a meal (Lee et al., 1989; Gries et al., 1996). During sleep, hepatic blood flow declines and nicotine clearance falls correspondingly. Nicotine levels during constant infusion rise at night. Thus, the day/night variation and meal effects of nicotine clearance result in circadian variations in plasma concentrations during constant dosing of nicotine (Gries et al., 1996).

Gender differences in drug metabolism have been observed for some drugs. In a small group of habitual heavy cigarette smokers, we found that the clearance of nicotine corrected for body weight was significantly higher in men than women (Benowitz and Jacob, 1984). However, in a later study we found that clearance normalized for body weight was similar in men and women, although total clearance was lower in women compared to men because they on average weigh less (Benowitz and Jacob, 1994). There were no differences between men and women in cotinine clearance.

Cigarette smoking itself may influence the rate of metabolism of nicotine. Cigarette smoking is known to accelerate the metabolism of some drugs. However, we found that the clearance of nicotine was significantly slower in cigarette smokers compared to nonsmokers (Benowitz and Jacob, 1993). In support of this observation was an earlier observation that cigarette smokers who had abstained from smoking for 7 days had a faster clearance of nicotine compared to overnight abstinence from cigarettes (Lee et al., 1987). These studies suggest that there are substances in tobacco smoke, as yet unidentified, that impede the metabolism of nicotine. Because nicotine and cotinine appear to be metabolized by the same hepatic enzymes, we asked whether cotinine might be responsible for the slowed metabolism of nicotine in smokers. However, in an experimental study in which nonsmokers received an intravenous infusion of nicotine with and without pretreatment with high doses of cotinine, there was no effect of cotinine on the clearance of nicotine (Zevin et al., 1997).

Genetic factors undoubtedly contribute to much of the individual variability in nicotine and cotinine metabolism. We have studied a woman with deficient C-oxidation of nicotine (Benowitz et al., 1995). She had very low clearance of nicotine, an unusually long half-life of nicotine, and converted only 12% of nicotine to cotinine. This individual has subsequently been determined to have a genetic defect, being a homozygous CYP2A6 v1 variant (unpublished data). The prevalence of this homozygous variant in the population is estimated to be about less than 5% based on v1 allele frequencies of 0 to 10% in different ethnic groups (Fernandez-Salguero et al., 1995).

Ethnic differences in nicotine and cotinine metabolism have also been observed. We compared nicotine and cotinine metabolism in 40 blacks and 39 whites of similar age and body weight (Benowitz and Jacob, 1997). The total and nonrenal clearance of cotinine was significantly lower in blacks vs. whites (total clearance 0.56 vs. 0.68 ml/min/kg; $p < 0.01$). The clearance of nicotine tended to be lower in blacks vs. whites (17.7 vs. 19.6 ml/min/kg), but this difference was not significant. Unpublished data from our laboratory indicate that the extent of glucuronide conjugation of both nicotine and cotinine is lower in blacks than whites. Whether slower conjugation explains the overall difference in clearances between blacks and whites remains to be determined.

RENAL EXCRETION

Nicotine is excreted by glomerular filtration and tubular secretion, with variable reabsorption depending on urinary pH. With uncontrolled urine pH, renal clearance averages about 100 ml/min, accounting for the elimination of about 10% of the daily intake of nicotine (Benowitz and Jacob, 1985). In acid urine, nicotine is mostly ionized and tubular reabsorption is minimized; renal clearance may be as high as 600 ml/min (urinary pH 4.4), depending on urinary flow rate. In alkaline urine, a larger fraction of nicotine is un-ionized, allowing net tubular reabsorption with renal clearance as low as 17 ml/min (urine pH 7.0).

Renal clearance of cotinine is much less than glomerular filtration rate (Benowitz et al., 1983b). Since cotinine is not appreciably protein bound, this indicates extensive tubular reabsorption. Renal clearance of cotinine can be enhanced by up to 50% with extreme urinary acidification. Cotinine excretion is less influenced by urinary pH than nicotine because it is less basic and, therefore, is primarily in the un-ionized form within the physiological pH range. As is the case for nicotine, the rate of excretion of cotinine is influenced by urinary flow rate. Renal excretion of cotinine is a minor route of elimination, averaging about 15% of total clearance. In contrast, 100% of nicotine-N-oxide is excreted unchanged in the urine.

NICOTINE AND COTININE BLOOD LEVELS DURING TOBACCO USE AND NICOTINE REPLACEMENT THERAPY

Blood or plasma nicotine concentrations sampled in the afternoon in smokers generally range from 10 to 50 ng/ml (Benowitz et al., 1990). The increment in venous blood nicotine concentration after smoking a single cigarette ranges from 5 to 30 ng/ml, depending on how a cigarette is smoked. Blood levels peak at the end of smoking a cigarette and decline rapidly over the next 20 minutes due to tissue distribution. The distribution half-life averages about 8 minutes.

Peak venous blood levels of nicotine are similar, although the rate of rise of nicotine is slower for cigar smokers and users of snuff and chewing tobacco compared to cigarette smokers (Benowitz et al., 1988; Armitage et al., 1978). Pipe smokers, particularly those who have previously smoked cigarettes, may have blood and urine levels of nicotine as high as cigarette smokers (Turner et al., 1977; Wald et al., 1981; McCusker et al., 1982). Primary pipe smokers who have not previously smoked cigarettes tend to have lower nicotine levels. Likewise, cigar smokers who have previously smoked cigarettes may inhale more deeply and achieve higher blood levels of nicotine than primary cigar smokers (Armitage et al., 1978), although on average, based on urinary cotinine levels, daily nicotine intake appears to be less for cigar compared to cigarette or pipe smokers (Wald et al., 1984).

The plasma half-life of nicotine after intravenous infusion or cigarette smoking averages about 2 hours (Benowitz and Jacob, 1994). However, when half-life is determined using the time course of urinary excretion of nicotine, which is more sensitive in detecting lower levels of nicotine in the body, the terminal half-life averages 11 hours (Jacob et al., 1998). The longer half-life detected at lower concentrations of nicotine is a consequence of slow release of nicotine from body tissues. Based on a half-life of 2 hours for nicotine, one would predict accumulation over 6 to 8 hours (3 to 4 half-lives) of regular smoking and persistence of significant levels for 6 to 8 hours after cessation of smoking. If a smoker smokes until bedtime, significant levels should persist all night. Studies of blood levels in regular cigarette smokers confirm these predictions (Fig. 12-9) (Benowitz et al., 1982b). Peak and trough levels follow each cigarette, but as the day progresses trough levels rise, and the influence of peak levels become less important. Thus, nicotine is not a drug to which smokers are exposed intermittently and which is eliminated rapidly from the body. To the contrary, smoking represents a multiple dosing situation with considerable accumulation while smoking and persistent levels for 24 hours of each day.

Plasma levels of nicotine from nicotine replacement therapies tend to be in the range of low-level cigarette smokers. Thus, typical steady-state plasma nicotine concentrations with nicotine patches range from 10 to 20 ng/ml, for nicotine gum and nicotine nasal spray from 5 to 15 ng/ml (Benowitz, 1995; Benowitz et al., 1987, 1997). For the sake of comparison, systemic doses from various nicotine delivery systems are as follows: Cigarette smoking—1 to 2 mg per cigarette; nicotine gum—1 mg for a 2 mg gum; transdermal nicotine—5 to 25 mg per day, depending on the patch; nicotine nasal spray—0.5 mg per dose of 1 spray in each nostril; oral snuff—3.6 mg for 2.5 grams held in the mouth for 30 minutes; and chewing tobacco—4.5 mg for 7.9 grams chewed for 30 minutes.

Cotinine is present in the blood of smokers in much higher concentrations than those of nicotine. Cotinine blood concentrations average about 250 to 300 ng/ml in groups of cigarette smokers (Gori and Lynch, 1985; Benowitz et al., 1983a). We have seen levels in tobacco users ranging from 10 to 900 ng/ml. After stopping smoking, levels decline in a log-linear fashion with an average half-life of about 16 hours (range, 11 to 36 hours) (Benowitz and Jacob, 1994). Because of the long half-life, there is much less fluctuation in cotinine concentrations through-

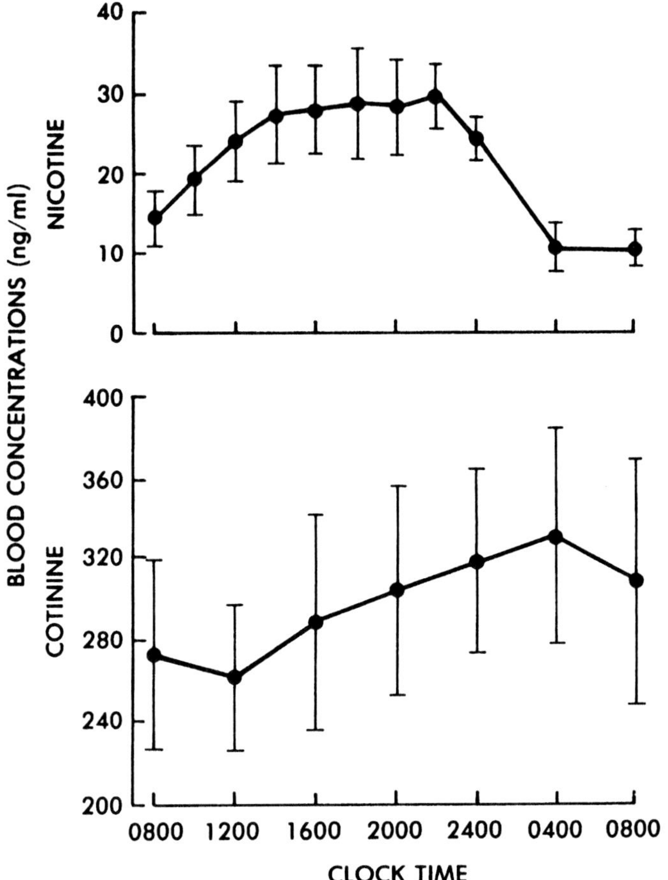

Figure 12.9. Circadian blood concentrations of nicotine and cotinine during unrestricted smoking. Data are mean ± S.E. for eight subjects. (Reprinted from Benowitz et al., 1983b.)

out the day compared to nicotine concentrations. As expected, there is a gradual rise in cotinine levels throughout the day, peaking at the end of smoking and persisting at high concentrations overnight (Fig. 12-9). Because of the long half-life of cotinine, it has been used as a biomarker for daily intake, both in cigarette smokers and in those exposed to environmental tobacco smoke (Benowitz, 1996). There is a high correlation among cotinine concentrations measured in plasma, saliva, and urine, and measurements in any one of these fluids can be used as a marker of nicotine intake.

PHARMACOKINETICS AND METABOLISM OF NICOTINE ANALOGS

Numerous nicotine analogs that are of potential therapeutic utility have been described in the literature (Holladay et al., 1997). These include (S)-3-methyl-5-(1-methyl-2-pyrrolidinyl) isoxazole hydrochloride (ABT-418); [2-methyl-3-(2-(S)-pyrrolidinylmethoxy) pyridine dihy-

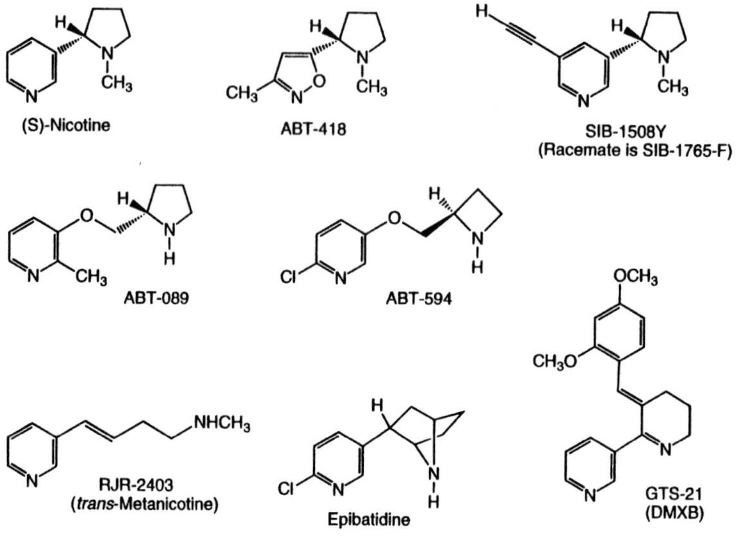

(S)-Nicotine

ABT-418

SIB-1508Y
(Racemate is SIB-1765-F)

ABT-089

ABT-594

RJR-2403
(*trans*-Metanicotine)

Epibatidine

GTS-21
(DMXB)

Figure 12.10. *Structures of nicotine analogs.*

drochloride] (ABT-089); [(R)-5-(2-azetidinyl-methoxy)-2-chloropyridine] (ABT-594); ([±]-5-ethynyl-3-(1-methyl-2-pyrrolidinyl) pyridine fumurate) (SIB-1765F); [S]-[-]-5-ethynyl- 3-(1-methyl-2-pyrrolidinyl) pyridine maleate (SIB-1508Y); (E)-N-methyl-4-(3-pyridinyl)-3-butene-1-amine (RJR-2403), 3′-(2,4-dimethoxybenzylidine)-anabaseine (GTS-21, DMXB); and epibatidine (Fig. 12-10) (Holladay et al., 1997; Bencherif et al., 1996; Menzaghi et al., 1997; Cosford et al., 1996; Sullivan et al., 1997; Davila-Garcia et al., 1997; Anderson et al., 1995; Bannon et al., 1998). Most of these compounds have similarities in either the pyridine or pyrrolidine ring to nicotine and might be expected to undergo some of the same metabolic reactions.

The metabolism of one of these compounds, ABT-418, has been studied in some detail, both in vitro in several animal species and in human liver slices (Fig. 12-11) (Rodrigues et al., 1994).

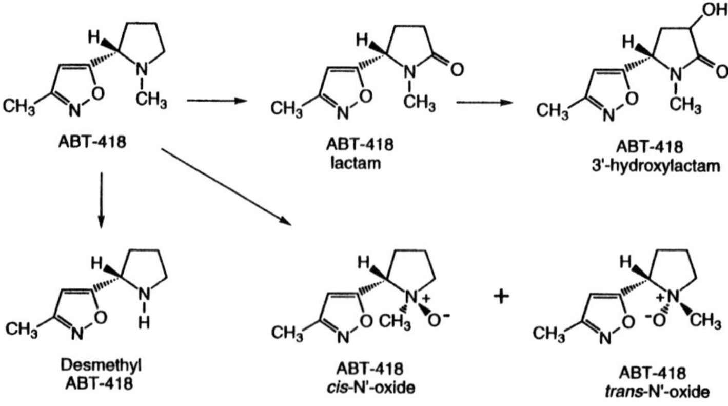

Figure 12.11. *ABT-418 metabolic pathways.*

Pathways of metabolism are similar to those of nicotine, including formation of a lactam (similar to cotinine), a 3′-hydroxy lactam, *cis*-and *trans*-N′-oxides, and N-desmethyl ABT-418. In human and monkey liver slices, the lactam metabolite predominates; in dog and rat liver, the *trans*-N′-oxide metabolite predominates. As for nicotine, the lactam formation is mediated by the CYP450 system, while N′-oxidation is mediated by the flavin—containing mono-oxygenase system. Cytosolic aldehyde oxidase was found to be a second enzyme involved in the metabolism of ABT-418 to its lactam, which is similar to the role of aldehyde oxidase as a second-step enzyme for the conversion of nicotine to cotinine. We are aware of no published systemic pharmacokinetic data. ABT-418 has been reported to have poor oral bioavailability in animals and has been delivered to humans in clinical trials via transdermal delivery systems.

CONCLUSION

In conclusion, our review has focused primarily on the pharmacokinetics of nicotine and related tobacco alkaloids. We have reviewed data on nicotine analogs that might be used therapeutically, but relatively little information has been published. However, because of similar structures, it is likely that many of the features of nicotine disposition will be shared by other nicotinic agonists. Pharmacokinetic considerations, including the importance of rate and route of dosing, are likely to influence the pharmacologic activity of these nicotine analogs, as they do for nicotine itself.

ACKNOWLEDGEMENTS

Much of the research described in this chapter was supported by U.S. Department of Health and Human Services grants DA02277, DA01696, and RR-00083. The authors thank Kaye Welch for editorial assistance.

REFERENCES

Aislaitner G, Li Y, Gorrod JW (1992): In vitro studies on (−)-nornicotine. Medical Sci. Res. 20: 897–899.

Anderson DJ, Williams M, Pauly JR, Raszkiewicz JL, Campbell JE, Rotert G, Surber B, Thomas SB, Wasicak J, Arneric SP, Sullivan JP (1995): Characterization of [^{3}H]ABT-418: a novel cholinergic channel ligand. J. Pharmacol. Exp. Ther. 273: 1434–1441.

Armitage AK, Dollery CT, Houseman TH, Kohner E, Lewis PJ, Turner DM (1978): Absorption of nicotine from small cigars. Clin. Pharmacol. Ther. 23: 143–151.

Bannon AW, Decker MW, Holladay MW, Curzon P, Donnelly-Roberts D, Puttfarcken PS, Bitner RS, Diaz A, Dickenson AH, Porsolt RD, Williams M, Arneric SP (1998): Broad-spectrum, non-opioid analgesic activity by selective modulation of neuronal nicotinic acetylcholine receptors. Science 279: 77–81.

Beckett AH, Sheikh AH (1973): In vitro metabolic N-oxidation of the minor tobacco alkaloids (−)-methylanabasine and (−)-anabasine to yield a hydroxylamine and a nitrone in lung and liver homogenates. J. Pharm. Pharmacol. 25:171P.

Beckett AH, Gorrod JW, Jenner P (1970): Absorption of nicotine-1′-N-oxide in man and its reduction in the gastrointestinal tract. J. Pharm. Pharmacol. 22:722–723.

Beckett AH, Gorrod JW, Jenner P (1972). A possible relation between pKa and lipid solubility and the amounts excreted in urine of some tobacco alkaloids given to man. J. Pharm. Pharmacol. 24:115–120.

Bencherif M, Lovette ME, Fowler KW, Arrington S, Reeves L, Caldwell WS, Lippiello PM (1996); RJR-

2403: A nicotinic agonist with CNS selectivity. I. In vitro characterization. J. Pharmacol. Exp. Ther. 279: 1413–1421.

Benowitz NL (1993): Nicotine replacement therapy. What has been accomplished—can we do better? Drugs 45: 157–170.

Benowitz NL (1995): Clinical pharmacology of transdermal nicotine. Eur. J. Pharm. Biopharm. 41: 168–174.

Benowitz NL (1996): Cotinine as a biomarker of environmental tobacco smoke exposure. Epidemiologic Rev. 18: 188–204.

Benowitz NL, Jacob P, III (1984): Daily intake of nicotine during cigarette smoking. Clin. Pharmacol. Ther. 35: 499–504.

Benowitz, NL, Jacob P III (1985). Nicotine renal excretion rate influences nicotine intake during cigarette smoking. J. Pharmacol. Exp. Ther. 234: 153–155.

Benowitz NL, Jacob P, III (1993): Nicotine and cotinine elimination pharmacokinetics in smokers and nonsmokers. Clin. Pharmacol. Ther. 53: 316–323.

Benowitz NL, Jacob P III (1994): Metabolism of nicotine to cotinine studied by a dual stable isotope method. Clin. Pharmacol. Ther. 56: 483–493.

Benowitz NL, Jacob P III (1997): Individual differences in nicotine kinetics and metabolism in humans. In Rapaka RS, Chiang N, Martin BR, editors. Pharmacokinetics, metabolism, and pharmaceutics of drugs of abuse, NIDA Research Monograph 173. Rockville, MD: U.S. Department of Health and Human Services, National Institutes of Health, pp. 48–64.

Benowitz NL, Jacob P III, Jones RT, Rosenberg J (1982a): Interindividual variability in the metabolism and cardiovascular effects of nicotine in man. J. Pharmacol. Exp. Ther. 221: 368–372.

Benowitz NL, Kuyt F, Jacob P III (1982b). Circadian blood nicotine concentrations during cigarette smoking. Clin. Pharmacol. Ther. 32: 758–764.

Benowitz NL, Hall SM, Herning RI, Jacob P, III, Jones RT, Osman AL (1983a): Smokers of low yield cigarettes do not consume less nicotine. N. Engl. J. Med. 309: 139–142.

Benowitz NL, Kuyt F, Jacob P III, Jones RT, Osman AL (1983b): Cotinine disposition and effects. Clin. Pharmacol. Ther. 309: 139–142.

Benowitz NL, Jacob P III, Savanapridi C (1987): Determinants of nicotine intake while chewing nicotine polacrilex gum. Clin. Pharmacol. Ther. 41: 467–473.

Benowitz NL, Porchet H, Sheiner L, Jacob P III (1988): Nicotine absorption and cardiovascular effects with smokeless tobacco use: comparison with cigarettes and nicotine gum. Clin. Pharmacol. Ther. 44: 23–28.

Benowitz NL, Porchet H, Jacob P III (1990): Pharmacokinetics, metabolism, and pharmacodynamics of nicotine. In Wonnacott S, Russell MAH, Stolerman IP, editors. Nicotine psychopharmacology: molecular, cellular and behavioral aspects. Oxford: Oxford University Press, pp. 112–157.

Benowitz NL, Jacob P III, Denaro C, Jenkins R (1991): Stable isotope studies of nicotine kinetics and bioavailability. Clin. Pharmacol. Ther. 49: 270–277.

Benowitz NL, Jacob P III, Fong I, Gupta S (1994): Nicotine metabolic profile in man: Comparison of cigarette smoking and transdermal nicotine. J. Pharmacol. Exp. Ther. 268: 296–303.

Benowitz NL, Jacob P III, Sachs DPL (1995): Deficient C-oxidation of nicotine. Clin. Pharmacol. Ther. 57: 590–594.

Benowitz NL, Jacob P III, Perez-Stable E (1996): CYP2D6 phenotype and the metabolism of nicotine and cotinine. Pharmacogenetics 6: 239–242.

Benowitz NL, Zevin S, Jacob P III (1997): Sources of variability in nicotine and cotinine levels with use of nicotine nasal spray, transdermal nicotine, and cigarette smoking. Br. J. Clin. Pharmacol. 43: 259–267.

Berkman CE, Park SB, Wrighton SA, Cashman JR (1995): *In vitro–in vivo* correlations of human (S)-nicotine metabolism. Biochem. Pharmacol. 50: 565–570.

Bowman ER, McKennis H (1962): Studies on the metabolism of (−)-cotinine in the human. J. Pharmacol. Exp. Ther. 135: 306–311.

Bowman ER, Turnbull LB, McKennis H Jr. (1959): Metabolism of nicotine in the human and excretion of pyridine compounds by smokers. J. Pharmacol. Exp. Ther. 127: 92–95.

Brandange S, Lindblom L (1979): The enzyme "aldehyde oxidase" is an iminium oxidase. Reaction with nicotine-Δ1′,5′-iminium ion. Biochem. Biophys. Res. Commun. 91: 991–996.

Byrd GD, Robinson JH, Caldwell WS, deBethizy JD (1995): Comparison of measured and FTC-predicted nicotine uptake in smokers. Psychopharmacology 122: 95–103.

Cashman JR, Park SB, Yang ZC, Wrighton SA, Jacob P III, Benowitz NL (1992): Metabolism of nicotine by human liver microsomes: stereoselective formation of *trans*-nicotine-N′-oxide. Chem. Res. Toxicol. 5: 639–646.

Clark MSG, Rand MJ, Vanov S (1965): Comparison of pharmacological activity of nicotine and related alkaloids occurring in cigarette smoke. Arch. Int. Pharmacodyn. 156: 363–379.

Cosford NDP, Bleicher L, Herbaut A, McCallum JS, Vernier J-M, Dawson H, Whitten JP, Adams P, Chavez-Noriega L, Correa LC, Crona JH, Mahaffy LS, Menzaghi F, Rao TS, Reid R, Sacaan AI, Santori E, Stauderman KA, Whelan K, Lloyd GK, McDonald IA (1996): (S)-(−)-5-Ethynyl-3-(1-methyl-2-pyrrolidinyl) pyridine maleate (SIB-1508Y): a novel anti-Parkinsonian agent with selectivity for neuronal nicotinic acetylcholine receptors. J. Med. Chem. 39: 3235–3237.

Crooks PA, Godin CS (1988): N-methylation of nicotine enantiomers by human liver cytosol. J. Pharm. Pharmacol. 40: 153–154.

Curvall M, Kazemi-Vala E, Englund G (1991): Conjugation pathways in nicotine metabolism. In Adlkofer F, Thurau K, editors. Effects of nicotine on biological systems. Basel: Birkhauser-Verlag, pp. 69–75.

Dagne E, Castagnoli N Jr. (1972a): Cotinine-N-oxide, a new metabolite of nicotine. J. Med. Chem. 15: 840–841.

Dagne E, Castagnoli N Jr. (1972b): Structure of hydroxycotinine, a nicotine metabolite. J. Med. Chem. 15: 356–360.

Dagne E, Castagnoli N Jr. (1974): Deuterium isotope effects in the vivo metabolism of cotinine. J. Med. Chem. 17: 1330–1333.

Dajani RM, Gorrod JW, Beckett AH (1975): Reduction *in vivo* of nicotine-1′-N-oxide by germ-free and conventional rats. Biochem. Pharmacol. 24: 648–650.

Davila-Garcia MI, Musachio JL, Perry DC, Xiao Y, Horti A, London ED, Dannals RF, Kellar KJ (1997): [^{125}I]IPH, an epibatidine analog, binds with high affinity to neuronal nicotinic cholinergic receptors. J. Pharmacol. Exp. Ther. 282: 445–451.

Dwoskin LP, Leibee LL, Jewell AL, Fang ZX, Crooks PA (1992): Inhibition of [^{3}H]dopamine uptake into rat brain striatal slices by quaternary N-methylated nicotine metabolites. Life Sci. 50: 233–237.

Fernandez-Salguero P, Hoffman SMG, Cholerton S, Mohrenweiser H, Raunio H, Rautio A, Pelkonen O, Huang J, Evans WE, Idle JR, Gonzalez FJ (1995): A genetic polymorphism in coumarin 7-hydroxylation: sequence of the human CYP2A genes and identification of variant CYP2A6 alleles. Am. J. Hum. Genet. 57: 651–660.

Flammang AM, Gelboin HV, Aoyama T, Gonzalez FJ, McCoy GD (1992): Nicotine metabolism by cDNA-expressed human cytochrome P-450s. Biochem. Archiv. 8: 1–8.

Gori GB, Benowitz NL, Lynch CJ (1986): Mouth versus deep airways absorption of nicotine in cigarette smokers. Pharmacol. Biochem. Behav. 25: 1181–1184.

Gori GB, Lynch CJ (1985): Analytical cigarette yields as predictors of smoke bioavailability. Regul. Toxicol. Pharmacol. 5: 314–326.

Gorrod JW, Aislaitner G (1994): The metabolism of alicyclic amines to reactive iminium ion intermediates. Eur. J. Drug Metab. Pharmacokin. 19: 209–217.

Gorrod JW, Hibberd AR (1982): The metabolism of nicotine-delta-1′(5′)-iminium ion, in vivo and in vitro. Eur. J. Drug Metab. Pharmacokin. 7: 293–298.

Gorrod JW, Jenner P (1975): The metabolism of tobacco alkaloids. In Hayes WJ Jr., editor. Essays in toxicology. New York: Academic Press, pp. 35–78.

Gourlay SG, Benowitz NL (1997): Arteriovenous differences in plasma concentration of nicotine and catecholamines and related cardiovascular effects after smoking, nicotine nasal spray, and intravenous nicotine. Clin. Pharmacol. Ther. 62: 453–463.

Gries JM, Benowitz NL, Verotta D (1996): Chronopharmacokinetics of nicotine. Clin. Pharmacol. Ther. 60: 385–395.

Harke HP, Schuller D, Frahm B. Mauch A (1974): Demethylation of nicotine and cotinine in pigs. Res. Commun. Chem. Pathol. Pharm. 9: 595–599.

Hecht SS, Hoffmann D (1989): The relevance of tobacco-specific nitrosamines to human cancer. Cancer Surveys 8: 273–294.

Henningfield JE, Keenan RM (1993): Nicotine delivery kinetics and abuse liability. J. Consult. Clin. Psychol. 61: 743–750.

Henningfield JE, Stapleton JM, Benowitz NL, Grayson RF, London ED (1993): Higher levels of nicotine in arterial than in venous blood after cigarette smoking. Drug Alcohol Depend. 33: 23–29.

Hibberd AR, Gorrod JW (1981): Nicotine-Δ1′,5′-iminium ion: a reactive intermediate in nicotine metabolism. Adv. Exp. Biol. Med. 136B: 1121–1131.

Holladay MW, Dart MJ, Lynch JK (1997): Neuronal nicotinic acetylcholine receptors as targets for drug discovery. J. Med. Chem. 40: 4169–4194.

Jacob P III, Benowitz NL (1991): Oxidative metabolism of nicotine in vivo. In Adlkofer F, Thurau K, editors. Effects of nicotine on biological systems. Basel: Birkhauser Verlag, pp. 35–44.

Jacob P III, Shulgin AT, Benowitz NL (1990): Synthesis of (3′R,5′S)-*trans*-3′-hydroxy-cotinine, a major metabolite of nicotine. Metabolic formation of 3′-hydroxycotinine in humans is highly steroselective. J. Med. Chem. 33: 1888–1891.

Jacob P III, Ulgen M, Gorrod JW (1997): Metabolism of (S)-(−)-nicotine by guinea pig and rat brain: identification of cotinine. Eur. J. Drug Metab. Pharmacokin. 22: 391–394.

Jacob P III, Yu L, Shulgin AT, Benowitz NL (1998): Minor tobacco alkaloids as biomarkers for tobacco use: comparison of cigarette, smokeless tobacco, cigar and pipe users. Am. J. Public Health (in press).

Jenner P, Gorrod JW (1973): Comparative *in vitro* hepatic metabolism of some tertiary N-methyl tobacco alkaloids in various species. Res. Commun. Clin. Path. Pharmacol. 6: 829–843.

Klus H, Kuhn H (1977): A study of the optical activity of smoke nicotines. Fachliche Mitt. Oesterr Tabakregie 17: 331–336.

Lee BL, Benowitz NL, Jacob P III (1987): Influence of tobacco abstinence on the disposition kinetics and effects of nicotine. Clin. Pharmacol. Ther. 41: 474–479.

Lee BL, Jarvik ME, Jacob P III, Benowitz NL (1989): Food and nicotine metabolism. Pharmacol. Biochem. Behav. 33: 621–625.

Leete E (1983): Biosynthesis and metabolism of the tobacco alkaloids. In: Pelletier SW, editor. Alkaloids: chemical and biological perspectives. New York: John Wiley & Sons, pp. 85–152.

Liu X, Jacob P III, Castagnoli N Jr. (1993): The metabolic fate of the minor tobacco alkaloids. In: Gorrod JW, Wahren J, editors. Nicotine and related alkaloids: Absorption, distribution, metabolism and excretion. London: Chapman and Hall, pp. 129–145.

McCracken NW, Cholerton S, Idle JR (1992): Cotinine formation by cDNA-expressed human cytochromes P450. Med. Sci. Res. 20: 877–878.

McCusker K, McNabb E, Bone R (1982): Plasma nicotine levels in pipe smokers. J. Am. Med. Assoc. 248: 577–578.

McKennis H Jr., Turnbull LB, Bowman ER (1962a): N-methylation of nicotine and cotinine in vivo. J. Biol. Chem. 238: 719–723.

McKennis H Jr., Turnbull LB, Bowman ER, Schwartz SL (1962b): The corrected structure of a ketoamide arising from the metabolism of (−)-nicotine. J. Amer. Chem. Soc. 84: 4598–4599.

McKennis H Jr., Turnbull LB, Schwartz SL, Tamaki E, Bowman ER (1962c): Demethylation in the metabolism of (−)-nicotine. J. Biol. Chem. 327: 541–546.

McKennis H Jr., Schwartz SL, Bowman ER (1964a): Alternate routes in the metabolic degradation of the pyrrolidine ring of nicotine. J. Biol. Chem. 239: 3990–3996.

McKennis H Jr., Schwartz SL, Turnbull LB, Tamaki E, Bowman ER (1964b): The metabolic formation of γ-(3-pyridyl)-γ-hydroxybutyric acid and its possible intermediary role in the metabolism of nicotine. J. Biol. Chem. 239: 3981.

Menzaghi F, Whelan KT, Risbrough VB, Rao TS, Lloyd GK (1997): Effects of a novel cholinergic ion channel agonist SIB-1765F on locomotor activity in rats. J. Pharmacol. Exp. Ther. 280: 384–392.

Murphy PJ (1973): Enzymatic oxidation of nicotine to nicotine-Δ1′,5′-iminium ion. J. Biol. Chem. 248: 2796–2800.

Nakajima M, Yamamoto T, Nunoya K, Yokoi T, Nagashima K, Inoue K, Funae Y, Shimada N, Kamataki T, Kuroiwa Y (1996a): Characterization of CYP2A6 involved in 3′-hydroxylation of cotinine in human liver microsomes. J. Pharmacol. Exp. Ther. 277: 1010–1015.

Nakajima M, Yamamoto T, Nunoya K, Yokoi T, Nagashima K, Inoue K, Funae Y, Shimada N, Kamataki T, Kuroiwa Y (1996b): Role of human cytochrome P4502A6 in C-oxidation of nicotine. Drug Metab. Disp. 24: 1212–1217.

Nasirov SK, Ryabchenko VP, Khalikova FR, Khazbievich IS, Kashkova EK (1978): Anabasine hydrochloride—a new antismoking agent. Khimiko-Farmatsevticheskii Zhurnal 12: 149–152.

Neurath GB, Orth D, Pein FG (1991): Detection of nornicotine in human urine after infusion of nicotine. In: Adlkofer F, Thurau K, editors. Advances in pharmacological sciences, effects of nicotine on biological systems. Basel: Birkhauser Verlag, pp. 45–49.

Nguyen TL (1976): Alpha-carbon hydroxylation in the metabolism of tobacco alkaloids, Ph.D. thesis, University of California, San Francisco.

Nguyen TL, Gruenke L, Castagnoli N Jr. (1979): Metabolic oxidation of nicotine to chemically reactive intermediates. J. Med. Chem. 22: 259–263.

Nguyen TL, Dagne E, Gruenke L, Bargava H, Castagnoli N Jr. (1981): The tautomeric structure of 5-hydroxycotinine, a secondary mammalian metabolite of nicotine. J. Med. Chem. 46: 758–760.

Nwosu CG, Crooks PA (1988): Species variation and stereoselectivity in the metabolism of nicotine enantiomers. Xenobiotica 18: 1361–1372.

Papadopoulos NM (1964): Formation of nornicotine and other metabolites from nicotine *in vitro* and *in vivo*. Can. J. Biochem. 42: 435.

Pool WF, Godin CS, Crooks PA (1985): Nicotine racemization during cigarette smoking. The Toxicologist 5: 232.

Rodrigues AD, Ferrero JL, Amann MT, Rotert GA, Cepa SP, Surber BW, Machinist JM, Tich NR, Sullivan JR, Garvey DS, Fitzgerald M, Arneric SP (1994): The in vitro hepatic metabolism of ABT-418, a cholinergic channel activator, in rats, dogs, cynomolgus monkeys, and humans. Drug Metab. Dispos. 22: 788–798.

Schmeltz I, Hoffmann D (1977): Nitrogen containing compounds in tobacco and tobacco smoke. Chem. Rev. 77: 295–311.

Schwartz SL, McKennis H Jr. (1963): Studies on the degradation of the pyrrolidine ring of (−)-nicotine in vivo. Formation of γ-(3-pyridyl)-γ-oxobutyric acid. J. Biol. Chem. 238: 1807–1812.

Shigenaga MK, Trevor AJ, Castagnoli N Jr. (1988): Metabolism dependent covalent binding of (S)-[5′-^{3}H]-nicotine to liver and lung microsomal macromolecules. Drug Metab. Disp. 16: 397–402.

Shigenaga MK, Kim BH, Caldera-Munoz P, Cairns T, Jacob P III, Trevor AJ, Castagnoli N Jr. (1989): Liver and lung microsomal metabolism of the tobacco alkaloid β-nicotyrine. Chem. Res. Toxicol. 2: 282–287.

Shulgin AT, Jacob P III, Benowitz NL, Lau D (1987): The identification and quantitative analysis of cotinine-N-oxide. J. Chromatogr. Biomed. Applic. 423: 365–372.

Stalhandski T, Slanina P (1982): Nicotyrine inhibits in vivo metabolism of nicotine without increasing its toxicity. Toxicol. Appl. Pharmacol. 65: 366–372.

Sullivan JP, Donnelly-Roberts D, Briggs CA, Anderson DJ, Gopalakrishnan M, Xue IC, Piattoni-Kaplan M, Molinari E, Campbell JE, McKenna DG, Gunn DE, Lin N, Ryther KB, He Y, Holladay MW, Wonnacott S, Williams M, Arneric SP (1997): ABT-089 [2-methyl-3-(2-(S)-pyrrolidinylmethoxy) pyridine]: I. A potent and selective cholinergic channel modulator with neuroprotective properties. J. Pharmacol. Exp. Ther. 283: 235–246.

Turner JAM, Sillett RW, McNicol MW (1977): Effect of cigar smoking on carboxy-haemoglobin and plasma nicotine concentrations in primary pipe and cigar smokers and ex-cigarette smokers. Br. Med. J. 2: 1387–1389.

Vahakangas K, Pelkonen O (1993): Extrahepatic metabolism of nicotine and related compounds by P-450. In: Gorrod JW, Wahren J, editors. Nicotine and related alkaloids: absorption-distribution-metabolism-excretion. London: Chapman and Hall, pp. 111–127.

Wada E, Bowman ER, Turnbull LB, McKennis H Jr. (1961): Norcotinine (desmethyl-cotinine) as a urinary metabolite of nornicotine. J. Med. Pharm. Chem. 4: 21–30.

Wald NJ, Idle M, Boreham J, Bailey A, Van Vunakis H (1981): Serum cotinine levels in pipe smokers: evidence against nicotine as cause of coronary heart disease. Lancet 2: 775–777.

Wald NJ, Idle M, Boreham J, Bailey A, Van Vunakis H (1984): Urinary nicotine concentrations in cigarette and pipe smokers. Thorax 39: 365–368.

Yi JM, Sprouse CT, Bowman ER, McKennis H Jr. (1977): The interrelationship between the metabolism of (S)-cotinine-N-oxide and (S)-cotinine. Drug Metab. Disp. 5: 355–362.

Zevin S, Jacob P III, Benowitz NL (1997): Cotinine effects on nicotine metabolism. Clin. Pharmacol. Ther. 61: 649–654.

13

Nicotine and Related Compounds as PET and SPECT Ligands

Victor L. Villemagne, MD, John L. Musachio, PhD, and Ursula Scheffel, ScD

Division of Nuclear Medicine
Department of Radiology
The Johns Hopkins Medical Institutions
Baltimore, Maryland

SYNOPSIS

In the past two decades, emission tomography has developed into a powerful technique for probing neuroreceptors in vivo. Neuroscientists, neuropharmacologists, and increasingly clinicians have utilized this noninvasive imaging modality to elucidate the localization, binding parameters, and alterations of neuroreceptors in the living human brain. Positron emission tomography (PET) and single photon emission computed tomography (SPECT) of high-affinity central nicotinic acetylcholine receptors (nAChRs) offer the potential to monitor human nAChRs in tobacco dependence/withdrawal, neurodegenerative disease, and other CNS disorders. Also, these noninvasive in vivo imaging techniques can aid in the development and testing of new cholinergic drugs. Examples of the current practice of emission tomography of human nAChRs can be found in the various PET studies that employ $[^{11}C](-)$-nicotine. This nAChR radioprobe, however, suffers from a high degree of nonspecific binding in vivo, which limits its utility in studies of nAChRs in living subjects. Current research efforts on the part of PET/SPECT radiochemists, therefore, have focused on development of new, highly specific

Neuronal Nicotinic Receptors: Pharmacology and Therapeutic Opportunities, Edited by S. P. Arneric and J. D. Brioni
ISBN 0-471-24743-x, pages 235–250. Copyright © 1998 by Wiley-Liss, Inc.

and highly selective nAChR radioligands that are able to localize high-affinity nAChRs in vivo. Key examples of new ^{11}C, ^{18}F, and ^{123}I labeled ligands that are derived from several different structural classes (e.g., radiolabeled (−)-nicotine, epibatidine, and 3-pyridyl ether analogs) are discussed along with a brief review of their in vitro and/or in vivo properties. In particular, the ^{18}F labeled analog of A-85380 is an example of a new type of PET nAChR radioligand that is able to bind to nAChRs in vivo with high specificity and selectivity; it also possesses a more favorable toxicity profile in comparison to radiolabeled, epibatidine-based probes. Based on the development of radioligands such as 2-[^{18}F]A-85380, we predict that a new era in human nAChR receptor imaging is dawning.

INTRODUCTION

Nicotinic acetylcholine receptors (nAChRs) are excitatory ligand-gated ion channels (Deneris et al., 1991; Lindstrom et al., 1995; Williams et al., 1994) constituted by two kinds of subunits (α and β) (Decker et al., 1995b). The pharmacological profile of nAChR subtypes is related to its subunit combination. Radioligand binding techniques have identified at least three subtypes of nAChRs in the central nervous system (Decker et al., 1995b, Nef et al., 1988): those with high affinity for (−)-nicotine, labeled by agonists such as ^{3}H-acetylcholine and ^{3}H-cytisine (nicotine/ACh) (Pabreza et al., 1991); those with high affinity for ^{125}I-α-bungarotoxin (α-Bgt) (Clarke et al., 1985); and those that selectively recognize neuronal bungarotoxin (n-Bgt) (Schulz et al., 1991). There is a good correlation between the distribution of nAChRs with the α4β2 subunit combination and the distribution of high-affinity nicotine/ACh binding sites (Flores et al., 1992), between the distribution of α7 mRNA and α-Bgt high-affinity binding sites (Clarke et al., 1985), and between α3 mRNA and n-Bgt binding sites (Schulz et al., 1991).

Development of in vivo radiotracers for the nAChR system has focused exclusively on probes for high-affinity, non-α7 sites. This can be ascribed in large part to the abundance of ligands that have been characterized by in vitro assay and shown to compete with nanomolar affinity for ^{3}H-(−)-nicotine and ^{3}H-(−)-cytisine binding sites (McDonald et al., 1995). These high-affinity binding sites are thought to be comprised predominantly (>90%) of the α4β2 subtype in mammalian brain (Lindstrom et al., 1995, Flores et al., 1992). This subtype is found throughout the mammalian CNS and is of relatively high abundance (Williams et al., 1994) in comparison to other nAChR subtypes. Thus, the current status of central nAChR imaging in all likelihood involves in vivo labeling of the α4β2 subtype.

EMISSION TOMOGRAPHY OF nAChRs: THE PROMISE

Neuroreceptor imaging via positron emission tomography (PET) and single photon emission tomography (SPECT) has been the focus of several reviews (Sedvall et al., 1986; Frost and Wagner Jr., 1990; Mazière and Mazière, 1990) and background information describing the basic principles of emission tomography is also available (Herscovitch, 1993; Bonne et al., 1992; Eriksson et al., 1990; Volkow et al., 1988). In this section, we present the rationale for current interest in PET and SPECT imaging of nAChRs, and address the question: How, in theory, might PET and SPECT imaging of high-affinity, central nAChRs be applied?

Neurodegenerative Diseases

The loss of cholinergic neurons in the basal forebrain has been associated with a variety of pathological disorders, such as senile dementia of the Alzheimer type (AD), Huntington's and

Parkinson's diseases, and progressive supranuclear palsy (Araujo et al., 1988; Kellar et al., 1987; London et al., 1989; Maelicke and Albuquerque, 1996; Nordberg and Winblad, 1986; Palacios et al., 1990; Whitehouse and Kalaria, 1995; Whitehouse et al., 1986; 1988a,b). Postmortem autoradiographic experiments on AD tissue using the nicotinic probes [^{3}H]-acetylcholine, [^{3}H]-(−) nicotine, and [^{3}H]-epibatidine consistently revealed a significant reduction of nAChRs in comparison to controls (Nordberg and Winblad, 1986; Whitehouse et al., 1986; Kellar et al., 1987; London et al., 1989; Warpman and Nordberg, 1995). Unlike such in vitro experiments, emission tomography offers the opportunity to monitor changes in human nAChRs in vivo. This may be of particular relevance in regard to AD since Court and Clementi (1995) suggest that significant reductions in nAChR binding site concentration may precede neuronal cell loss in AD. Thus, PET/SPECT studies of nAChRs would offer a means to examine the temporal and spatial sequence of decline of the nicotinic system in AD at the early stages of the disease process.

PET/SPECT imaging of nAChRs could also play an important role in monitoring the effects of new and established cholinergic drugs on nAChR binding site concentration. Emission tomography might prove useful in identifying patients who respond to AD cholinergic replacement therapy and in establishing individualized doses that prevent or slow down nAChR decline.

Drug Development

A variety of therapeutic targets for nAChR ligands have been identified (Williams et al., 1994), and synthetic chemists have now prepared subtype-selective nicotine agonists (Abreo et al., 1996; Arneric et al., 1996; Holladay et al., 1997) that might prove useful in the treatment of neurodegenerative disorders. PET, in particular, offers an excellent opportunity to expedite the drug development process of new nAChR-based agents. Often it is possible to label drug candidates isotopically with the short-lived, cyclotron-produced ^{11}C ($t_{1/2}$ = 20.4 min) as a substitute for the naturally occuring ^{12}C. The result is a high specific activity PET tracer of the drug that has very little mass associated with it (typically less than 10 μg). The PET tracer of the therapeutic drug candidate can then be studied in humans with less extensive safety and toxicity documentation (Farde, 1996). In a fairly rapid time frame, then, information concerning the drug's ability to bind to its target receptor (in this case nAChRs), temporal biodistribution, and metabolism in humans can be obtained.

Alternatively, a radiolabeled analog of the drug candidate (e.g., ^{18}F [$t_{1/2}$ = 109.8 min] for PET) can be produced, administered, and then competed against the "cold" test drug that is given in therapeutic dosage (Farde, 1996). This approach will demonstrate how the "cold" drug specifically inhibits radiotracer binding. Through the use of kinetic modeling and PET it may be possible to measure nAChR receptor occupancy of nicotinic drugs. Such receptor occupancy studies, as previously demonstrated for antipsychotic drugs and dopamine D_2 PET radioligands (Kapur et al., 1996; Roemer et al., 1996; Farde et al., 1992), would prove of value in establishing thresholds for clinically effective doses of nAChR drugs.

Smoking Cessation

Neuronal nAChRs have been implicated in tobacco dependence. For example, the addictive nature of cigarette smoking can be attributed to the reinforcing properties of the nAChR agonist, nicotine (Corrigall et al., 1992; Henningfield and Woodson, 1989). Thus, current smoking cessation strategies have focused on nicotine replacement therapies (Williams et al., 1994; Balfour and Fagerström, 1996). PET and SPECT imaging of nAChRs offers a noninvasive means to monitor the effects of nicotine replacement drugs (e.g., the nicotine patch and nicotine mouthpiece) on radiotracer binding. Such imaging studies would assist in assessing the therapeutic efficacy of smoking cessation drugs by examining the specific site of pharmacological action. Emission tomography of high-affinity nAChRs might also help determine in vivo the time course of smoking-induced nAChR upregulation (Benwell et al., 1988; Breese et al., 1997; Marks et al., 1985; Pauly et al., 1991; Wonnacot, 1990; Yates et al., 1995), as well as receptor changes during withdrawal in human subjects.

EMISSION TOMOGRAPHY OF nAChRs: THE PRACTICE

Some of the theorized applications of imaging central nAChRs have already been realized. As briefly outlined below, several research groups from Sweden have utilized [^{11}C](−)-nicotine and PET in an effort to visualize nAChRs, particularly in AD patients. Although the appropriateness of [^{11}C](−)-nicotine as an in vivo radiotracer for nAChRs can be questioned (see the next section), the methodologic approach used in these noninvasive imaging studies illustrates the potential of emission tomography to elucidate nAChR function.

The radiosynthesis and preliminary distribution kinetics of [^{11}C](−)-nicotine in rabbits was reported by Mazière et al. in 1976. More recently, [^{11}C](−)-nicotine has been administered to humans (Nyback et al., 1989; Sedvall et al., 1990), and the in vivo accumulation of the radioligand correlated well with the distribution of nAChRs measured by in vitro binding techniques in autopsy brain tissue (Nordberg, 1993). This radioligand has been employed for PET studies of nAChRs in Alzheimer's patients. Decreased uptake and binding of [^{11}C](−)-nicotine was reported in AD subjects (Nordberg et al., 1990). [^{11}C](−)-Nicotine and PET also have been used to monitor the effects of the anti-AD drug, tacrine, on nAChRs in AD patients (Nordberg et al., 1992b, 1997). The results showed increased cholinergic activity (measured as increased [^{11}C](−)-nicotine binding) after tacrine treatment. Moreover, the PET data obtained 3 weeks after tacrine treatment paralleled improvements in neuropsychological performance (Nordberg et al., 1992b). PET and [^{11}C](−)-nicotine were also used to monitor intracranial infusion of purified nerve growth factor to an Alzheimer patient (Seiger et al., 1993). These imaging studies revealed transient increases in [^{11}C](−)-nicotine uptake and binding and are suggestive of the ability of nerve growth factor to improve cholinergic brain function. These pioneering PET studies point toward the feasibility of using imaging techniques to customize and monitor treatment regimes in AD patients.

[^{11}C](−)-Nicotine and PET were also used to compare the nicotine distribution pattern of a smoking cessation device, the nicotine inhaler, with that of a conventional cigarette in a healthy male smoker (Lunnell et al., 1996). This tracer study, which spiked the inhaler and cigarette with [^{11}C](−)-nicotine, showed differential deposition of radioactivity. [^{11}C](−)-nicotine from cigarette smoking showed a markedly higher lung uptake while the radioactivity from the vapor inhaler was deposited bucally or in the upper respiratory tract. The above study offers support for using PET in the examination of the pharmacokinetics of nicotine replacement therapies and in the selection of nicotine delivery systems.

Currently [^{11}C](−)-nicotine has been the only PET radiotracer used in humans in attempts to visualize nAChRs. Comparable SPECT agents labeled with ^{123}I have yet to be developed and would prove complementary to PET ligands (Bourguigon et al., 1997). Such SPECT ligands do not require an on-site cyclotron for their production and, due to the longer half-life of ^{123}I ($t_{1/2}$ = 13.2 h), would afford a longer imaging window.

PET STUDIES WITH [^{11}C](−)-NICOTINE: THE PITFALLS

It is a difficult task to use [^{11}C](−)-nicotine as a ligand for measurement of neuronal nAChRs in vivo via PET. This is a result of the uptake of the radiotracer being mediated principally by regional cerebral blood flow (rCBF) (Nybäck et al., 1994, Grunwald et al., 1996). The fact that $[^{11}C](-)$-nicotine does not fulfill the receptor-based biodistribution criterion of saturability (Sedvall et al., 1986) also suggests the difficulty in using this radiotracer to map nAChRs in vivo (Nybäck et al., 1994). Finally, the short drug-receptor interaction time (on-off rate) is another limiting factor for the use of radiolabeled nicotine (Nybäck et al., 1994) or its radiolabeled analogs (Kämpfer et al., 1996; Sorger et al., 1996) for nAChR imaging.

Despite all of the drawbacks of using $[^{11}C](-)$-nicotine to localize nAChRs via PET, methodology has been developed to compensate for the flow dependence of $[^{11}C](-)$-nicotine uptake (Nordberg et al., 1995; Lundqvist et al., 1997). This methodology consist of a dual-tracer two-scan paradigm: administration of $[^{15}O]$-water (used to compensate for the influence of rCBF) closely followed by $[^{11}C](-)$-nicotine. With this protocol and the use of a two-compartment model, the authors report that in vivo quantification of nAChRs is achievable (Nordberg et al., 1995). Nonetheless, Lundqvist and colleagues (1997) recognize the need for new in vivo nAChR radioligands that can overcome the difficulties posed by $[^{11}C](-)$-nicotine.

NEW RADIOTRACERS FOR TOMOGRAPHIC IMAGING OF nAChRs

Design Requirements

The process of selecting a particular neuroreceptor ligand to be radiolabeled with an imaging radionuclide (e.g., ^{11}C or ^{18}F for PET or ^{123}I for SPECT) is not simple and has been discussed in greater detail elsewhere (Dannals et al., 1993; Eckelman and Gibson, 1993). Briefly, it is highly desirable to know the in vivo behavior of the compound prior to devoting resources toward its preparation in radiolabeled form. Often, however, one must rely principally on in vitro binding data. In such cases, target ligands that possess high affinity (typically <1 nM) and selectivity for the neuroreceptor site of interest are desired. In terms of in vivo properties, an ideal radioprobe for a neuroreceptor site should demonstrate: (1) good blood-brain barrier penetration and rapid clearance from the blood; (2) high specificity and selectivity for the binding site; (3) an absence of radiolabeled metabolites capable of crossing the blood-brain barrier; (4) appropriate kinetics so that high-contrast images of specific binding regions can be obtained within the time window afforded with the radionuclide (typically 90 min maximum for a ^{11}C radiotracer, 3 h for a ^{18}F radioligand, and 24 h for an ^{123}I compound); (5) low toxicity so that the radiotracer can be administered with a high margin of safety.

The tracer selection process must also consider the constraints imposed by the radiochemistry. For development of short-lived radiotracers such as ^{11}C ($t_{1/2}$ = 20.4 min) and ^{18}F ($t_{1/2}$ = 109.7), time limits the radiosynthetic approaches that can be employed. Specific activity of the radioligand, which often is dependent on the time of radiosynthesis, is also a factor to be considered. It is desirable for imaging radioligands to be of high specific activity for two reasons. First, since neuroreceptors such as high-affinity nAChRs are in relatively low density (in the range of 23 and 16 fmol/mg in human and rodent thalamus, respectively (Nordberg et al., 1992a; London et al., 1985), a high specific activity radioligand is required so as not to saturate the binding sites. Second, high specific activity PET/SPECT radiotracers are more likely to ensure that the radiolabeled compound does not illicit any pharmacologic effects when administered.

This second point is highly pertinent to the nAChR system since nAChR agonists such as epibatidine are known to produce severe pharmacologic effects and death at low doses (Sullivan et al., 1994; Badio et al., 1997). Therefore, an ideal in vivo radiotracer for the nAChR system should be an antagonist, as this would lessen concerns about eliciting pharmacological responses. However, a high-affinity (subnanomolar) nAChR antagonist has yet to be identified. In fact, the most potent nAChR antagonist to date, erysodine (Decker et al., 1995a), is approximately 5-fold less potent than cytisine, and already it has been shown that [^{3}H]-cytisine possesses only a moderate degree of specific binding in vivo in mouse brain (Flesher et al., 1994). Thus, until a high-affinity nAChR antagonist is identified, radiotracer development will more than likely involve preparation of labeled nAChR agonists.

PET Radioligands Based on (−)-Nicotine

In an attempt to develop a ^{11}C PET radioligand with improved in vivo properties as compared to [^{11}C](−)-nicotine, **1** (Fig. 13-1), the closely related analog, **2,** was prepared as a racemate (Sihver et al., 1997) (Fig. 13-1). The in vivo properties of **2** have yet to be published, but in vitro saturation binding experiments showed this radioligand to possess high affinity (Kd = 0.1 nM) and low nonspecific binding in rat brain. The cognition-enhancing ABT-418 (Newhouse et al., 1996), **3,** an isoxazole analog of (−)-nicotine, has also been labeled with ^{11}C (Fig. 13-1) but has proven disappointing in localizing nAChRs in baboon brain via PET (Valette et al., 1997). The inability of [^{11}C]-**3** to bind to nAChRs in vivo may be a result of the compound's affinity (Ki = 4.2 vs. [^{3}H]-cytisine), which is comparable to that of **1** (Garvey et al., 1994). Higher-affinity nAChR ligands that have been labeled with positron emitters or radioiodine and are able to image nAChRs in primates can be found in the epibatidine series.

PET/SPECT Radioligands Based on Epibatidine

With the discovery and subsequent pharmacological characterization of epibatidine, **4** (Fig. 13-1) (Badio and Daly, 1994; Houghtling et al., 1995; Spande et al., 1992), a new probe to study

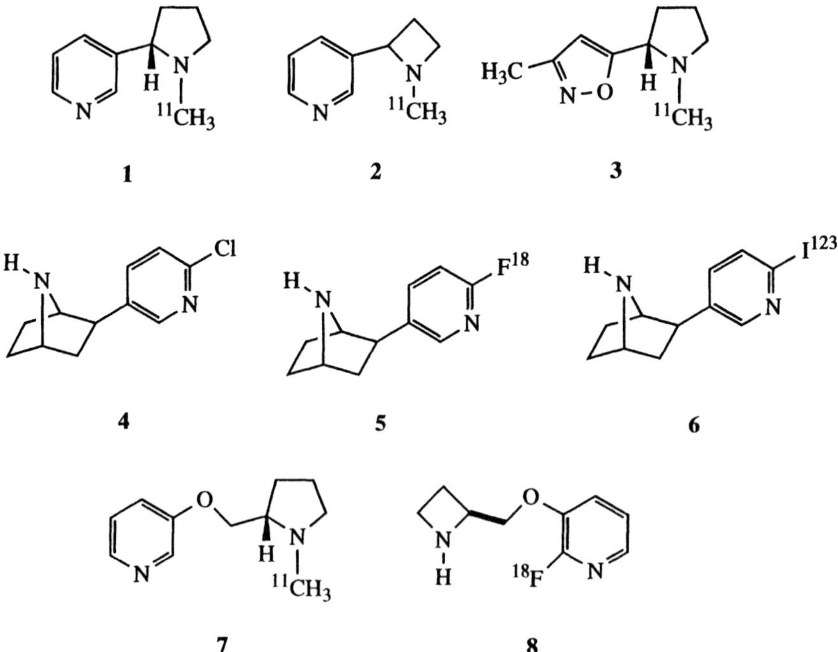

Figure 13-1. Ligands developed for in vivo studies of nAChRs: [^{11}C](−)-nicotine, **1**; *[^{11}C] MPA,* **2**; *[^{11}C] ABT-418,* **3**; *epibatidine,* **4**; *[^{18}F] FPH,* **5**; *[^{123}I] IPH,* **6**; *and 3-pyridyl ethers [^{11}C] A-84543* **7**; *and 2-[^{18}F]fluoro-A-85380,* **8**.

nAChRs has become available. Both enantiomers of epibatidine display subnanomolar affinity for nAChRs in the brain (Badio and Daly, 1994), and this compound in tritiated form has proven useful for autoradiographic (Perry and Kellar, 1995) and in vivo studies of neuronal nAChRs (London et al., 1995). Similar studies with [^{3}H]-norchloroepibatidine have demonstrated selective and specific binding to nAChRs in rodent brain (Scheffel et al., 1995). These results have prompted development of several PET and SPECT analogs of the parent structure.

For example, the [^{18}F] labeled 2-fluoro-pyridyl analog, [^{18}F] FPH, **5** (Fig. 13-1), has been synthesized and evaluated as an imaging agent for nAChRs at two different PET centers (Horti et al., 1996a; Villemagne et al., 1996, 1997; Ding et al., 1996; Liang et al., 1997). Other positron labeled epibatidine analogs have been reported including a Br-76 tracer (Bottlaender et al., 1997) and ^{11}C N-methyl derivatives (Patt et al., 1995; Horti et al., 1996b). Since [^{18}F] FPH has been the most thoroughly studied member in this series, a summary of our PET imaging studies (Villemagne et al., 1996, 1997) with this compound in baboon brain is provided.

Figure 13-2 shows the PET-generated time-radioactivity curves for [^{18}F] FPH in various brain regions. Brain radioactivity reached a plateau approximately 60 min postinjection in the thalamus. Clearance was relatively rapid from the cerebellum ($t_{1/2}$ = 3 h), intermediate from the hypothalamus/midbrain ($t_{1/2}$ = 7 h), and slow from the thalamus ($t_{1/2}$ = 16 h). Uptake of [^{18}F] FPH at 130 min postinjection was highest in the thalamus and hypothalamus/midbrain; intermediate in the neocortex and hippocampus; and lowest in the cerebellum (Fig. 13-2, left column). Similar results were reported by Ding et al. (1996). In order to establish specificity of the binding, a subcutaneous injection of 1 mg/kg cytisine was administered at 45 minutes after the injection of [^{18}F] FPH. This challenge dose using a well-characterized nAChR agonist reduced brain radioactivity at 130 min by 67%, 64%, and 52% of control values in the thalamus, hypothalamus/midbrain, and cerebellum, respectively (Fig. 13-2). Radioactivity uptake was highly correlated with the known densities of nAChRs measured in vitro by ^{3}H-nicotine (r = 0.97), ^{3}H-cytisine (r = 0.97), ^{3}H-epibatidine (r = 0.91) in rat brain, and by ^{3}H-nicotine in human brain (r = 0.81) (Perry and Kellar, 1995; London et al., 1985; Nordberg et al., 1992a). Thus, our results and those of Ding et al. show [^{18}F] FPH to be an exquisite in vivo PET probe for high-affinity nAChRs.

Similarly, [^{123}I] IPH (**6**), has been developed (Musachio et al., 1997a,b) and was the first SPECT agent used to image nAChRs in primate brain (Musachio, 1997c). [^{123}I] IPH displayed the appropriate regional localization for a high-affinity nAChR ligand (thalamus > frontal cortex > cerebellum), and radioactivity was displaced (35–45% displacement) by a challenge dose of cytisine. IPH labeled with the longer-lived ^{125}I ($t_{1/2}$ = 59.7 days) has also proven useful for external monitoring of central nAChRs in living animals via a simple probe system (Liu et al., 1997) and for rapid autoradiographic studies of nAChRs in brain and peripheral ganglia (Dávila-García et al., 1997).

Although **5** and **6** are highly sensitive in vivo probes for nAChRs, they are also highly toxic (Molina et al., 1997; Horti et al., in press). It is believed that the potent toxicity of epibatidine and its analogs is related to activation of peripheral ganglionic nAChRs (Lin et al., 1997). The acute toxicity of **5** has a reported LD_{50} of 1.5 μg/kg via intravenous injection in unanesthetized rat, and even at the lowest dose of FPH studied (0.25 μg/kg), plasma catecholamine levels increased significantly (Molina et al., 1997). If a standard 5 mCi injection of [^{18}F] FPH (conservatively projected specific radioactivity of 1000 mCi/μmol) were given (hypothetically) to a 70 kg human subject, a dose of 0.01 μg/kg of the tracer would be administered. This dose is only 25-fold lower than the pharmacologic threshold. Thus, the usually wide safety margin afforded with a PET radiopharmaceutical (Farde et al., 1996) is compromised in the case of [^{18}F] FPH, and it has been concluded that human PET studies with [^{18}F] FPH, and it has been concluded that human PET studies with [^{18}F] FPH are inadvisable (Molina et al., 1997). Nonetheless, [^{18}F] FPH and the SPECT analog, [^{123}I] IPH, have advanced nAChR re-

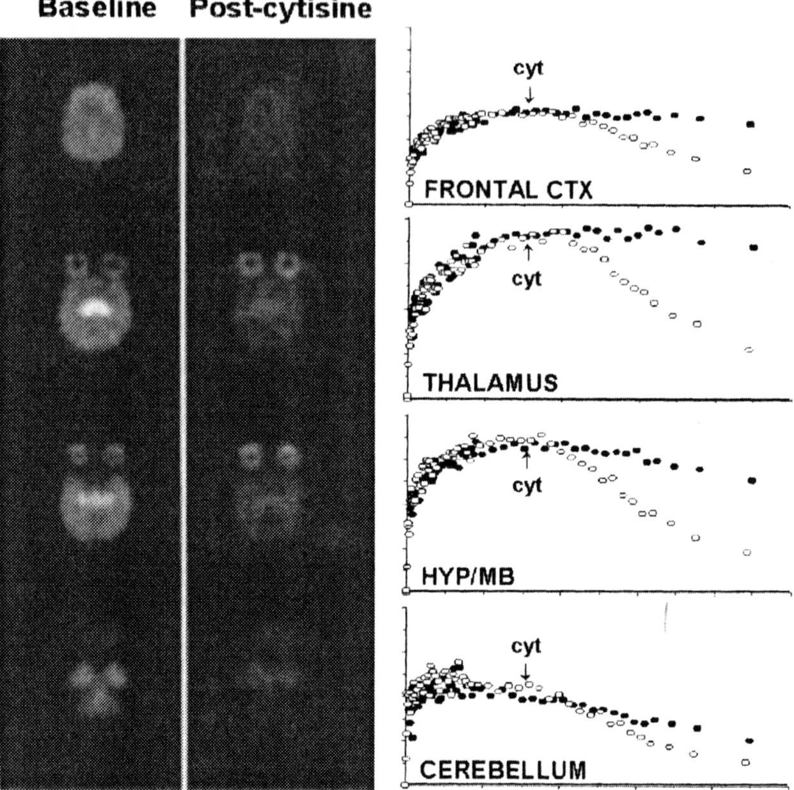

Figure 13-2. Black and white coded transforms of PET scans showing ^{18}F-FPH binding in the baboon brain at 130 mpi in the control study (left column) and after displacement with 1 mg/kg cytisine (right column). Corresponding time-activity curves for each region are shown. In the control study (left column), there is high uptake of the radiotracer in the thalamus and hypothalamus/midbrain regions, intermediate uptake in the frontal cortex, and low uptake in the cerebellum. Time activity curves from the frontal cortex, thalamus, hypothalamus/midbrain (HYP/MB), and cerebellum in the control study (full symbols ●) show the radioactivity reaching a plateau approximately 50 min after radiotracer injection followed by a slow dissociation. The administration of 1 mg/kg cytisine (cyt) at 45 min after tracer injection (open symbols ○) markedly decreased the ^{18}F-FPH binding in all areas, showing specificity of the binding of ^{18}F-FPH to nAChRs.

ceptor imaging greatly by showing that high specific binding of a nAChR radioligand in vivo can be achieved.

PET Radioligands Based on 3-Pyridyl Ethers

In an effort to circumvent the side-effect liability seen with (−)-nicotine, Abbott Laboratories has developed a series of 3-pyridyl ethers that possess subnanomolar affinity for brain nAChRs and differentially activate subtypes of neuronal nAChRs (Abreo et al., 1996). This series shows promise for development of therapeutic agents for treatment of AD (Arneric et al., 1996) and has been the basis for development of an exciting new class of nAChR imaging agents. The N-methyl pyrolidinyl ether, A-84543, **7** (Fig. 13-1), was of particular interest since it possesses subnanomolar affinity for [^{3}H]-cytisine binding sites (Ki = 0.15 nM) and was 84-fold more potent in stimulating the $\alpha4\beta2$ subtype than ganglionic-type nAChRs (Abreo et al., 1996). A-84543 has been recently labeled with ^{11}C methyl iodide (Kassiou et al., 1997) and evaluated for its in vivo binding properties to central nAChRs in mice (Kassiou et al., 1998). The results indicate that [^{11}C] A-84543 is able to label nAChRs in vivo, although blocking studies suggested a higher degree of nonspecific binding for this radiotracer as compared to radiolabeled epibatidine analogs. Perhaps the most encouraging result with the high-affinity A-84543 was its reduced toxicity liability. Whereas the LD_{50} of epibatidine and its analogs is in the low μg/kg range in mice, the LD_{50} for A-84543 (via intravenous injection) was greater than 1 mg/kg. This suggests that the toxicity of 3-pyridyl ethers is sufficiently low that PET and SPECT tracers from this series could ultimately be used for human studies of nAChRs provided that the radiotracer exhibits high in vivo specificity.

In an attempt to increase further the in vivo specific binding of radiolabeled 3-pyridyl ethers, a PET ligand based on the structure of A-85380 has been prepared. A-85380 is an (*S*)-azetidinylmethoxy pyridine ether that possesses ca. 50 pM affinity for rat brain [^{3}H]-cytisine binding sites, which is comparable to that of ($\pm$)-epibatidine. This compound displays roughly 3-fold higher affinity than that of A-84543 (Abreo et al., 1996), and it is highly selective for nAChRs over other receptor systems (Sullivan et al., 1996). Unlike A-84543, however, A-85380 cannot be readily labeled with the positron-emitting ^{11}C. Thus, the fluorinated analog of A-85380 (2-fluoro-A-85380), **8** (Fig. 13-1), was synthesized (Horti et al., 1997) and found to retain subnanomolar affinity (Ki = 0.052 nM vs. [^{3}H]-cytisine) for nAChR binding sites in rat brain homogenates. Radiolabeling conditions have been established to prepare 2-[^{18}F]fluoro-A-85380 in high specific radioactivity (Horti et al., 1998; Dollé et al., 1998). In vivo, 2-[^{18}F]fluoro-A-85380 showed adequate brain penetration, and the radioligand exhibited the appropriate regional distribution (thalamus, superior colliculi > striatum, cortex > cerebellum) for a high-affinity nAChR probe. The high specificity and selectivity of 2-[^{18}F]fluoro-A-85380 binding to nAChRs was also confirmed by pharmacologic blocking studies in mouse brain (Horti et al., 1997) and baboon (Vallette et al., 1998). In summary, the biodistribution profile of 2-[^{18}F]fluoro-A-85380 was similar to that of the ^{18}F labeled analog of epibatidine: Both are highly specific in vivo probes for nAChRs. However, unlike the epibatidine series, the fluoro analog of A-85380 is much less toxic (LD_{50} > 7.3 μmol/kg i.v. in mice). Thus, it appears that development of 2-[^{18}F]fluoro-A-85380 has overcome the stumbling block of toxicity seen in radiotracers of the epibatidine series.

Of course, several other criteria must be met before 2-[^{18}F]fluoro-A-85380 can be used to image nAChRs in humans via PET. First, metabolism studies in rodents and nonhuman primates should be performed to test whether there are radiolabeled metabolites formed and if they are able to cross the blood-brain barrier. Second, a more detailed determination of the dose threshold(s) at which 2-fluoro-A-85380 first elicits pharmacological effects (e.g., cardiovascular effects, hypomotility, hypothermia) is in order. Such measurements would more accurately reflect the safety margin afforded with 2-[^{18}F]fluoro-A-85380 as compared to gross lethality determinations. Finally, radiation dosimetry estimates need to be calculated according to the Medical Internal Radiation Dose method so as to determine maximal permissible amounts of radioactivity that can be administered to humans.

Notwithstanding these additional criteria that must be met, the ^{18}F labeled analog of A-85380 appears most promising as an in vivo PET radiotracer of human neuronal nAChRs as a result of its high specificity and selectivity and relatively low toxicity. Undoubtedly, other PET and SPECT radioligands will be developed from the 3-pyridyl ether series. Indeed, our laboratory has recently reported a radioiodinated analog of A-85380 that is able to label nAChRs in vivo in mice (Musachio et al., 1998) and image baboon nAChRs via SPECT (Musachio et al., in press). These new radiotracers should offer a new era in nAChR receptor imaging that is no longer hampered by specificity or toxicity concerns.

SUMMARY

For years, the only tracer available for study of high-affinity nAChRs via emission tomography was [^{11}C](−)-nicotine. This is not longer the case. However, much remains to be accomplished before a high-affinity, selective, nontoxic radiotracer for neuronal nAChRs is incorporated into routine clinical studies (e.g., monitoring response to cholinergic drugs in AD).

Development of new tracers for emission tomography has focused exclusively on the $\alpha 4\beta 2$ nAChR subtype. Imaging of the $\alpha 7$ subtype, although yet to be accomplished, is another promising area of study. This is true because the $\alpha 7$ subtype has been implicated in working memory and learning (Gray et al., 1996; Granon et al., 1995; Blozovski, 1985), and appears to play a key role in schizophrenia (Freedman et al., 1995; Leonard et al., 1996).

Advances in the field of therapeutic nicotinic drugs will translate into development of novel nAChR radioprobes. These radiotracers, in turn, can assist in shaping the development, validation, and approval of cholinergic modulators. In vivo imaging techniques are unique in their ability to provide pharmacokinetic information concerning nicotinic drugs on their specific binding site. Such an approach will not only help in the development of new nAChR-based therapeutic agents but will also elucidate the physiological mechanisms of nAChRs in health and disease.

ACKNOWLEDGMENTS

The authors would like to acknowledge the contributions made by Robert F. Dannals, Edythe D. London, Andrew G, Horti, Andreí O. Koren, Michael Kassiou, Hayden T. Ravert, William B. Mathews, Paige Finley, Dave Clough, Robert Smoot, Karen Edmonds, Yougen Zhan, and Henry N. Wagner Jr.

REFERENCES

Abreo MA, Lin Nan-Horng, Garvey DS, Gunn DE, Hettinger AM, Wasicak JT, Pavlik PA, Martin YC, Donnelly-Roberts DL, Anderson DJ, Sullivan JP, Williams M, Arneric SP, Holladay MW (1996): Novel 3-pyridyl ethers with subnanomolar affinity for central neuronal nicotinic acetylcholine receptors. J. Med. Chem. 39: 817–825.

Araujo DM, Lapchak PA, Robitaille Y, Gauthier S, Quirion RJ (1988): Differential alteration of various cholinergic markers in cortical and subcortical regions of human brain in Alzheimer's disease. Neurochemistry 50: 1914–1923.

Arneric SP, Bannon AW, Briggs CA, Brioni JD, Buccafusco JJ, Decker MW, Donnelly-Roberts DK, Gopalkrishnan M, Holladay MW, Kyncy J, Lin NH, Marsh KC, Qui Y, Radek R, Sullivan JP, Williams M (1996): ABT-089: an orally active cholinergic channel modulator with cognition enhancing and neuroprotective activity. In: Becker R, Giacobini E, Robert P, editors. Alzheimer disease: from molecular biology to therapy. New York: Birkhauser Boston Inc., pp. 287–292.

Badio B, Daly JW (1994): Epibatidine, a potent analgetic and nicotinic agonist. Mol. Pharmacol. 45: 563–569.

Badio B, Garraffo HM, Plummer CV, Padgett WL, Daly JW (1997): Synthesis and nicotinic activity of epiboxidine: an isoxazole analogue of epibatidine. Eur. J. Pharmacol. 321: 189–194.

Balfour DJK, Fagerström KO (1996): Pharmacology of nicotine and its therapeutic use in smoking cessation and neurodegenerative disorders. Pharmacol. Ther. 72: 51–81.

Benwell MEM, Balfour DJK, Anderson JM (1988): Evidence that tobacco smoking increases the density of (−)-[^{3}H]nicotine binding sites in human brain. J. Neurochem. 50: 1243–1247.

Blozovski D (1985): Mediation of passive avoidance learning by nicotinic hippocampo-entorhinal components in young rats. Dev. Psychobiol. 18: 355–366.

Bonne O, Krausz Y, Lerer B (1992): SPECT imaging in psychiatry. A review. Gen. Hosp. Psychiatry 14: 296–306.

Bottlaender M, Loc'h C, Kassiou M, Ottaviani M, Coulon C, Tavitian B, London E, Musachio J, Dannals R, Mazière B (1997): In vivo PET study of nicotinic acetylcholine receptors (nAChR) with [Br-76]-bromoepibatidine. Soc. Neurosci. Abstr. Book 23: 382.

Bourguignon MH, Pauwels EKJ, Loc'h C, Mazière B (1997): Iodine-123 labelled radiopharmaceuticals and single-photon emission tomography: a natural liaison. Eur. J. Nucl. Med. 24: 331–344.

Breese CR, Marks MJ, Logel J, Adams CE, Sullivan B, Collins AC, Leonard S (1997): Effect of smoking history on [^{3}H]nicotine binding in human postmortem brain. J. Pharmacol. Exp. Ther. 282: 7–13.

Clarke PB, Schwartz RD, Paul SM, Pert CB, Pert A (1985): Nicotinic binding in rat brain: autoradiographic comparison of [^{3}H]acetylcholine, [^{3}H]nicotine, and [^{125}I]-alpha-bungarotoxin. J. Neurosci. 5: 1307–1315.

Corrigall WA, Franklin KBJ, Coen KM, Clarke PBS (1992): The mesolimbic dopaminergic system is implicated in the reinforcing properties of nicotine. Psychopharmacology 107: 285–289.

Court J, Clementi F (1995): Distribution of nicotinic subtypes in human brain. Alzheimer disease and associated disorders 9, Suppl. 2: 6–14.

Dannals RF, Ravert HT, Wilson AA (1993): Chemistry of tracers for positron emission tomography. In: Burns HD, Gibson RE, Dannals RF, Siegl PKS, editors. Nuclear imaging in drug discovery, development, and approval. Boston: Birkhäuser, pp. 55–74.

Dávila-García MI, Musachio JL, Perry DC, Xiao Y, Horti A, London ED, Dannals RF, and Kellar KJ (1997): [^{125}I]IPH, an epibatidine analog, binds with high affinity to neuronal nicotinic cholinergic receptors. J. Pharmacol. Exp. Ther. 282: 445–451.

Decker MW, Anderson DJ, Brioni JD, Donnelly-Roberts DL, Kang CH, O'Neill AB, Piattoni-Kaplan M, Swanson S, and Sullivan JP (1995a): Erysodine, a competitive antagonist at neuronal nicotinic acetylcholine receptors. Eur. J. Pharmacol. 280: 79–89.

Decker MW, Brioni JD, Bannon A W, Arneric SP (1995b): Diversity of neuronal nicotinic acetylcholine receptors: lessons from behavior and implications for CNS therapeutics. Life Sci. 56: 545–570.

Deneris ES, Connolly J, Rogers SW, Duvoisin R (1991): Pharmacological and functional diversity of neuronal nicotinic acetylcholine receptors. Trends Pharmacol. Sci. 12: 34–40.

Ding YS, Gatley SJ, Fowler JS, Volkow ND, Aggarwal D, Logan J, Dewey SL, Liang SL, Carroll FI, Kuhar MJ (1996): Mapping nicotinic acetycholine receptors with PET. Synapse 24: 403–407.

Dollé F, Valette H, Hinnen F, Vaufrey F, Guenther I, Bottlaender M, Crouzel C (1998): Synthesis of 2-[^{18}F]fluoro-3-[(S)-2-azetidinylmethoxy]pyridine, a novel highly potent radioligand for in vivo imaging central nicotinic acetylcholine receptors. J. Nucl. Med. 39: 21P.

Eckelmann WC, Gibson RE (1993): The design of site-directed radiopharmaceuticals for use in drug discovery. In: Burns HD, Gibson RE, Dannals RF, Siegl PKS, editors. Nuclear imaging in drug discovery, development, and approval. Boston: Birkhäuser, pp. 113–134.

Eriksson L, Dahlbom M, Widen L (1990): Positron emission tomography—a new technique for studies of the central nervous system. J. Microsc. 157: 305–333.

Farde L (1996): The advantage of using positron emission tomography in drug research. Trends Neurosci. 19: 211–214.

Farde L, Nordstrom AL, Wielsel FA, Pauli S, Halldin C, Sedvall G (1992): Positron emission tomographic analysis of central D1 and D2 dopamine receptor occupancy in patients treated with classical neuroleptics and clozapine. Relation to extrapyramidal side effects. Arch. Gen. Psychiatry 49: 538–544.

Flesher JE, Scheffel U, London ED, Frost JJ (1994): *In vivo* labeling of nicotinic cholinergic receptors in brain with [^{3}H]cytisine. Life Sci. 54: 1883–1890.

Flores CM, Rogers SW, Pabreza LA, Wolfe B, Kellar KJ (1992): A subtype of nicotinic cholinergic receptor in rat brain is composed of alpha 4 and beta 2 subunits and is up-regulated by chronic nicotine treatment. Mol. Pharmacol. 41: 31–37.

Freedman R, Hall M, Adler LE, Leonard S (1995): Evidence in postmortem brain tissue for decreased number of hippocampal nicotinic receptors in schizophrenia. Biol. Psychiatry 38: 22–33.

Frost JJ, Wagner HN Jr, editors (1990): Quantitative imaging: neuroreceptors, neurotransmitters, and enzymes. New York: Raven Press.

Garvey DS, Wasicak JT, Decker MW, Brioni JD, Buckley MJ, Sullivan JP, Carrera GM, Holladay MW, Arneric SP, Williams M (1994): Novel isoxazoles which interact with brain cholinergic channel receptors have intrinsic cognitive enhancing and anxiolytic activities. J. Med. Chem. 37: 1055–1059.

Granon S, Poucet B, Thinus-Blanc C, Changeux JP, Vidal C (1995): Nicotinic and muscarinic receptors in the rat prefrontal cortex: differential roles in working memory, response selection and effortful processing. Psychopharmacology 119: 139–144.

Gray R, Rajan AS, Radcliffe KA, Yakehiro M, Dani JA (1996): Hippocampal synaptic transmission enhanced by low concentrations of nicotine. Nature 383: 713–716.

Grünwald F, Biersack H-J, Kuschinsky W (1996): Nicotine receptor mapping. Eur. J. Nucl. Med. 23: 1012–1016.

Henningfield JE, Woodson PP (1989): Dose-related actions of nicotine on behavior and physiology: review and implications for replacement therapy for nicotine dependence. J. Substance Abuse 1: 301–317.

Herscovitch P (1993): Evaluation of the brain by positron emission tomography. Rheum. Dis. Clin. North Am. 19: 765–794.

Holladay MW, Dart MJ, Lynch JK (1997): Neuronal nicotinic acetylcholine receptors as targets for drug discovery. J. Med. Chem. 40: 4169–4194.

Horti AG, Ravert HT, London ED, Dannals RF (1996a): Synthesis of a radiotracer for studying nicotinic acetylcholine receptors by positron emission tomography: (±)-exo-2-(2-[^{18}F]fluoro-5-pyridyl)-7-azabicyclo [2.2.1] heptane. J. Labelled, Compd. Radiopharm. 38: 355–366.

Horti AG, Ravert HT, Mathews WB, Musachio JL, Kimes A, London ED, Dannals RF (1996b): Synthesis of high specific activity carbon-11 N-methylated analogs of epibatidine for imaging nAChRs. J. Nucl. Med. 37S: 192P (abs).

Horti AG, Koren AO, Ravert HT, Musachio JL, Mathews WB, London ED, Dannals RF. (1998): Synthesis of a radiotracer for studying nicotinic acetylcholine receptors: 2-[18F]fluoro-3-(2(*S*)-azetidinylmethoxy) pyridine (2-[^{18}F]A-85380). J. Labelled Compd. Radiopharm. 41: 309–318.

Horti AG, Scheffel UA, Kimes AS, Musachio JL, Ravert HT, Mathews WB, Zhan Y, Finley PA, London ED, Dannals RF. Synthesis and evaluation of *N*-[C-11]methylated analogs of epibatidine as tracers for positron emission tomographic studies of nicotinic acetycholine receptors. J. Med. Chem. in press.

Houghtling RA, Davila-Garcia MI, Kellar KJ (1995): Characterization of (±)-[^{3}H]epibatidine binding to nicotinic cholinergic receptors in rat and human brain. Mol. Pharmacol. 48: 280–287.

Kämpfer I, Sorger D, Schliebs R, Kärger W, Günther K, Schulze K, Knapp WH (1996): Radiodination of nicotine with specific activity high enough for mapping nicotinic acetylcholine receptors. Eur. J. Nucl. Med. 23: 157–162.

Kapur S, Zipursky RB, Jones C, Remington GJ, Wilson AA, DaSilvia J, Houle S (1996): The D2 receptor occupancy profile of loxapine determined using PET. Neuropsychopharmacology 15: 562–566.

Kassiou M, Ravert HT, Mathews WB, Musachio JL, London ED, Dannals RF (1997): Synthesis of 3[(1-[^{11}C]methyl-2(S)-pyrrolidinylmethoxy]pyridine and 3-[1-[^{11}C]methyl-2-(R)-pyrrolidinyl) methoxy] pyridine: radioligands for *in vivo* studies of neuronal nicotinic acetylcholine receptors. J. Labelled Compd. Radiopharm. 38: 425–431.

Kassiou M, Scheffel UA, Ravert HT, Mathews WB, Musachio JL, London ED, Dannals RF (1998): Pharmacological evaluation of [^{11}C]A-84543: an enantioselective ligand for *in vivo* studies of neuronal nicotinic acetylcholine receptors. Life Sci. 63: PL13–PL18.

Kellar KJ, Whitehouse PJ, Martino-Barrows AM, Marcus K, Price DL (1987): Muscarinic and nicotinic cholinergic binding sites in Alzheimer's disease cerebral cortex. Brain Res. 436: 62–68.

Leonard S, Adams C, Breese CR, Adler LE, Bickford P, Byerley W, Coon H, Griffith JM, Miller C, Myles-Worsley M, Nagamoto HT, Rollins Y, Stevens KE, Waldo M, Freedman R (1996): Nicotinic receptor function in schizophrenia. Schizophren. Bull. 22: 431–445.

Liang F, Navarro HA, Abraham P, Kotian P, Ding Y-S, Fowler J, Volkow N, Kuhar MJ, Carrol FI (1997): Synthesis and nicotinic acetylcholine receptor binding properties of exo-2-(2′-fluoro-5′-pyridinyl)-7-azabicyclo-[2.2.1]heptane: a new positron emission tomography ligand for nicotinic receptors. J. Med. Chem. 40: 2293–2295.

Lin, N-H, Gunn DE, Ryther KB, Garvey DS, Donnelly-Roberts DL, Decker MW, Brioni JD, Buckley MJ, Rodrigues AD, Marsh KG, Anderson DJ, Buccafusco JJ, Prendergast MA, Sullivan JP, Williams M, Arneric SP, Hollady MW (1997): Structure-activity studies on 2-methyl-3-(2(S)-pyrrolidinyl-methoxy)pyridine (ABT-089): an orally bioavailable 3-pyridyl ether nicotinic acetylcholine receptor ligand with cognition-enhancing properties. J. Med. Chem. 40: 385–390.

Lindstrom J, Anand R, Peng X, Gerzanich V, Wang F, Li Y (1995): Neuronal nicotinic receptor subtypes. Ann. NY Acad. Sci. 757: 100–116.

Liu X, Musachio JL, Wagner HN Jr, Mochizuki T, Dannals RF, London ED (1997): External monitoring of cerebral nicotinic acetylcholine receptors in living mice. Synapse 27: 378–380.

London ED, Ball MJ, Waller SB (1989): Nicotinic binding sites in cerebral cortex and hippocampus in Alzheimer's dementia. Neurochem. Res. 14: 745–750.

London ED, Scheffel U, Kimes AS, Kellar KJ (1995): *In vivo* labeling of nicotinic acetylcholine receptors in brain with [^{3}H]epibatidine. Eur. J. Pharm. 278: R1–2.

London ED, Waller SB, Wamsley JK (1985): Autoradiographic localization of [^{3}H]nicotine binding sites in the rat brain. Neurosci. Lett. 53:179–184.

Lundqvist H, Langstrom B, Nordberg A (1997): Use of carbon-11 nicotine in PET studies. Eur. J. Nucl. Med. 24: 825–826.

Lunnel E, Berström M, Antoni G, Långström B, Nordberg A (1996): Nicotine deposition and body distribution form a nicotine inhaler and a cigarette studied with positron emission tomography. Clin. Pharmacol. Ther. 59: 593–594.

Maelicke A, Albuquerque E (1996): New approach to drug therapy in Alzheimer's dementia. Drug Discovery Today 1: 53–59.

Marks MJ, Stitzel JA, Collins AC (1985): Time course of the effects of chronic nicotine infusion on drug response and brain receptors. J. Pharmacol. Exp. Ther. 235: 619–628.

Mazière M (1995): Cholinergic neurotransmission studied *in vivo* using positron emission tomography or single photon emission computerized tomography. Pharmacol. Ther. 66: 83–101.

Mazière B, Mazière M (1990): Where have we got to with neuroreceptor mapping of the human brain? Eur. J. Nucl. Med. 16: 817–835.

Mazière M, Comar D, Marazano C, Berger G (1976): Nicotine-11C: synthesis and distribution kinetics in animals. Eur. J. Nucl. Med. 1: 255–258.

McDonald IA, Cosford N, Vernier J-M (1995): Nicotinic acetylcholine receptors: molecular biology, chemistry and pharmacology. Annu. Rept. Med. Chem. 30: 41–50.

Molina PE, Ding Y-S, Carroll FI, Liang F, Volkow ND, Pappas N, Kuhar M, Abumrad N, Gatley SJ, Fowler JS (1997): Fluoro-norchloroepibatidine: preclinical assesment of acute toxicity. Nucl. Med. Biol. 24: 743–747.

Mulholland GK, Kilbourn MR, Sherman P, Carey JE, Frey KA, Koeppe RA, Kuhl DE (1995): Synthesis, *in vivo* biodistribution and dosimetry of [^{11}C]N-methylpiperidyl benzilate ([^{11}C]NMPB), a muscarinic acetylcholine receptor antagonist. Nucl. Med. Biol. 22: 13–17.

Musachio JL, Horti A, London ED, Dannals RF (1997a): Synthesis of a radioiodinated analog of epibatidine: ($\pm$)-exo-2-(2-iodo-5-pyridyl)-7-azabicylco[2.2.1]heptane for *in vitro* and *in vivo* studies of nicotinic acetylcholine receptors. J. Labelled Compd. Radiopharm. 39: 39–48.

Musachio JL, Scheffel U, Villemagne V, Horti A, Finley P, Xiao Y, Dávila-García M, Keller KJ, Ravert HT, Mathews WB, London ED, Dannals RF (1997b): [^{11}C] and ^{123}I/125] (+)- and (−)-MeIPH: analogs of epibatidine for *in vivo* studies of nicotinic acetylcholine receptors. J. Nucl. Med. 38: 65P.

Musachio JL, Villemagne VL, Scheffel U, Stathis M, Finley P, Horti A, London ED, Dannals RF (1997c): [$^{125/123}$I]IPH: a radioiodinated analog of epibatidine for *in vivo* studies of nicotinic acetylcholine receptors. Synapse 26: 392–399.

Musachio JL, Scheffel U, Finley PA, Zhan Y, Mochizuki T, Wagner HN Jr, Dannals RF (1998): 5-[I-125/123]iodo-3-(2(*S*)-azetidinylmethoxy)pyridine, a radioiodinated analog of A-85380 for *in vivo* studies of central nicotinic acetylcholine receptors. Life Sci. 62: PL351–357.

Musachio JL, Villemagne VL, Scheffel UA, Dannals RF, Dogan AS, Yokoi F, Wong DF. Synthesis of an I-123 analog of A-85380 and preliminary SPECT imaging of nicotinic receptors in baboon. Nucl. Med. Biol. in press.

Nef P, Oneyser C, Alliod C, Couturier S, Ballivet M (1988): Genes expressed in the brain define three distinct neuronal nicotinic acetylcholine receptors. EMBO J. 7: 595–601.

Newhouse PA, Potter A, Corwin J (1996): Acute administration of the cholinergic channel activator ABT-418 improves learning in Alzheimer's disease. 2nd Annual Society for Research on Nicotine and Tobacco Conference, Washington DC, 1996, Abstr. A39.

Nordberg A (1993): *In vivo* detection of neurotransmitter changes in Alzheimer's disease. Ann. NY Acad. Sci. 695: 27–33.

Nordberg A, Winblad B. (1986): Reduced number of [^{3}H]nicotine and [^{3}H]acetylcholine binding sites in the frontal cortex of Alzheimer brains. Neurosci. Lett. 72: 115–119.

Nordberg A, Hartvig P, Lilja A, Viitanen M, Amberla K, Lundqvist H, Andersson Y, Ulin J, Winblad B, Langstrom B (1990): Decreased uptake and binding of ^{11}C-nicotine in brain of Alzheimer patients as visualized by positron emission tomography. J. Nueral. Transm. Park. Dis. Dement. Sect. 2: 215–224.

Nordberg A, Alafuzoff I, Winblad B (1992a): Nicotinic and muscarinic subtypes in the human brain: changes with aging and dementia. J. Neurosci, Res. 31: 103–111.

Nordberg A, Lilja A, Lundqvist H, Hartvig P, Amberla K, Viitanen M, Warpman U, Johansson M, Hellström-Lindahl E, Bjurling P, Fasth K-J, Långström B, Winblad B (1992b): Tacrine restores cholinergic nicotinic receptors and glucose metabolism in Alzheimer patients as visualized by positron emission tomography. Neurobiol. Aging 13: 747–758.

Nordberg A, Lundqvist H, Hartvig P, Lilja A, Långström B (1995): Kinetic analysis of regional (S)(−)^{11}C-nicotine binding in normal and Alzheimer brains—*in vivo* assessment using positron emission tomography. Alzheimer Dis. Assoc. Discord. 9: 21–27.

Nordberg A, Lundqvist H, Hartvig P, Andersson J, Johansson M, Hellstrom-Lindahi E, Långström B (1997): Imaging of nicotinic and muscarinic receptors in Alzheimer's disease: effect of tacrine treatment. Dement. Geriatr. Cogn. Disord. 8: 78–84.

Nybäck H, Nordberg A, Långström B, Halldin C, Hartvig P, Ahlin A, Swahn CG, Sedvall G (1989): Attempts to visualize nicotinic receptors in the brain of monkey and man by positron emission tomography. Prog. Brain Res. 79: 313–319.

Nybäck H, Halldin C, Ahlin A, Curvall M, Eriksson L (1994): PET studies of the uptake of (S)- and (R)-[^{11}C]nicotine in the human brain: difficulties in visualizing specific receptor binding *in vivo.* Psychopharmacol. Berl. 115: 31–36.

Pabreza LA, Dhawan S, Kellar KJ (1991): [^{3}H]cytisine binding to nicotinic cholinergic receptors in brain. Mol. Pharmacol. 39: 9–12.

Palacios JM, Mengod G, Vilaro MT, Wiederhold KH, Boddeke H, Alvarez FJ, Chinaglia G, Probst A (1990): Cholinergic receptors in the rat and human brain: microscopic visualization. Prog. Brain Res. 84: 243–253.

Patt JT, Westera G, Buck A, Fletcher SR, Schubiger PA (1995) [^{11}C]-N-methyl- and [^{18}F]fluoroethylepibatidine: ligands for the neuronal nicotinic receptor. *J.* Labelled Compd. Radiopharm. 37: 355–356.

Pauly JR, Marks MJ, Gross SD, Collins AC (1991): An autoradiographic analysis of cholinergic receptors in mouse brain after chronic nicotine treatment. J. Pharmacol. Exp. Ther. 258: 1127–1136.

Perry DC, Kellar KJ (1995): ^{3}H-epibatidine labels nicotinic receptors in rat brain: an autoradiographic study. J. Pharmacol. Exp. Ther. 275: 1030–1034.

Roemer RA, Richelson E, Shagass C, Leventhal L (1996): A method to estimate *in vivo* D2 receptor occupancy by antipsychotic drugs. J. Psychiatry Neurosci. 21: 325–333.

Scheffel U, Taylor GF, Kepler JA, Carroll FI, Kuhar MJ (1995): *In vivo* labeling of neuronal nicotinic acetylcholine receptors with radiolabeled isomers of norchloroepibatidine. NeuroReport 6: 2483–2488.

Schulz DW, Loring RH, Aizenman E, Zigmond RE (1991): Autoradiographic localization of putative nicotinic receptors in the rat brain using ^{125}I-neuronal bungarotoxin. J. Neurosci. 11: 287–297.

Sedvall G, Farde L, Persson A, Wiesel F-A (1986): Imaging of neurotransmitter receptors in the living human brain. Arch. Gen. Psychiatry 43: 995–1005.

Sedvall G, Farde L, Nyback H, Pauli S, Persson A, Savic I, Wiesel FA (1990): Recent advances in psychiatric brain imaging. Acta Radiol. Suppl. 374: 113–115.

Seiger Å, Nordberg A, van Holst H, Bäckman L, Ebendal T, Alafuzoff I, Amberla K, Hartvig P, Herlitz A, Lilja A, Lundqvist H, et al. (1993): Intracranial infusion of purified nerve growth factor to an Alzheimer patient: the first attempt of a possible future treatment strategy. Behav. Brain Res. 57: 255–261.

Sihver W, Fasth KJ, Ögren M, Nordberg A, Watanabe Y, Långström B (1997): Development of ^{11}C-nicotinic receptor ligands to characterize nicotinic acetycholine receptors in the central nervous system. XII international symposium on radiopharmaceutical chemistry. Uppsala, Sweden (June 15–19).

Sorger D, Kampfer I, Knapp WH (1996): Nicotine receptor mapping—reply. Eur. J. Nucl. Med. 23: 1013.

Spande TF, Garraffo HM, Yeh HJ, Pu QL, Pannell LK, Daly JW (1992): A new class of alkaloids from a dendrobatid poison frog: a structure for alkaloid 251F. J. Nat. Prod. 55: 707–722.

Sullivan JP, Decker MW, Brioni JD, Donnelly-Roberts D, Anderson DJ, Bannon AW, Kang C-H, Adams P, Piattoni-Kaplan M, Buckley MJ, Gopalakrishnan M, Williams M, Arneric SP (1994): (±)-Epibatidine elicits a diversity of *in vitro* and *in vivo* effects mediated by nicotinic acetylcholine receptors. J. Pharmacol. Exp. Ther. 271: 624–631.

Sullivan JP, Donnelly-Roberts D, Briggs CA, Anderson DJ, Gopalkrishnan M, Piattoni-Kaplan M, Campbell JE, McKenna DG, Molinari E, Hettinger A-M, Garvey DS, Wasicak JT, Holladay MW, Williams M, Arneric SP (1996): A-85380 [3-(2(S)-azitidinemethoxy)pyridine]: *in vitro* pharmacological properties of a novel, high affinity $\alpha 4\beta 2$ acetylcholine receptor ligand. Neuropharmacology 35: 725–734.

Valette H, Bottlaender M, Dollé F, Dolci L, Syrota A, Crouzel C (1997): An attempt to visualize baboon brain nicotinic receptors with N-[^{11}C]ABT-418 and N-[^{11}C]methyl-cytisine. Nucl. Med. Commun. 18: 164–168.

Valette H, Bottlaender M, Dollé F, Fuseau C, Coulon C, Guenther I, Crouzel C (1998): In vivo study of central nicotinic acetylcholine receptors using 2-[^{18}F]fluoro-3-[(S)-2-azetidinylmethoxy]pyridine and PET in baboons J. Nucl. Med. 39: 72P.

Villemagne VL, Horti A, Scheffel A, Ravert HT, London ED, Dannals RF (1996): Imaging nicotinic acetylcholine receptors in baboon brain by PET. J. Nucl. Med. 37S: 11P (abs.).

Villemagne VL, Horti A, Scheffel A, Ravert HT, Finely P, Clough DJ, London ED, Wagner HN Jr. Dannals RF (1997): Imaging nicotinic acetylcholine receptors with PET and [^{18}F]-FPH, a [^{18}F]-labeled analog of epibatidine. J. Nucl. Med. 38: 1737–1741.

Volkow ND, Mullani NA, Bendriem B (1988): Positron emission tomography instrumentation: an overview. Am. J. Physiol. Imaging 3: 142–153.

Warpman U, Nordberg A (1995): Epibatidine and ABT 418 reveal selective losses of $\alpha 4\beta 2$ nicotinic receptors in Alzheimer brains. Neuroreport 6: 2419–2423.

Whitehouse PJ, Kalaria RN (1995): Nicotinic receptors and neurodegenerative dementing diseases: basic research and clinical implications. Alzheimer Dis. Associated Disorders 9 (suppl. 2): 3–5.

Whitehouse PJ, Martino AM, Antuono PG, Lowenstein PR, Coyle JT, Price DL, Kellar KJ (1986): Nicotinic acetylcholine binding sites in Alzheimer's disease. Brain Res. 371: 146–151.

Whitehouse PJ, Martino AM, Marcus KA, Zweig RM, Singer HS, Price DL, Kellar KJ (1988a): Reductions in acetylcholine and nicotine binding in several degenerative diseases. Arch. Neurol. 45: 722–724.

Whitehouse PJ, Martino AM, Wagster MV, Price DL, Mayeux L, Atack JR, Kellar KJ (1988b): Reductions in [^{3}H]-nicotinic acetylcholine binding in Alzheimer's disease and Parkinson's disease: an autoradiographic study. Neurology 38: 720–723.

Williams M, Sullivan JP, Arneric SP (1994): Neuronal nicotinic acetylcholine receptors. Drug News and Perspectives 7: 205–223.

Wonnacott S (1990): The paradox of nicotinic acetylcholine receptor upregulation by nicotine. Trends Pharmacol. Sci. 11: 216–219.

Yates SL, Bencherif M, Fluhler EN, Lippiello PM (1995): Upregulation of nicotinic acetylcholine receptors following chronic exposure of rats to mainstream cigarette smoke or α4β2 receptors to nicotine. Biochem. Pharmacol. 50: 2001–2008.

Part IV

Medicinal Chemistry

14

Natural Products as a Source of Nicotinic Acetylcholine Receptor Modulators and Leads for Drug Discovery

Mark W. Holladay, Ph.D.
Neurological and Urological Diseases Research
Pharmaceutical Products Division
Abbott Laboratories
Abbott Park, Illinois

Nicholas D. P. Cosford Ph.D. and Ian A. McDonald Ph.D.
SIBIA Neurosciences Inc.
La Jolla, California

The neurotransmitter acetylcholine (**1**) activates nicotinic acetylcholine receptors (nAChRs), which opens the associated ion channel and permits the flow of cations, principally sodium and calcium, across the cell membrane. (S)-(−)-nicotine (**2,** (Fig 14-1) hereafter simply referred to as nicotine), an alkaloid found in tobacco plants, is the prototypical agent that pharmacologically distinguishes between nAChRs and the other major class of acetylcholine-activated receptors, the metabotropic muscarinic receptors (mAChRs).

Neuronal Nicotinic Receptors: Pharmacology and Therapeutic Opportunities, Edited by S. P. Arneric and J. D. Brioni
ISBN 0-471-24743-x, pages 253–270. Copyright © 1998 by Wiley-Liss, Inc.

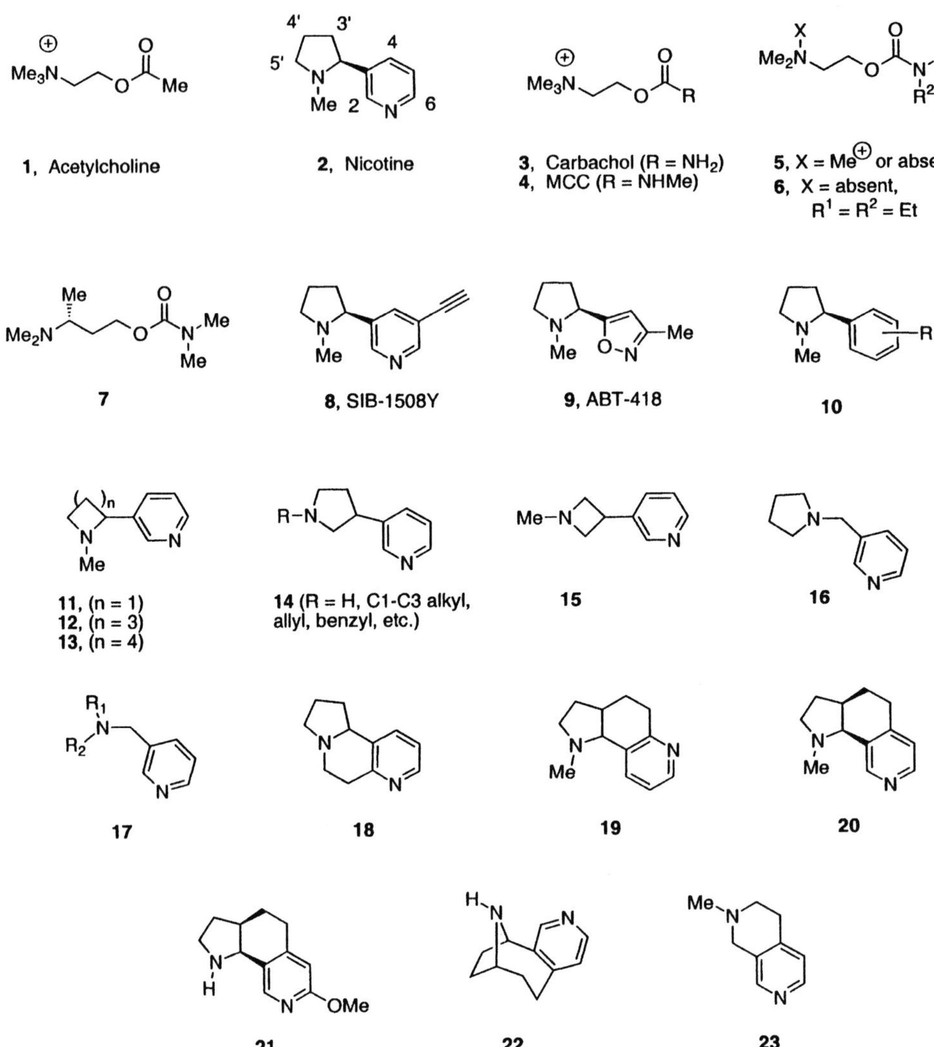

Figure 14-1. *Structures 1-23.*

Recent advances in biochemistry and molecular biology have led to the isolation of genes that encode neuronal protein subunits (α_2–α_9 and β_2–β_4). These subunits can assemble in various combinations to form functional pentameric neuronal nAChRs. Among the known and putative neuronal nAChR subtypes, the most prominent include α4β2 nAChRs, which comprise the majority of the high-affinity nicotine binding sites in rat brain; α7 nAChRs, which are labeled by α-bungarotoxin and exhibit a different pattern of distribution than α4β2; and heterogeneous populations of α3-containing nAChRs found in autonomic ganglia and in the adrenal gland (reviewed in Sargent, 1993; Lindstrom et al., 1995; Holladay et al., 1997). Experiments on the pharmacology of recombinant and native receptors suggest that modulation of specific nAChR subtypes is related to certain physiological effects. For example, activation of α3-containing nAChRs in the adrenal medulla ultimately leads to undesirable side effects, while activation of specific nAChRs in the CNS relates to the ability of nAChR modulators to exert effects on motor and cognitive performance, which has implications in the treatment of Parkinson's disease and Alzheimer's disease, respectively. A contemporary challenge in drug discovery is to identify selective nAChR modulators to maintain or enhance the desirable therapeutic properties while eliminating undesirable side effects.

In addition to nicotine and acetylcholine, numerous other substances of natural origin have been discovered that either activate or inhibit nAChRs, and these compounds have provided valuable pharmacological tools to understand the function of nAChRs. A number of these substances have served as important lead compounds in medicinal chemistry efforts to discover agents that selectively modulate nAChR subtypes (Holladay et al., 1995; McDonald et al., 1995; Swanson et al., 1995; McDonald et al., 1996; Glennon and Dukat, 1996; Holladay et al., 1997). In this chapter, several important natural nAChR modulators will be discussed, together with selected examples that have served as leads for the discovery of potential therapeutic agents.

ACETYLCHOLINE

Stabilization of the ester moiety of acetylcholine as a carbamate yields carbachol (**3**), which is poorly selective for brain nicotinic vs. muscarinic receptors (throughout this chapter, unless otherwise specified, binding to brain nAChRs refers to displacement of a radioligand, e.g., [^{3}H]-nicotine or [^{3}H]-cytisine, that interacts predominantly with the α4β2 subtype). However, the *N*-methyl derivative (*N*-methylcarbamoylcholine, MCC, **4**) shows binding affinity in rat brain comparable to that of nicotine and is more than 100-fold selective for binding to CNS nicotinic vs. muscarinic receptors (Abood and Grassi, 1986; Abood et al., 1993; Anderson and Arneric, 1994). Among a variety of analogs based on general structure **5,** quaternary compounds, including MCC, showed uniformly higher affinity for central nAChR sites than the corresponding tertiary amines (Abood, et al., 1993). However, several tertiary amine compounds in which R^1 and R^2 = C_2-C_3 alkyl, e.g., **6,** had binding constants in the mid-nanomolar range. Recent further explorations in this series (Søkilde et al., 1996; Mikkelson et al., 1997) have led to the identification of compound **7,** a tertiary amine. Compared to the MCC analogs, compound **7** has one additional carbon separating the basic nitrogen atom from the carbamate oxygen and an additional methyl substituent. This analog is reported to possess low nanomolar binding affinity and 900-fold selectivity for nicotinic vs. muscarinic receptors, and thus represents a significant advance in this series.

Recently, choline, the breakdown product of acetylcholine from the action of acetylcholinesterase, has been shown to be a comparatively weak but selective activator of the α7 nAChR subtype (Papke et al., 1996).

NICOTINE

Pyridine Ring Modified Analogs

Early pyridine ring–modified analogs, including alkyl-substituted derivatives (Catka and Leete, 1978; Leete and Leete, 1978; Chavdarian et al., 1982; Haglid, 1967; Secor et al., 1981; Seeman et al., 1983; Seeman et al., 1985a,b), halogenated derivatives (Karrer and Takahashi, 1926; Lowry and Gore, 1931; Leete et al., 1971; Rondahl, 1977), and amino derivatives

(Tschitschibabin and Kirssanow, 1924; Shibagaki and Matsushita, 1985; Rondahl, 1977) preceded the advent of routine characterization in standardized CNS assays.

5-Fluoronicotine was one of several nAChR modulators shown to differentially stimulate the release of the neurotransmitters acetylcholine, norepinephrine, dopamine, and serotonin using in vivo microdialysis (Summers et al., 1995). 6-Substituted fluoro-, chloro-, bromo-, and methylnicotine derivatives were shown to possess comparable [^{3}H]-nicotine binding affinities to nicotine, whereas activity in the rat tail flick assay were either comparable (Me, F) or roughly an order of magnitude more potent (Cl, Br) than nicotine (Dukat et al., 1996). The 6-methoxy derivative, on the other hand, was considerably weaker than nicotine in both assays.

(S)-(−)-5-ethynyl-3-(1-methyl-2-pyrrolidinyl)pyridine maleate (SIB-1508Y, **8**), a compound that is entering Phase II clinical trials for the treatment of Parkinson's disease, was the result of a drug discovery program focused on the synthesis and characterization of subtype-selective nAChR agonists (Cosford et al., 1996; Menzaghi et al., 1997a,b; Sacaan et al., 1997). During initial SAR studies it was noted that substitution of the nicotine pyridyl ring, especially at the 5-position, provided analogs that retained potency at the $\alpha 4\beta 2$ nAChR subtype but exhibited a diminished response at the peripheral and ganglionic ($\alpha 3\beta 4$) subtypes. This was determined by the effect of compounds in a fluorescence-based whole cell assay that measures changes in intracellular calcium concentrations in cells that express specific, recombinant human nAChR subtypes. Furthermore, **8** was found to be more efficacious than nicotine in an in vitro assay that measures drug-stimulated release of [^{3}H]-dopamine from a rat striatal slice preparation. Subsequent profiling of **8** in a battery of in vivo assays, including the 6-OHDA-lesioned rat turning model and a delayed matching to position model in MPTP-treated primates, demonstrated the potential of **8** for the treatment of both the motor and cognitive deficits associated with Parkinson's disease.

Isosteric replacements of the pyridine ring of nicotine have been reported also. The nAChR agonist ABT-418 (**9**), which contains a 3-methylisoxazole isostere of pyridine (Garvey et al., 1994a,b) has been extensively characterized in in vitro and in vivo assays (Arneric et al., 1994; Decker et al., 1994). Beneficial effects of this drug in patients suffering from Alzheimer's disease have been reported (Newhouse et al., 1996). A series of phenyl pyrrolidine analogs (**10**) of nicotine has been prepared and evaluated in a CNS binding assay with K_i values ranging from 46 nM to > 10 000 nM (Elliott et al., 1995). An analog of nornicotine in which the pyridine ring is replaced by a thiazole moiety was reported in 1946 (Erlenmeyer and Marbet, 1946).

Pyrrolidine Ring Modified Analogs

The pyrrolidine ring of nicotine has been systematically substituted (for numbering convention, see structure **2**) to provide a variety of nicotine derivatives, and the binding affinites of these compounds versus [^{3}H]-cytisine- labeled whole rat brain were measured (Lin et al., 1994). It was found that the 4′-position was in general tolerant of substituents (K_i range 4.23–510 nM) whereas substitution at the 5′-position led to analogs with low affinity, with the notable exceptions of the (β)-Me and (β)-n-Bu derivatives, which showed K_i values of 34.9 and 125.2 nM respectively. Variation of the pyrrolidine *N*′-substituent has also been studied (Glassco et al., 1993). Both removal and homologation of the *N*-methyl substituent led to decreases in binding affinity. However, these studies showed that certain in vivo activities (rat tail flick and disruption of locomotor activity) did not correlate well with binding affinity. Other pyrrolidine-modified nicotine analogs, including unsaturated pyrrolidine derivatives, have been prepared for various purposes prior to the routine characterization of compounds in CNS assays (Rueppel and Rapoport, 1971; Cushman and Catagnoli, 1972; Shibagaki et al., 1986; Frank et al., 1942; Maeda et al., 1978; Brandange and Lindblom, 1979; Chavdarian, 1983).

Nicotine analogs have been synthesized in which the pyrrolidine ring has been replaced by an isostere. Contraction of the ring to the four-membered azetidine analog led to the synthesis of racemic 2-(3-pyridyl)azetidine (**11**), which shows binding affinity approximately 10-fold greater than that of nicotine, whereas the corresponding piperidine (**12**) and azepine (**13**) analogs were less potent than nicotine by at least an order of magnitude (Abood et al., 1993). The isomeric azetidine **14** and the corresponding desmethyl derivative have been prepared also, but activity in nAChR assays was not reported (Secor and Edwards, 1979). Nicotine analogs in which the pyridine ring is attached at the 3′-position of the pyrrolidine nucleus (isonicotines, **15**) were prepared as racemates and evaluated for binding affinity and in a number of in vivo assays (Dukat et al., 1996; Damaj et al., 1996). These analogs were active but less potent than the corresponding nicotine derivatives. The nicotine isomer *N*-(3-pyridylmethyl) pyrrolidine (**16**) had a binding affinity of 49 nM (IC_{50}) for CNS [^{3}H]nicotine sites (Caldwell and Lippiello, 1993), and stimulated release of acetylcholine and norepinephrine in an in vivo brain microdialysis study (Summers et al., 1995). Replacement of the pyrrolidine ring with *N′*,*N*-dialkyl substituents (**17**) has been investigated (Dukat et al., 1996; Damaj et al., 1996). In this series, the most potent compound tested (R_1 = Me, R_2 = Et) had a binding affinity about 20-fold weaker than that of nicotine.

Conformationally Restricted Analogs

There have been several attempts to explore the active conformation of nicotine by the synthesis and biological testing of torsionally constrained nicotine derivatives. Thus, the bridged nicotine analogs **18** and **19,** prepared as racemates, were reported to have no appreciable biological activity (Catka and Leete, 1978; Chavdarian et al., 1983), whereas the enantiomerically pure conformationally constrained isoquinoline **20** bound to brain nAChRs with modest affinity (Ki = 0.6 μM) (Glassco et al., 1993). Compound **21,** which is structurally related to **20,** has been prepared as a pure enantiomer and found to possess functional activity at β4-containing nAChRs, as well as the ability to stimulate striatal dopamine release and activity in vivo in models of Parkinson's disease and pain (McDonald et al., 1996). The racemic pyridine ring-fused azabicyclo[4.2.1]octane **22,** a compound that encompasses features of both nornicotine and anatoxin-a, is a potent, conformationally locked nicotine analog (Kanne and Abood, 1988; Kanne et al., 1986). A series of fused bicyclic nicotine-like compounds has been reported, of which the most potent (**23**) had a K_i of 18 nM in a CNS binding assay (Damaj et al., 1996; Dukat et al., 1996).

ANABASINE AND ANABASEINE

Anabasine (**24** (Fig. 14-2)), a homolog of nornicotine isolated from tobacco, and anabaseine (**25**), found naturally in a marine worm (Kem, 1985), differ structurally only in the bond order at the 1,2-position. Anabasine has ca. 30-fold lower affinity than nicotine for mouse brain agonist binding sites, and ca. 40% of the efficacy of nicotine in stimulating ion flux in mouse midbrain (Marks et al., 1993). Anabaseine exhibits approximately 20-fold weaker affinity than nicotine for agonist binding sites in rat brain and ca. 10% of the efficacy of nicotine at stimulating α4β2 receptors in oocytes (de Fiebre et al., 1995). In contrast, anabaseine is highly efficacious at stimulating α7 homomeric channels in oocytes, and thus is functionally selective for this subtype.

Reaction of anabaseine with aldehydes affords 3-substituted derivatives (Zoltewicz et al., 1993). One such analog, GTS-21 (**26,** also known as DMXB), has been extensively character-

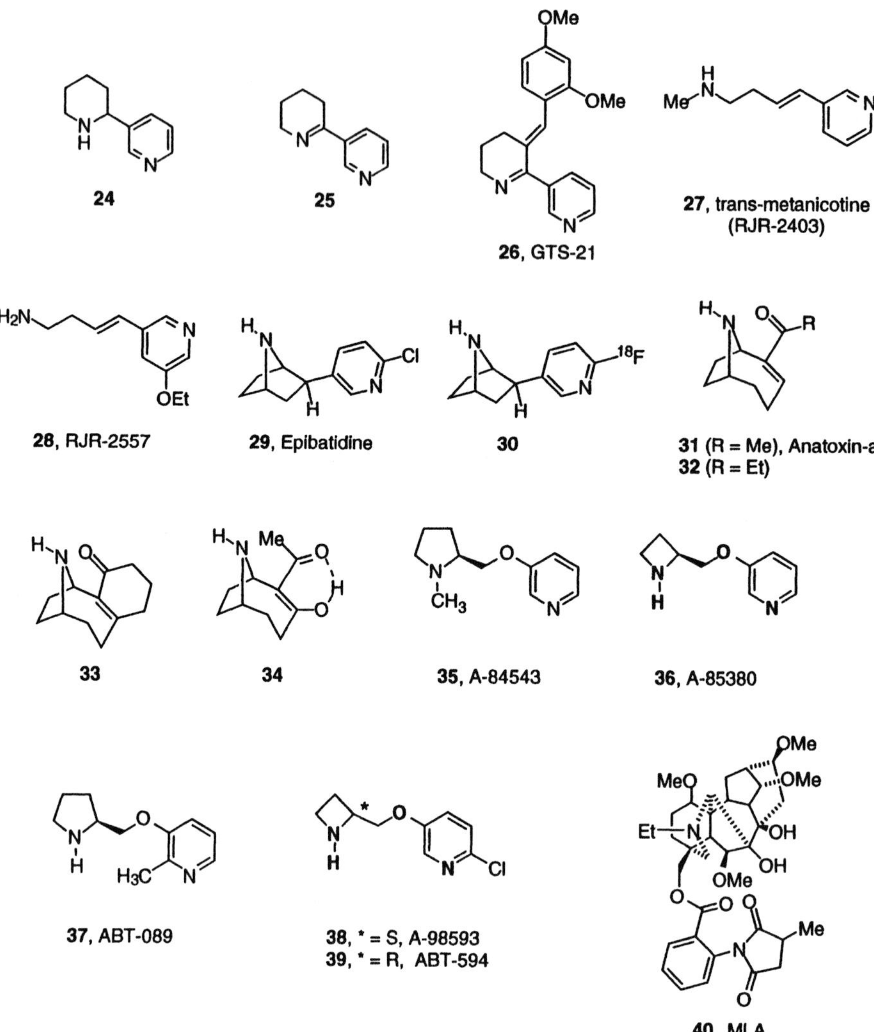

Figure 14-2. Structures 24-40.

ized in vitro and in vivo, and has reportedly been in clinical trials for treatment of Alzheimer's disease (Kem et al., 1996). GTS-21 has shown positive effects in measures of cognition in rats (Meyer et al., 1994) and rabbits (Woodruff-Pak et al., 1994), and has shown cytoprotective effects in cells in culture and in vivo (Martin, *et al.*, 1994). GTS-21 is effective following chronic administration in several assays of learning and memory (Arendash et al., 1995), and also has been shown to demonstrate protective effects against Aβ-induced neurotoxicity (Shimohama, 1996). The synthesis of some rigidified 2-azabicyclo[2.2.2]octane anabasine analogs has been reported (Szczepanski and Anouna, 1996).

TRANS-METANICOTINE

Trans-meta nicotine (**27**) is a metabolite of nicotine that is synthesized via a ring-opening reaction on the pyrrolidine ring of nicotine (Acheson et al., 1980). Although nicotinic properties of **27** have been recognized for some time (Wilson et al., 1976), recent investigations have demonstrated **27** (now designated RJR-2403) has potential for the treatment of Alzheimer's disease (Bencherif et al., 1996; Lippiello et al., 1996). Based on ion flux experiments in rat thalamic tissue, which is believed to reflect activation of the $\alpha4\beta2$ nAChR subtype, and PC12 cells, which contain the $\alpha3$-containing ganglionic subtype, RJR-2403 appears to be one of only few agents (others include ABT-418, A-84543, and SIB-1508Y, discussed elsewhere in this chapter) that show functional selectivity for the $\alpha4\beta2$ subtype. Recently, a related analog RJR-2557 (**28**) has been disclosed, which has improved selectivity for activating central vs. peripheral nAChRs, and also appears to show some intra-CNS selectivity (Bane et al., 1997).

EPIBATIDINE

The skin from the poisonous frog, *Epipedobates tricolor*, has furnished the exquisitely potent nAChR agonist, epibatidine (**29**) (Spande et al., 1992). This conformationally constrained nicotine analog binds with subnanomolar affinity to rat brain preparations (Badio and Daly, 1994) and to many of the recombinant nAChR subtypes (Gerzanich et al., 1995). As a consequence, epibatidine binding can be used to determine whether a particular recombinant receptor combination in fact has been expressed.

Both enantiomers of epibatidine have been prepared (Fletcher et al., 1993, 1994; Huang and Shen, 1993) and have been shown to have similar affinity for nAChRs (Badio and Daly, 1994; Gerzanich et al., 1995). Examination of molecular models, viewed end on, allows one to reconcile that the nAChR binding site does not select one enantiomer over the other, since each isomer presents similar lipophilic residues in the same region of space while the heteroatoms can occupy identical positions.

Epibatidine is a very toxic compound and serves the role in nature to protect the frog from potential predators. It has also been shown to be a potent analgesic agent in a number of animal models (Badio and Daly, 1994; Senokuchi et al., 1994; Spande et al., 1992). Mechanistically, both the toxicity and analgesic activity are a consequence of activation of nAChRs (Qian et al., 1993; Rupniak et al., 1994). A detailed evaluation of the antinociceptive effects of nAChR agonists in the rat tail flick assay revealed that the effect is mediated by both peripheral and central nAChRs (Rao et al., 1996). Although the receptor subtype responsible for this activity was not determined, the $\alpha7$ subtype could be eliminated as a likely possibility because the antinociception could not be antagonized with the $\alpha7$ antagonist methyllycaconitine (MLA).

Since the seminal publication disclosing the structure of epibatidine (Spande et al., 1992), numerous syntheses of either racemic or optically pure epibatidine have appeared in the literature. A comprehensive review summarizing these approaches has appeared (Chen and Trudell, 1996). In conjunction with this synthetic effort, a number of analogs were found to be synthetically accessibly (Table 14.1). Many of these compounds have been evaluated for analgesic properties and/or for the ability to displace [^{3}H]-nicotine from rat brain membranes. Although the number of compounds is relatively small, it appears that so long as the general bicyclic amine structure is maintained with the aromatic moiety being attached in the 2-position, the potency and functional efficacy remains high. It would be interesting to evaluate all analogs for their ability to activate specific, recombinant nAChR combinations. Thus, the true value of these compounds may be revealed as they are used to define the pharmacophore for the different nAChR subtypes.

Recently, the ^{18}F-derivative **30** has been described (Liang et al., 1997). This compound has been used in positron emission tomography studies to map nAChR in the brain. Unfortunately, due to the toxicity and high affinity of epibatidine for most nAChRs, this approach has little utility in mapping the distribution of specific receptor subtypes.

ANATOXIN

First isolated from mass cultures of the freshwater blue-green algae *Anabaena flos-aquae* (Devlin et al., 1977), (+)-anatoxin-a (**31**) is a highly potent nAChR agonist. This alkaloid, which possesses a 9-azabicyclo[4.2.1]nonane skeleton, was initially synthesized by a procedure that utilized cocaine as the starting material (Campbell et al., 1977). Since then anatoxin-a has been the target of numerous total syntheses that have been reviewed (Mansell, 1996). The biosynthesis of anatoxin-a has also been investigated (Hemscheidt et al., 1995).

Because of its properties as an extremely potent nAChR agonist, anatoxin-a has been characterized in many biological assays (Aronstam and Witkop, 1981; Carmichael et al., 1979; Macallan et al., 1988; Swanson et al., 1986; Thomas et al., 1993). The natural enantiomer, (+)-anatoxin-a, is a much more potent nAChR agonist than its enantiomer (Swanson et al., 1986). This stereospecificity is in marked contrast to the lack of selectivity exhibited by the enantiomers of epibatidine and reflects subtle effects exerted by the structural differences between these two bicyclic alkaloids.

Attempts to determine the structural features that contribute to the potency and selectivity of the parent compound for nAChRs have focused on the preparation of various anatoxin-a analogs. Methylation of the nitrogen results in a large decrease in potency (Kofuji et al., 1990), while dihydroanatoxin, in which the alkene functionality has been reduced, is an order of magnitude less potent (Swanson et al., 1991; Wonnacott et al., 1991). Homoanatoxin (**32**), a homologue synthesized to study the effects of extending the side chain, was subsequently isolated as a natural product in *Oscillatoria formosa* (Skulberg et al., 1992) and is roughly equipotent with anatoxin-a in various assays (Wonnacott et al., 1992). A series of alkyl-modified side-chain analogs of anatoxin-a were prepared and evaluated in nAChR binding assays and in voltage clamp recordings of neuronal $\alpha 7$ receptors expressed in *Xenopus* oocytes (Thomas et al., 1994). Other modifications of the side chain have been investigated, including a series of analogs in which the methyl ketone moiety was systematically replaced with carboxylic acid, ester, alkanol, amide, aldehyde, or alkoxime functional groups (Howard et al., 1990; Swanson et al., 1989, 1991; Wonnacott et al., 1991). When these compounds were evaluated in several peripheral and neuronal nAChR binding and functional assays, none was found to be as potent as the parent anatoxin-a.

Efforts to determine the active conformation of anatoxin-a have focused on whether the molecule adopts the *s-cis* or the *s-trans* conformation when it is bound to the active site of the receptor. To address this question, conformationally constrained analogs that mimic the molecular in each of the conformations have been prepared. The *s-trans*-locked compound **33** (Hernandez and Rapport, 1994) was found to have binding affinity in the low nanomolar range (Holladay et al., 1997), whereas an *s-cis* mimic **34** was found to be more than a thousand-fold less potent than the parent compound (Brough et al., 1992).

TABLE 14.1. Epibatidine Analogs

Structure	nAChR Binding	Analgesic Activity	Reference
H-N (7-azabicyclo[2.2.1]heptane) – 3-pyridyl	Ki = 0.03 nM; rat brain membranes	Less effective than epibatidine	Badio and Daly, 1994; Senokuchi et al., 1994
H-N (7-azabicyclo[2.2.1]heptane) – 3-pyridyl, Me	Ki = 0.13 nM; rat brain membranes	NT	Badio and Daly, 1994
H-N (7-azabicyclo[2.2.1]heptane) – 3-pyridyl, I	Ki = 0.48 nM; rat brain membranes	NT	Badio and Daly, 1994
R-N bicyclic – 3-pyridyl, Cl	NT	R = H, potent R = Me, potent R = i-Pr, less potent	Bai et al.,1996; Malpass et al., 1996
H-N bicyclic – 3-pyridyl, Cl	NT	NT	Malpass et al., 1996
H-N bicyclic – 3-pyridyl, Cl, H	30-fold less potent than epibatidine	30-fold less potent than epibatidine	Zhang et al., 1997
H-N (7-azabicyclo[2.2.1]heptane) – isoxazole, Me Epiboxidine	10-fold less potent than epibatidine	10-fold less potent than epibatidine	Badio et al., 1997

NT = not tested

PYRIDYL ETHERS: HYBRID COMPOUNDS OF ACETYLCHOLINE AND NICOTINE

A-84543 (**35**) is the prototypical member of a series of compounds that originated as a hybrid structure based on acetylcholine and nicotine (Abreo et al., 1996). Interestingly, A-84543 (Ki = 150 pM) has higher affinity for brain nAChR sites than either acetylcholine or nicotine, and structure-activity studies with respect to pyrrolidine *N*-substitution suggest a divergence in structure-activity relationships compared to nicotine. Several other compounds of interest have been derived from this series. A-85380 (**36**) possesses 50 pM affinity for brain binding sites and is a potent agonist at neuronal nAChRs (Abreo et al., 1996; Sullivan et al., 1996). ABT-089 (**37**) possesses weak or partial agonist activity at several subtypes of nAChRs but nevertheless is effective as a cognition-enhancing agent, whereas the low activity at peripheral ganglionic-like nAChRs contributes to the improved safety profile of ABT-089 (Arneric et al., 1997; Decker et al., 1997; Lin et al., 1997; Sullivan et al., 1997). Although A-85380 possesses weak analgesic properties in mice, the corresponding (S)-chloropyridine analog **38** (A-98593) and its (R)-enantiomer **39** (ABT-594) are potent analgesic compounds (Holladay et al., 1998; Decker et al., 1998). ABT-594 shows reduced activation of ganglionic-like nAChRs compared to (±)-epibatidine, which likely accounts for its lower cardiovascular liability (Holladay et al., 1998). The demonstration that ABT-594 also possesses activity in models of neuropathic pain serves to further enhance the potential of this compound as a useful therapeutic substance (Bannon et al., 1998).

METHYLLYCACONITINE (MLA)

The tertiary diterpenoid MLA (**40**) is a natural product isolated from a poisonous plant found in western Canada, *Delphinium brownii* (Aiyar et al., 1979). MLA is selective for the putative α7 nAChR subtype compared with numerous other subtypes (Drasdo et al., 1992; Vijayaraghavan et al., 1992; Yum et al., 1996; Alkondon et al., 1992; Quik et al., 1996). It has recently been demonstrated that low concentrations of MLA can enter the CNS following peripheral administration, which may permit its use to deduce what, if any, behavioral actions are mediated by central α-Bgt-sensitive nAChRs (Turek et al., 1995). Based on studies with MLA, neither the nicotine discriminative stimulus effect (Brioni et al., 1996) nor the tail flick analgesia response (Rao et al., 1996) appears to be mediated by nAChRs containing the α7 subunit.

PEPTIDE TOXINS

Although peptide toxins are not generally regarded as useful leads for drug discovery, they have proven to be invaluable as pharmacological tools, and a brief discussion of these agents is therefore included. The foremost member of this group is α-bungarotoxin (α-Bgt), a 75 amino acid peptide isolated from an East Asian snake *Bungarus multicinctus* (Lee, 1972). α-Bgt is a potent blocker of the nAChR subtype found on skeletal muscle (Colquhoun and Rang, 1976). High affinity for α-Bgt also is a principal distinguishing pharmacological property of the α7 class of nAChRs in the brain, in that it labels this subtype with much higher affinity than it does the other prominent CNS subtype, α4β2. Neuronal bungarotoxin (n-Bgt) (Lindstrom et al., 1987) has also been referred to as κ-bungarotoxin (Chiappinelli, 1983), toxin F (Loring and Zigmond, 1988), and bungarotoxin 3.1 (Ravdin and Berg, 1979), and is a minor component of

Bungarus multicinctus venom. n-Bgt was originally characterized by its ability to block cholinergic transmission in peripheral autonomic ganglia and was subsequently found to potently antagonize several discrete responses in the CNS, including nicotine-stimulated release of [^{3}H]-dopamine (DA) from rodent striatal tissues (Schulz et al., 1991; Grady et al., 1992). Autoradiography studies in brain demonstrate that n-Bgt labels both α-Bgt-sensitive sites as well as a population of sites displaced by a high concentration of nicotine but distributed much less widely than are α4β2 receptors (Schulz et al., 1991). The subunit composition of this unique site is not known, but indirect evidence suggests the possibility that it may be an α3-containing nAChR.

Recently, certain α-conotoxins, isolated from *Conus* marine snails, have been characterized as selective ligands for specific subtypes of nAChRs. The α-conotoxins are small peptides generally having 12–16 amino acids with two disulfide bridges. α-Conotoxin MI and α-conotoxin GI selectively target skeletal muscle nAChRs. α-Conotoxin ImI selectively blocks α7 homomeric channels, with somewhat lower potency at α9 homomeric channels and essentially no activity at numerous other nAChR subunit combinations, including that corresponding to the skeletal muscle receptor (Johnson et al., 1995). α-Conotoxin MII possesses specificity for the α3β2 subunit combination (Cartier et al., 1996) and partially blocks nicotine-stimulated striatal dopamine release (Kulak et al., 1997). Thus, following much speculation on the identity of the nAChR subtype responsible for dopamine release (El-Bizri and Clarke, 1994; Wonnacott, 1997), studies with this peptide have permitted the conclusions that this response is probably mediated by more than one nAChR subtype and strongly suggest that one of these contains α3 and β2 subunits (Kulak et al., 1997).

OTHER NATURAL PRODUCTS

A number of other natural products have been found to interact with one or more nAChR subtypes. Prominent examples include cytisine (Barlow and McLeod, 1969), lobeline (reviewed in Holladay et al., 1995), dihydro-β-erythroidine (Hider et al., 1986), lophotoxin (Sorenson et al., 1987), histrionicotoxin (Aronstam et al., 1985), neosurugatoxin (Wada et al., 1992), and strychnine (Elgoyhen et al., 1994). So far, little or no structure-activity work on these compounds has been reported.

SUMMARY

A rich diversity of natural products has provided a source of important pharmacological tools to study nAChR function, as well as lead compounds for the development of potential medicinal agents. Progress in the identification of more specific agents will be aided greatly by continued medicinal chemistry efforts combined with utilization of advanced biological characterization tools, such as functional evaluation in cell lines containing defined nAChR subtypes.

REFERENCES

Abood LG, Grassi S (1986): [^{3}H]Methylcarbamylcholine, a new radioligand for studying brain nicotinic receptors. Biochem. Pharmacol. 35: 4199–4202.

Abood LG, Lerner-Marmarosh N, Wang D, Saraswati M (1993): Structure-activity relationships of various nicotinoids and N-substituted carbamate esters of choline and other amino alcohols. Med. Chem. Res. 2: 552–563.

Abreo MA, Lin N-H, Garvey DS, Gunn DE, Hettinger A-M, Wasicak JT, Pavlik PA, Martin YC, Donnelly-Roberts DL, Anderson DJ, Sullivan JP, Williams M, Arneric SP, Holladay MW (1996): Novel 3-pyridyl ethers with subnanomolar affinity for central neuronal nicotinic acetylcholine receptors. J. Med. Chem. 39: 817–825.

Acheson RM, Ferris MJ, Sinclair NM (1980): Transformations involving the pyrrolidine ring of nicotine. J. Chem. Soc. Perkin Trans. 1: 579–585.

Aiyar VN, Benn MH, Hanna T, Jacyno J, Roth SH, Wilkens JL (1979): The principal toxin of *Delphinium brownii,* and its mode of action. Experientia 35: 1367–1368.

Alkondon M, Pereira EFR, Wonnacott S, Albuquerque EX (1992): Blockade of nicotinic currents in hippocampal neurons defines methyllycaconitine as a potent and specific receptor antagonist. Mol. Pharm. 41: 802–808.

Anderson DJ, Arneric SP (1994): Nicotinic receptor binding of [^{3}H]cytisine, [^{3}H]nicotine and [^{3}H]methylcarbamylcholine in rat brain. Eur. J. Pharmacol. 253: 261–267.

Arendash GW, Sengstock GJ, Sanberg PR, Kem WR (1995): Improved learning and memory in aged rats with chronic administration of the nicotinic receptor agonist GTS-21. Brain Res. 674: 252–259.

Arneric SP, Sullivan JP, Briggs CA, Donnelly-Roberts D, Anderson DJ, Raszkiewicz JL, Hughes ML, Cadman ED, Adams P, Garvey DS, Wasicak JT, Williams M (1994): (S)-3-Methyl-5-1-(1-methyl-2-pyrrolidinyl) isoxazole (ABT 418): a novel cholinergic ligand with cognition-enhancing and anxiolytic activities: I. In vitro characterization. J. Pharm. Exp. Ther. 270: 310–318.

Arneric SP, Campbell JE, Carroll S, Daanen JF, Holladay MW, Johnson P, Lin N-H, Marsh KC, Peterson B, Qui Y, Roberts EM, Rodrigues AD, Sullivan JP, Trivedi J, Williams M (1997): ABT-089 [3-(2(S)-Pyrrolidinylmethoxy)-2-methylpyridine]: An orally effective cholinergic channel modulator with potential once-a-day dosing and cardiovascular safety. Drug Devel. Res. 41: 31–43.

Aronstam RS, Witkop B (1981): Anatoxin-a interactions with cholinergic synaptic molecules. Proc. Natl. Acad. Sci. USA 78: 4639–4643.

Aronstam RS, King CT, Albuquerque EX, Daly JW, Feigl DM (1985): Binding of [^{3}H]perhydrohistrionicotoxin and [^{3}H]phencyclidine to the nicotinic receptor-ion channel complex of Torpedo electroplax. Biochem. Pharmacol. 34: 3037–3047.

Badio B, Daly JW (1994): Epibatidine, a potent analgetic and nicotinic agonist. Mol. Pharm. 45: 563–569.

Badio B, Garraffo M, Plummer CV, Padgett WL, Daly JW (1997): Synthesis and nicotinic activity of epiboxidine: an isoxazole analogue of epibatidine. Eur. J. Pharmacol. 321: 189–194.

Bai D, Xu R, Chu G, Zhu X (1996): Synthesis of (±)-epibatidine and its analogues. J. Org. Chem. 61: 4600–4606.

Bane AJ, Bencherif M, Lippiello P (1997): The effects of the novel nicotinic agonist RJR-1557 on spontaneous locomotor activity in the rat. Soc. Neurosci. Abstr. 23: 669, Abstr. 266.4.

Bannon AW, Decker MW, Holladay MW, Curzon P, Donnelly-Roberts D, Puttfarcken PS, Bitner RS, Pauly JL, Diaz A, Porsolt R, Dickenson AH, Williams M, Arneric SP (1998): Broad-spectrum, non-opioid analgesic activity by selective modulation of neuronal nicotinic acetylcholine receptors. Science 279: 77–81.

Barlow RB, McLeod LJ (1969): Some studies on cytisine and its methylated derivatives. Br. J. Pharmacol. 35: 161–174.

Bencherif M, Lovette ME, Fowler KW, Arrington S, Reeves L, Caldwell WS, Lippiello PM (1996): RJR-2403: a nicotinic agonist with CNS selectivity I. *In vitro* characterization. J. Pharmacol. Exp. Ther. 279: 1413–1421.

Brandange S, Lindblom L (1979): Synthesis, structure and stability of nicotine $\delta^{1'(5')}$ iminium ion, an intermediary metabolite of nicotine. Acta Chem. Sc. (Ser. B) 33: 187–191.

Brioni JD, Kim DJB, O'Neill AB (1996): Nicotine cue: Lack of effect of the α7 nicotinic receptor antagonist methyllycaconitine. Eur. J. Pharmacol. 301: 1–5.

Brough PA, Gallagher T, Thomas P, Wonnacott S, Baker R, Malik KMA, Hursthouse, MB (1992): Synthesis and X-ray crystal structure of 2-acetyl-9-azabicyclo[4.2.1]nonan-3-one. A conformationally locked *s-cis* analogue of anatoxin-a. J. Chem. Soc. Chem. Comm. 1087–1089.

Caldwell WS, Lippiello PM (1993): Method for the treatment of neurodegenerative diseases, US 5,214,060 (RJ Reynolds Tobacco Company).

Campbell HF, Edwards OE, Kolt R (1977): Synthesis of nor-anatoxin-a and anatoxin-a. Can. J. Chem. 55: 1372–1379.

Carmichael WW, Biggs DF, Peterson MA (1979): Pharmacology of anatoxin-a, produced by the freshwater cyanophyte *Anabaena flos-aquae* NRC-44-1. Toxicon 17: 229–236.

Cartier GE, Yoshikami D, Gray WR, Luo S, Olivera BM, McIntosh JM (1996): A new α-conotoxin which targets α3β2 nicotinic acetylcholine receptors. J. Biol. Chem. 271: 7522–7528.

Catka T, Leete E (1978): Synthesis of a "bridged nicotine": 1,2,3,5,6,10b-hexahydropyrido[2,3-g]indolizine. J. Org. Chem. 43: 2125–2126.

Chavdarian CG (1983): Optically active nicotine analogues. Synthesis of (S)-(−)-2,5-dihydro-1-methyl-2-(3-pyridyl)pyrrole ((S)-(−)-3′,4′-dehydronicotine). J. Org. Chem. 48:1529–1531.

Chavdarian CG, Sanders EB, Bassfield RL (1982): Synthesis of optically active nicotinoids. J. Org. Chem. 47: 1069–1073.

Chavdarian CG, Seeman JI, Wooten JB (1983): Bridged nicotines. Synthesis of cis-2,3,3a,4,5,9b-hexahydro-1-methyl-1H-pyrrolo[2,3-f]quinoline. J. Org. Chem. 48: 492–494.

Chen Z, Trudell ML (1996): Chemistry of 7-azabicyclo[2.2.1]hepta-2,5-dienes, 7-azabicyclo[2.2.1]hept-2-enes, and 7-azabicyclo[2.2.1]heptanes. Chem. Rev. 96: 1179–1193.

Chiappinelli VA (1983): Kappa bungarotoxin: a probe for the neuronal nicotinic receptor in the avian ciliary ganglion. Brain Res. 277: 9–21.

Colquhoun D, Rang HP (1976): Effects of inhibitors on the binding of iodinated α-bungarotoxin to acetylcholine receptors in rat muscle. Mol. Pharm. 12: 519–535.

Cosford NDP, Bleicher L, Herbaut A, McCallum JS, Vernier J-M, Dawson H, Whitten JP, Adams P, Chavez-Noriega L, Correa LD, Crona JH, Mahaffy LS, Menzaghi LS, Rao TS, Reid R, Sacaan AI, Santori E, Stauderman KA, Whelan K, Lloyd GK, McDonald IA (1996): (S)-(−)-5-Ethynyl-3-(1-methyl-2-pyrrolidinyl)pyridine maleate (SIB-1508Y): A novel anti-parkinsonian agent with selectivity for neuronal nicotinic acetylcholine receptors. J. Med. Chem. 39: 3235–3237.

Cushman M, Castagnoli N (1972): The synthesis of *trans*-3′-methylnicotine. J. Org. Chem. 37: 1268–1271.

Damaj MI, Glassco W, Dukat M, May EL, Glennon RA, Martin BR (1996): Pharmacology of novel nicotinic analogs. Drug Dev. Res. 38:177–187.

Decker MW, Brioni JD, Sullivan JP, Buckley MJ, Radek RJ, Raszkiewicz JL, Kang CH, Kim DJB, Giardina WJ, Wasicak JT, Garvey DS, Williams M, Arneric SP (1994): (S)-3-Methyl-5-1-(1-methyl-2-pyrrolidinyl)isoxazole (ABT 418): A novel cholinergic ligand with cognition-enhancing and anxiolytic activities: II. In vivo characterization. J. Pharmacol. Exp. Ther. 270: 319–328.

Decker MW, Bannon AW, Curzon P, Gunther KL, Brioni JD, Holladay MW, Lin N-H, Li Y, Daanen JF, Buccafusco JF, Prendergast MA, Jackson W, Arneric SP (1997): ABT-089 [2-Methyl-3-(2-(S)-pyrrolidinylmethoxy)pyridine dihydrochloride]: II. A novel cholinergic channel modulator with effects on cognitive performance in rats and monkeys. J. Pharmacol. Exp. Ther. 283: 247–258.

Decker MW, Bannon AW, Buckley MJ, Kim DJB, Holladay MW, Ryther KB, Lin N-H, Wasicak JT, Williams M, Arneric SP (1998): Antinociceptive effects of the novel neuronal nicotinic acetylycholine receptor agonist. ABT-594, in mice. Eur. J. Pharmacol. 346: 23–33.

de Fiebre CM, Meyer EM, Henry JC, Muraskin SI, Kem WR, Papke RL (1995): Characterization of a series of anabaseine-derived compounds reveals that the 3-(4)-dimethylaminocinnamylidine derivative (DMAC) is a selective agonist at neuronal α7/[^{125}I]α-bungarotoxin receptor subtypes. Mol. Pharm. 47: 164–171.

Devlin JP, Edwards OE, Gorham PR, Hunter NR, Pike RK, Stavric B (1977): Anatoxin-a, a toxic alkaloid from *Anabaena flos-aquae*. Can. J. Chem. 55: 1367–1371.

Drasdo A, Caulfield M, Bertrand D, Bertrand S, Wonnacott S (1992): Methyllycaconitine: a novel nicotinic antagonist. Mol. Cell Neurosci. 3: 237–243.

Dukat M, Fielder W, Dumas D, Damaj I, Martin BR, Rosecrans JA, James JR, Glennon RA (1996): Pyrrolidine-modified and 6-substituted analogs of nicotine: a structure-affinity investigation. Eur. J. Med. Chem. 31: 875–888.

El-Bizri H, Clarke PBS (1994): Blockade of nicotine receptor-mediated release of dopamine from striatal synaptosomes by chlorisondamine and other nicotinic antagonists administered in vitro. Br. J. Pharmacol. 111: 406–413.

Elgoyhen AB, Johnson DS, Boulter, J, Vetter DE, Heinemann S (1994): α9: an acetylcholine receptor with novel pharmacological properties expressed in rat cochlear hair cells. Cell 79: 705–715.

Elliott RL, Ryther KB, Anderson DJ, Raszkiewicz JL, Campbell JE, Sullivan JP, Garvey DS (1995): Phenyl pyrrolidine analogues as potent nicotinic acetylcholine receptor (nAChR) ligands. Bioorg. Med. Chem. Lett. 5: 991–996.

Erlenmeyer H, Marbet R (1946): Zur Kenntnis des α-[Thiazolyl-(5)]-pyrrolidins. Helv. Chim. Acta 29: 1946–1947.

Fletcher SR, Baker R, Chambers MS, Hobbs SC, Mitchell PJ (1993): The synthesis of (+)-and (−)-epibatidine. J. Chem. Soc. Chem. Comm. 1216–1218.

Fletcher SR, Baker R, Chambers MS, Herbert RH, Hobbs SC, Thomas SR, Verrier HM, Watt AP, Ball RG (1994): Total synthesis and determination of the absolute configuration of epibatidine. J. Org. Chem. 59: 1771–1778.

Frank RL, Holley RW, Wikholm DM (1942): 3,2′-Nicotyrine. Insecticidal properties of certain azo derivatives. J. Am Chem. Soc. 64: 2835–2838.

Garvey DS, Wasicak JT, Elliott RL, Lebold SA, Hettinger A-M, Carrera GM, Lin N-H, He Y, Holladay MW, Anderson DJ, Cadman ED, Raszkiewicz JL, Sullivan JP, Arneric SP (1994a): Ligands for brain cholinergic channel receptors: Synthesis and in vitro characterization of novel isoxazoles and isothiazoles as bioisosteric replacements for the pyridine ring in nicotine. J. Med. Chem. 37: 4455–4463.

Garvey DS, Wasicak JT, Decker MW, Brioni JD, Buckley MJ, Sullivan JP, Carrera GM, Holladay MW, Arneric SP, Williams M (1994b): Novel isoxazoles which interact with brain cholinergic channel receptors have intrinsic cognitive enhancing and anxiolytic activities. J. Med. Chem. 37: 1055–1059.

Gerzanich V, Peng X, Wang F, Wells G, Anand R, Fletcher S, Lindstrom J (1995): Comparative pharmacology of epibatidine: a potent agonist for neuronal nicotinic acetylcholine receptors. Mol. Pharmacol. 48: 774–782.

Glassco W, Suchocki J, George C, Martin BR, May EL (1993): Synthesis, optical resolution, absolute configuration, and preliminary pharmacology of (+) and (−)-*cis*-2,3,3a,4,5,9b-hexahydro-1-methyl-1H-pyrrolo[3,2-h]isoquinoline, a structural analog of nicotine. J. Med. Chem. 36: 3381–3385.

Glennon RA, Dukat M (1996): Nicotine receptor ligands. Med. Chem. Res. 6: 465–486.

Grady S, Marks MJ, Wonnacott S, Collins AC (1992): Characterization of nicotinic receptor-mediated [^{3}H]dopamine release from synaptosomes prepared from mouse striatum. J. Neurochem. 59: 848–856.

Haglid F (1967): The methylation of nicotine with methyllithium. Acta Chem. Sc. 21: 329–334.

Hemscheidt T, Rapala J, Sivonen K, Skulberg OM (1995): Biosynthesis of anatoxin-a in anabaena flosaquae and homoanatoxin-a in *Oscillatoria formosa*. J. Chem. Soc. Chem. Comm. 1361–1362.

Hernandez A, Rapoport H (1994): Conformationally constrained analogues of anatoxin. Chriospecific synthesis of s trans carbonyl ring fused analogues. J. Org. Chem. 59: 1058–1066.

Hider RC,, Walkinshaw MD, Saenger W (1986): Erythrina alkaloid nicotinic antagonists: structure-activity relationships. Eur. J. Med. Chem.-Chim. Ther. 21: 231–234.

Holladay MW, Lebold SA, Lin N-H (1995): Structure-activity relationships of nicotinic acetylcholine receptor agonists as potential treatments for dementia. Drug Dev. Res. 35: 191–213.

Holladay MW, Dart MJ, Lynch JK (1997): Neuronal nicotinic acetylcholine receptors as targets for drug discovery. J. Med. Chem. 40: 4169–4194.

Holladay MW, Wasicak JT, Lin N-H, He Y, Ryther KB, Bannon AW, Buckley MJ, Kim DJB, Decker MW, Anderson DJ, Campbell JE, Kuntzweiler TA, Donnelly-Roberts DL, Piattoni-Kaplan M, Briggs CA, Williams M, Arneric SP (1998): Identification and structure-activity relationships of (*R*-5-(2-azetidinylmethoxy)-2-chloropyridine (ABT-594), a potent orally active analgesic agent acting via neuronal nicotinic acetylcholine receptors. J. Med. Chem. 41: 407–412.

Howard MH, Sardina FJ, Rapoport H (1990): Chirospecific synthesis of nitrogen and side-chain modified anatoxin analogues. Synthesis of (1R)-anatoxinal and (1R)-anatoxinic acid derivatives. J. Org. Chem. 55: 2829–2838.

Huang DF, Shen TY (1993): A versatile total synthesis of epibatidine and analogs. Tetrahedron Lett. 34: 4477–4480.

Johnson DS, Martinez J, Elgoyhen AB, Heinemann SF, McIntosh JM (1995): α-Conotoxin ImI exhibits subtype-specific nicotinic acetylcholine receptor blockade: preferential inhibition of homomeric α7 and α9 receptors. Mol. Pharm. 48: 194–199.

Kanne DB, Abood LG (1988): Synthesis and biological characterization of pyridohomotropanes. Structure-activity relationships of conformationally restricted nicotinoids. J. Med. Chem. 31: 506–509.

Kanne DB, Ashworth DJ, Cheng MT, Mutter LC (1986): Synthesis of the first highly potent bridged nicotinoid. 9-azabicyclo[4.2.1]nona[2,3-c]pyridine (pyrido[3,4-b]homotropane). J. Am. Chem. Soc. 108: 7864–7865.

Karrer P, Takahashi T (1926): Uber nicotine. Helv. Chim. Acta 9: 458–461.

Kem WR (1985): Structure and action of nemertine toxins. Amer. Zool. 25: 99–111.

Kem WR, Mahnir VM, Lin B, Prokai-Tartrai K (1996): Two primary GTS-21 metabolites are potent partial agonists at alpha 7 nicotinic receptors expressed in the *Xenopus* oocyte. Soc. Neurosci. Abstr. 22: 268 Abstr. 110.7.

Kofuji P, Aracava Y, Swanson KL, Aronstam RS, Rapoport H, Albuquerque EX (1990): Activation and blockade of the acetylcholine receptor-ion channel by the agonists (+)-anatoxin-a, the N-methyl derivative and the enantiomer. J. Pharm. Exp. Ther. 252: 517–525.

Kulak JM, Nguyen TA, Olivera BM, McIntosh JM (1997): A-conotoxin MII blocks nicotine-stimulated dopamine release in rat striatal synaptosomes. J. Neurosci. 17: 5263–5270.

Lee CY (1972): Chemistry and pharmacology of polypeptide toxins in snake venoms. Annu. Rev. Pharmacol. 12: 265–281.

Leete E, Leete SAS (1978): Synthesis of 4-methylnicotine and an examination of its possible biosynthesis from 4-methylnicotinic acid in *Nicotinia tabacum*. J. Org. Chem. 43: 2122–2124.

Leete E, Bodem GB, Manuel MF (1971): Formation of 5-fluoronicotine from 5-fluoronicotinic acid. Phytochemistry 10: 2687–2692.

Liang F, Navarro HA, Abraham P, Kotian P, King Y-S, Fowler J, Volkow N, Kuhar MJ, Carroll FI (1997): Synthesis and nicotinic acetylcholine receptor binding properties of exo-2-(2′-fluoro-5′-pyridinyl)-7-azabicyclo[2.2.1]heptane: a new positron emission tomography ligand for nicotinic receptors. J. Med. Chem. 40: 2293–2295.

Lin N-H, Carrera GM Jr, Anderson DJ (1994): Synthesis and evaluation of nicotine analogs as neuronal nicotinic acetylcholine receptor ligands. J. Med. Chem. 37: 3542–3553.

Lin N-H, Gunn DE, Ryther KB, Garvey DS, Donnelly-Roberts DL, Decker MW, Brioni JD, Buckley MJ, Rodrigues AD, Marsh KG, Anderson DJ, Buccafusco JJ, Pendergast MA, Sullivan JP, Williams M, Arneric SP, Holladay MW (1997): Structure-activity studies on ABT-089: an orally bioavailable 3-pyridyl ether nicotinic acetylcholine receptor (nAChR) ligand with cognition enhancing properties. J. Med. Chem. 40: 385–390.

Lindstrom J, Schoepfer R, Whiting P (1987): Molecular studies of the neuronal nicotinic acetylcholine receptor family. Mol. Neurobiol. 1: 281–337.

Lindstrom J, Anand R, Peng X, Gerzanich V, Wang F, Li Y (1995): Neuronal receptor subtypes. Ann. NY Acad. Sci. 757: 100–116.

Lippiello PM, Bencherif M, Gray JA, Peters S, Grigoryan G, Hodges H, Collins AC (1996): RJR-2403: a nicotinic agonist with CNS selectivity II. *In vivo* characterization. J. Pharmacol. Exp. Ther. 279: 1422–1429.

Loring RH, Zigmond RE (1988): Characterization of neuronal nicotinic receptors by snake venom neurotoxins. Trends Neurosci. 11: 73–78.

Lowry TM, Gore HM (1931): The properties of nicotine and its derivatives. Part III. Chloronicotine and methylnicotine. J. Chem. Soc. 319–323.

Macallan DRE, Lunt GG, Wonnacott S, Swanson KL, Rapoport H, Albuquerque EX (1988): Methylycaconitine and (+)-anatoxin-a differentiate between nicotinic receptors in vertebrate and invertebrate nervous systems. FEBS Lett. 226: 357–363.

Maeda S, Matsushita H, Mikami Y, Kisaki T (1978): Synthesis and characterization of N-methylmyosmine. Agric. Biol. Chem. 42: 2177–2178.

Malpass JR, Hemmings DA, Wallis AL (1996): Synthesis of epibatidine homologues: homoepibatidine and bis- homoepibatidine. Tetrahedron Lett. 37: 3911–3914.

Mansell HL (1996): Synthetic approaches to anatoxin-a. Tetrahedron 52: 6025–6061.

Marks MJ, Farnham DA, Grady SR, Collins AC (1993): Nicotinic receptor function determined by stimulation of rubidium efflux from mouse brain synaptosomes. J. Pharmacol. Exp. Ther. 264: 542–552.

Martin EJ, Panickar KS, King MA, Deyrup M, Hunter BE, Wang G, Meyer EM (1994): Cytoprotective actions of 2,4-dimethoxybenzylidene anabaseine in differentiated PC12 cells and septal cholinergic neurons. Drug Devel. Res. 31:135–141.

McDonald IA, Cosford N, Vernier J-M (1995): Nicotinic acetylcholine receptors: molecular biology, chemistry and pharmacology. Ann. Rep. Med. Chem. 30: 41–50.

McDonald IA, Vernier J-M, Cosford N, Corey-Naeve J (1996): Neuronal nicotinic acetylcholine receptor agonists. Curr. Pharmaceutical Design 2: 357–366.

Menzaghi F, Whelan KT, Risborough VT, Rao TS, Lloyd GK (1997a): Effects of a novel cholinergic ion channel agonist SIB-1765F on locomotor activity in rats. J. Pharmacol. Exp. Ther. 280: 384–392.

Menzaghi F, Whelan KT, Risborough VT, Rao TS, Lloyd GK (1997b): Interactions between a novel cholinergic ion channel agonist, SIB-1765F and L-DOPA in the reserpine model of Parkinson's disease in rats. J. Pharmacol. Exp. Ther. 280: 393–401.

Meyer EM, de Fiebre CM, Hunter BE, Simpkins CE, Frauworth N, de Fiebre NEC (1994): Effects of anabaseine-related analogs on rat brain nicotinic receptor binding and on avoidance behaviors. Drug Devel. Res. 31: 127–134.

Mikkelson I, Falch E, Krogsgaard-Larsen P, Frederiksen K, Lenz S (1997): Amino carbamates as high-affinity stereoselective ligands for neuronal nicotinic acetylcholine receptors (nAChRs). Abstr. 214th ACS Meeting, Las Vegas, NV, Abstr. MEDI-001.

Newhouse P, Potter A, Corwin J (1996): Acute administration of the cholinergic channel activator ABT-418 improves learning in Alzheimer's disease. Society for Research on Nicotine and Tobacco, Washington, DC, Abstract A39.

Papke RL, Bencherif M, Lippiello P (1996): An evaluation of neuronal nicotinic acetylcholine receptor activation by quaternary nitrogen compounds indicates that choline is selective for the α7 subtype. Neurosci. Lett. 213: 201–204.

Qian C, Li T, Shen TY, Libertine-Garahan L, Eckman J, Biftu T, Ip S (1993): Epibatidine is a nicotinic analgesic. Eur. J. Pharmacol. 250: R13–R14.

Quik M, Choremis J, Komourian J, Lukas RJ, Puchacz E (1996): Similarity between rat brain nicotinic α-bungarotoxin receptors and stably expressed α-bungarotoxin binding sites. J. Neurochem. 67: 145–154.

Rao TS, Correa LD, Reid RT, Lloyd GK (1996): Evaluation of anti-nociceptive effects of neuronal nicotinic acetylcholine receptor (nAChR) ligands in the rat tail-flick assay. Neuropharmacology 35: 393–405.

Ravdin PM, Berg DK (1979): Inhibition of neuronal acetylcholine sensitivity by α-toxins from *Bungarus multicinctus* venom. Proc. Natl. Acad. Sci. USA 76: 2072–2076.

Rondahl L (1977): Synthetic analogues of nicotine VI. Nicotine substituted in the 5-position. Acta Pharm. Suec. 14: 113–118.

Rueppel ML, Rapoport H (1971): Aberrant alkaloid biosynthesis. Formation of nicotine analogs from unnatural precursors in Nicotinia glutinosa. J. Am. Chem. Soc. 93: 7021–7028.

Rupniak NMJ, Patel S, Marwood R, Webb J, Traynor JR, Elliott J, Freedman SB, Fletcher SR, Hill RG (1994): Antinociceptive and toxic effects of (+)-epibatidine oxalate attributable to nicotinic agonist activity. Br. J. Pharmacol. 113: 1487–1493.

Sacaan AI, Reid RT, Santori EM, Adams P, Correa LD, Mahaffy LS, Bleicher L, Cosford NDP, Stauderman KA, McDonald IA, Rao TS, Lloyd GK (1997): Pharmacological characterization of SIB-1765F: a novel cholinergic ion channel agonist. J. Pharmacol. Exp. Ther. 280: 373–383.

Sargent PB (1993): The diversity of neuronal nicotinic acetylcholine receptors. Ann. Rev. Neurosci. 16: 403–443.

Schulz DW, Loring RH, Aizenmann E, Zigmond RE (1991): Autoradiographic localization of putative nicotinic receptors in the rat brain using ^{125}I-neuronal bungarotoxin. J. Neurosci. 11: 287–297.

Secor HV, Edwards WB (1979): Nicotine analogues: synthesis of pyridylazetidines. J. Org. Chem. 44: 3136–3140.

Secor HV, Chavdarian CG, Seeman JI (1981): The radical and organometallic methylation of nicotine and nicotine N-oxide. Tetrahedron Lett. 22: 3151–3154.

Seeman JI, Secor HV, Howe CR, Chavdarian CG, Morgan LW (1983): Organometallic methylation of nicotine and nicotine N-oxide. Reaction pathways and racemization mechanisms. J. Org. Chem. 48: 4899–4904.

Seeman JI, Chavdarian CG, Kornfeld RA, Naworal JD (1985a): Nicotine chemistry. The addition of organolithium reagents to (−)-nicotine. Tetrahedron 41: 595–602.

Seeman JI, Clawson LE, Secor HV (1985b): Nicotine chemistry. The addition of alkyl radicals to (S)-(−) nicotine: synthesis of optically active 6-alkyl nicotines. Synthesis 953–955.

Senokuchi K, Nakai H, Kawamura M, Katsube N, Nonaka S, Sawaragi H, Hamanaka N (1994): Synthesis and biological evaluation of (+/−)-epibatidine and the congeners. Syn. Lett. 343–344.

Shibagaki M, Matsushita H (1985): The synthesis of 4-aminonicotine and 4-aminocotinine. Heterocycles 23: 1681–1684.

Shibagaki, M, Matsushita H, Kaneko H (1986): The synthesis of 5′-alkylnicotines. Heterocycles 24: 423–428.

Shimohama S (1996): Nicotinic agonists prevent neuronal cell death by glutamate and amyloid beta protein. Neurobiol. Aging 17: S40.

Skulberg O, Carmichael WW, Anderson RA, Matsunaga RE, Moore RE, Skulberg R (1992): Investigations of a neurotoxic oscillatorialean strain (Cyanophyceae) and its toxin. Isolation and characterization of homoanatoxin-a. Environ. Toxicol. Chem. 11: 321.

Søkilde B, Mikkelson I, Stensbøl TB, Andersen B, Ebdrup S, Krogsgaard-Larsen P, Falch E (1996): Analogues of carbacholine: Synthesis and relationship between structure and affinity for muscarinic and nicotinic acetylcholine receptors. Arch. Pharm. 329: 95–104.

Sorenson EM, Culver P, Chiapinelli VA (1987): Lophotoxin: selective blockade of nicotinic transmission in autonomic ganglia by a coral neurotoxin. Neuroscience 20: 875–884.

Spande TF, Garraffo HG, Edwards MW, Yeh HJC, Pannell L, Daly JW (1992): Epibatidine: a novel (chloropyridyl)azabicycloheptane with potent analgesic activity from an Ecuadoran poison frog. J. Am. Chem. Soc. 114: 3475–3478.

Sullivan JP, Donnelly-Roberts D, Briggs CA, Anderson DJ, Gopalakrishnan M, Piattoni-Kaplan M, Campbell JE, McKenna DG, Molinari E, Hettinger A-M, Garvey DS, Wasicak JT, Holladay MW, Williams M, Arneric SP (1996): A-85380 [3-(2(S)-azetidinylmethoxy)pyridine]: In vitro pharmacological properties of a novel, high affinity $\alpha 4\beta 2$ nicotinic acetylcholine receptor ligand. Neuropharmacology 35: 725–734.

Sullivan JP, Donnelly-Roberts D, Briggs CA, Gopalakrishnan M, Hu I, Campbell JE, Anderson DJ, Piattoni-Kaplan M, Molinari E, McKenna DG, Gunn DE, Lin N-H, Ryther KB, He Y, Holladay MW, Williams M, Arneric SP (1997): ABT-089 [2-methyl-3-(2-(S)-pyrrolidinylmethoxy)pyridine dihydrochloride]: a potent and selective cholinergic channel modulator with cytoprotective properties. J. Pharmacol. Exp. Ther. 283: 235–246.

Summers KL, Lippiello P, Verhulst S, Giacobini E (1995): 5-Fluoronicotine, noranhydroecgonine, and pyridyl-methylpyrrolidine release acetylcholine and biogenic amines in rat cortex in vivo. Neurochem. Res. 20: 1089–1094.

Swanson KL, Allen CN, Aronstam RS, Rapoport H, Albuquerque EX (1986): Molecular mechanisms of the potent and stereospecific nicotinic receptor agonist(+)-anatoxin-a. Mol. Pharmacol. 29: 250–257.

Swanson KL, Aracava Y, Sardina FJ, Rapoport H, Aronstam RS, Albuquerque EX (1989): N-methylanatoxinol isomers: derivatives of the agonist (+)-anatoxin-a block the nicotinic acetylcholine receptor ion channel. Mol. Pharmacol. 35: 223–231.

Swanson KL, Aronstam RS, Wonnacott S, Rapoport H, Albuquerque EX (1991): Nicotinic pharmacology of anatoxin analogs. I. Side chain structure-activity relationships at peripheral agonist and noncompetitive antagonist sites. J. Pharmacol. Exp. Ther. 259: 377–386.

Swanson KL, Alkondon M, Pereira EFR, Albuquerque EX (1995): The search for subtype-selective nicotinic acetylcholine receptor agonists. In Blum MS, editor. The toxic action of marine and terrestrial alkaloids. Fort Collins, CO: Alaken, Inc, pp. 191–280.

Szczepanski SW, Anouna KG (1996): Synthesis of semi-rigid analogs of anabasine. Tetrahedron Lett. 37: 8841–8844.

Thomas P, Stephens M, Wilkie G, Amar M, Lunt GG, Whiting P, Gallagher T, Pereira E, Alkondon M, Albuquerque EX, Wonnacott S (1993): (+)-Anatoxin-a is a potent agonist at neuronal nicotinic acetylcholine receptors. J. Neurochem. 60: 2308–2311.

Thomas P, Brough PA, Gallagher T, Wonnacott S (1994): Alkyl-modified side chain variants of anatoxin-a: a series of potent nicotinic agonists. Drug Dev. Res. 31: 147–156.

Tschitschibabin AE, Kirssanow AW (1924): Aminierung des Nicotins mit Natrium und Kaliumamid. Chem. Ber. 57: 1163–1169.

Turek JW, Kang Ch, Campbell JE, Arneric SP, Sullivan JP (1995): A sensitive technique for the detection of the α7 neuronal nicotinic acetylcholine receptor antagonist, methyllycaconitine, in rat plasma and brain. J. Neurosci. Methods 61: 113–118.

Vijayaraghavan S, Pugh PC, Zhang Z-W, Rathouz MM, Berg DK (1992): Nicotinic receptors that bind α-bungarotoxin on neurons raise intracellular free Ca^{2+}. Neuron 8: 353–362.

Wada A, Uezono Y, Arita M, Tsuji K, Yanagihara N, Kobayashi H, Izumi F (1992): Neosurugatoxin: a probe for neuronal nicotinic receptors in adrenal medulla, brain, and ganglia. Methods Neurosci. 8: 311–322.

Wilson KL Jr, Chang RSL, Bowman ER, McKennis H Jr. (1976): Nicotine-like actions of *cis*-metanicotine and *trans*-metanicotine. J. Pharmacol. Exp. Ther. 196: 685–696.

Wonnacott S (1997): Presynaptic nicotinic ACh receptors. Trends Neurosci. 20: 92–98.

Wonnacott S, Jackman S, Swanson KL, Rapoport H, Albuquerque EX (1991): Nicotinic pharmacology of anatoxin analogs. II. Side chain structure-activity relationships at neuronal nicotinic ligand binding sites. J. Pharmacol. Exp. Ther. 259: 387–391.

Wonnacott S, Swanson KL, Albuquerque EX, Huby NJS, Thompson P, Gallagher T (1992): Homoanatoxin: a potent analogue of anatoxin-a. Biochem. Pharmacol. 43: 419–423.

Woodruff-Pak DS, Li Y-T, Kem WR (1994): A nicotinic agonist (GTS-21), eyeblink classical conditioning, and nicotinic receptor binding in rabbit brain. Brain Res. 645: 309–317.

Yum L, Wolf KM, Chiappinelli VA (1996): Nicotinic acetylcholine receptors in separate brain regions exhibit different affinities for methyllycaconitine. Neuroscience 72: 545–555.

Zhang C, Gyermek L, Trudell ML (1997): Synthesis of optically pure epibatidine analogs: (1R, 2R, 5S)-2β-(2-chloro-5-pyridinyl)-8-azabicyclo[3.2.1]octane and (1R, 2R, 5S)-2α-(2-chloro-5-pyridinyl)-8-azabicyclo[3.2.1]octane from (−)-cocaine. Tetrahedron Lett. 38: 5619–5622.

Zoltewicz JA, Prokai-Tetrai K, Bloom LB, Kem WR (1993): Long range transmission of polar effects in cholinergic 3-arylidene anabaseines. Conformations calculated by molecular modelling. Heterocycles 35: 171–179.

15

Nicotinic Cholinergic Receptor Pharmacophores

Richard A. Glennon Ph.D. and Małgorzata Dukat Ph.D.
Department of Medicinal Chemistry
Virginia Commonwealth University
Richmond, Virginia

DEVELOPMENT OF nAChR PHARMACOPHORIC MODELS

In its broadest sense a pharmacophore is a minimal ensemble of structural features (atoms or physicochemical properties associated with given atoms), common to a series of agents, seemingly responsible for a specific effect. Therefore, a nicotinic pharmacophore or nicotinic acetylcholinergic receptor (nAChR) pharmacophore should represent those features necessary or essential for nicotinic cholinergic activity (e.g., nicotine-like activity, nAChR affinity). Why is identification of a nicotinic cholinergic pharmacophore a useful endeavor? If pharmacophoric features can be reliably identified it should be theoretically possible to design novel agents on the basis of the pharmacophore.

Historical Considerations

Hey (1952) suggested that nicotinic cholinergic activity is dependent upon molecules with a partial positive charge a suitable distance from a quaternary amine. Later, Barlow and Hamilton (1962) proposed that nicotinic activity requires a positively charged site situated 3.4 to 4.5 Å from an onium site. Concurrent with this hypothesis, Holland and co-workers (Sekul and Holland, 1961; Coleman et al., 1965) proposed that a partial negative charge was required at some distance from an onium site, and this concept was seemingly supported by the results of

Neuronal Nicotinic Receptors: Pharmacology and Therapeutic Opportunities, Edited by S. P. Arneric and J. D. Brioni
ISBN 0-471-24743-x, pages 271–284. Copyright © 1998 by Wiley-Liss, Inc.

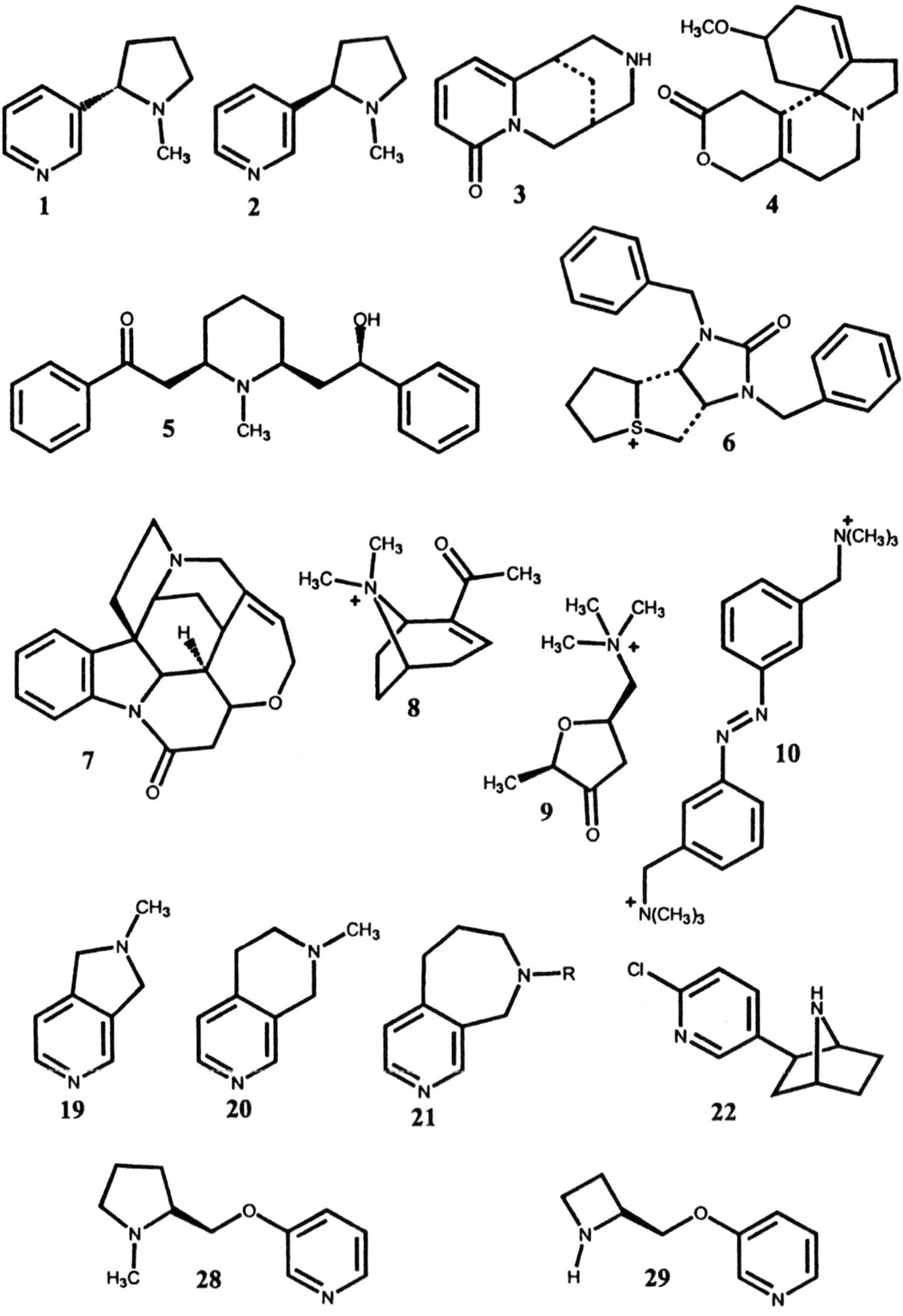

Figure 15-1. Chemical structures of some agents discussed in this chapter.

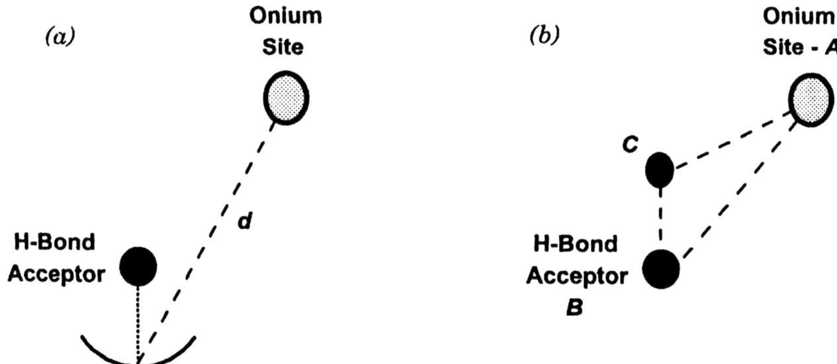

Figure 15-2. *The Beers and Reich (1970) pharmacophore model. (a): Distance d represents the distance between the onium site and the van der Waals surface of a hydrogen bond acceptor atom; optimal d = 5.9 Å. A representation of the pharmacophore produced by Sheridan et al. (1986). (b): Idealized distances are A-B: 4.8 Å, B-C: 1.2 Å, and C-A: 4.0 Å.*

Ormerod (1956) on a series of benzylcholine esters. Triggle (1965) explained these divergent concepts by invoking the need for both a partially positive and a partially negative secondary site. Kier (1968) attempted to better define interatom distances by comparing the structures of nicotine and acetylcholine. Using extended Huckel theory molecular orbital calculations Kier found that (−)-nicotine (**1;** see Fig. 15-1 for structure) can exist in either one of two preferred conformations that were consistent with then available NMR-predicted solution conformations of various nicotine analogs. It was concluded from these studies that a secondary, negatively charged site situated 4.85 Å from an onium site is a key feature for nicotinic-acting molecules.

Several years later, Beers and Reich (1970) examined a series of agents using Dreiding and CPK space-filling models to identify common structural features necessary for nicotinic cholinergic activity; parallel studies were conducted on muscarinic cholinergic agents. Nicotinic activity (both agonist and antagonist activity) was found to be mediated by two elements: (1) a coulombic interaction involving an alkylammonium moiety, and (2) a hydrogen bond that depends upon the presence of an acceptor moiety in the nicotinic agent at a distance of 5.9 Å from the center of positive charge (Beers and Reich, 1970). Kier (1968) had previously commented on the ambiguity introduced by not considering conformationally restricted molecules in studying pharmacophoric components of nicotinic agents. The conformationally restricted agent cytisine (**3**) was a key consideration of Beers and Reich, who found that this agent perfectly met the 5.9 Å requirement. The Beers and Reich model (shown in Fig. 15-2) did not define the internitrogen distance in (−)-nicotine (**1**). Rather, it defined a distance from the onium site (i.e., a site that accounts for the interaction of the pyrrolidine nitrogen atom) and a point on the van der Waals surface of the hydrogen bond acceptor. Consequently, this distance would be longer than the internitrogen distance itself and is not inconsistent with the 4.85 Å distance proposed by Kier. Dihydro-β-erythroidine (DHβE; **4**) is a competitive nicotinic antagonist. Beers and Reich (1970) reported that DHβE fits the model regardless of whether the distance is measured from the basic nitrogen atom to the van der Waals surface of either the lactone oxygen or the ether oxygen atom. Other agents considered in developing the model included (+)nicotine (**2**), lobeline (**5**), acetylcholine, trimethaphan (**6**), and strychnine (**7**).

Wasserman, Bartels, and Erlanger (1979) later reported that a distance of 5.2 Å between a cationic center and an electronegative atom 1.5 Å above the plane represents an important nicotinic pharmacophoric feature. Utilizing a frog nAChR preparation Spivak et al. (1983) examined the agonist activity of some of the agents used by Beers and Reich as well as several other nicotinic cholinergic agents. They suggested that the receptor donates a hydrogen bond to the agonist and recognizes a plane defined by the hydrogen bond acceptor (i.e., a carbonyl group and its substituents); the anionic site of the receptor is positioned out of the plane off the C=O axis. Such conditions reportedly suffice to establish chirality at the recognition site. Furthermore, they suggested that certain antagonists might block the action of nicotinic agonists by interaction at sites distinct from the recognition site of agonists.

Sheridan and co-workers (1986) subsequently employed an ensemble distance geometry approach to better define features identified by Beers and Reich. Figure 15-2 shows the Sheridan pharmacophore model. The four agents used in developing this model were the "agonists" (−)-nicotine (**1**), (−)-cytisine (**3**), (−)-ferruginine (**8**), and muscarone (**9**). Several other agents were used to check the consistency of the model: DHβE (**4**), trimethaphan (**6**), strychnine (**7**), and *trans*-3,3′-bisQ (**10**). Features found to be important were (1) a distance between a basic (e.g., pyrrolidine) nitrogen atom, point ***A***, and a hydrogen bonding atom (e.g., the pyridine nitrogen atom of nicotine or the carbonyl oxygen atom of cytisine), point ***B***, of 4.4 to 5.0 Å (with an idealized distance of 4.8 ± 0.3 Å); (2) a distance between point ***B*** and a dummy atom (representing the near-centroid of the pyridine ring of nicotine) or the carbonyl carbon atom of cytisine, point ***C***, fixed at 1.2 Å; and (3) a distance between point ***A*** and point ***C*** of 3.7 to 4.3 Å (with an idealized distance of 4.0 ± 0.3 Å). The 5.9 Å distance between the onium site and van der Waals surface of the Beers and Reich model is said to be geometrically compatible with the ***A–B*** distance of the Sheridan model, and the former distance is represented by a point ***D*** in the latter model (not shown in Fig. 15-1). The ***A–D*** distance was found to be identical to that suggested by Beers and Reich (i.e., 5.9 Å). Sheridan and co-workers pointed out that there is nothing in their model that distinguishes between agonists and antagonists, and that agonists and antagonists may or may not bind in a similar manner sharing a common amine site; to account for this they suggested that agonists might share a certain common spatial volume, and that this volume is related to the handedness of the molecules.

Waters et al. (1988) later examined a series of semirigid nicotinic cholinergic agonists using frog rectus abdominus muscle and found that although the agents easily conformed to the Beers and Reich, and Sheridan, pharmacophore models agonist potencies spanned nearly a 10,000-fold range. They concluded that the pharmacophore models, while probably correct, may be incomplete.

Barlow and Johnson (1989) have challenged the concept of the Sheridan pharmacophore model. On the basis of x-ray crystallographic analysis they found that the crystal structure of (−)-nicotine (**1**) and (−)-cytisine (**3**) could be overlayed but that the hydrogen-bonding pyridone cabonyl group of cytisine is on the side of the ring opposite that of the pyridine nitrogen atom. They acknowledged that crystal structures may not accurately represent the solution conformation of conformationally flexible molecules; nevertheless, they provided examples of other agents with nicotinic cholinergic activity that lacked a hydrogen-bond acceptor site. They found that an out-of-plane onium group 4.5 to 6.5 Å distance from an aromatic ring is a feature common to many agents with nicotinic cholinergic activity. They proposed that agonist activity may be associated with a charged nitrogen atom site (i.e., an onium site) and an area of planarity on the receptor, such as a tyrosine or phenylalanine residue, that would accommodate an aromatic ring or double bond. Binding at the latter site was suggested to involve a hydrophobic or π-electron interaction. This "point plus flat area" concept seemed to better account than the Sheridan "three-point pharmacophore model" for the actions of a wide variety of nicotinic agonists.

Factors Confounding Pharmacophoric Studies

Much of the confusion, or controversy, surrounding the early pharmacophoric studies might be attributed to the nature and small number of agents investigated, which differed from study to study, and to the different tissue/receptor preparations used to evaluate activity and define the agents. Indeed, examination of a set of agents in different tissue preparations resulted in differences in potencies (Barlow and Hamilton, 1962), and investigation of the optical isomers of nicotine in various isolated tissue preparations resulted in some nonparallel stereoselectivity effects indicating that differences likely exist among nicotine receptors from the diverse preparations (Barlow and Hamilton, 1965). Another factor complicating the interpretation of results could be related to differences in the efficacies of agents from preparation to preparation (Barlow and Hamilton, 1965), or even differences in efficacies among a series of agents within the same receptor preparation (Waters et al., 1988). In addition, the observed effects of some nAChR ligands may be secondary (direct or indirect) actions mediated via other neurotransmitter systems (Shacka and Robinson, 1996). Furthermore, most of the pharmacophoric studies failed to account for data in a quantitative sense (i.e., agents utilized in various studies were typically included because they were nicotine-like in action and/or in structure).

It is now recognized that the composition of nicotinic receptors can vary widely (especially those from peripheral versus central sources). Insect nicotinic receptors represent an extreme example; structure-activity relationships being derived from insect receptors appear quite different than those from mammalian sources (see Buckingham et al., 1995; Tomizawa et al., 1996, and references therein). Given the paucity yet structural diversity of agents previously available and the different preparations used in the past to characterize nicotinic cholinergic ligands, identified pharmacophores may either be invalid, complicated by efficacy issues, or more likely, limited to certain specific actions of the agents mediated by the receptors present in the preparations being investigated. It is likely that multiple pharmacophores will be identified in the future, with different populations of nicotine cholinergic receptors possessing somewhat different pharmacophoric requirements.

REEXAMINATION OF PHARMACOPHORIC MODELS

With the development of radioligand binding techniques and the commercial availability of suitable radioligands, it became possible to investigate the binding requirements of nicotinic agents for central nicotinic cholinergic receptors. Using affinity as a measure, the concept of efficacy could be eliminated. This has advantages and disadvantages. Nonconsideration of efficacy allows pharmacophore models to be developed on the basis of those structural features seemingly important solely for binding. However, if agonists and antagonists bind in a dissimilar manner, any formulated structure-affinity relationships (SAFIRs) may be confounded by the possibility that some of the examined compounds are potential agonists and others are antagonists.

With the above caveat in mind, we began an investigation of nAChR SAFIR in order to identify the minimal structural requirements for binding. The nicotine molecule was dissected into fragments and the affinities of these fragments for central nAChRs were measured. The studies employed rodent brain homogenates, which are though to consist primarily of $\alpha_4\beta_2$-type nicotinic cholinergic receptors. Thus, studies of this sort should assist in the identification of an $\alpha_4\beta_2$ pharmacophore model. Preliminary findings were published in 1993 with follow-up studies published in 1994 and 1996; some of these findings are reviewed below (Glennon et al., 1993, 1994; Dukat et al., 1996).

TABLE 15.1. Affinity of Some Simple Aminomethylpyridines for Nicotinic Receptors.[a]

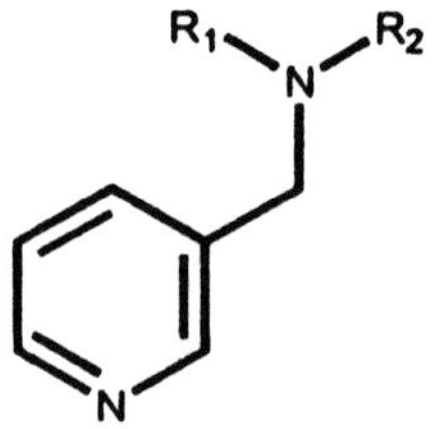

	R_1	R_2	Ki, nM
11	H	H	>10,000
12	H	Me	>10,000
13	H	Et	970
14	Me	Me	540
15	Me	Et	28
16	Me	*n*Pr	1,140
17	Me	*n*Bu	>10,000
18	Et	Et	122

[a]For purposes of comparison, *S*-(−)-nicotine (**1**) binds with an affinity (Ki value) of 2.3 nM, and *R*-(+)-nicotine (**2**) with a KI = 30nM. See Dukat et al. (1996) and references therein for additional examples.

One of the smallest structural components of nicotine to retain affinity was the aminomethylpyridine (AMP) moiety. The simple aminomethylpyridines, which differed only with respect to the nature of the terminal amine substituents, varied in affinity from 28 to more than 10,000 nM (see Table 15-1 for some representative examples). Optimal affinity was associated with a tertiary amine where one of the amine substituents was a methyl group; the N-ethyl-N-methyl AMP derivative **15** (Ki = 28 nM; Table 15-1) displayed an affinity 12 times less than that of (−)-nicotine but similar to that of (+)-nicotine.

The N-N distance in **15** (4.79 Å) is comparable to that for (−)-nicotine (**1**). Although N-N distances were not calculated for all of the compounds in Table 15-1, it is evident from inspection that they should not vary significantly. And yet the affinities of the compounds in Table 15-1 vary widely, indicating that terminal amine substituents play a role in receptor binding. Further support comes from investigations with 1′-substituted analogs of nornicotine that reached similar conclusions (Glassco et al., 1993). Likewise, Lin et al. (1994) investigated a series of 3′-, 4′- and 5′-substituted derivatives of nicotine and found that ring substituents can also dramatically influence affinity. Thus, even when the N-N distance is held relatively constant, substituent effects can have a substantial effect on receptor affinity. This is an important issue because many of the agents previously used in pharmacophoric definition represented isolated molecules and features such as N-N distance may have been overshadowed by the presence or absence of substituent groups that could have inadvertently resulted in a given agent being classified as "active" or "inactive."

Nicotine likely exists in multiple low-energy conformations and the exact pharmacologically relevant conformation is unknown. Two possible conformational extremes in the AMP series are those in which the amine-containing aminomethyl group is fully extended in an "up conformation" and those where it is bent in a "down conformation" (Dukat et al., 1996a). Although neither may represent the preferred conformation of nicotine itself, we prepared and examined several analogs in the AMP series to test the possibilities. For example, the 5- and 6-membered ring analogs **19** and **20** (Ki = 85 and 18 nM, respectively) bind with higher affinity than the simple conformationally flexible **14** (Ki = 540 nM). Consistent with the Sheridan pharmacophore, **14** and **20** both possess internitrogen distances of 4.8 to 4.9 Å. A 7-membered ring analog, **21** (Ki ca. 1,000 nM where R = H or Me), binds with lower affinity. The calculated internitrogen distance in **21** is somewhat shorter (4.6 Å) (Cheng et al., 1995).

EPIBATIDINE AND ITS INFLUENCE ON nAChR PHARMACOPHORIC MODELS

Daly and co-workers (Spande et al., 1992) reported the isolation, characterization, and potent analgesic activity of an alkaloid, (−)-epibatidine (**22**), they had isolated from the skin of an Ecuadorian frog; for review, see Dukat (1994). The actions of this agent were not antagonized by opiate antagonists, suggesting a novel mechanism of antinociceptive action. Although there are some structural similarities between (−)-epibatidine and nicotine, there also exist some obvious structural differences. Intrigued by a possible relationship between these two agents, we conducted molecular modeling studies and found that the two structures were superimposable but that the internitrogen (N-N) distance in (−)-epibatidine (5.51 Å) exceeds that found in (−)-nicotine (4.87 Å) by approximately 0.6 Å (Dukat, 1994); furthermore, the N-N distance in (−)-epibatidine exceeds both the idealized distance and variability of the Sheridan pharmacophore (i.e., 4.8 ± 0.3 Å). On this basis alone, it was assumed that (−)-epibatidine might not bind with high affinity at nicotine receptors. Interestingly, (−)-epibatidine (Ki = 0.055 nM) was found to bind with approximately 30-fold higher affinity than (−)-nicotine itself (Ki = 1.5 nM) (Dukat, 1994) and remains one of the highest affinity nAChR agents yet reported. Qian et al. (1993) independently reported a similarly high affinity for (−)-epibatidine. Others have since confirmed this high affinity, and tritiated (e.g., Houghtling et al., 1994) and radioiodinated (e.g., Davila-Garcia et al., 1997) analogs of epibatidine now have been demonstrated to label nicotine receptors. The later agent, [^{125}I]IPH, represents an analog of epibatidine where the chloro group has been replaced by ^{125}I. Although labeling more than one type of receptor, in rat forebrain [^{125}I]IPH behaves like [^{3}H]epibatidine and binding appears to be primarily to $\alpha_4\beta_2$-type receptors (Davila-Garcia et al., 1997), validating the use of epibatidine for comparative purposes in molecular modeling studies.

Epibatidine, then, is a very high-affinity nicotine receptor ligand with an internitrogen distance that is greater than that found in nicotine. Due to the competitive nature of their displacement kinetics, it can be assumed that the two agents share some binding characteristics (e.g., a common onium site); at this time, however, it is unknown if these agents share other common binding features on the receptor or whether they bind somewhat differently. For the time being we have made a second assumption that both agents also share a common hydrogen-bonding site with the pyridine nitrogen atoms binding in like manner. It might be noted, however, that Abreo et al. (1996) have provided some alternate possibilities. Nevertheless, using the common overlap, we have investigated the differential volumes of the epibatidine isomers and nicotine in order to shed light on the binding requirements of nAChR agents (Dukat et al., 1996). We have conducted QSAR studies on a limited series of structurally related agents and have also found that a parabolic relationship seems to exist between N-N distance and affinity, with an N-N distance of 5.1–5.5 Å being optimal (Glennon et al., 1994). However, see the next section.

CHAIN-EXTENDED AGENTS: PHARMACOPHORIC IMPLICATIONS

Attempts to identify particular substituent groups necessary for nicotinic cholinergic activity, other than the onium nitrogen atom, have always been accompanied by argument (Barlow and Johnson, 1989). Investigations with newer agents have only further fueled the argument.

The 7-membered ring analogs of the AMPs were found to bind with low affinity; (e.g., **21** where R = H, Ki = 1,100 nM). We reasoned that this could be due to their short N-N distance and that moving the nitrogen atom (as shown in **23**, Fig. 15-3; calculated N-N distance = 5.5 Å) might result in enhanced affinity. Compound **23** (R = H; Ki = 46 nM) was synthesized and found to bind with 10-fold higher affinity than the simple AMP **14** (Ki = 540 nM) (Cheng et al., 1995). The affinity of **23**, although enhanced relative to that of **21**, is still low compared to that of (−)-nicotine (**1**). Because this low affinity might reflect the presence of the added bulk introduced by the 7-membered ring, ring-opened analogs of **23** were examined: aminoethylpyridines (AEPs) **24**. The AEPs (calculated N-N distance ≈5.9 Å) displayed enhanced affinity, but no greater than that of **23**. Nevertheless, the affinity of, for example, N,N-dimethyl AEP (**24**, $R_1 = R_2$ = Me; Ki = 47 nM) (Fig. 15-3) is still 10-fold higher than that of **14**. Further extension of chain length, such as to the aminopropylpyridines (APPs, **25**; N-N distance ≈6.9 Å) resulted in compounds devoid of affinity (i.e., Ki > 10,000 nM) regardless of the nature of R_1 and R_2. Unsaturation (see **26**, Fig. 15-3) was introduced in an effort to slight-

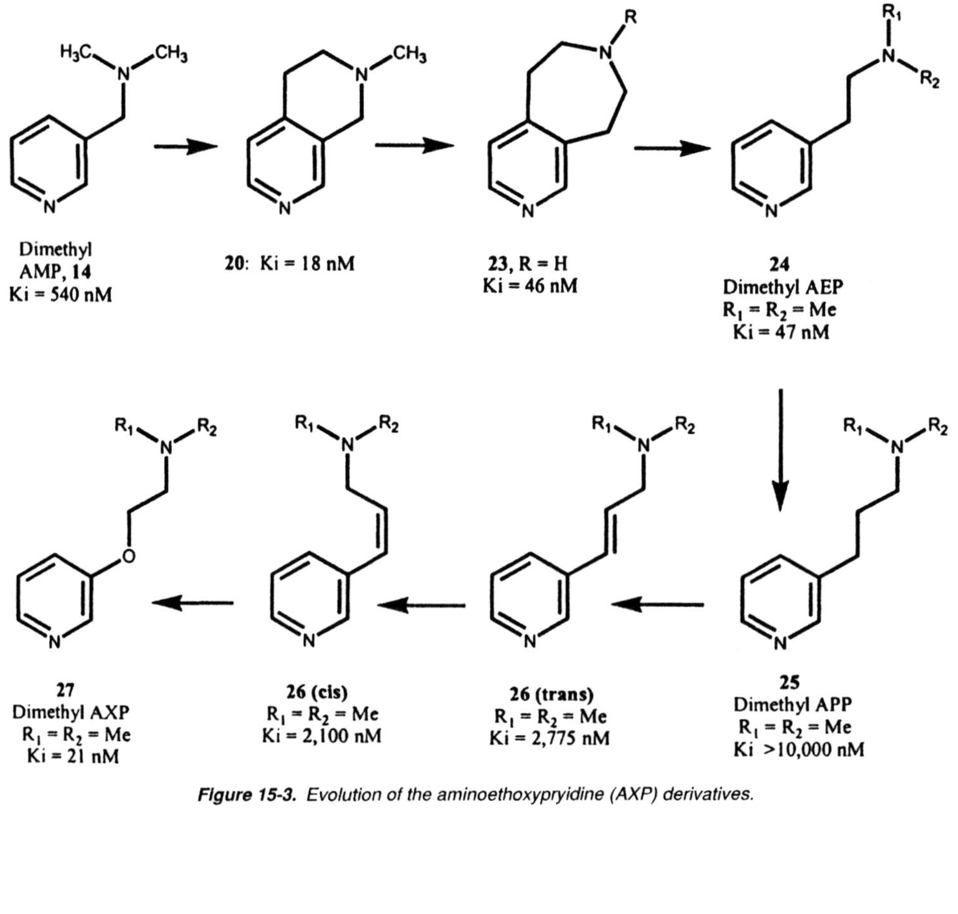

Figure 15-3. Evolution of the aminoethoxypryidine (AXP) derivatives.

ly shorten chain length and reduce conformational flexibility; both *cis* and *trans* isomers were examined. Although these alkenes displayed low affinity, they possessed higher affinity than the corresponding APP (i.e., **25**) analog where $R_1 = R_2 =$ Me. Interestingly, stereochemistry did not seem to play a major role, suggesting that the modest affinity-enhancing effect might be due more to an electronic effect rather than a steric effect.

Alkyne analogs were also examined (data not shown) and displayed even higher affinity than the corresponding alkene. Consequently, a series of aminoethoxypyridine (AXP, **27;** N-N distance ≈6.9 Å) analogs was examined. These AXP analogs (Fig. 15-3) were found to bind with higher affinity than their corresponding AMP derivatives or AEP (**24**) derivatives. The higher affinity of the AXP analogs over the corresponding APP derivatives **25** suggests that the ether oxygen atom may contribute to binding via an electronic effect. Of interest, however, is that parallel changes in the alkylamine substituents of the AMP, AEP (**24**), and AXP (**27**) series did not result in parallel changes in affinity (unpublished results), suggesting that differences may exist in the manner in which these agents bind at nAChRs.

While it might be argued that these chain-extended analogs bind, they bind with lower affinity than (−)-nicotine (**1**) and this lower affinity might be due to N-N distances that are longer than that found in nicotine or epibatidine (i.e., 4.8–5.5 Å). Is this a case where substituent effects overshadow N-N distance? While our work was in progress, the Abbott group (Abreo et al., 1996) disclosed a series of pyridyl ethers with close structural similarity to our AXP derivatives. These agents, such as A-84543 (**28**) and A-85380 (**29**), bind with subnanomolar affinity. A significant difference between the two series is that **28** and **29** bind with 140- to 400-fold higher affinity than AXP **27** where, for example, $R_1 = R_2 =$ Me, suggesting that either the branching or the cyclic structures associated with the former contribute to their very high affinity. It would seem, then, that substituent groups can play a significant role in binding. Of interest is that unlike what is seen with nicotine, the R(+)-isomer of A-84543 binds with 130-fold lower affinity than its S(−)-isomer (i.e., A-84543), and N-demethylation of A-84543 did not result in decreased affinity; both optical isomers of the N-desmethyl compound bind with comparably high affinity (Ki = 0.14–0.16 nM) (Abreo et al., 1996). R(+)-nicotine (**2**) typically binds with 10- to 30-fold lower affinity than S(−)-nicotine (**1**) and N-demethylation of nicotine results in about a 30-fold decrease in affinity (Glennon and Dukat, 1996).

RJR-2403 (**30;** Ki = 26 nM) was found to be the highest affinity member of a series of chain-extended alkene derivatives (Caldwell et al., 1997). These agents are similar in structure to **26** but are longer by one additional methylene group in the side chain. The results of a structure-activity study have been reported and some of the findings are presented in Fig. 15-4. Of particular interest is that (1) secondary amines possess greater affinity than primary or tertiary amines, (2) the double bond and its geometry are important for binding, (3) a four-membered chain is preferred over a three- or five-membered chain, and (4) bulky groups on the alkylamine decrease affinity (Caldwell et al., 1997).

Do agents such as (−)-nicotine (**1**), (−)-epibatidine (**22**), the AEP (**24**) and AXP (**25**) analogs, A-85380 (**29**), and RJR-2403 (**30**) analogs bind at nicotine receptors in the same fashion? This question remains to be answered. Being a fairly rigid structure, (−)-epibatidine would seem to be a suitable template for modeling studies. As discussed above, nicotine approximates the N-N distance of epibatidine, and the shorter distance of nicotine may account for its lower affinity. Agents with N-N distances longer than that found in nicotine or epibatidine may or may not interact with the receptors in a comparable manner. Agents with longer N-N distances, if conformationally flexible, may fold back to accommodate this distance. Abreo et al. (1996), for example, have found that the minimum energy conformation of A-85380 (Ki = 0.052 nM) results in an N-N distance of 6.1 Å. They have examined several models in an attempt to overlay critical features of this agent with those of (−)-nicotine and (−)-

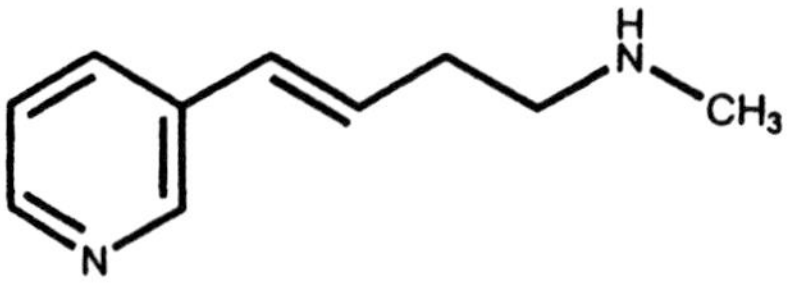

30: Ki = 26 nM

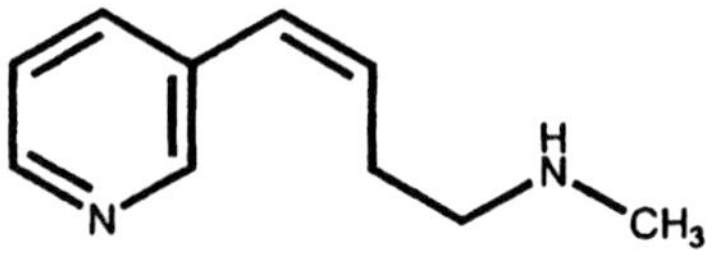

31: Ki = 354 nM

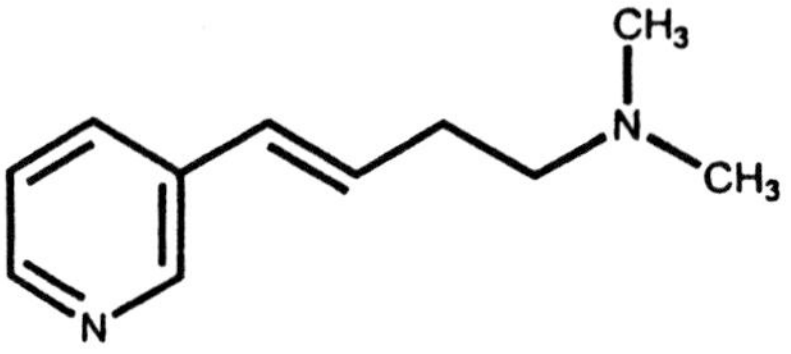

32: Ki = 4,500 nM

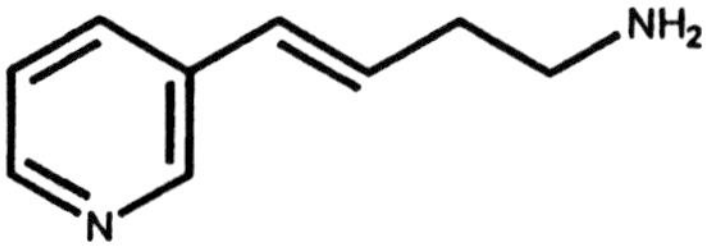

33: Ki = 119 nM

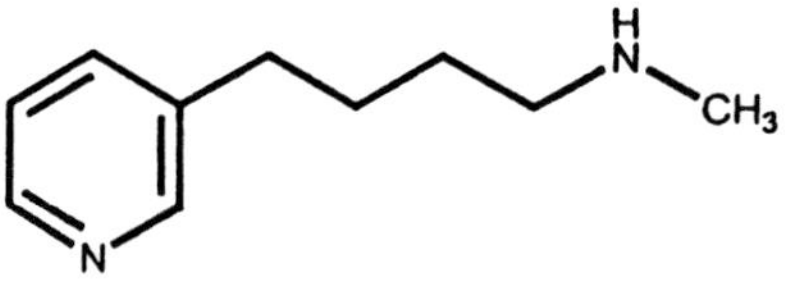

34: Ki = 910 nM

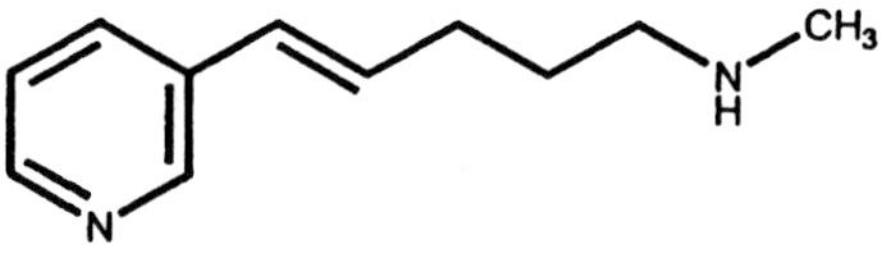

35: Ki = 1,955 nM

36: Ki = 58 nM

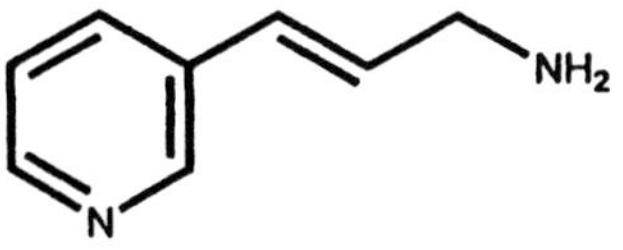

37: Ki = 46,619 nM

Figure 15-4. *Structures of RJR-2403 (30) and related agents with associated nicotine receptor affinities (Ki values) (Caldwell et al., 1997).*

epibatidine, and have been able to do so by using conformations of A-85380 that are somewhat energetically less favorable than the lowest energy conformation. It would seem possible, then, that some agents with longer N-N distances may be accommodated in a common manner.

Alternatively, agents with longer distances (indeed, even agents with shorter distances) (1) may bind in a manner proposed by Barlow and Johnson (1989) using the "point plus flat area" concept, or (2) may bind in such a manner that they utilize only one common feature

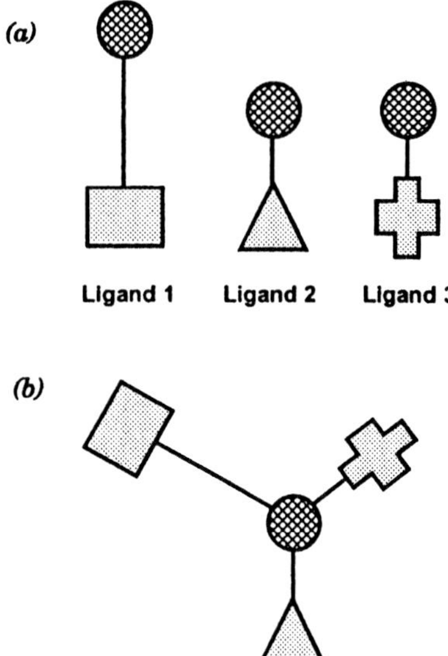

Figure 15-5. *Hypothetical nonsuperimposable modes of binding of Ligands 1–3 at nicotine receptors. (a): The hatched circles represent terminal amines with different or similar amine substituents, and the geometrical designs represent other structural aspects of the ligands. (b): The three different ligands of varying internitrogen distance may share a common onium binding site (hatched circle site) but otherwise use different sites to accommodate other structural features.*

(e.g., the onium site) of the receptor but use different binding features for other portions of the molecule (this is dramatized in Fig. 15-5). The "point plus flat area" concept, particularly given the corollary that a hydrogen-bonding acceptor portion of the molecule may not be important on the basis of the argument concerning the binding of (−)-nicotine and cytisine (see above), seems unattractive due to the low affinity of deazanicotine (Ki = 860 nM) (Glennon et al., 1993), where the pyridine nitrogen atom of nicotine was replaced by an sp^2-hybridized carbon atom. That is, the presence of the pyridine nitrogen atom, whether because of its hydrogen-bonding acceptor character or because of some other property, contributes to the affinity of nicotine.

Is there any evidence that would support the nonsuperimposable mode of binding shown in Figure 15-5? As mentioned above, structural changes on the alkylamine portions of nicotine, the AEP (**24**), AXP (**27**), A-84543 (**30**), and RJR-2403 (**31**) series of compounds do not result in parallel shifts in binding affinity. There are several possible explanations. The agents may bind similarly but do not share a common onium site. Or, the agents share a common onium site, but the amines are oriented in somewhat different directions (Glennon et al., 1994). With the latter possibility, the remaining portions of the molecules may or may not utilize common binding features. As mentioned previously, flexible long-chain compounds may bend back to be accommodated by the same binding features used by, for example, nicotine. However, it is difficult to envision the amine and pyridine portions of agents such as the AMP **15** and the alkyne **36** both being capable of simultaneously interacting at the same onium site and hydrogen-bonding site.

Certain agents, in particular the AXP derivatives, A-84543, and A-85380, may be using an affinity-enhancing auxiliary binding site, not used by nicotine, that accommodates the pyridyl ether oxygen atom. It is also possible that agents such as **36** utilize the same or a similar binding feature to account for their enhanced affinity over derivatives such as **34** that lack the electronic character associated with an alkenyl or alkynyl function.

For the time being, it would appear that the model shown in Figure 15-5 best explains much of the present $\alpha_4\beta_2$ binding data. The model is unattractive from a design standpoint because SAFIR data obtained from one series of agents (e.g., the Ligand 1 series; see Fig. 15-5) may not be applicable to another series of agents (e.g., the Ligand 2 series) due to different modes of binding. On the other hand, there are a number of receptor features that would serve as likely candidates to accommodate the various nononium portions of diverse nicotinic cholinergic ligands (e.g., see Barlow and Johnson, 1989; Westkaemper, 1996). The synthesis and evaluation of additional agents of varying N-N distance with parallel substituent modifications and/or with greater conformational restriction will be necessary to test these possibilities.

SUMMARY

Attempts have been made for 50 years to define a nicotinic cholinergic pharmacophore. Early studies were hampered by a lack of agents and little understanding of the structural differences among different types of nicotinic cholinergic receptors. Furthermore, early studies typically focused on agents with nicotinic activity with little regard for the quantitative data associated with activity and/or used diverse types of receptor preparations and may have been confounded by issues such as efficacy. With the advent of radioligand binding techniques, the identification of nAChR subtypes, and the development or discovery or newer high-affinity ligands, significant progress has been and continues to be made. Nevertheless, most recent studies have been devoted to identification of what might be more correctly defined as a $\alpha_4\beta_2$ nAChR pharmacophore using radioligand binding data. The reasons are that these receptors are most accessible to investigators, the efficacy factor has been eliminated, and a significant number of agents now have been studied. What emerges is that neither the Beers and Reich nor the Sheridan pharmacophores satisfactorily account for the binding of newer agents. A single pharmacophoric model explaining the binding of various agents has yet to be identified, and might not be identified. The future will likely witness the identification of more than a single pharmacophore for $\alpha_4\beta_2$ binding and will certainly result in the identification of multiple pharmacophoric models to explain binding at the different subtypes of nicotinic cholinergic receptors.

ACKNOWLEDGMENTS

Work from the authors' laboratories was supported in part by US PHS grant DA 05274 (RAG) and an A. D. Williams grant (MD).

REFERENCES

Abreo MA, Lin NH, Garvey DS, Gunn DE, Hettinger AM, Wasicak JT, Pavlik PA, Martin YC, Donnelly-Roberts DL, Anderson DJ, Sullivan JP, Williams M, Arneric SP, Holladay MW (1996): Novel 3-pyridyl ethers with subnanomolar affinity for central neuronal nicotinic acetylcholine receptors. J. Med. Chem. 39: 817–825.

Barlow RB, Hamilton JT (1962): Effects of some isomers and analogues of nicotine on junctional transmission. Br. J. Pharmacol. 18: 510–542.

Barlow RB, Hamilton JT (1965): The stereoselectivity of nicotine. Br. J. Pharmacol. 25: 206–212.

Barlow RB, Johnson O (1989): Relations between structure and nicotine-like activity: X-ray crystal structure analysis of (−)-cytisine and (−)-lobeline hydrochloride and a comparison with (−)-nicotine and other nicotine-like compounds. Br. J. Pharmacol. 98: 799–808.

Beers WH, Reich E (1970): Structure and activity of acetylcholine. Nature 225: 917–922.

Buckingham SD, Balk ML, Lummis SCR, Jewess P, Sattelle DB (1995): Actions of nitromethylenes on an α-bungarotoxin-sensitive neuronal nicotinic acetylcholine receptor. Neuropharmacology 34: 591–597.

Cadwell WS, Benchenif M, Bhatti BS, Deo NM, Dobson GP, Dull GM, Lipiello PM, Lovette ME, Miller CH, Ravard A, Schmitt JD, Crooks PA (1997): Synthesis and structure-activity relationships of analogs of RJR-2403, a CNS-selective nicotinic agonist. Abstracts of International Business Communications Symposium on Nicotinic Acetylcholine Receptors as Pharmaceutical Targets, Washington, DC, July 24–25.

Cheng YX, Fiedler W, Dukat M, Damaj I, Martin B, Glennon RA (1995): Conformationally restricted aminomethylpyridine derivatives as novel nicotine receptor ligands. VA J. Sci. 46: 135.

Coleman ME, Hume AH, Holland WC (1965): Nicotine-like stimulant actions of several substituted phenylcholine ethers. J. Pharmacol. Exp. Ther. 148: 66–70.

Davila-Garcia MI, Musachio JL, Perry DC, Xiao YX, Horti A, London ED, Dannals RF, Kellar KJ (1997): $[^{125}I]$IPH, an epibatidine analog, binds with high affinity to neuronal nicotinic cholinergic receptors. J. Pharmacol. Exp. Ther. 282: 445–451.

Dukat M (1994): 208/210 a.d.a. Epibatidine. Med. Chem. Res. 4: 433–439.

Dukat M, Fiedler W, Dumas D, Damaj I, Martin BR, Rosecrans JA, James JR, Glennon RA (1996a): Pyrrolidine-modified and 6-substituted analogs of nicotine: a structure-affinity investigation. Eur. J. Med. Chem. 31: 875–888.

Dukat M, Herndon JL, Glennon RA (1996b): Epibatidine: reconsideration of the nicotine receptor pharmacophore. NIDA Res. Mono. 162: 286.

Glassco W, May EL, Damai MI, Mortin BR (1993): In vivo and in vitro activity of some N-substituted (±)-nornicotine analogs. Med. Chem. Res. 4: 273–282.

Glennon RA, Dukat M (1996): Nicotine receptor ligands. Med. Chem. Res. 6: 465–486.

Glennon RA, Maarouf A, Fahmy S, Martin B, Fan F, Yousif Y, Shafik RM, Dukat M (1993): Structure-activity relationships of simple nicotine analogs. Med. Chem. Res. 2: 546–551.

Glennon RA, Herndon JL, Dukat M (1994): Epibatidine-aided studies toward definition of a nicotinic receptor pharmacophore. Med. Chem. Res. 4: 461–473.

Hey P (1952): On relationships between structure and nicotine-like stimulant activity in choline esters and ethers. Br. J. Pharmacol. 7: 117–129.

Houghtling RA, Davila-Garcia MI, Hurt SD, Kellar KJ (1994): $[^{3}H]$Epibatidine binding to nicotinic cholinergic receptors in brain. Med. Chem. Res. 4: 538–546.

Kier LB (1968): A molecular orbital calculation of the preferred conformation of nicotine. Mol. Pharmacol. 4: 70–76.

Lin N-H, Carrera GM, Anderson DJ (1994): Synthesis and evaluation of nicotine analogs as neuronal nicotinic acetylcholine receptor ligands. J. Med. Chem. 37: 3542–3553.

Ormerod WE (1956): The pharmacology of benzylcholine derivatives and the nature of carbonyl receptors. Br. J. Pharmacol. 11: 267–272.

Qian C, Li T, Shen TY, Libertine-Garahan L, Eckman J, Biftu T, Ip S (1993): Epibatidine is a nicotinic analgesic. Eur. J. Pharmacol. 250: R13–R14.

Sekul AA, Holland WC (1961): Pharmacology of senecioylcholine. J. Pharmacol. Exp. Ther. 132: 171–175.

Shacka JJ, Robinson SE (1996): Central and peripheral anatomy of nicotine sites. Med. Chem. Res. 6: 444–464.

Sheridan RP, Nilakantan R, Dixon JS, Venkataraghavan R (1986): The ensemble approach to distance geometry: application to the nicotinic pharmacophore. J. Med. Chem. 29: 899–906.

Spande TF, Garraffo HM, Edwards MW, Yeh HJC, Pannell L, Daly JW (1992): Epibatidine: a novel (chloropyridyl) azabicycloheptane with potent analgesic activity from an Ecuadoran poison frog. J. Am. Chem. Soc. 114: 3475–3478.

Spivak CE, Waters J, Witkop B, Albuquerque EX (1983): Potencies and channel properties induced by semirigid agonists at frog nicotinic acetylcholine receptors. Mol. Pharmacol. 23: 337–343.

Tomizawa M, Latli B, Casida JE (1996): Novel neonicotinoid-agarose affinity column for Drosophila and Musca nicotinic acetylcholine receptors. J. Neurochem. 67: 1667–1676.

Triggle DJ (1965): Chemical aspects of the autonomic nervous system. Chapter XII: The cholinergic receptor. London: Academic Press, pp. 159–165.

Wassermann NH, Bartels E, Erlanger BF (1979): Conformational properties of the acetylcholine receptor as revealed by studies with constrained depolarizing ligands. Proc. Natl. Acad. Sci. USA 76: 256–259.

Waters JA, Spivak CE, Hermsmeier M, Yadav JS, Liang RF, Gund TM (1988): Synthesis, pharmacology, and molecular modeling studies of semirigid, nicotinic agonists. J. Med. Chem. 31: 545–554.

Westkaemper RB (1996): Construction of 3-helix models of the nicotinic receptor ligand binding site. Med. Chem. Res. 6: 511–525.

Part V

Current and Potential Therapeutic Applications

16

Neuronal Nicotinic Receptor Channel Dysfunction in a Human Epilepsy Syndrome

Sam Berkovic, M.D., Ph.D

Department of Medecine (Neurology)
University of Melbourne, Austin
and Repatriation Medical Centre, Heidelberg
(Melbourne)
Australia

Logos Curtis, M.D.

Department of Physiology
Faculty of Medicine
University of Geneva, Switzerland

Daniel Bertrand, Ph.D

Department of Physiology
Faculty of Medicine
University of Geneva, Switzerland

The unraveling of the mysteries of the functions of neuronal nicotinic acetylcholine receptors (nAChRs) may well be aided by understanding the effects of causative disease mutations. Recently mutations in the α4 subunit of the neuronal nicotinic acetylcholine receptor (CHRNA4) have been associated with a specific form of epilepsy and there is considerable interest in the possible role of CHRNA4 and other nAChRs in disease of the nervous system.

Neuronal Nicotinic Receptors: Pharmacology and Therapeutic Opportunities, Edited by S. P. Arneric and J. D. Brioni
ISBN 0-471-24743-x, pages 287–306. Copyright © 1998 by Wiley-Liss, Inc.

Epilepsy, or the tendency to have recurrent unprovoked epileptic seizures, affects at least 2% of the population at some time in life (Hauser et al., 1993). Epilepsy is not one condition but is divided into numerous syndromes that probably have different biological bases. The fundamental nature of epilepsy is of episodic neuronal hyperexcitability and hypersynchronization leading to an abnormal excessive neuronal discharge that leads to the symptoms of seizures.

Broadly, epilepsies are divided into generalized and partial (focal) forms (Commission on Classification, 1989). In the generalized epilepsies the fundamental defect is believed to be an abnormality of thalamo-cortical synchronization that occurs diffusely throughout the forebrain (Gloor and Fariello, 1988). The generalized epilepsies have long been known to have an inherited component. In contrast, the partial epilepsies are due to a relatively localized cortical abnormality. The clinical manifestations of the seizure will depend on the anatomical site of the seizure focus. Focal epilepsy can be caused by acquired neuronal damage such as head injury, tumors, and a variety of other local insults to the brain. It is emerging, however, that certain focal epilepsies are genetically determined and an increasing number of these have been recognized in recent years (Berkovic and Scheffer, 1997).

INHERITED FORMS OF EPILEPSY

Like virtually all common diseases, the genetics of the common epilepsies is complex with evidence of multiple genetic components, with or without additional environmental factors. Detection of genetic linkage and subsequently identifying defective genes has been frustratingly difficult in these epilepsies (for review, see Berkovic, 1997).

A few idiopathic epilepsies have, however, been recognized where there is simple (Mendelian) inheritance. Such epilepsies provide a much better opportunity for successful mapping studies and subsequent identification of the abnormal gene product. These disorders are relatively infrequent. Molecular genetic study of such epilepsies is likely to yield insights into the basic mechanisms and may well give clues as to the genetic mechanisms of the epilepsies with complex inheritance (Berkovic, 1997).

Surprisingly, many but not all of these idiopathic epilepsies with simple inheritance are focal epilepsies. This raises the perplexing biological issue as to how a single gene mutation, which presumably affects gene products in all or many neurons, can lead to seizure focus in a localized part of the brain. The best understood form of this newly emerging group of epilepsies is autosomal-dominant nocturnal frontal lobe epilepsy where defects in CHRNA4 have been found (Berkovic and Scheffer, 1997).

AUTOSOMAL DOMINANT NOCTURNAL FRONTAL LOBE EPILEPSY (ADNFLE)

ADNFLE is an idiopathic partial epilepsy syndrome with a mean age of onset of 12 years and a median of 8 years. Individuals have clusters of brief seizures during sleep. They generally wake up and, in the older patients, describe a nonspecific aura, perhaps of a feeling in their body or limbs followed by motor activity characteristically associated with preservation of consciousness. The motor activity varies from brief episodes of tonic posturing to hyperkinetic thrashing movements characteristic of seizures coming from the anterior and mesial parts of the frontal lobes (Scheffer et al., 1995, 1996).

Recognition of this disorder as an epilepsy has only been recent as electroencephalographic (EEG) studies are generally normal in between seizures, and daytime attacks are uncommon. Many such patients were misdiagnosed as having nightmares or other forms of nonepileptic sleep disturbance (Scheffer et al., 1995, 1996; Oldani et al., 1996).

Neuroimaging studies of the brain show no structural abnormalities and these individuals are intellectually and neurologically normal. Brain tissue from patients with ADNFLE has not been examined, so the possibility of a subtle structural abnormality cannot be ruled out. Studies with positron emission tomography (PET) may show defects in glucose metabolism in the frontal lobe, whereas studies during seizures using single photon emission computed tomography (SPECT) show a focal increase in cerebral blood flow during attacks (Hayman et al., 1997).

GENETICS OF ADNFLE

Pedigree analysis revealed that ADNFLE is an autosomal dominant disorder with a penetrance of approximately 75%. The largest family described to date has 27 affected individuals and was identified in South Australia (Scheffer et al., 1996). Linkage studies performed at the Women's and Children's Hospital, Adelaide, South Australia, by Phillips et al. (1995) showed that the disorder mapped to the distal end of the long arm of chromosome 20 (20q13.2–13.3) in this family.

CHRNA4 was known to be localized at 20q13.2–13.3. In collaboration with Dr. Ortrud Steinlein, Institute of Human Genetics, Bonn, a missense mutation in CHRNA4 converting serine 248 to a phenylalanine was found in this large Australian family (Steinlein et al., 1995). There was strong evidence that this was a pathogenic mutation. All affected individuals and obligate carriers had the mutation, whereas it was not found in individuals marrying into the pedigree or in a large panel of blood donors in Germany and Australia (Steinlein et al., 1995). The serine 248 was known to be conserved during evolution in vertebrates and invertebrates and it was known to be central to the ion channel and forming the binding site for chlorpromazine, a noncompetitive antagonist of the receptor. Studies of the receptor in the open and closed state had also shown that serine 248 was critical to the normal opening and closing of the receptor (Devillers-Thiéry et al., 1993; Unwin, 1995).

None of the other four families initially reported by us with ADNFLE had a mutation in CHRNA4, but subsequently a Norwegian family with the same clinical phenotype was identified with an insertion mutation (776ins3) in the same region (Magnusson et al., 1996; Steinlein et al., 1997).

In a number of other families, however, there is good evidence that the disorder does not link to 20q (Berkovic et al., 1995, and unpublished data). The underlying cause of other families with ADNFLE is not known, but other nAChR subunits are strong candidates. Thus, it appears that ADNFLE is a genetically heterogeneous disorder with the disease in some families being associated with abnormalities in CHRNA4 and the other nAChR subunits being candidate regions for other families.

HOW CAN A MUTATION IN THE α4 SUBUNIT LEAD TO EPILEPSY?

The identification of an association between one form of genetically transmissible epilepsy and a mutation in the gene coding for the α4 subunit of the nAChR suggests that this mutation either predisposes to, or is at the origin of, epileptic discharges. In this section we examine the possible natures of dysfunction in light of our current knowledge of the structure, function, and anatomic localization of nAChRs, discuss possible neuronal circuits involved, and examine the properties of control and known mutant subunits.

Focal epileptogenesis such as that seen in ADNFLE results from the generation of abnormal synchronous discharges in a neuronal population. The factors predisposing a neuronal population to this sort of abnormal activity may include (1) altered membrane properties in a group of neurons, which may lead to a pacemaker rhythm; (2) increased excitatory coupling between cells of an epileptogenic region; and (3) reduced inhibitory control within these neuronal circuits. Assuming that an altered α4 subunit plays a causal role in the seizures of ADNFLE, the linked mutation must affect one or several of these factors in critical groups of neurons. Although it is known that in ADNFLE epileptic activity is manifest in frontal cortex (Hayman et al., 1997), the neuronal population(s) implicated in epileptogenesis still remain(s) to be determined. Likewise, which specific factors, apparently under hereditary influence of the α4 nAChR subunit gene, predispose to epileptic seizures is presently only a matter of speculation. A better understanding of the involved neuronal populations and mechanisms of epileptogenesis may provide further insight into the pathogenesis of seizures as well as the functional role of the nAChR in the human central nervous system.

THE NEURONAL NICOTINIC ACETYLCHOLINE RECEPTOR IS A CATIONIC CHANNEL

Although it was known for years that acetylcholine activates ACh ligand-gated channels at the neuromuscular junction or in ganglia (Sakmann et al., 1985; Bertrand and Changeux, 1995), little or nothing was known about their possible distribution and function in the central nervous system. The identification of a large family of closely related genes originating from a common ancestral gene, which encodes for the nAChR subunits, opened new ways to investigate brain functions. To date, 11 nAChR subunits have been identified in vertebrates: α2–9 and β2–4 (McGehee and Role, 1995).

It is widely accepted that, as are the muscular nAChRs, the neuronal nAChRs are oligomeric proteins composed of five subunits that form at the same time the binding site for the neurotransmitter and the ionic pore (Fig. 16-1). Depending upon the subunit considered, receptors can be further subdivided into heteromeric and homomeric in function of their known assembly properties. Each subunit is a protein spanning the membrane four times with both its N- and C-terminal domain facing the extracellular space. From numerous studies it was further deduced that the ligand-binding site resides at the interface between two adjacent subunits and that the alpha subunits contribute to the major binding domain while the adjacent (nonalpha) subunits produce the complementary binding site. In the case of homomeric receptors, such as α7, it was shown that a single subunit possesses both the main and complementary ligand-binding sites (Corringer et al., 1997). The aqueous pore through which ions can flow when the receptor is in its active state lies at the center of the pseudosymmetrical arrangement of the protein. Moreover, it is known that when activated, these channels are permeable to cations and thereby cause a depolarization of the cell membrane in which they are inserted. Although it is beyond the scope of this section to review in detail the basic structure of the nAChRs, it is of value to recall that the walls of the ionic pore are made by the juxtaposition of second transmembrane segments (TMII) from the five subunits that form a receptor (Bertrand et al., 1993). Consequently, it can be expected that mutations within TMII can affect channel opening and/or permeability as well as other receptor properties by allosteric interactions (reviewed in Bertrand and Changeux, 1995).

Determination of the distribution of neuronal nAChRs revealed that these proteins are widely expressed in many areas of the central nervous system (Clarke, 1993). In addition, when examined at the cellular level it was found that nAChRs can be expressed either pre- and/or postsynaptically. Although surprising at first, these findings are of high relevance when considering the potential role of neuronal nAChRs. First, and most easily understood, postsynaptic receptors can be viewed as participating in the transmission of electrical information across synapses and, given their microsecond activation time scale, can be considered as fast wiring systems. Second, and more difficult to apprehend, presynaptic receptors may even play a more important role

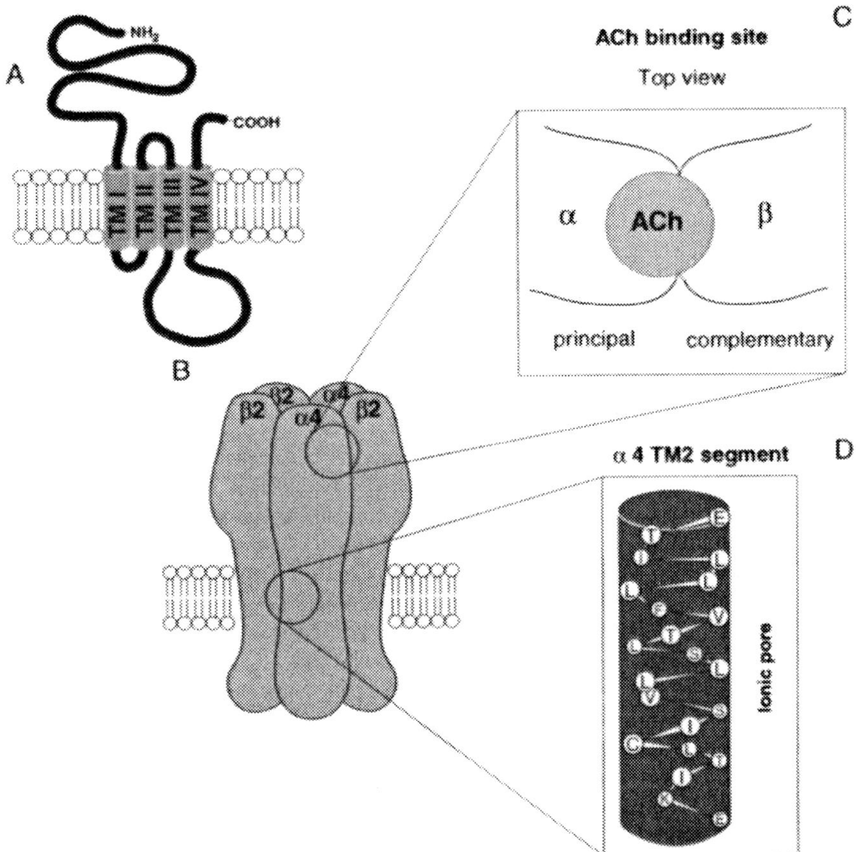

Figure 16-1. *Structure of nAChRs.* ***A:*** *Transmembrane topology of a single subunit.* ***B:*** *Pentameric disposition of combined subunits constituting a nAChR (in this case, a heteromeric $\alpha 4\beta 2$ receptor).* ***C:*** *ACh binding site at the extracellular interface between two adjacent subunits. In the $\alpha 4\beta 2$ receptor, the principal component is formed by the $\alpha 4$ subunit, and the complementary by $\beta 2$.* ***D:*** *Amino acid sequence of the pore-lining second transmembrane domain (TMII). The amino acids form an α-helix with specific residues facing the ionic pore.*

in the modulation of neurotransmission (Wonnacott, 1997). Evidence for the participation of these presynaptic receptors in mechanisms was first obtained when examining the effects of nicotine on synaptosomes. In these brain preparations, which mostly result from closure of presynaptic boutons in small spherical-shaped bodies, it was shown that application of nicotine or other nicotinic agonists could trigger the release of the dopamine contained in these small vesicles—a release that was readily blocked by nicotinic antagonist and was shown to be independent of voltage-activated calcium channels (Rapier et al., 1988; Grady et al., 1992). In agreement with these observations, electrophysiological studies done on tissue slices or primary cell cultures also revealed that low nicotine concentrations could trigger the release of the neurotransmitter glutamate or gamma-amino-butyric acid (GABA) (Léna et al., 1993; Gray et al., 1996; Role and Berg, 1996).

To understand this presynaptic modulatory role of nicotinic receptors it is necessary to recall that, in contrast to their muscle counterpart, these channels are highly permeable to divalent cations and their activation can induce a significant calcium influx in neurons (Bertrand et al., 1993). Since it is also known that the fusion of the neurotransmitter containing vesicles with the presynaptic membrane strongly depends upon the intracellular calcium concentration, it was readily suggested that activation of nAChRs at the presynaptic bouton causes a calcium influx sufficient to trigger the release of another neurotransmitter. It was also shown this calcium influx can trigger internal cell processes affecting cellular functions other than neurotransmission (Zheng et al., 1994).

In view of these latest findings it is therefore necessary to take into account that neuronal nAChRs may exert either a direct role in excitatory neurotransmission, or an indirect and modulatory role in excitatory or inhibitory neurotransmission, or possibly even a role in other internal cellular processes.

NEURONAL nAChRs AND BRAIN DEVELOPMENT

In addition to the effects they may exert on neurotransmission, it has been suggested that nAChRs play a role in brain development. Brain in situ hybridization studies, using mRNA probes directed against the different subunits of the neuronal nAChRs, have demonstrated that proteins coding for these receptors are expressed very early in embryonic life (Couturier et al., 1990; Court and Clementi, 1995; Schröder et al., 1996; Aghulon, 1998). Moreover, it was found that expression of the nAChR subunits often precedes innervation and that mRNA levels display specific developmental patterns. These results were further supported by other studies in which nAChR protein levels were quantified using immunoprecipitations (Schröder, 1992; Liu et al., 1997). In view of these findings it was proposed that neuronal nAChRs may play a role in brain development. Thus, it follows that perturbation of a given receptor function may cause a modification of the establishment of appropriate neuron connections. The distinction between the effects of a receptor malfunctioning in an otherwise correctly wired neuronal circuitry versus an altered pattern of development must, however, still await the development of other possibilities of investigation such as PET, which could allow visualization of the level of expression of a given receptor in vivo (Ding et al., 1996).

DISTRIBUTION OF NEURONAL nAChRs IN RELATIONSHIP TO BRAIN STRUCTURES POTENTIALLY INVOLVED IN SEIZURE GENERATION/PROPAGATION

Several approaches have been used to determine the localization and constitution of different types of nAChRs in various structures of the central nervous system of the rodent. Radioligand binding studies initially established the distribution of two populations of nAChRs: one type preferentially binding [^{3}H]-labeled nicotine and another preferentially binding [^{125}I]-αBTX (α-bungarotoxin, a snake toxin) (Clarke, 1987, 1993). Subsequent immunohistochemical studies revealed the [^{3}H]-nicotine binding sites to be largely associated with nAChRs containing α4 and β2 subunits, whereas the [^{125}I]-αBTX binding sites corresponded to nAChRs mainly or exclusively composed of α7 subunits (Whiting and Lindstrom, 1986; Flores et al., 1992; Schröder, 1992; Séguéla et al., 1993; Dominguez et al., 1994).

Rat central nervous system mRNA expression patterns, as revealed by in situ hybridization, have been determined for five of the α subunits (2, 3, 4, 5, and 7) and all three known neuronal β subunits (2, 3, and 4) (Deneris et al., 1989; Duvoisin et al., 1989; Wada et al., 1989; Boulter et al., 1990; Wada et al., 1990). While there are no specific radioligands available for the individual subunits, α4 and β2 mRNA expression often parallels [^{3}H]-nicotine binding just as α7 mRNA expression does [^{125}I]-αBTX binding, thus further contributing to the view that most nAChRs of the central nervous system contain either α4 and β2 subunits or α7 subunits. However, it must be stressed that mRNA expression does not consistently parallel binding or immunohistochemical localization and the following points must be considered before drawing conclusions: (1) mRNA transcript levels can have little or no correlation with the corresponding protein expression; (2) for proteins such as nAChR subunits, which may demonstrate presynaptic localization, mRNA in the neuronal somas of a particular area may correspond to proteins expressed in distant areas receiving axonal projections from the expressing neurons. Finally, functional studies employing additional electrophysiological approaches have identified sites of probable nicotinic transmission including the medial habenula (McCormick and Prince, 1987; Mulle et al., 1991), the interpeduncular nucleus (Brown et al., 1983; Mulle et al., 1991), the substantia nigra (Lichtensteiger et al., 1982), the ventral tegmental area (VTA) (Calabresi et al., 1989; Brodie, 1991), the thalamus (McCormick and Prince, 1987), and the cortex (Rowell and Winkler, 1984).

While these studies have led to a relatively extensive description of nAChR localization in the rodent brain, studies of the human nervous system are still at a preliminary stage. Labeling studies have revealed [^{3}H]-nicotine binding in areas including human cortex, thalamus, striatum, hippocampal structures, substantia nigra, and the nucleus basalis of Meynert (NBM) (Adem et al., 1988, 1989; Rubboli et al., 1994). nAChR mRNA distribution in humans has been determined for only some subunits and often only in limited areas of the brain (Sugaya et al., 1990; Wevers et al., 1994; Court and Clementi, 1995; Schröder et al., 1996). Although these initial results suggest a nAChR distribution similar to that found in rodents, several significant differences (for example, β2 mRNA shows a different distribution in humans and rodents) preclude systematic inference of human distribution based on rodent studies. Human α4 mRNA expression in adults has so far only been studied in the frontal cortex, where it is highly expressed. Recent in situ hybridization analysis of human fetal brain has shown high levels of α4 mRNA in the cortex, striatum, thalamus, and hippocampus. Clearly, much remains to be determined concerning α4 subunit distribution in the human brain. However, currently available data suggest that α4-containing nAChRs are widely expressed in the human brain in patterns similar to those seen in rodents and in particular in several structures often implicated in seizure generation (either clinically or in animal models) (Fig. 16-2). These structures, which may contain neuronal circuits either capable of epileptogenesis or responsible for seizure suppression under certain conditions and which are potential areas where dysfunctional nAChR-mediated transmission or modulation may cause the generation of synchronous oscillations leading to seizure, are as follows:

- The hippocampus and related structures of the limbic system: Although ADNFLE seizures probably do not originate in hippocampal structures (Scheffer et al., 1995), the frequent involvement of areas of the limbic system in the origination and the propagation of complex partial seizures (Lothman et al., 1991; Lothman, 1992; Luciano, 1993) warrants a brief mention of nAChRs in these structures. [^{3}H]-nicotine binding has been demonstrated throughout structures of the human hippocampal formation and is particularly strong in the entorhinal cortex, subiculum, parasubiculum, and amygdala (Rubboli et al., 1994; Court and Clementi, 1995). In the rat, α4 mRNA expression parallels this pattern (Wada et al., 1989).
- The basal forebrain and related structures of the ascending activating system: Several crucial cholinergic systems in the basal forebrain and brainstem are known to contribute to a diffuse group of structures related to the level of arousal and include the NBM, areas of the reticular formation and the VTA, the pedunculopontine nucleus (PPN), and the lat-

A

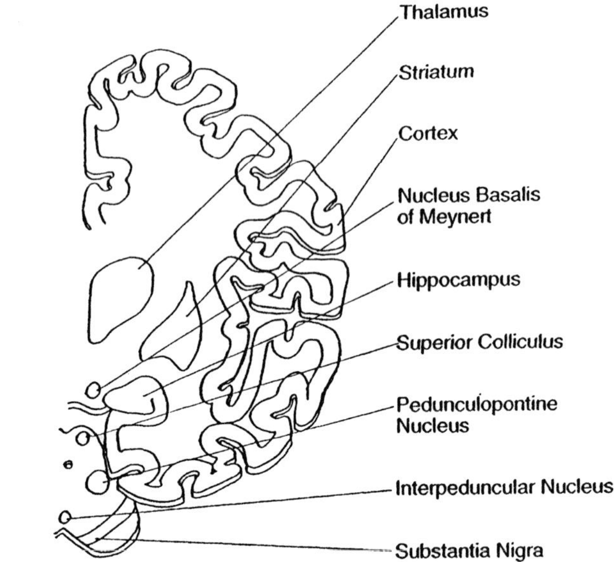

B

	^{3}H-Nicotine binding	α4 mRNA (adult rat)	α4 mRNA (adult human)	α4 mRNA (fetal human)
Cortex	++	++	++	++
Thalamus[a]	++	++	?	++
Nucleus Basalis M.	++	?	?	?
Striatum	++	(+)	?	+
Hippocampus	++	++	?	++
Pedunculo-pontine N.	+	+	?	?
Interpeduncular Nucleus	++	++	?	?
Substantia Nigra (PR[b])	+	+	?	?
Superior Colliculus	+	(+)	?	?

Figure 16-2. *Anatomic localization of α4 nAChR subunits.* ***A:*** *Schematic cerebral hemisphere and midbrain sections localizing structures of potential α4 nAChR mediated nicotinic transmission.* ***B:*** *[^{3}]H-Nicotine binding and α4 mRNA expression in illustrated structures. (Key: ++ = strong signal, + = moderate signal, (+) = weak signal and/or signal restricted to limited areas of the structure [see text]; a: including thalamic relay nuclei and reticular nucleus, b: pars reticulata.)*

erodorsal tegmental nucleus (LDTN) (Jones, 1993; Szymusiak, 1995). All of the mentioned structures show relatively high levels of [^{3}H]-nicotine binding, and in the rat the VTA and most areas of the reticular formation show high levels of α4 mRNA expression, whereas only weak expression has been seen in the PPN and LDTN (Wada et al., 1989). Furthermore, nAChR-mediated transmission has been demonstrated in cells of the VTA (Calabresi et al., 1989; Brodie, 1991). Rhythmic activity within these structures is linked to transitions between states of sleep and arousal (Jones, 1993). As seizures in ADNFLE mainly occur during these transitions (Oldani et al., 1996), and it has recently been shown that cholinergic activity within these circuits can have seizure-suppressing effects (Danober et al., 1995), it is tempting to suspect that nAChR transmission dysfunction in these structures contributes to the symptomatology of ADNFLE.

- The cerebellum: [^{3}H]-nicotine binds in most areas of the cerebellum in humans as in rats; however, mRNA studies in rats have revealed α4 mRNA to be mostly expressed in the deep cerebellar nuclei. Lesions of the cerebellum have been shown to facilitate focal cortical seizures (Paz et al., 1985) and electrical stimulation of either cerebellar cortex or deep cerebellar nuclei inhibits focal seizures (Hutton et al., 1972; Babb et al., 1974), suggesting a seizure-suppressing capacity of certain cerebellar structures (Gale, 1992).
- Cortex and thalamus, substantia nigra, and related basal areas. Several studies have implicated thalamocortical circuits and the neuronal circuits related to the substantia nigra and striatum in epileptogenesis. In order to illustrate possible mechanisms by which a molecular defect in the α4 subunit can engender cellular and ultracellular dysfunction leading to epileptogenesis, these two circuits are further discussed.

THALAMOCORTICAL CIRCUITS

The thalamocortical circuits illustrated in Figure 16-3A are typical of the connections between neurons of sensory and motor cortex and thalamic relay nuclei. They play a critical role in the physiology of sleep as well as the pathogenesis of certain forms of seizure (Stériade et al., 1993; Snead, 1995; McCormick, 1996). As they are richly populated by nAChRs, they present an attractive candidate for circuits involved in ADNFLE.

In rats, α4 and β2 mRNAs are highly expressed in cortical neurons as are—but to a lesser degree—α2, α3 and α5 mRNAs (Wada et al., 1989, 1990). The pattern of expression of α7 mRNA is significantly different as it is expressed in deep and superficial cortical layers (Séguéla et al., 1993). Not surprisingly, [^{3}H]-nicotine binding is high throughout the cortex and [^{125}I]-αBTX binds mainly in deep and superficial layers (Clarke et al., 1985; Sugaya et al., 1990). While the cortex receives important cholinergic afferents from the basal forebrain, in particular the nucleus basalis of Meynert (Rye et al., 1987), there is but little evidence that they terminate on nicotinic rather than muscarinic receptors (Clarke, 1993). However, certain studies tentatively suggest that at least some cortical nAChRs are located presynaptically on thalamocortical afferents from thalamic relay nuclei (Parkinson et al., 1988; Vidal et Changeux, 1989). All thalamic nuclei show strong expression of α3, α4, and β2 mRNAs (Wada et al., 1989) and in particular the relay nuclei that project to cortical sensory and motor areas. This expression pattern is paralleled by [^{3}H]-nicotine binding (Swanson et al., 1987; Adem et al., 1988). Contrary to the cortex, neurons of thalamic nuclei have more readily demonstrated fast currents in response to nicotine exposure, suggestive of the presence of postsynaptic receptors (McCormick and Prince, 1987). Cholinergic input to the thalamus has been shown to originate principally in the laterodorsal tegmental nucleus (LDTN) and pedunculopontine nucleus (PPN) and innervates thalamocortical relay cells as well as inhibitory neurons of the thalamic reticular nucleus (TRN) (Hallanger et al., 1987; Stériade et al., 1988).

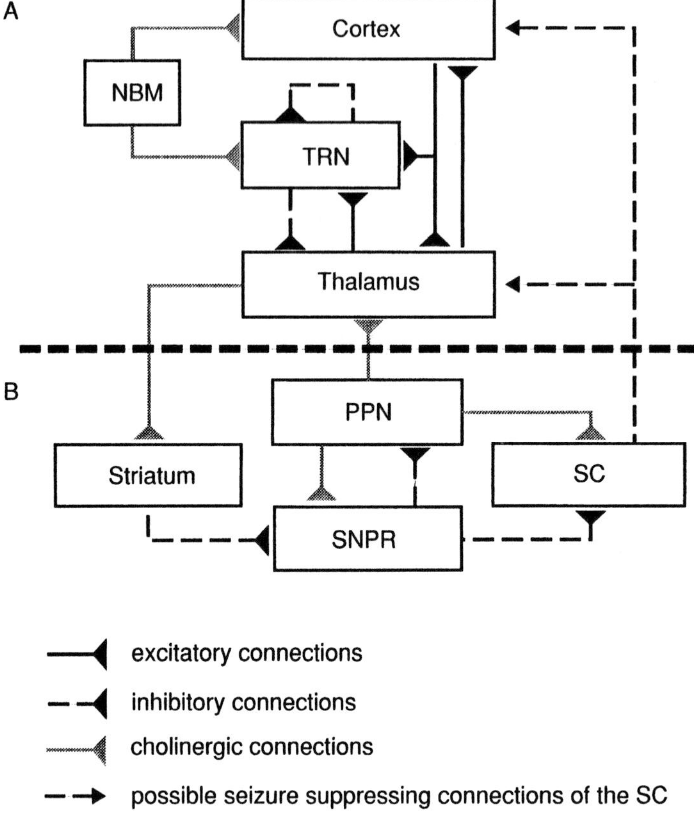

NBM: Nucleus Basalis of Meynert
TRN: Thalamic Reticular Nucleus
PPN: Pendunculopontine Nucleus
SC: Superior Colliculus
SNPR: Substantia Nigra Pars Reticulata

Figure 16-3. *Cholinergic influence on neuronal circuits implicated in epileptogenesis. Schematic diagram of thalamocortical circuits (**A**) and seizure suppressing circuits of basal structures (**B**). These circuits illustrate excitatory and inhibitory couplings that may be influenced by nAChRs innervated by cholinergic afferents. The inhibitory connections are thought to be GABAergic and the excitatory connections mainly glutamate mediated.*

Figure 16-3A illustrates a schematic view of the thalamocortical neuronal circuitry and in particular the relationship between the cortex and thalamic nuclei and the inhibitory neurons of the TRN (which surrounds the thalamus). The cortex and thalamic relay nuclei have reciprocal excitatory couplings and are modulated by inhibitory neurons of the TRN (Stériade et al., 1993). The GABAergic inhibitory neurons of the TRN are capable of generating synchronized oscillations that are typically observed in cortical networks during sleep and certain forms of seizures (Stériade et al., 1993; Meierkord, 1994; Snead, 1995). The capacity to generate these oscillations is dependent upon the intrinsic properties of the TRN neurons but also the various interconnections between these neurons and thalamic relay neurons and cortical neurons (Stériade et al., 1993; von Krosigk et al., 1993; Bal and McCormick, 1996; Contreras et al., 1997).

Taking into account these observations, one can speculate as to several levels at which nAChR dysfunction could impair the normal neuronal activity of these networks and render them susceptible of either generating epileptic discharges or propagating epileptic discharges in a facilitated way, particularly during sleep (when seizures occur in ADNFLE): Fast synaptic transmission mediated by nAChRs is possible on thalamic neurons innervated by cholinergic afferents from the PPN and LDTN. If these nAChRs over- or underactivate certain of these neurons, in particular during crucial transitions between arousal and sleep, this can conceivably lead to an imbalance between excitatory and inhibitory signaling within the thalamic circuitry, ultimately causing an epileptic focus. Considering the possibility of nAChRs functioning as presynaptic modulators of neurotransmitter release extends the possible involved neurons to those responding to noncholinergic transmitters. For example, dysfunction of presynaptic nAChRs on GABAergic boutons could impede the capacity of inhibitory neurons to release increased amounts of GABA, which may be necessary during transition to the oscillatory modes characteristic of sleep. Finally, as discussed earlier, nAChR dysfunction may affect the development of thalamocortical structures. In situ hybridization studies in the human fetal brain have revealed that thalamic and cortical neurons show very high levels of α4 mRNA expression even before innervation. If nAChRs play an important role in the development of certain of these circuits, the α4 mutation could cause abnormal connections to form, causing seizure-prone circuits that later in life could be independent of cholinergic mechanisms.

SEIZURE-GATING CIRCUITS OF THE BASAL GANGLIA AND RELATED STRUCTURES

Neurons of the substantia nigra pars reticulata (SNPR), striatum, PPN, and superior colliculus (SC) participate in a neuronal circuit (Fig. 16-3B) with probable seizure-suppressing effects. The seizure-suppressing effects of this circuit are generated by neurons of the SC via as yet unknown mechanisms (it is, however, interesting to note that the SC neurons project to thalamic relay nuclei and frontal cortex). As for the thalamocortical circuits discussed previously, this seizure-suppressing circuit may be modulated by nAChR-mediated transmission and thus prone to aberrant activity in ADNFLE.

The substantia nigra demonstrates very high levels of [^{3}H]-nicotine binding and α3, α4, α5, β2, and β3 mRNAs. The expression of nAChR subunits is particularly high in the substantia nigra pars compacta (SNPC) and is expressed at a lower level in the SNPR. Nicotinic transmission on dopaminergic neurons of the SNPC that project to the striatum has been described and implicated in reward mechanisms involved in tobacco addiction (Pidoplichko et al., 1997; Nestler, 1992; Dani, 1996). No such transmission has yet been reported in the SNPR; however, the expression of nAChR mRNA and relatively high levels of [^{3}H]-nicotine binding suggest the presence of nAChRs, the function of which has yet to be determined. [^{3}H]-nicotine binding is also moderate to high in the striatum, PPN, and SC, and α4 mRNA is expressed in these areas, although not at levels that consistently parallel binding, suggesting that some binding sites are either formed by non-α4 receptors or by presynaptic receptors. There is little nAChR mRNA expression in the striatum except for α4 and β2 transcripts in certain parts of the globus pallidus.

The SC shows nAChR mRNA only in the deep layers, which are innervated by the SNPR. The seizure-suppressing neurons of the SC are under inhibitory influence of nigral GABAergic neurons, which in turn are inhibited by GABAergic neurons of the striatum. Excitatory cholinergic neurons of the PPN innervate both the SNPR and the SC and are under inhibitory influence of the GABAergic neurons of the SNPR. Thus, excitation of striatal inhibitory neurons results in the activation of seizure-suppressing effects of the superior colliculus and these effects may be modulated by nAChRs responding to cholinergic afferents from the PPN. As for the thalamocortical circuits described previously, one may reasonably speculate that nAChR dysfunction in this circuit can lead to an imbalance of inhibitory and excitatory signaling, particularly in the SC, resulting in the impairment of an endogenous seizure control mechanism.

RECONSTITUTION OF FUNCTIONAL nAChRs

Further study of the anatomical and cellular distribution of nAChR subunits, in particular human α4, will undoubtedly provide further insight into the neuronal populations involved in epileptogenesis in ADNFLE. However, the in vitro reconstitution of mutant nAChRs has provided a method of investigating the altered physiological and pharmacological properties of a mutated subunit. At present two methods can be employed to reconstitute functional receptors by expression of exogenous DNAs: (1) reconstitution in *Xenopus* oocytes or (2) transfection into a competent cell line. Briefly, reconstitution in oocytes consists of injecting the desired cDNA (or mRNA) in appropriate combination with other subunits required to assemble functional channels in the nucleus of a *Xenopus* oocyte and to investigate it electrophysiologically a couple of days later. Upon application of ACh, or appropriate ligands, oocytes expressing functional receptors will display a significant current that can be investigated using standard methods. Contrary to the oocyte method, expression into cell lines requires transient or stable transfection with the appropriate genes (Buisson et al., 1998). Using the so-called voltage-clamp technique, electrophysiological investigations can be carried out at the macroscopic level. In this technique the voltage across the membrane of the cell is kept constant by an electronic circuit and the measure is the current needed to compensate the ionic flux that flows through the membrane. Although slightly more complex, measurements of the activity of a single protein are also possible by employing the so-called patch-clamp technique. Both of these methods are applicable to *Xenopus* oocytes as well as to cell lines.

EFFECTS OF ADNFLE MUTANTS ON THE nAChR FUNCTIONS

Up to now, two distinct mutations have been found to be linked with ADNFLE (Steinlein et al., 1995, 1997). Although significantly different, these two mutations both result in a modification in the TMII segment. The first identified S248F mutation consists of the substitution of a serine (an amino acid with a short lateral chain), located close to the putative cytoplasmic mouth of the channel, with a phenylalanine (a residue that displays a longer side chain). Introduction of the equivalent mutation in the chick α4 nAChR produced three characteristic modifications of receptors reconstituted in *Xenopus* oocytes with this altered subunit. These modifications, respectively are: (1) an increase of the desensitization to acetylcholine, (2) a small shift of the dose-response curve to ACh toward a lower sensitivity, and (3) a reduction of the mean currents evoked by a saturating ACh concentration (Fig. 16-4B–D). In agreement with these findings it was later reported that reconstitution of human α4β2 with the S248F mutated subunit displayed too fast a desensitization (Weiland et al., 1997). Moreover, recent findings have confirmed that the single-channel conductance of receptors containing the mutation S248F is reduced (Kuryatov et al., 1997). In view of these findings it was concluded that mutation S248F produces a loss of function of the neuronal nAChRs.

The identification of a linkage between another mutation and ADNFLE in a Norwegian family prompted some speculations about a possible loss of function of the corresponding α4-containing nAChR. However, reconstitution experiments conducted in *Xenopus* oocytes with

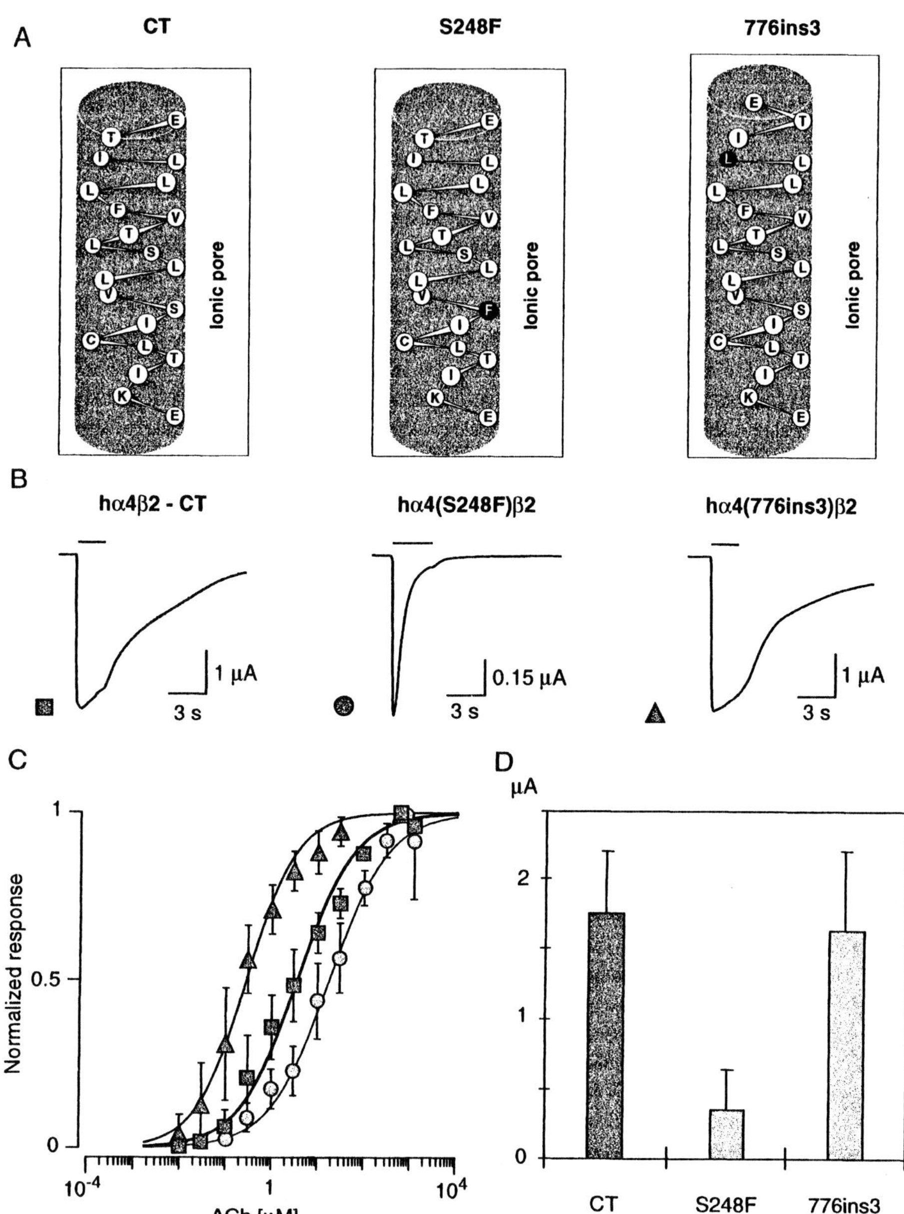

Figure 16-4. *Reconstitution of ADNFLE α4 mutant receptors.* ***A:*** *TMII domains of the common type (CT) α4 subunit, S248F and L776ins3 mutants. In S248F, a phenylalanine residue replaces a serine lining the ionic pore. In L776ins3, a leucine residue is inserted near the top of the α-helix, possibly distorting this region of the protein.* ***B:*** *Current response curves of the three receptor types to acetylcholine pulses. Acetylcholine pulses were as follows: hα4β2-CT: 600 μM ACh 2 sec; hα4(S248F)β2: 1mM ACh 3 sec; hα4(776ins3)β2: 600 μM ACh 2 sec.* ***C:*** *Dose-response curves of hα4β2-CT and mutants. hα4β2-CT displays an EC50 3.13±2 μM. hα4(S248F)β2: EC50 20±8 μM, hα4(776ins3)β2: EC50 0.69±0.08 μM* ***D:*** *Maximal elicited currents from oocytes expressing CT and mutant nAChRs.*

the corresponding leucine insertion mutant (776ins3) yielded surprising results. Oocytes coinjected with the 776ins3 mutant in combination with the control β2 subunit displayed, in response to ACh, robust currents that could not readily be distinguished from those of the control either in their time course or amplitude (Steinlein et al., 1997). Furthermore, determination of their sensitivity to ACh revealed that mutation 776ins3 causes an apparent increase of the receptor affinity by about 10-fold (Fig. 16-C). Although results obtained with the S248F and 776ins3 mutants seem difficult to reconcile, determination of the effects of calcium substitution brought a more comprehensible interpretation. Namely, it was found that mutant 776ins3 displays a higher sensitivity to external calcium, which was interpreted as a reduction of the permeability of this receptor to divalent cations (Steinlein et al., 1997). This hypothesis is further reinforced when comparing results previously obtained by point mutations of leucine residues in the upper segment of the α7 TMII (Bertrand et al., 1993). Taken together these data suggest that, although acting by different mechanisms, both mutation S248F and 776ins3 impair the basic properties of the neuronal nAChRs. Moreover, in light of the 776ins3 mutation and its sensitivity to calcium, it could be postulated that these receptors may essentially act at the presynaptic level rather than the postsynaptic level.

OTHER MUTATIONS

Point mutagenesis experiments employed for the structure function relationship studies have demonstrated that, when produced in a critical domain, the modification of a single amino acid can profoundly alter the receptor function. This is best illustrated by the pleiotropic effects caused by the mutation of a single leucine residue into a threonine (L247T) in the second transmembrane segment of the α7 nAChR (Revah et al., 1991; Bertrand et al., 1992, 1993a,b). The leucine residue mutated in this work is an amino acid that is extremely well conserved throughout the entire family of ligand-gated channels and was shown, using biochemical methods, to be pointing toward the ionic pore (Giraudat et al., 1985). Effects induced by this single mutation include a shift of about 200-fold toward a higher affinity, suppression of desensitization, and modification of the pharmacological profile with competitive antagonists becoming agonists (Bertrand et al., 1992). Moreover, it was found that introduction of the comparable mutation in the serotoninergic (5-HT_3) as well as muscle and more recently $GABA_A$ receptors also causes profound modifications of the receptor functions (Yakel et al., 1993; Labarca et al., 1995; Pan et al., 1997). In contrast, however, mutation of another leucine residue adjacent to L247 causes no significant changes in the receptor function and therefore illustrates the critical part played by determinant residues and the minor role of others.

Numerous point mutations as well as microchimera were also effectuated in the ligand binding domain, and similarly to the previous observation, mutations of determinant amino acids produce important changes in the receptor affinity for a given ligand whereas replacement of other residues has little or no effects. Determinant amino acids in the ligand binding domain have been recently reviewed by Galzi and Changeux (Galzi et al., 1996). More recently, it was also shown that a specific amino acid segment, or loop C, participates in the determination of the desensitization properties (Corringer et al., 1997). This work illustrates that desensitization is defined by the entire protein structure and cannot be restricted exclusively either to the channel domain or to the ligand binding site.

Studies of muscular nAChR mutants bring further insight into the mechanisms through which mutations can alter protein function: Several hereditary mutations have been described in these receptor subunits (which share common lineage with neuronal nicotinic receptor subunits) that cause congenital myasthenic syndromes. While one mutant localizes to the critical

TMII region, others have been described that alter the signal peptide region or a glycosylation consensus site (Ohno et al., 1995, 1996).

Since a single subunit of nAChR comprises roughly 600 amino acids, at present it is still impossible to predict the effects of any given mutation. However, the identification of functional domains that participate in the formation of either the binding site, the channel pore, or the modulatory sites, including the muscular and intracellular loops, represents an important step toward possible prediction of the putative effects of a given mutation. Nonetheless, at present the reconstitution methods described herein remain the best approaches to evaluate the physiological and pharmacological modifications induced by spontaneous gene alterations.

FROM MUTATION TO CELLULAR FUNCTION

The studies of mutated subunits expressed in vitro offer valuable insight into possible functional alterations of nAChRs in individuals with ADNFLE. However, how closely these in vitro systems reflect in vivo consequences is debatable. First, the in vitro experiments performed so far have examined properties of receptors expressed in a haploid manner—that is, one copy of each gene. In individuals with ADNFLE (an autosomal dominantly transmitted disease), one of every two α4 subunit alleles in every cell is presumably not mutated, and if expressed in vitro it would exhibit normal receptor characteristics. This complicates the possible alterations that nAChRs may show in ADNFLE; their properties depend on how an individual cell uses its mutated and "healthy" alleles to constitute α4-containing nAChRs and it may be that certain nAChR-expressing cell populations within the nervous system suffer more than others from the presence of one mutated allele. Second, as witnessed by many knockout mice phenotypes, living organisms are capable of important developmental compensations in response to genetic defects. This is relevant for nicotinic subunits, as α4 is part of a large family ofhomologous subunits that may be able to partially compensate loss of function within the family. Supporting this notion, a recent report suggests that in oocytes expressing S248F-α4 and-β2, the addition of α5 subunits partially restores the functionality of expressed nAChRs (Kuryatov et al., 1997).

PROBABILITY OF MUTATION IN THE α4 GENE

Although mutations such as S248F or L776ins in α4 may seem to be chance events, their occurrence and expression in a human population are at least partially nonrandom and dependent on several relatively constant factors. These factors, which may differ between species and from gene to gene, can usually be summarized as pertaining to either (1) the variation in the rate of mutation, (2) the rate of DNA repair, or (3) the "tolerance" of a gene's mutation by the organism.

- The rate of mutation, i.e., the rate of occurrence of actual physical alterations of the DNA sequence of a gene, depends in part on the base pair sequences themselves. Often certain particular types of mutations occur at a greatly increased frequency at or near "hot spots," which are sequence patterns constituting regions of relatively high mutation rate. In humans, for example, CpG islands have been associated with increased occurrences of base transitions (Youssoufian et al., 1988), and short sequence repeats can be a cause of deletions within genes (Krawczak and Cooper, 1991).
- The rate of DNA repair reflects the capability of an organism to correct mutations and varies considerably between organisms. Furthermore, in higher animals, certain chromosomal regions or genes seem to be more actively repaired than others (Bohr, 1985).

- The functional impact of a mutation will affect its observed frequency in a population. Genes for which even slight variations are deleterious or fatal will show little variation in a population even if their mutation rate is high.

Determining how these factors apply to the α4 gene awaits further analysis of the α4 gene region and sequence and studies of its genetic variation (for example, the isolation of a dinucleotide repeat polymorphism in the first intron of the α4 gene [Weiland and Steinlein, 1996]). However, in order to roughly approximate the rate of mutation of several nAChR subunits, we have examined the sequence identity between nAChR subunits α3, α4, α5, α7, β2, and β4 of three vertebrate species (chicken, rat, human) (Fig. 16-5). Such an analysis reveals the degree of conservation of a given protein between species. It is well known that certain genes, for instance, cytochrome C, are highly conserved between species, whereas others, such as hemoglobin, are not. Figure 16-5 readily reveals that the α4 subunit, one of the most highly expressed subunits in vertebrate brains and known to contribute in large part to the functional nAChRs of the central nervous system, displays a higher degree of difference amongst these species than other nAChR subunits. In comparison, α7 and β2 nAChR subunits, both also highly expressed in vertebrate brains, show strong homology between species, comparable to cytochrome C.

Interpretation of these comparisons requires some caution, as many factors can conceivably cause protein sequences to vary between species. However, the degree of protein sequence divergence for a given protein can be taken as a rough estimate of the rate of evolution of this protein. This led us to the highly speculative hypothesis that perhaps the α4 subunit is an evolving subunit among higher vertebrates compared to other important subunits such as α7 or β2, which are relatively stable. Consequently, one would expect the α4 gene to be relatively variable even within one species (such as human), leading to a higher proportion of mutant alleles but also a greater variability in functional ("healthy") alleles. Considering this variability in a two-allele system might explain why roughly 25% of individuals carrying a mutant α4 allele remain seizure-free.

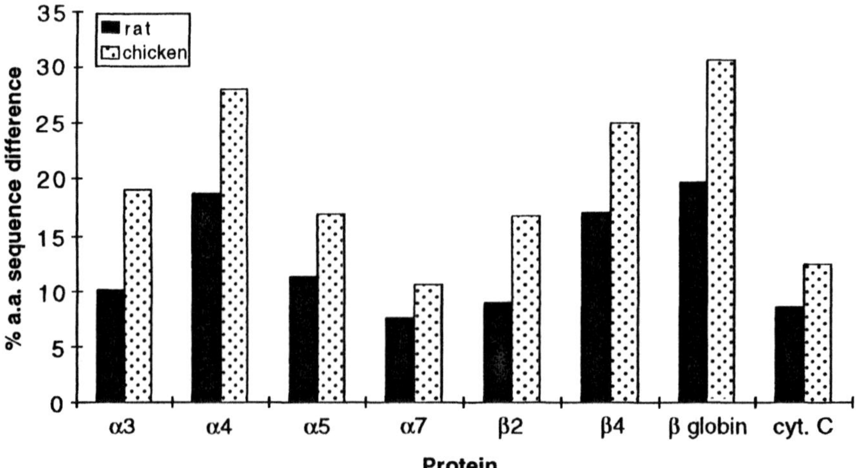

Figure 16-5. Comparison of interspecies sequence diversity of nAChR subunits. The percentage of differing amino acids between homologous subunits of rat and human or chick and human is shown for six nAChR subunits. Comparative values for β-globin and cytochrome C are given as examples of proteins with high and low sequence variation.

REFERENCES

Adem A, et al. (1988): Distribution of nicotinic receptors in human thalamus as visualized by ^{3}H-nicotine and ^{3}H-acetylcholine receptor autoradiography. J. Neural Transm. 73: 77–83.

Adem A, et al. (1989): Quantitative autoradiography of nicotinic receptors in large cryosections of human brain hemispheres. Neuroscience Lett. 101: 247–252.

Aguthon C, et al. (1998) Distribution of mRNA for the α4 subunit of the nicotinic acetylcholine receptor in the human fetal brain. Brain Res. Mol. Brain Res. 58(1-2): 123–131.

Babb TL, et al. (1974): Fastigiobulbar and dentatothalamic influences on hippocampal cobalt epilepsy in the cat. Electroencephalogr. Clin. Neurophysiol. 36: 141–154.

Bal T, McCormick DA (1996): What stops synchronized thalamocortical oscillations? Neuron 17: 297–308.

Berkovic SF (1997): Genetics of epilepsy syndromes. In: Engel J Jr., Pedley TA, editors. Epilepsy: a comprehensive textbook. Philadelphia: Lippincott-Raven, pp. 217–224.

Berkovic SF, Scheffer IE (1997): Genetics of human partial epilepsy. Curr. Opin. Neurol. 10: 110–114.

Berkovic SF, Phillips HA, Scheffer IE, et al. (1995): Genetic heterogeneity in autosomal dominant nocturnal frontal lobe epilepsy. Epilepsia 36 (suppl. 4): 147.

Bertrand D, Changeux J-P (1995): Nicotinic receptor: an allosteric protein specialized for intercellular communication. Sem. Neurosci. 7: 75–90.

Bertrand D, et al. (1992): Unconventional pharmacology of a neuronal nicotinic receptor mutated in the channel domain. Proc. Natl. Acad. Sci. USA 89(4): 1261–1265.

Bertrand D, et al. (1993a): Mutations at two distinct sites within the channel domain M2 alter calcium permeability of neuronal alpha 7 nicotinic receptor. Proc. Natl. Acad. Sci. USA 90(15): 6971–6975.

Bertrand D, et al. (1993b): Stratification of the channel domain in neurotransmitter receptors. Curr. Op. Cell Biol. 5: 688–693.

Bohr VA, et al. (1985). DNA repair in an active gene: removal of pyrimidine dimers from the DHFR gene of CHO cells is much more efficient than in the genome overall. Cell 40(2): 359–369.

Boulter J, et al. (1990): Alpha3, alpha5, and beta4: three members of the rat neuronal nicotinic acetylcholine receptor-related gene family form a gene cluster. J. Biol. Chem. 265(8): 4472–4482.

Brown DA, et al. (1983): Chemical transmission in the rat interpeduncular nucleus in vitro. J. Physiol. (London) 341: 655–670.

Buisson B, et al. (1998): Stable expression of human neuronal nicotinic receptors in human embryonic kidney-293 cells. Chapter 6.

Calabresi P, et al. (1989): Nicotinic excitation of rat ventral tegmental neurones in vitro studied by intracellular recording. Br. J. Pharmacol. 98: 135–140.

Clarke PB (1993): Nicotinic receptors in mammalian brain: localization and relation to cholinergic innervation. Prog. Brain Res. 98(77): 77–83.

Clarke PBS (1987): Recent progress in identifying nicotinic cholinoceptors in mammilian brain. TIPS 8: 32–35.

Clarke PBS et al. (1985): Nicotinic binding in rat brain: autoradiographic comparison of 3H-acetylcholine, 3H-nicotine and 125I-alpha-bungarotoxin. J. Neurosci. 5: 1307–1315.

Commission on Classification and Terminology of the International League Against Epilepsy (1989): Proposal for revised classification of epilepsies and epileptic syndromes. Epilepsia 30: 389–399.

Contreras D, et al. (1997): Spatiotemporal patterns of spindle oscillations in cortex and thalamus. J. Neurosci. 17(3): 1179–1196.

Corringer P-J, et al. (1997): Critical elements determining diversity in agonist binding and desensitization of neuronal nAChR. J. Neurosci. 18(2): 648–657.

Court J, Clementi F (1995): Distribution of nicotinic subtypes in human brain. Alzheimer's Dis. Assoc. Disorders 9 (suppl. 2): 6–14.

Couturier S, et al. (1990): A neuronal nicotinic acetylcholine receptor subunit (alpha7) is developmentally regulated and forms a homo-oligomeric channel blocked by alpha-BTX. Neuron 5: 847–856.

Danober L, et al. (1995): Mesopontine cholinergic control over generalized nonconvulsive seizures in a genetic model of absence epilepsy in the rat. Neuroscience 69(4): 1183–1193.

Deneris ES, et al. (1989): Beta 3: a new member of the nicotinic acetylcholine receptor gene family is expressed in brain. J. Biol. Chem. 264: 6268–6272.

Devillers-Thiéry A, Galzi JL, Eiselé JL, Bertrand S, Bertrand D, Changeux JP (1993): Functional architecture of the nicotinic acetylcholine receptor: a prototype of ligand-gated ion channels. J. Membrane Biol. 36: 97–112.

Ding YS, et al. (1996): Mapping nicotinic acetylcholine receptors with PET. Synapse 24: 403–407.

Dominguez DTE, et al. (1994): Immunocytochemical localization of the alpha 7 subunit of the nicotinic acetylcholine receptor in the rat central nervous system. J. Comp. Neurol. 349(3): 325–342.

Duvoisin RM, et al. (1989): The functional diversity of the neuronal nicotinic acetylcholine receptors is increased by a novel subunit: beta 4. Neuron 3: 487–496.

Flores CM, et al. (1992): A subtype of nicotinic cholinergic receptor in rat brain is composed of $\alpha4$ and $\beta2$ subunits and is up-regulated by chronic nicotine treatment. Mol. Pharmacol. 41: 31–37.

Gale K (1992): Subcortical structures and pathways involved in convulsive seizure generation. J. Clin. Neurophysiol. 9(2): 264–277.

Galzi J-L, et al. (1996): The multiple phenotypes of allosteric receptor mutants. Proc. Natl. Acad. Sci. USA 93: 1853–1858.

Giraudat J, et al. (1985): Transmembrane topology of acetylcholine receptor subunits probed with photoreactive phospholipids. Biochemistry 24: 3121–3127.

Gloor P, Fariello RG (1988): Generalized epilepsy: some of its cellular mechanisms differ from those of focal epilepsy. TINS 11: 63–68.

Grady S, et al. (1992): Characterization of nicotinic receptor-mediated [^{3}H]dopamine release from synaptosomes prepared from mouse striatum. J. Neurochem. 59: 848–856.

Gray R, et al. (1996): Hippocampal synaptic transmission enhanced by low concentrations of nicotine [see comments]. Nature 383(6602): 713–716.

Hallanger AE, et al. (1987): The origins of cholinergic and other subcortical afferents to the thalamus in the rat. J. Comp. Neurol. 262: 105–124.

Hauser WA, Annegers JF, Kurland LT (1993): Incidence of epilepsy and unprovoked seizures in Rochester, Minnesota: 1935–1984. Epilepsia 34: 453–468.

Hayman M, Scheffer IE, Chinvarun Y, Berlangieri SU, Berkovic SF (1997): Autosomal dominant nocturnal frontal lobe epilepsy: demonstration of focal frontal onset and intrafamilial variation. Neurology 49: 969–975.

Hutton JT, et al. (1972): The influence of the cerebellum in cat penicillin epilepsy. Epilepsia 13: 401–408.

Jones BE (1993): The organization of central cholinergic systems and their functional importance in sleep-waking cycles. Prog. Brain Res. 98: 61–71.

Krawczak M, Cooper DN, (1991): Gene deletions causing human genetic disease: mechanisms of mutagenesis and the role of the local DNA sequence environment. Hum. Genet. 86: 425.

Kuryatov A, et al. (1997): Mutation causing autosomal dominant nocturnal frontal lobe epilepsy alters Ca2+ permeability, conductance, and gating of human $\alpha4\beta2$ nicotinic acetylcholine receptors. J. Neurosci. 17(23): 9035–9047.

Labarca C, et al. (1995): Channel gating governed symmetrically by conserved leucine residues in the M2 domain of nicotinic receptors. Nature 376: 514–516.

Léna C, et al. (1993): Evidence for "preterminal" nicotinic receptors on GABAergic axons in the rat interpeduncular nucleus. J. Neurosci. 13(6): 2680–2688.

Lichtensteiger W, et al. (1982): Stimulation of nigrostriatal dopamine neurones by nicotine. Neuropharmacology 21: 963–968.

Liu W, et al. (1997): α4 nicotinic acetylcholine receptors (nAChRs) in cortical areas of the fetal human brain. Soc. Neurosci. Abstr. 23: 40.1.

Lothman EW (1992): Basic mechanisms of the epilepsies. Curr. Op. Neurobiol. Neurosurg. 5: 216–223.

Lothman EW et al. (1991): Functional anatomy of hippocampal seizures. Prog. Neurobiol. 37: 1–82.

Luciano D (1993): Partial seizures of frontal and temporal origin. Neurologic Clinics 11(4): 805–822.

Magnusson A, Nakken KO, Brubakk E (1996): Autosomal dominant frontal epilepsy. Lancet 347: 1191–1192.

McCormick DA, Prince DA (1987a): Acetylcholine causes rapid nicotinic excitation in the medial habenular nucleus of guinea pig, in vitro. J. Neurosci. 7: 742–752.

McCormick DA, Prince DA (1987b): Actions of acetylcholine in the guinea-pig and cat medial and laterla geniculate nuclei, in vitro. J. Physiol. (London) 392: 147–165.

McGehee DS, Role LW (1995): Physiological diversity of nicotinic acetylcholine receptors expressed by vertebrate neurons. Annu. Rev. Physiol. 57(521): 521–546.

Meierkord, H (1994): Epilepsy and sleep. Curr. Op. Neurol. 7: 107–112.

Mulle C, et al. (1991): Existence of different subtypes of nicotinic acetylcholine receptors in the rat habenulo-interpeduncular system. J. Neurosci. 11: 2588–2597.

Nestler EJ (1992): Molecular mechanisms of drug addiction. J. Neurosci. 12: 2439–2450.

Ohno K, et al. (1995): Congenital myasthenic syndrome caused by prolonged acetylcholine receptor channel openings due to a mutation in the M2 domain of the ε subunit. Proc. Natl. Acad. Sci. 92: 758–762.

Ohno K, et al. (1996): Congenital myasthenic syndrome caused by decreased agonist binding affinity due to a mutation in the acetylcholine receptor ε subunit. Neuron 17: 157–170.

Oldani A, Zucconi M, Ferini-Strambi L, Bizzozero D, Smirne S (1996): Autosomal dominant nocturnal frontal lobe epilepsy: electroclinical picture. Epilepsia 37: 964–976.

Pan A-H, et al. (1997): Agonist-induced closure of constitutively open γ-aminobutyric acid channels with mutated M2 domains. Proc. Natl. Acad. Sci. USA 94: 6490–6495.

Parkinson D, et al. (1988): Evidence for a nicotinic component to the actions of acetylcholine in cat visual cortex. Exp. Brain Res. 73: 553–568.

Paz C, et al. (1985): Amygdala kindling in chronically cerebellectomized cats. Exo. Neurol. 88: 418–424.

Phillips HA, Scheffer IE, Berkovic SF, Hollway GE, Sutherland GR, Mulley JC (1995): Localization of a gene for autosomal dominant nocturnal frontal lobe epilepsy to chromosome 20q13.2. Nature Genet. 10: 117–118.

Pidoplichko VI, et al. (1997): Nicotine activates and desensitizes midbrain dopamine neurons. Nature 390: 401–404.

Rapier C, et al. (1988): Stereoselective nicotine induced release of dopamine from striatal synaptosomes: concentration dependence and repetitive stimulation. J. Neurochem. 50: 1123–1130.

Revah F, et al. (1991): Mutations in the channel domain alter desensitization of a neuronal nicotinic receptor. Nature 353: 846–849.

Role LW, Berg DK (1996): Nicotinic receptors in the development and modulation of CNS synapses. Neuron 16: 1077–1085.

Rowell PP, Winkler DL (1984): Nicotinic stimulation of [^{3}H]acetylcholine release from mouse cerebral cortical synaptosomes. J. Neurochem. 43: 1593–1598.

Rubboli F, et al. (1994): Distribution of nicotinic receptors in the human hippocampus and thalamus. Eur. J. Neurosci. 6: 1596–1604.

Rye DB, et al. (1987): Cortical projections arising from the basal forebrain: a study of cholinergic and noncholinergic components employing combined retrograde tracing and immunohistochemical localization of choline acetyltransferase. Neuroscience 13: 627–643.

Sakmann B, et al. (1985): Role of acetylcholine receptor subunits in gating of the channel. Nature 318: 538–543.

Scheffer IE, Bhatia KP, Lopes-Cendes I, et al. (1994): Autosomal dominant frontal epilepsy misdiagnosed as sleep disorder. Lancet 343: 515–517.

Scheffer IE, Bhatia KP, Lopes-Cendes I, Fish DR, Marsden CD, Andermann E, Andermann F, Desbiens R, Keene D, Cendes F, Manson JI, Constantinou J, McIntosh A, Berkovic SF (1995): Autosomal dominant nocturnal frontal epilepsy: a distinctive clinical disorder. Brain 118: 61–73.

Schröder H (1992): Immunohistochemistry of cholinergic receptors. Anta. Embryol. Berl. 186(5): 407–429.

Séguéla P, et al. (1993): Molecular cloning, functional properties, and distribution of rat brain alpha 7: a nicotinic cation channel highly permeable to calcium. J. Neurosci. 13(2): 596–604.

Snead OCI (1995): Basic mechanisms of generalized absence seizures. Ann. Neurol. 37: 146–157.

Steinlein OK, et al. (1995): A missense mutation in the neuronal nicotinic acetylcholine receptor alpha 4 subunit is associated with autosomal dominant nocturnal frontal lobe epilepsy. Nat. Genet. 11(2): 201–203.

Steinlein OK, Magnusson A, Stoodt J, Bertrand S, Weiland S, Berkovic SF, Nakken K, Propping P, Bertrand D (1997): An insertion mutation of the CHRNA4 gene in a family with autosomal dominant nocturnal frontal lobe epilepsy. Hum. Mol. Genet. 6: 943–948.

Stériade M, et al. (1988): Projections of cholinergic and non-cholinergic neurons of the brainstem core to relay and associational thalamic nuclei in the cat and macaque monkey. Neuroscience 25: 47–67.

Stériade M, et al. (1993): Thalamocortical oscillations in the sleeping and aroused brain. Science 262(5134): 670–685.

Sugaya K, et al. (1990): Nicotinic acetylcholine receptor subtypes in human frontal cortex: changes in Alzheimer's disease. J. Neurosci. Res. 27: 349–359.

Swanson LW, et al. (1987): Immunohistochemical localization of neuronal nicotinic receptors in the rodent central nervous system. J. Neurosci. 7: 3334–3342.

Szymusiak, R (1995): Magnocellular nuclei of the basal forebrain: substrates of sleep and arousal regulation. Sleep 18(6): 478–500.

Unwin N (1995): Acetylcholine receptor channel imaged in the open state. Nature 373: 37–43.

Vidal C, Changeux J-P (1989): Pharmacological profile of nicotinic acetylcholine receptors in the rat prefrontal cortex: an electrophysiological study in slice preparation. Neuroscience 29: 261–270.

von Krosigk M, et al. (1993): Cellular mechanisms of a synchronized oscillation in the thalamus. Science 261(5119): 361–364.

Wada E, et al. (1989): Distribution of alpha2, alpha3, alpha4, and beta2 neuronal nicotinic receptor subunit mRNAs in the central nervous system: a hybridization histochemical study in the rat. J. Comp. Neurol. 284: 314–335.

Wada E, et al. (1990): The distribution of mRNA encoded by a new member of the neuronal nicotinic acetylcholine receptor gene family (alpha5) in the rat central nervous system. Brain Res. 526: 45–53.

Weiland S, Steinlein O (1996): Dinucleotide polymorphism in the first intron of the human neuronal nicotinic acetylcholine receptor $\alpha 4$ subunit gene (CHRNA4). Clin. Genet. 50: 433–434.

Weiland S, et al. (1997): An amino acid exchange in the second transmembrane segment of a neuronal nicotinic receptor causes partial epilepsy by altering its desensitization kinetics. FEBS Lett. 398: 91–96.

Wevers A, et al. (1994): Cellular distribution of nicotinic acetylcholine receptor subunit mRNAs in the human cerebral cortex as revealed by non-isotopic in situ hybridization. Brain Res. Mol. Brain Res. 25(1–2): 122–128.

Whiting P, Lindstrom J (1986): Pharmacological properties of immunoisolated neuronal nicotinic receptors. J. Neurosci. 6: 3061–3069.

Wonnacott S. (1997): Presynaptic nicotinic ACh receptors. Trends Neurosci. 20(2): 92–98.

Yakel JL, et al. (1993): Single amino acid substitution affects desensitization of the 5-hydroxytryptamine type 3 receptor expressed in *Xenopus* oocytes. Proc. Natl. Acad. Sci. USA 90(11): 5030–5033.

Youssoufian H, et al. (1988): Nonsense and missense mutations in hemophilia A: estimate of the relative mutation rate at CG dinucleotides. Am. J. Hum. Genet. 42: 718.

Zheng JQ, et al. (1994): Turning of nerve growth cones induced by neurotransmitters. Nature 368(6467): 140–144.

17

The Role of Nicotine and Nicotinic Receptors in Psychopathology

Sherry Leonard, PhD, Lawrence E. Adler, MD, Ann Olincy, MD, Charles R. Breese, PhD, Judith Gault, PhD, Randal G. Ross, MD, Michael Lee, BS, Ellen Cawthra, RN, MPA, Herbert T. Nagamoto, MD, and Robert Freedman, MD

University of Colorado Health Science Center
Denver, Colorado

The role of nicotine in psychopathology is a relatively new interest for neuroscientists but may not be news to those in our population who suffer from mental illness, as many are heavy users of tobacco products. This review focuses on schizophrenia, with some comments on nicotine use in bipolar disorder and major depression. These diseases constitute, in large part, the principal psychotic diagnoses that require chronic hospitalization in the United States and throughout the world (Roberts, 1990; Capenter and Buchanan, 1994). Additionally, they usually present in early adulthood and, with poor prognoses, represent the largest burden on families and society at large (Roberts, 1990; Rugg and Keith, 1993; Carpenter and Buchanan, 1994).

SYMPTOMS AND PREVALENCE OF SCHIZOPHRENIA AND AFFECTIVE DISORDERS

The characteristic symptoms of schizophrenia include delusions, hallucinations (most often these are auditory), disorganized thinking and speech, and a flattening of affect and volition

Neuronal Nicotinic Receptors: Pharmacology and Therapeutic Opportunities, Edited by S. P. Arneric and J. D. Brioni
ISBN 0-471-24743-x, pages 307–322. Copyright © 1998 by Wiley-Liss, Inc.

(DSM-IV). In bipolar disorder, subjects suffer from cycling mania with elevated mood and hyperactivity, and may also have periods of major depression. In contrast, unipolar depression includes prolonged periods of depressive mood, lack of pleasure, insomnia, and thoughts of suicide. Although schizophrenia, bipolar disorder, and major depression can present at any age, they are most often diagnosed in early adulthood (DSM-IV; Carpenter and Buchanan, 1994; Ridley and Baker, 1990). Ethnically, schizophrenia exists uniformly throughout the world, although there are isolated populations in which higher incidences have been noted (Hammond et al., 1983). Sex biases, although different in each disease, have been recorded for all three disorders. There is a higher incidence of both bipolar disorder and major depression in females than in males, the lifetime risk for major depression being 10% to 25% for women and 5% to 12% for men (DSM-IV). Schizophrenia, however, appears to present earlier in males and with more severe symptoms, and has a lifetime risk of 1.3% in the general population (DSM-IV).

All of these diseases have a presumed genetic component, where higher risk occurs for individuals who come from families with another member who is ill (Kendler and Davis, 1981; Kendler, 1988; Straube and Oades, 1992; Kendler et al., 1983). There is also a high concordance in monozygotic twins (McGue and Gottesman, 1989; Goldberg et al., 1990). The strong likelihood of familial transmission suggests that these disorders can be studied at both the genetic and biochemical level.

USE OF TOBACCO PRODUCTS IN SCHIZOPHRENIA AND RELATED DISORDERS

A commonality in schizophrenia, bipolar disorder, and major depression is the high incidence of smoking (Masterson and O'Shea, 1984; Hughes et al., 1986; Goff et al., 1992). In the general population of the United States today, smokers account for approximately 30% and this fraction is declining (Hughes et al., 1986). The number of subjects suffering from schizophrenia, manic depression, or major depression constitutes more than 10% of the population (DSM-IV). Of the total number of smokers in the United States, then, more than 25% are mentally ill, and this fraction is likely to increase as smoking among the non–mentally ill decreases (Glassman, 1993).

Distribution of smoking histories in subjects from whom postmortem brain tissue or lymphocytes were collected in the Denver Schizophrenia Center is shown in Figure 17-1. Schizophrenics have the highest incidence of smoking, approaching 80% of those affected. This group often smokes high-tar cigarettes and sometimes uses multiple forms of tobacco (Hughes et al., 1986). They have also been found to extract more nicotine from cigarettes compared to normal smokers (Olincy et al., 1997b). Smoking cessation programs are less successful among all of these groups than in the normal population and, indeed, psychiatric wards are finding it necessary to provide a smoking room in nonsmoking hospitals (Glassman, 1993). Of these three disorders, the role of nicotine has been most fully characterized in schizophrenia (Freedman et al., 1994; Leonard et al., 1996).

SENSORY PROCESSING DEFICITS IN SCHIZOPHRENIA

Schizophrenics are known to have deficits in the processing of sensory information, several of which can be assayed in both humans and laboratory animals. These patients have difficulty with inhibitory mechanisms regulating extraneous visual and auditory stimuli. Two measures that have been used to characterize these deficits are smooth pursuit eye movement (Holzman,

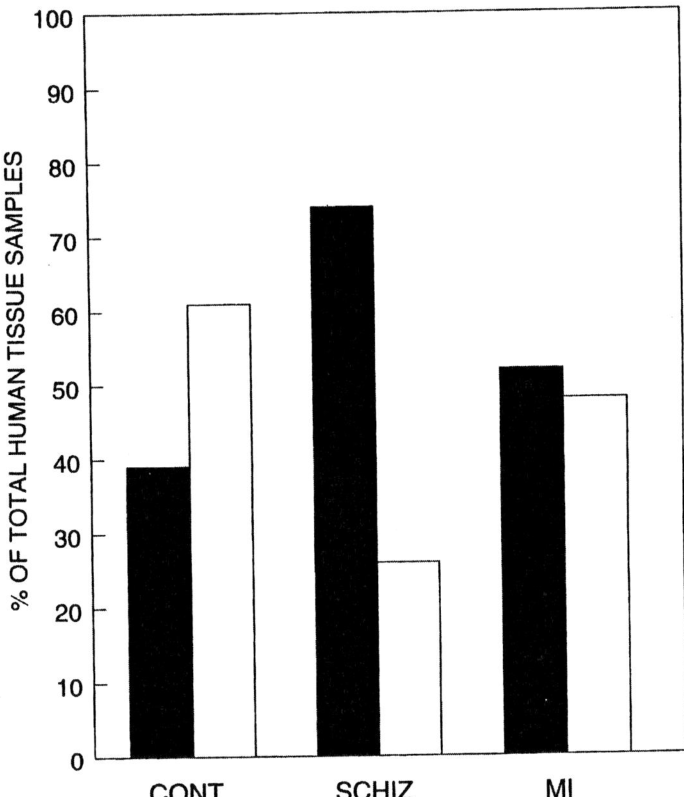

Figure 17.1. Group prevalence of smoking in subjects used for neuronal nicotinic receptor studies. Samples for studies on neuronal nicotinic receptors include postmortem brain (43 controls with no history of mental illness, 44 schizophrenics, and 24 bipolar and depressive subjects) and immortalized lymphocytes (73 controls, 16 schizophrenics, and 43 bipolar or depressive subjects). The percent of total tissue samples (smokers, filled bars; nonsmokers, open bars) for each group is shown. CONT, control subjects; SCHIZ, schizophrenics; MI, other mentally ill subjects.

1983; Friedman et al., 1991; Radant and Hommer, 1992; Clementz and Sweeney, 1990) and gating of auditory-evoked potentials (Baker et al., 1987; Freedman et al., 1987, 1991).

Tracking of eye movement is measured using an infrared photoelectrode limbus detection eye-tracking device in a smooth pursuit task (Radant and Hommer, 1992). Auditory gating is assessed using an auditory-evoked potential combined with a conditioning/testing paradigm. Electrodes, placed on the skull surface, record a wave with a 50 ms latency (P50) following paired auditory stimuli delivered 0.5 s apart (Freedman et al., 1991). Normal controls decrement the amplitude of the second of the two paired-click stimuli through the action of an inhibitory neuronal pathway, activated in response to the first stimulus.

For both of these measures most schizophrenic subjects have uniformly abnormal responses. Abnormal auditory gating and eye tracking are found in approximately 10% of the normal population (Holzman et al., 1973, 1988; Waldo et al., 1991). However, both abnormal eye tracking and the P50 deficit are inherited in families of schizophrenics in an apparent autosomal-dominant pattern (Holzman et al., 1984; 1988; Grove et al., 1992; Freedman et al., 1987, 1996). Thus, 50% of family members have aberrant gating of the P50 auditory-evoked potential and abnormal smooth pursuit eye movement whether or not they also have the disease. This suggests that both of these deficits represent factors that predispose to schizophrenia. Further, since they are more penetrant than the disease itself, they provide assayable traits for which the genetics and biochemistry can be studied directly.

Subjects with bipolar disorder also have abnormal auditory gating during their manic phases, although gating normalizes when they are euthymic (Baker et al., 1987). Eye tracking abnormalities in mania also occur but may be due to medications such as lithium, used to treat the illness, and are not found in family members of mania patients (Holzman et al., 19873: Iacono and Koenig, 1983; Levy et al., 1983; Abel et al., 1991). A further difference between loss of gating in bipolar mania and schizophrenia is that the typical neuroleptic haloperidol normalizes the gating deficit in bipolar mania but does not normalize gating in schizophrenia (Baker et al., 1987). Typical neuroleptics also have no effect on smooth pursuit eye tracking (Levy et al., 1983; Litman et al., 1994).

NORMALIZATION OF SENSORY DEFICITS BY NICOTINE

Deficits in both smooth pursuit eye movement (Olincy et al., 1997a) and in gating of the P50 auditory-evoked potential (Adler et al., 1992, 1993) are normalized by nicotine in schizophrenics. The effect of nicotine on the gating deficit in a single schizophrenic subject is shown in Figure 17-2. He was originally on molindone but was changed to perphenazine when he did not respond well to treatment. However, he was not able to tolerate conventional neuroleptics due to severe extrapyramidal side effects and clozapine was eventually prescribed. The subject was a smoker and was asked to come in for recording before his first cigarettes of the day. The subject had a test-to-conditioning ratio (T/C) of 133%, indicating failure to inhibit the second stimulus. He then smoked several cigarettes and was recorded again, where the normalization of the test response can be seen. Clozapine, prescribed for this nonresponder to typical neuroleptics, also normalized his gating deficit (Figure 17-2), as has been found in other non-gating subjects (Nagamoto et al., 1996).

Normalization of the smooth pursuit eye movement deficit in a schizophrenic by nicotine is shown in Figure 17-3. A schizophrenic smoker and nonschizophrenic smoker, after abstaining from cigarettes for at least 2 hours, performed a baseline constant velocity, trapezoidal, smooth pursuit task. At baseline, the schizophrenic showed an increase in leading saccades compared to the nonsmoker. The leading saccades were significantly decreased in the patient after smoking, while smoking had no effect on the nonschizophrenic subject. Clozapine does not normalize leading saccades in schizophrenic subjects (Litman et al., 1994).

Normalization of both the P50 gating deficit and abnormal eye tracking by nicotine suggests that nicotinic receptors may play a role in neuronal pathways regulating the function of these sensorimoter processing mechanisms in the brain. However, differential effects of the atypical neuroleptic, clozapine (normalization of the P50 deficit; no effect on eye tracking abnormalities), indicate that the neuronal circuits controlling these phenotypes are not likely to be identical.

It is important to note that cigarette smoking is known to increase the metabolism of typical neuroleptics, such as haloperidol (Jann et al., 1986; Miller et al., 1990), leading to increased smoking in schizophrenic patients using these medications (McEvoy et al., 1995a). The atypical neuroleptic, clozapine, actually seems to decrease the smoking levels in patients (McEvoy

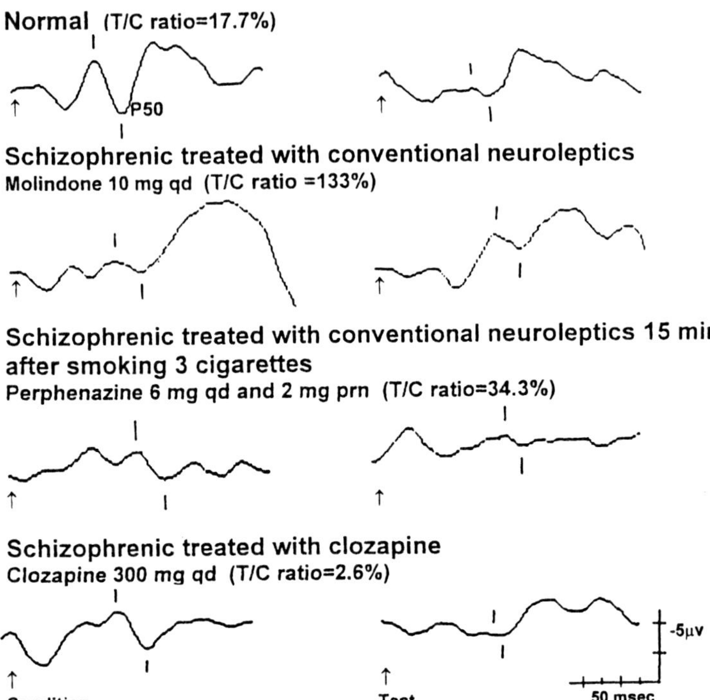

Figure 17.2. Effects of nicotine and clozapine on gating of the P50 deficit in a single subject. Evoked potentials in a conditioning testing paradigm, with stimuli delivered 0.5 s apart. The normal (top) shows suppression of the test P50 wave. The percentage ratios of test-to-conditioning amplitude are shown in parentheses. The same schizophrenic patient was recorded during conventional neuroleptic treatment, during conventional neuroleptic treatment after nicotine administration via cigarette smoking, and during clozapine treatment. The patient's lack of suppression of the test P50 response was not normalized by conventional neuroleptics but was normalized by nicotine and clozapine. The patient was treated with more than one conventional neuroleptic because of lack of efficacy. The P50 waveform is indicated by tick marks. Positive polarity is down.

et al., 1995b). It has been suggested that clozapine blockade of 5HT3 receptors might lead to increased release of acetylcholine (Ashby et al., 1989; Barnes et al., 1989; Wang et al., 1994) and, further, that this release of acetylcholine may account for normalization of the P50 deficit through stimulation of nicotinic receptors (Nagamoto et al., 1996).

Rodents also have the capacity to filter out extraneous auditory information, measured by gating of a similar wave, the N40 (Adler et al., 1986). This animal model provides the opportunity to study the pharmacological aspects of the gating pathway, using invasive techniques. The effects of cholinergic agonists and antagonists on gating of auditory stimuli have been extensively studied in both anesthetized and awake-behaving rat paradigms (Adler et al., 1986, 1988; Bickford-Wimer et al., 1990). It was found that antagonists of the α7 nicotinic receptor (α-bungarotoxin, methyllycaconitine, and antisense oligonuclèotides complementary to the α7 translation start site) efficiently blocked the capacity to gate or filter the N40 wave, an animal model for the auditory gating deficit in schizophrenia (Luntz-Leybman et al., 1992; Rollins et al., 1993, 1996). Neither mecamylamine nor dihydro-β-erythroidin, antagonists of the high-affinity and ganglionic types of nicotinic receptors, had any effect (Luntz-Leybman et al., 1992;

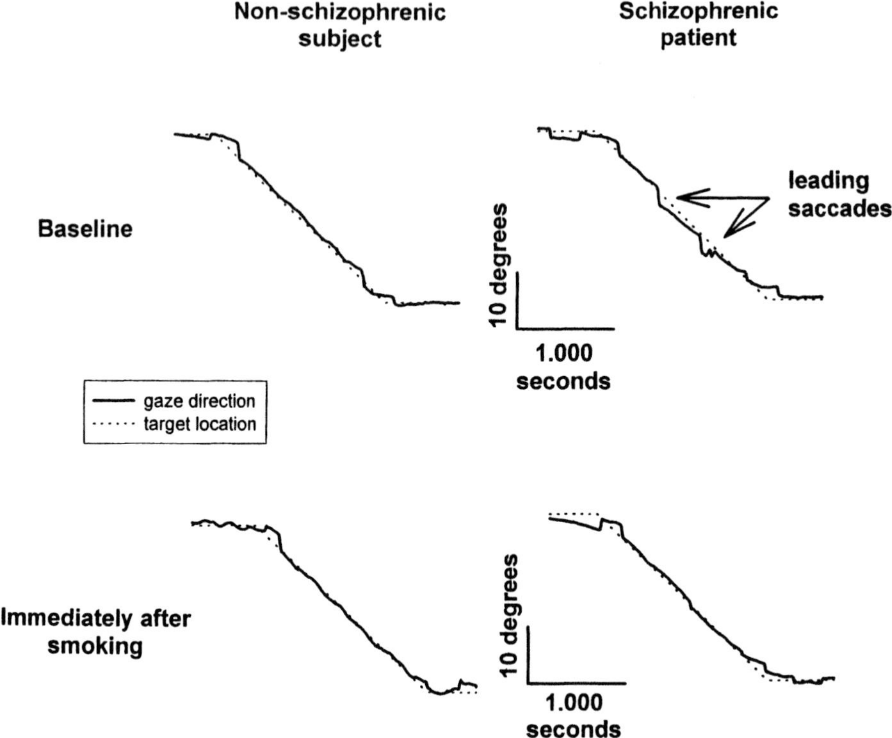

Figure 17.3. Normalization of smooth pursuit eye movements in a schizophrenic by smoking. Example of a portion of a trapezoidal smooth pursuit task with a constant velocity of 16.7°/second in a schizophrenic patient and nonschizophrenic subject. Before smoking, the schizophrenic subject has 3.6% of the total eye movements due to leading saccades, where the eye jumps ahead of the target and then the subject has to wait for the target to catch up. After smoking 2 cigarettes, the schizophrenic subject decreased the percentage of eye movements due to leading saccades to 0.6%, with eye velocity matching target velocity. The nonschizophrenic subject accurately tracks the target before smoking, with 0.5% of total eye movements due to leading saccades. After smoking 1.5 cigarettes, the percentage of eye movements due to leading saccades is virtually unchanged at 0.7%.

Rollins et al., 1996). Scopalamine, a muscarinic blocker, likewise had no effect on auditory gating in the rat (Luntz-Leybman et al., 1992). This pharmacological evidence suggests a role for the α7 neuronal nicotinic receptor in the processing of auditory information.

A mouse model for the auditory gating deficit has been found in the DBA/2 line. These mice have a schizophrenia-like deficit in the inhibition of auditory-evoked potentials compared to other mouse strains such as C3H (Stevens et al., 1996). The gating deficit in these animals is normalized by nicotine and also by a partial agonist of the α7 nicotinic receptor, GTS-21 (Stevens et al., 1997). Interestingly, the DBA/2 mouse line also exhibits a 40% decrease in levels of α-bungarotoxin binding receptors in the brain compared to other strains (Marks et al., 1986, 1989). These results further support the hypothesis that the α7 neuronal nicotinic receptor lies in the inhibitory pathway regulating the gating of auditory stimuli.

EXPRESSION OF THE α7 NICOTINIC RECEPTOR SUBUNIT IN HUMAN TISSUES

Regional studies of the localization of α7-containing receptors indicate a wide but relatively low level of distribution in both rat and human brain. A recent study comparing α7 mRNA and α-bungarotoxin binding at the cellular level in human brain showed that α7 is most highly expressed in nuclei involved in processing of sensory information, such as the hippocampus, the lateral and medial geniculates, and the reticular nucleus of the thalamus (RTN; Breese et al., 1997a). The RTN, a capsular structure surrounding the thalamus, provides the principal inhibitory input into the dorsal area of the thalamus (Steriade et al., 1987; Mitrofanis and Guillery, 1993; Lozsádi, 1994; Gonzalo-Ruiz and Lieberman, 1995).

Localization of α7 receptors appears to be mainly on nonprincipal cells, where labeling with α-bungarotoxin was often seen on dendrites as well as on the cell somata. Expression of α7 mRNA and receptor binding in human hippocampus and RTN is shown in Figure 17-4. In the RTN apparent α7 receptors are present on cell somata on a dense network of intertwined dendritic fibers. In the hippocampus, binding was mainly seen on cell bodies and dendrites of several types of interneurons, as described by Amaral (1978). The RTN in the rat is much less heavily labeled by α-bungarotoxin. Another difference in rat and humans appears to be in the CA3c region of the hippocampus. In the rat, a dense and diffuse binding of αBTX is seen, which is suggestive of an axonal localization (Breese et al., 1997a). In human hippocampus, the CA3c binding was not observed, although postmortem effects could not be ruled out.

The α7-containing receptor is also found in the periphery, having been extensively studied in ganglionic tissue from the chick (McGehee and Role, 1995). The receptor has been found

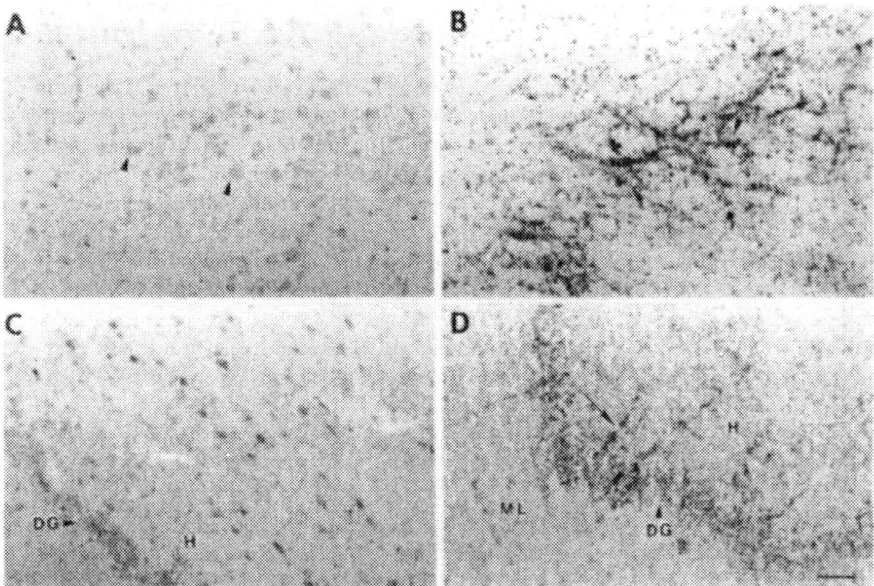

*Figure 17.4 Expression of α7 mRNA and [^{125}I]-α-bungarotoxin binding in human reticular thalamic nucleus and hippocampus. In situ hybridization with an α7-specific cRNA probe and [^{125}I]-α-bungarotoxin autoradiography in postmortem brain of a control subject, as described in Breese et al., 1997a (**A** and **B**): reticular thalamic nucleus;(**C** and **D**): hilus of the dentate gyrus.*

in lung tumor cells (Quik et al., 1994) and a recent study in tissues of neural crest origin shows that α7 receptors are expressed in almost all cells derived from these progenitors (Breese et al., 1997b). The latter results suggest that α7 may have nonsynaptic roles in the periphery, possibly involving peptide secretion.

COMPARISON OF EXPRESSION OF LOW- AND HIGH-AFFINITY NICOTINIC RECEPTORS IN POSTMORTEM BRAIN OF SCHIZOPHRENIC AND CONTROL SUBJECTS

Expression of both high- and low-affinity nicotinic receptors has been studied in postmortem brain from both schizophrenics and control subjects. We found that α-bungarotoxin binding was decreased by about 40% in the brains of subjects who were schizophrenic in life compared to subjects who were not mentally ill (Freedman et al., 1995). In this study, we carefully matched the tissue for age, sex, cause of death, postmortem interval, and smoking history. Most of the schizophrenics were heavy smokers and were matched with like controls. This is the same decrease in α7 receptor expression that was found to result in a loss of auditory gating in the rat and mouse models (Rollin et al., 1993; Stevens et al., 1996).

High-affinity nicotinic receptors appear to be affected in schizophrenia as well. We studied the upregulation of nicotine binding in postmortem brain of smokers and nonsmokers. In normal subjects, the number of high-affinity nicotinic receptors in both hippocampus and thalamus is correlated with the number of cigarettes smoked per day (Breese et al., 1997c). In smokers who had quit for varying periods before death, we found receptor levels in the nonsmoking range. However, when we examined the effect of smoking history on high-affinity receptor levels in postmortem brain of schizophrenics, we found that the schizophrenics had fewer receptors than comparable normal smokers (Breese et al., 1997) (Fig. 17-5, top). In nonsmoking schizophrenic subjects, there was no significant difference in receptor numbers from those of controls, but at every smoking level the schizophrenics had lower levels (Fig. 17-5, bottom). We have also examined regulation of high-affinity receptors in a few subjects who suffered from either bipolar disorder or major depression in life. Preliminary results suggest that their receptor numbers do not differ significantly from the control subjects when compared for smoking history. Thus, the failure to upregulate either low- or high-affinity receptors in response to heavy smoking may be specific for the schizophrenic samples.

To investigate any effects of typical neuroleptics on regulation of nicotinic receptor numbers, we studied [^{3}H]-nicotine binding in rats treated with either nicotine, haloperidol, or haloperidol and nicotine together. As can be seen in Figure 17-6, the typical neuroleptic haloperidol does not appear to affect the normal upregulation of nicotinic receptor numbers seen with nicotine dosing in the rat. As the postmortem brain tissue used in our experiments was from sub-

Figure 17.5 Scatterplots and group means of single-point [^{3}H]-nicotine binding data for the hippocampus (top) and correlational analysis of [^{3}H]-nicotine binding with the number of packs smoked per day at the time of death (bottom). (Top): In control smokers, a significant difference was found in mean [^{3}H]-nicotine binding between nonsmokers, lifelong smokers, and smokers who had quit (, $p < 0.001$). In schizophrenic subjects, there was no significant difference between any of the groups (all $p < 0.45$), although there was a significant reduction in the nicotinic receptor numbers in schizophrenic smokers compared to control smokers (**, $p < 0.02$). (Bottom): In control subjects (black line) receptor number showed an increase with smoking history (hippocampus: $r = 0.65$, $p < 0.001$), indicating a dose-dependent increase in [^{3}H]-nicotine binding levels in the brains of human smokers. While schizophrenic subjects demonstrated a dose-dependent increase in hippocampal [^{3}H]-nicotine binding ($r = 0.516$, $p < 0.02$), the slope of the regression line was decreased to 43% of the rate of normal smokers (normals, $m = 6.58$; schizophrenics, $m = 3.77$).*

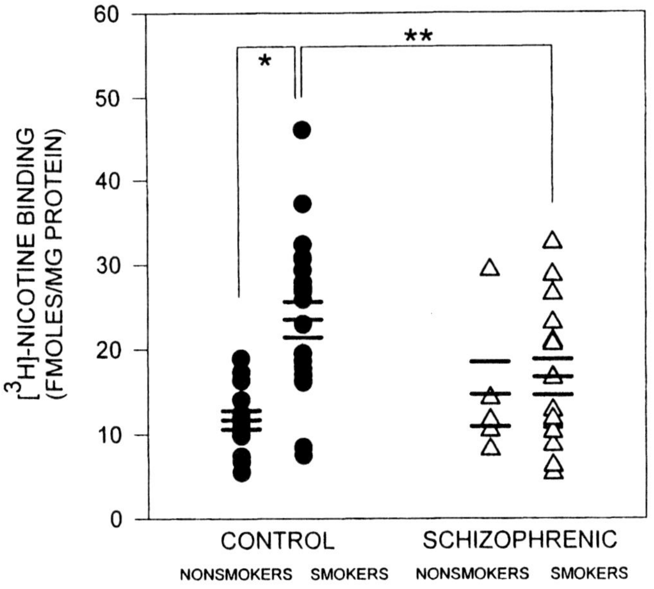

60
50
40
30
20
10
0
*
**
[3H]-NICOTINE BINDING
(FMOLES/MG PROTEIN)
CONTROL
SCHIZOPHRENIC
NONSMOKERS
SMOKERS
NONSMOKERS
SMOKERS

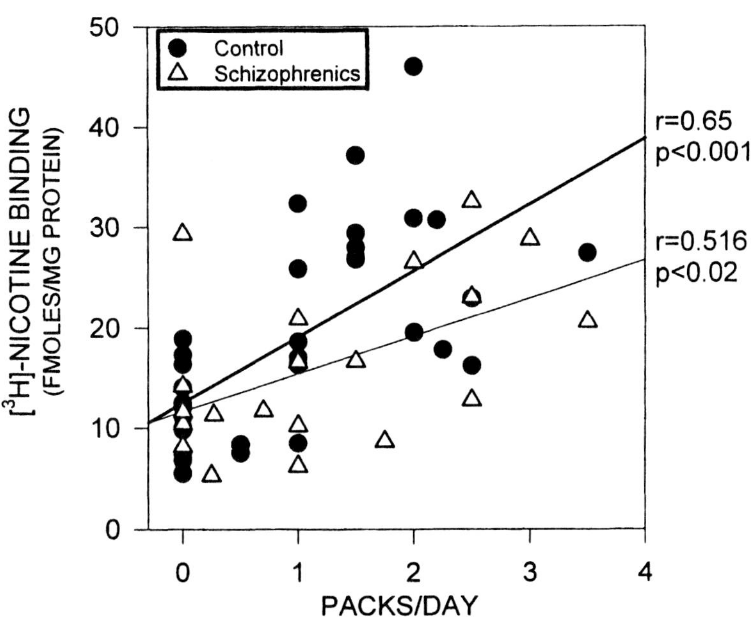

Control
Schizophrenics
r=0.65
p<0.001
r=0.516
p<0.02
[3H]-NICOTINE BINDING
(FMOLES/MG PROTEIN)
PACKS/DAY
0
1
2
3
4
10
20
30
40
50

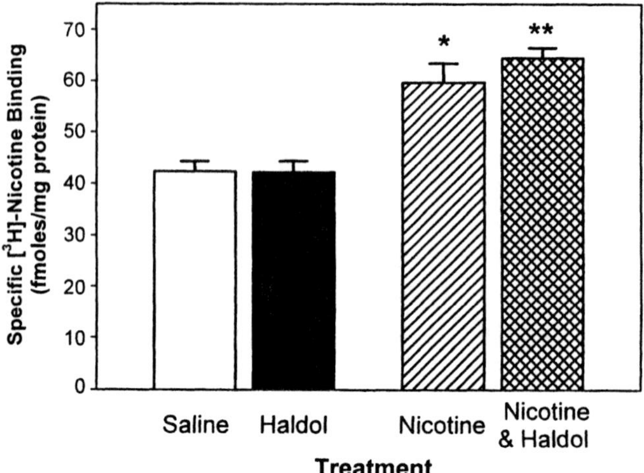

Figure 17.6. *Effect of chronic nicotine, haloperidol, and haloperidol and nicotine treatment on [^{3}H]-nicotine receptor binding in the rat cortex. Rats treated with chronic haloperidol showed no difference from saline-treated control animals. Chronic nicotine treatment significantly increased nicotinic receptor binding in the rat cortex (*, $p < 0.05$). Chronic haloperidol and nicotine cotreatment also induced an increase in [^{3}H]-nicotine receptor binding in rat cortex similar to that seen for nicotine treatment alone (**, $p < 0.05$). These data would suggest that the changes observed in schizophrenics are not related to chronic neuroleptic treatment.*

jects who had taken only conventional neuroleptics in life, our results suggest that regulation of receptor levels in schizophrenics may, indeed, be aberrant. Additionally, since nicotine addiction is thought to be partially characterized by nicotinic receptor number (Breese et al., 1997c), the low levels of receptor in schizophrenics who are heavy smokers suggest that their addiction level and/or need for nicotine may be different than in normal smoking subjects.

GENETIC EVIDENCE FOR INVOLVEMENT OF THE α7 NICOTINIC RECEPTOR IN THE P50 DEFICIT IN SCHIZOPHRENIA

Genetic studies in schizophrenic pedigrees also support an involvement of the α7 nicotinic receptor in schizophrenia. The α7 nicotinic receptor subunit maps to human chromosome 15q14 (Chini et al., 1994; Orr-Urtreger et al., 1995). Nine pedigrees were ascertained with apparent familial schizophrenia and gating of the P50 auditory-evoked potential was recorded in all family members. A full genome scan was performed using more than 500 polymorphic markers at a 10 cM resolution, including a polymorphic dinucleotide repeat D15S1360 that lies within 120 kb of the α7 gene. Linkage was found to the D15S1360 marker with a lod of 5.3 at zero recombination, $p < 0.001$ (Freedman et al., 1996). This means there are 200,000 to 1 odds that the chromosomal locus of the α7 gene is genetically linked to the P50 deficit in the nine schizophrenic pedigrees studied. We have recently repeated linkage analysis in a second schizophrenic cohort ascertained through the NIMH Schizophrenia Genetics Initiative, in which we used schizophrenia as the phenotype instead of the P50 deficit. We also found linkage of schizophrenia to the α7 marker D15S1360, with a $p < 0.005$ (Leonard et al., 1997b). The genetic studies complement the biological data supporting the involvement of the α7 receptor in a schizophrenic endophenotype.

ISOLATION OF GENOMIC CLONES FOR THE $\alpha7$ NICOTINIC RECEPTOR SUBUNIT

We have recently isolated genomic clones for the human $\alpha7$ receptor subunit. The gene is similar in structure to that of the chick (Couturier et al., 1990), having 10 exons and exon/intron borders in the same locations. However, we have found that the human gene is partially duplicated. Exons 5 to 10 have been duplicated, with intervening introns, and lie proximal to the full-length $\alpha7$ gene at chromosome 15q14 (Leonard et al., 1997a; Gault et al., 1998). Although the DNA sequence of the duplicated exons is the same as in the full-length gene, we do not yet know if the introns are duplicated in their entirety. Additionally, there are novel exons located 5′ of the duplicated exon 5. These novel exons are expressed with the duplicated sequences (exons 5–10) as mRNA. The function of these duplicated and expressed $\alpha7$ sequences is not yet known, but alternative transcripts in at least two other mamalian nicotinic receptor subunits have been shown to have a regulatory function (Garcia-Guzman et al., 1995; Newland et al., 1995). The duplication of the $\alpha7$ gene and expression of these sequences will most certainly complicate mutation screening. Not only would mutations have to be localized to the full-length or partially duplicated gene but expression of each would also have to be quantified.

CONCLUSIONS

The body of data summarized here provides support for the role of nicotinic receptors in schizophrenia. At the present time, evidence is strongest for involvement of $\alpha7$ subunit–containing receptors. However, the finding that schizophrenics do not increase receptor numbers of either the low- or high-affinity nicotine binding subtypes with nicotine dose suggests that the regulation and function of nicotinic receptors in human brain may be more complicated than previously thought. Schizophrenia is a complex disease, with a multifactorial etiology. Indeed, genetic linkage has not only been found at 15q14 but at multiple other chromosomal loci, including chromosomes 6p (Straub et al., 1995), 8p (Kendler et al., 1996; Pulver et al., 1995), and 22q (Gill et al., 1996). One of the high-affinity nicotinic receptor subunits, $\alpha2$, maps to the linkage region on chromosome 8p21–25 (Wood et al., 1995; Anand and Lindstrom J., 1990; Kendler et al., 1996; Pulver et al., 1995). Whether this linkage will prove to be as informative as was the linkage to the $\alpha7$ subunit remains to be determined.

Additionally, we know little at present about the role of nicotinic receptors in the other two principal psychotic diseases. The incidence of smoking is also high in bipolar disorder and major depression. Certainly continued efforts will be made to understand the biological meaning of nicotine use in all of these diseases and to direct treatments based on this contribution.

REFERENCES

Abel LA, Friedman L, Jesberger J, Malki A, Meltzer HY (1991): Quantitative assessment of smooth pursuit gain and catch-up saccades in schizophrenia and effective disorders. Biol. Psych. 29: 1063–1072.

Adler LE, Rose G, Freedman R (1986): Neurophysiological studies of sensory gating in rats: Effects of amphetamine, phencyclidine, and haloperidol. Biol. Psych. 21: 787–798.

Adler LE, Pang K, Gerhardt G, Rose GM (1988): Modulation of the gating of auditory evoked potentials by norepinephrine: Pharmacological evidence obtained using a selective neurotoxin. Biol. Psych. 24: 179–190.

Adler LE, Hoffer LJ, Griffith J, Waldo MC, Freedman R (1992): Normalization by nicotine of deficient auditory sensory gating in the relatives of schizophrenics. Biol. Psych. 32: 607–616.

Adler LE, Hoffer LD, Wiser A, Freedman R (1993): Normalization of auditory physiology by cigarette smoking in schizophrenic patients. Am. J. Psych. 150: 1856–1861.

Amaral DG (1978): A Golgi study of cell types in the hilar region of the hippocampus in the rat. J. Comp. Neurol. 183: 851–914.

Anand R, Lindstrom J (1990): Nucleotide sequence of the human nicotinic acetylcholine receptor α2 subunit gene. Nucleic Acids Res. 18: 4272.

Ashby CRJ, Edwards E, Harkins KL, Wang RY (1989): Differential effect of typical and atypical antipsychotic drugs on the suppressant action of 2-methylserotonin on medical prefrontal cortical cells: a microiontophoretic study. Eur. J. Pharm. 166: 583–584.

Baker N, Adler LE, Franks RD, Waldo MC, Berry S, Nagamoto H, Muckle A, Freedman R (1987): Neurophysiological assessment of sensory gating in psychiatric inpatients: comparison between schizophrenia and other diagnoses. Biol. Psych. 22: 603–617.

Barnes JM, Barnes NM, Costall B, Naylor, RJ, Tyers MB (1989): 5-HT3 receptors mediate inhibition of acetylcholine release in cortical tissue. Nature 338: 762–763.

Bickford-Wimer PC, Nagamoto H, Johnson R, Adler L, Egan M, Rose GM, Freedman R (1990): Auditory sensory gating in hippocampal neurons: A model system in the rat. Biol. Psych. 27: 183–192.

Breese CR, Adams C, Logel J, Drebing C, Rollins Y, Barnhart M, Sullivan B, DeMasters BK, Freedman R, Leonard S (1997a): Comparison of the regional expression of nicotinic acetylcholine receptor α7 mRNA and [^{125}I]-α-bungarotoxin binding in human postmortem brain. J. Comp. Neurol. 387: 385–398.

Breese CR, Gillen KM, Adams CE, Leonard S (1997b): Expression of high and low affinity neuronal nicotinic receptors in tissues of neural crest origin. Soc. Neurosci. Abstr. 23: 382.

Breese CR, Marks MJ, Logel J, Adams CE, Sullivan B, Collins AC, Leonard S (1997c): Effect of smoking history on [^{3}H]nicotine binding in human postmortem brain. J. Pharm. Exp. Ther. 282: 7–13.

Breese C, Adams CE, Sullivan B, Logel J, Drebing C, Gillen KM, Marks MJ, Collins AC, Leonard S (1997d): Abnormal regulation of high affinity nicotinic receptor binding in schizophrenics. Soc. Neurosci. Abstr. 23: 2202.

Carpenter WT, Buchanan RW (1994): Schizophrenia. Med. Prog. 330: 681–689.

Chini B, Raimond E, Elgoyhen AB, Moralli D, Balzaretti M, Heinemann S (1994): Molecular cloning and chromosomal localization of the human α7- nicotinic receptor subunit gene (CHRNA7). Genomics 19: 379–381.

Clementz BA, Sweeney JA (1990): Is eye movement dysfunction a biological marker for schizophrenia? A methodological review. Psychol. Bull. 108: 77–92.

Couturier S, Bertrand D, Matter J-M, Hernandez M-C, Bertrand S, Millar N, Valera S, Barkas T, Ballivet M (1990): A neuronal nicotinic acetylcholine receptor subunit α7 is developmentally regulated and forms a homo-oligomeric channel blocked by α-BTX. Neuron 5: 847–856.

DSM-IV, American Psychiatric Association (1994): Diagnostic and statistical manual of mental disorders, 4th ed. Washington, DC: American Psychiatric Association.

Freedman R, Adler LE, Baker N, Waldo M, Mizner G (1987): Candidate for inherited neurobiological dysfunction in schizophrenia. Somat. Cell Mol. Gen. 13: 479–484.

Freedman R, Waldo M, Bickford-Wimer P, Nagamoto H (1991): Elementary neuronal dysfunctions in schizophrenia. Schiz. Res. 4: 233–243.

Freedman R, Adler LE, Bickford P, Byerley W, Coon H, Cullum CM, Griffith JM, Harris JG, Leonard S, Miller C, Myles-Worsley M, Nagamoto HT, Rose G, Waldo MC (1994): Schizophrenia and nicotinic receptors. Harvard Rev. Psych. 2: 179–192.

Freedman R, Hall M, Adler LE, Leonard S (1995): Evidence in postmortem brain tissue for decreased numbers of hippocampal nicotinic receptors in schizophrenia. Biol. Psych. 38: 22–33.

Freedman R, Coon H, Myles-Worsley M, Orr-Urtreger A, Olincy A, Davis A, Polymeropoulos M, Holik J, Hopkins J, Hoff M, Rosenthal J, Waldo MC, Reimherr R, Wender P, Yaw J, Young DA, Breese DR, Adams C, Patterson D, Adler LE, Kruglyak L, Leonard S, Byerley W (1996): Linkage of a neurophysiological deficit in schizophrenia to a chromosome 15 locus. Proc. Natl. Acad. Sci. USA 94: 587–592.

Friedman L, Jesberger JA, Meltzer HY (1991): A model of smooth pursuit performance illustrates the relationship between gain, catch-up saccade rate, and catch-up saccade amplitude in normal controls and patients with schizophrenia. Biol. Psych. 30: 537–556.

Garcia-Guzman M, Sala R, Sala S, Campos-Caro A, Stuhmer W, Gutierrez LM, and Criado M (1995): Alpha-bungarotoxin-sensitive nicotinic receptors on bovine chromaffin cells: molecular cloning, functional expression and alternative splicing of the alpha 7 subunit. Eur. J. Neurosci. 7: 647–655.

Gault J, Berger R, Robinson M, Drebing C, Logel J, Hopkins J, Moore T, Jacobs S, Meriwether J, Choi MJ, Kim EJ, Walton K, Buiting K, Davis A, Breese CR, Freedman R, Leonard S (1998): Structural analysis of the human $\alpha 7$ neuronal nicotinic acetylcholine receptor gene and its partial duplication. Genomics (In press).

Gill M, Vallada H, Collier D, Sham P, Holmans P, Murray R, McGuffin P, Nanko S, Owen M, Antonarakis S, Housman D, Kazazian H, Nestadt G, Pulver AE, Straub RE, MacLean CJ, Walsh D, Kendler KS, DeLisi L, Polymeropoulos M, Coon H, Byerley W, Lofthouse R, Gershon E, Read CM (1996): A combined analysis of D22S278 marker alleles in affected sib-pairs: support for a susceptibility locus for schizophrenia at chromosome 22q12. Schizophrenia Collaborative Linkage Group (Chromosome 22). Am. J. Med. Genet. 67: 40–45.

Glassman AH (1993): Cigarette smoking: implications for psychiatric illness. Am. J. Psych. 150: 546–553.

Goff DC, Henderson DC, Amico E (1992): Cigarette smoking in schizophrenia: relationship to psychopathology and medication side effects. Am. J. Psych. 149: 1189–1194.

Goldberg TE, Ragland JD, Torrey F, Gold JM, Bigelow LB, Weinberger DR (1990): Neuropsychological assessment of monozygotic twins discordant for schizophrenia. Arch. Gen. Psych. 47: 1066–1072.

Gonzalo-Ruiz A, Lieberman AR (1995): GABAergic projections from the thalamic reticular nucleus to the anteroventral and anterodorsal thalamic nuclei of the rat. J. Chem. Neuroanat. 9: 165–174.

Grove WM, Clementz BA, Iacono WG, Katsanis J (1992): Smooth pursuit ocular motor dysfunction in schizophrenia: evidence for a major gene. Am. J. Psych. 149: 1362–1368.

Hammond KW, Kauders FR, MacMurray JP (1983): Schizophrenia in Palau: a descriptive study. Int. J. Soc. Psych. 29: 161–170.

Holzman PS (1983): Smooth pursuit eye movements in psychopathology. Schiz. Bull. 9: 33–36.

Holzman PS, Proctor LR, Hughes DW (1973): Eye-tracking patterns in schizophrenia. Science 181: 179–181.

Holzman PS, Solomon CM, Levin S, Waternaux CS (1984): Pursuit eye movement dysfunctions in schizophrenia. Family evidence for specificity. Arch. Gen. Psych. 41: 136–139.

Holzman PS, Kringlen E, Matthysse S, Flanagan SD, Lipton RB, Cramer G, Levin S, Lange K, Levy DL (1988): A single dominant gene can account for eye tracking dysfunctions and schizophrenia in offspring of discordant twins. Arch. Gen. Psych. 45: 641–647.

Hughes JR, Hatsukami DK, Mitchell JE, Dahlgren IA (1986): Prevalence of smoking among psychiatric outpatients. Am. J. Psych. 143: 993–997.

Iacono WG, Koenig WG (1983): Features that distinguish the smooth-pursuit eye-tracking performance of schizophrenic, affective-disorder, and normal individuals. J. Abnormal Psychol. 92: 29–41.

Jann MW, Saklad SR, Ereshefsky L, Richards AL, Harrington CA, Davis DM (1986): Effects of smoking on haloperidol and reduced haloperidol plasma concentrations and haloperidol clearance. Psychopharmacology 90: 468–470.

Kendler KS (1988): The genetics of schizophrenia and related disorders: a review. In: Dunner DL, Gershon ES, Barrett JE, Editors, Relatives at risk for mental disorder. New York: Raven Press, pp. 247–266.

Kendler KS, Davis KL (1981): The genetics and biochemistry of paranoid schizophrenia and other paranoid psychoses. Schiz. Bull. 7: 689–709.

Kendler KS, Neale MC, MacLean CJ, Heath AC, Eaves LJ, Kessler RC (1993): Smoking and major depression. A causal analysis. Arch. Gen. Psych. 50: 36–43.

Kendler KS, MacLean CJ, O'Neill FA, Burke J, Murphy B, Duke F, Shinkwin, Easter SM, Webb BT, Zhang J, Walsh D, Straub RE (1996): Evidence for a schizophrenia vulnerability locus on chromosome 8p in the Irish Study of High-Density Schizophrenia Families. Am. J. Psych. 153: 1534–1540.

Leonard S, Adams C, Breese CR, Adler LE, Bickford, P, Byerley W, Coon H, Griffith JM, Miller C, Myles-Worsley M, Nagamoto HT, Rollins Y, Stevens KE, Waldo M, Freedman R (1996): Nicotinic receptor function in schizophrenia. Schiz. Bull. 22: 431–445.

Leonard S, Gault J, Logel J, Drebing C, Berger R, Robinson M, Hopkins J, Moore T, Kim EJ, Davis A, Barnhart M, Meriwether J, Breese C, Freedman R (1997a): Genomic structure of the human α7 neuronal nicotinic receptor subunit. Soc. Neurosci. Abstr. 23: 391.

Leonard S, Moore T, Gault J, Hopkins J, Robinson M, Olincy A, Adler LE, Cloninger CR, Kaufmann CA, Tsuang MT, Faraone SV, Malaspina D, Svrakic DM, and Freedman R (1998): Further investigation of a chromosome 15 locus in schizophrenia: Analysis of affected sibpairs from the NIMH Genetics Initiative. Am. J. Med. Gen. (Neuropsych. Gen.) 81: 308–312.

Levy DL, Yasillo NJ, Dorus E, Shaughnessy R, Gibbons RD, Peterson J, Janicak PG, Gaviria M, and Davis JM (1983): Relatives of unipolar and bipolar patients have normal pursuit. Psych. Res. 10: 285–293.

Litman RE, Hommer DW, Radant A, Clem T, and Pickar D (1994): Quantitative effects of typical and atypical neuroleptics on smooth pursuit eye tracking in schizophrenia. Schiz. Res. 12: 107–120.

Lozsádi DA (1994): Organization of cortical afferents to the rostral, limbic sector of the rat thalamic reticular nucleus. J. Comp. Neurol. 341: 520–533.

Luntz-Leybman V, Bickford PC, Freedman R (1992): Cholinergic gating of response to auditory stimuli in rat hippocampus. Brain Res. 587: 130–136.

Marks MJ, Romm E, Gaffney DK, Collins AC (1986): Nicotine-induced tolerance and receptor changes in four mouse strains. J. Pharm. Exp. Ther. 237: 809–819.

Marks MJ, Romm E, Campbell SM, Collins AC (1989): Variation of nicotinic binding sites among inbred strains. Pharm. Biochem. Behav. 33: 679–689.

Masterson E, O'Shea B (1984): Smoking and malignancy in schizophrenia. Br. J. Psych. 145: 429–432.

McEvoy JP, Rose JE, Levin ED, Vanderzwaag C, and McGee M (1994): Clozapine decreases drive to smoke. Schiz. Res. [Abstract] 11: 104.

McEvoy JP, Freudenreich O, Levin ED, and Rose JE (1995): Haloperidol increases smoking in patients with schizophrenia. Psychopharmacology 119: 124–126.

McGehee DS, Role LW (1995): Physiological diversity of nicotinic acetylcholine receptors expressed by vertebrate neurons. Ann. Rev. Physiol. 57: 521–546.

McGue M, Gottesman II (1989): Genetic linkage in schizophrenia: perspectives from genetic epidemiology. Schiz. Bull. 15: 453–464.

Miller DD, Kelly MW, Perry PJ, Coryell WH (1990): The influence of cigarette smoking on haloperidol pharmacokinetics. Biol. Psych. 28: 529–531.

Mitrofanis J. Guillery RW (1993): New views of the thalamic reticular nucleus in the adult and the developing brain. Trends Neurosci. 16: 240–245.

Nagamoto HT, Adler LE, Hea RA Griffith JM, McRae KA, Freedman R (1996): Gating of auditory P50 in schizophrenics: unique effects of clozapine. Biol. Psych. 40: 181–188.

Newland CF, Beeson D, Vincent A, Newsom-Davis J (1995): Functional and non-functional isoforms of the human muscle acetylcholine receptor. J. Physiol. 489: 767–778.

Olincy A, Ross RG, Young DA, Roath M, and Freedman R (1998): Improvement in smooth pursuit eye movements after cigarette smoking in schizophrenic patients. Neuropsychopharmacology 18: 175–185.

Olincy A, Young DA, Freedman R (1997b): Increased levels of the nicotine metabolite cotinine in schizophrenic smokers compared to other smokers. Biol. Psych. 42: 1–5.

Orr-Urtreger A, Seldin MF, Baldini A, Beaudet AL (1995): Cloning and mapping of the mouse alpha 7-neuronal nicotinic acetylcholine receptor. Genomics 26: 399–402.

Pulver AE, Lasseter VK, Kasch L, Wolyniec P, Nestadt G, Blouin JL, Kimberland M, Babb R, Vourlis S, Chen H (1995): Schizophrenia: a genome scan targets chromosomes 3p and 8p as potential sites of susceptibility genes. Am. J. Med. Genet. 60: 252–260.

Quik M, Chan J, Patrick J (1994): α-Bungarotoxin blocks the nicotinic receptor mediated increase in cell number in a neuroendocrine cell line. Brain Res. 655: 161–167.

Radant AD, Hommer DW (1992): A quantitative analysis of saccades and smooth pursuit during visual pursuit tracking. A comparison of schizophrenics with normals and substance abusing controls. Schiz. Res. 6: 225–235.

Ridley RM, Baker HF (1990): Implications of age of onset for the genetics of schizophrenia. Biol. Psych. 28: 455–458.

Roberts GW (1990): Schizophrenia: the cellular biology of a functional psychosis. TINS 13: 207–211.

Rollins YD, Steven KE, Harris KR, Hall ME, Rose GM, Leonard S (1993): Reduction in auditory gating following intracerebroventricular application of α-bungarotoxin binding site ligands and α7 antisense oligonucleotides. Soc. Neurosci. Abstr. 19: 837.

Rollins YD, Breese CR, Adams C, Drebing C, Rose RM, Leonard S (1996): Cellular localization of α-bungarotoxin binding and α7 mRNA in the hippocampus related to auditory gating in the awake-behaving rat. Soc. Neurosci. Abstr. 22: 1272.

Rupp A, Keith SJ (1993): The costs of schizphrenia: Assessing the burden. Psychiatr. Clin. North. Am. 16: 413–423.

Steriade M, Parent A, Pare D, Smith Y (1987): Cholinergic and noncholinergic neurons of cat basal forebrain project to reticular and mediodorsal thalamic nuclei. Brain Res. 408: 372–376.

Stevens KE, Freedman R, Collins AC, Hall M, Leonard S, Marks, MJ, Rose GM (1996): Genetic correlation of inhibitory gating of hippocampal auditory evoked response and alpha-bungarotoxin-binding nicotinic cholinergic receptors in inbred mouse strains. Brain Res. 15: 152–162.

Stevens KE, Kem WR, Mahnir VM, Freedman R (1998): Selective α7-nicotinic agonists normalize inhibition of auditory response in DBA mice. Psychopharmacology 136: 320–327.

Straub RE, MacLean CJ, O'Neill FA, Burke, J, Murphy B, Duke F, Shinkwin R, Webb BT, Zhang J, Walsh D (1995): A potential vulnerability locus for schizophrenia on chromosome 6p24-22: evidence for genetic heterogeneity. Nature Gen. 11: 287–293.

Straube ER, Oades RD (1992): Genetic studies. In: Anonymous schizophrenia: empirical research and findings. San Diego: Academic Press, pp. 361–384.

Waldo MC, Carey G, Myles-Worsley M, Cawthra E, Adler LE, Nagamoto HT, Wender P, Byerley W, Plaetke R, Freedman R (1991): Codistribution of a sensory gating deficit and schizophrenia in multi-affected families. Psych. Res. 39: 257–268.

Wang RY, Ashby CRJ, Edwards E, Zhang JY (1994): The role of 5-HT3-like receptors in the action of clozapine. J. Clin. Psych. 55 (suppl B): 23–26.

Wood S, Schertzer M, Yaremko ML (1995): Identification of the human neuronal nicotinic cholinergic alpha 2 receptor locus, (CHRNA2), within an 8p21 mapped locus, by sequence homology with rat DNA. Somat. Cell Mol. Genet. 21: 147–150.

18

Nicotine and Non-Nicotine Formulations for Smoking Cessation

Scott J. Leischow, PhD, and Gretchen Cook MS
Arizona Program for Nicotine and Tobacco Research
University of Arizona
Tucson, Arizona

Nearly all ex-smokers (95%) report quitting on their own or with no formal therapy (Fiore et al., 1990). However, fewer than 5% of the smokers who attempted to quit without behavioral or pharmacological support achieve abstinence for 1 year (Cohen et al., 1989). These data suggest that smoking cessation on any one occasion is exceedingly difficult for most smokers; however, since smokers typically attempt to quit on many occasions, the probability of eventually achieving success increases with each attempt. Because most smokers want to quit and make attempts to quit, it is critical that the probability of quitting be improved. In fact, currently approved pharmacological treatments, in combination with behavioral support, can increase 1-year quit rates to 20 to 24% (Fiore et al., 1990). Given the dramatically improved abstinence rates that occur with pharmacological therapies and the public health benefit of increasing the proportion of ex-smokers to smokers, continued intensive evaluation of pharmacological treatments for smoking is well justified. This chapter reviews the safety and efficacy of the currently available and promising new pharmacological interventions for the treatment of nicotine dependence.

Neuronal Nicotinic Receptors: Pharmacology and Therapeutic Opportunities, Edited by S. P. Arneric and J. D. Brioni
ISBN 0-471-24743-x, pages 323–336. Copyright © 1998 by Wiley-Liss, Inc.

NICOTINE REPLACEMENT THERAPIES FOR SMOKING CESSATION

Rationale and Mechanisms of Action

The nicotine replacement therapies (NRTs) currently approved by the Food and Drug Administration (FDA) for use in the United States are nicotine polacrilex, transdermal nicotine system, nicotine nasal spray, and the oral nicotine inhaler. The nicotine inhaler, though FDA approved, is not currently on the market. A nicotine tablet and an oral transmucosal nicotine stick are currently being tested. All NRT delivery systems provide nicotine in a manner that is different than a cigarette delivers nicotine. Specifically, the delivery of the nicotine is both slower and results in lower plasma and arterial blood nicotine levels than when smoking (Henningfield et al., 1990b; Henningfield, 1995; Schneider et al., 1996b). This substantially reduces nicotine toxicity and eliminates the arterial "boli," or the rapid delivery of nicotine to the brain, that occurs when the nicotine is delivered via smoking (Russell and Feyerabend, 1978).

It is generally accepted that NRT successfully enhances smoking cessation efforts in three ways. First, nicotine replacement serves to reduce the severity of the withdrawal symptoms normally experienced with quitting "cold turkey" (Hughes et al., 1991; Schneider et al., 1984; Schneider, 1994a). This then allows the smoker to focus on making behavioral changes necessary for a successful quit attempt. Second, nicotine provided by these medications replaces some of the nicotine previously provided by the cigarette, which at least partially enhances mood and cognitive changes (Resnick, 1993). And third, use of NRT may result in the reduction in the reinforcing effects of nicotine by making nicotine intake independent of environmental events (Hughes, 1993a).

Nicotine Replacement Therapy Medications

Nicotine Polacrilex (Gum). Nicotine gum was approved for prescription use in 1984 for the 2 mg form and in 1991 for the 4 mg form and for over-the-counter (OTC) use in 1996. The nicotine is bound to an ion-exchange resin and incorporated into a gum base. When the gum is chewed properly it slowly releases nicotine, which is absorbed across the mucosal surfaces in the mouth (Sachs, 1989). It takes 15 to 30 minutes to reach peak nicotine levels and venous nicotine levels are ⅓ to ⅔ of between cigarette nicotine levels (Benowitz, 1988) and arterial nicotine levels are 1/10 to 1/20 of between cigarette nicotine levels (Henningfield et al., 1990a, 1993).

Nicotine gum reduces the severity of several tobacco withdrawal symptoms (Hughes et al., 1990; Hughes and Hatsukami, 1992). However, it often results in minor nicotine toxicity effects caused by improper chewing of the nicotine gum, including hiccups, belching, sore jaw, sore throat, nausea, and vomiting (Hughes, 1986).

The efficacy of nicotine gum has been well established (Cepeda-Benito, 1993; Gourlay and McNeil, 1990; Hughes, 1991a, 1993a; Lam et al., 1987; Silagy et al., 1994; Tang et al., 1994). These meta-analyses reported that the odds of abstinence at long-term follow-up (at least 6 months) were between 1.4 and 2.1 higher for gum users than for placebo, independent of whether more intense behavioral intervention was provided. However, when the absolute increases in quit rates are examined, the use of nicotine gum increased quit rates by 0 to 7% and by 7 to 15%, when minimal and more intense behavioral therapy were provided, respectively (Hughes, 1995). Quit rates when using nicotine gum can also be influenced by the level of nicotine dependence and the dosage of nicotine gum used. For instance, the 4 mg nicotine gum appears to be more effective than the 2 mg in highly dependent smokers (Herrera et al., 1995; Jarvik and Henningfield, 1993; Kornitzer et al., 1987; Sachs, 1995; Tonnesen et al., 1988).

Transdermal Nicotine System (Nicotine Patch). In 1991 four transdermal nicotine delivery systems became available in the United States by prescription for the treatment of nicotine dependence. Two brands of transdermal nicotine recently became available without prescription (Nicotrol and Nicoderm), one is still available by prescription (Habitrol), and one is no longer marketed (ProStep). With the fixed initial dose, peak nicotine blood levels are reached within 4 to 9 hours after application, 15 to 22 mg of nicotine/day is delivered, and plasma nicotine levels are in the range of 6 to 23 μg/L (Palmer and Faulds, 1992).

The use of transdermal nicotine produces very few side effects. The one problem experienced by a significant proportion of patients is mild to moderate dermatological reactions (Sachs et al., 1993). Other less frequent side effects are symptoms of nicotine toxicity (Hughes, 1993b). Transdermal nicotine decreases nicotine withdrawal after smoking cessation, though variability of change occurs (Fagerstrom et al., 1992, 1993; Palmer and Faulds, 1992).

The efficacy of transdermal nicotine as an aid in smoking cessation has been firmly established. It has been demonstrated that the use of transdermal nicotine doubles the odds of being abstinent at 6 and 12 months, independent of the intensity of the behavioral therapy provided, when compared to the use of the placebo patch (Fiore et al., 1994; Hughes, 1995; Silagy et al., 1994; Tang, 1994). Fiore's meta-analysis found that across the 17 studies examined, the 6-month absolute pooled abstinence rates for the active patch were 27% and 13% for placebo, and at 1 year were 22% for the active patch and 9% for the placebo patch.

Some products gradually reduce nicotine intake by encouraging the use of smaller patches, or "weaning," while one product recommends abrupt medication termination. Although one of the supposed benefits of the "weaning" strategy is that it will reduce the nicotine withdrawal experienced with the abrupt strategy, there is little evidence that the abrupt strategy results in significant withdrawal when treatment is terminated (Hurt et al., 1994). There is no difference between the abrupt and weaning programs in the rates of relapse to smoking (Fiore et al., 1994; Silagy et al., 1994).

Transdermal nicotine is available in two administration systems: 16 hours and 24 hours. There are pros and cons for both administrations (Hughes, 1996). For example, 24-hour administration seems to decrease morning craving (Leischow et al., 1994a) while 16-hour administration results in less sleep disturbance (Fagerstrom et al., 1993; Leischow et al., 1994a). There is no evidence that either approach is more efficacious (Daughton et al., 1991; Fagerstrom et al., 1993; Fiore et al., 1994).

Nicotine Nasal Spray (NNS). Nicotine nasal spray is available by prescription only. The nicotine nasal spray consists of a nicotine solution contained in a small bottle fitted for insertion into the nostril. By pressing the pump mechanism in the bottle, a fine spray of nicotine (0.5 mg) is released into the nostril. The patient squirts once in each nostril for a total of 1.0 mg of nicotine. Typically, a patient will use 13 to 20 doses per day (Schneider et al., 1996b; Sutherland et al., 1992). Nicotine concentrations in the blood with *ad libitum* use are moderate and in the same range as low-to-moderate patches and 2 mg gum (Schneider et al., 1996b).

Nicotine nasal spray has been shown to decrease withdrawal symptom intensity, decrease craving for cigarettes, and decrease weight gain in abstinent patients in the first 48 hours of use, but not after a week (Sutherland et al., 1992). Nasal irritation, runny nose, watery eyes, throat irritation, sneezing, and coughing are the most common side effects reported (Sutherland et al., 1992; Schneider et al., 1995; Tonnesen et al., 1996).

Several placebo-controlled clinical trials have demonstrated that the nicotine nasal spray is an efficacious pharmacological treatment for smoking cessation (Hjalmarson et al., 1994; Schneider et al., 1995; Sutherland et al., 1992). For instance, Schneider and colleagues reported that the use of the nicotine nasal spray showed significantly higher quit rates over placebo overall and at the short-term and long-term time points (Schneider et al., 1995). The 6-week quit rates were 43% vs. 20%, at 6 months the rates were 25% vs. 10%, and at 1 year the rates were 18% vs. 8% for active versus placebo, respectively. The results of another study indicated that heavier smokers receive the greatest benefit from the nicotine nasal spray (Sutherland et al., 1992).

Nicotine Inhaler (or Inhaler). Nicotine inhalers consist of a nicotine-impregnated plug that is inserted into a cigarette-like tube or mouthpiece. The patient "puffs" on the mouthpiece, which draws warm air through the plug. The air becomes saturated with nicotine and the vaporized nicotine is inhaled by the user. The nicotine is absorbed mainly in the upper respiratory tract via the buccal mucosa (Schneider, 1994b). Each plug contains 10 mg of porous nicotine and delivers a maximum of 400 puffs of vaporized nicotine. The number of puffs, the volume of air passing through the plug, and the environmental temperature affect the amount of nicotine obtained with each puff. At room temperature, a single puff delivers approximately 13 micrograms of nicotine and it takes a minimum of 80 puffs to obtain the same nicotine delivery as one would receive from 8 to 12 puffs on a cigarette (Henningfield, 1995; Leischow, 1994b).

The nicotine inhaler delivers nicotine relatively slowly, in part because of the small amount of nicotine delivered by the inhaler but also because buccal absorption is slow. The primary benefit of the inhaler is its use may minimize the problems associated with quitting by partially mimicking the hand-to-mouth activity of smoking and thereby replace some of the sensory reinforcements of the smoking behavior (Lieschow, 1994b; Rose, 1988).

The most commonly reported adverse effects are mild, transient, and did not affect continued use of the inhaler. They included mild mouth and throat irritation and mild coughing (Hjalmarson et al., 1997; Leischow et al., 1996; Schneider et al., 1996c; Tonnesen et al., 1993). Other side effects that are reported include nausea and oral burning/smarting (Schneider et al., 1996c).

The ability of the nicotine inhaler to consistently reduce postcessation withdrawal symptoms has not been demonstrated. The results of one study indicated that craving was relieved with the use of the nicotine inhaler at day 3 and week 1 only. It did not provide relief at any of the other time points throughout the study (Schneider et al., 1996c).

Four randomized, placebo-controlled, double-blind clinical trials have investigated the efficacy of the nicotine inhaler for smoking cessation (Leischow et al., 1996; Hjalmarson et al., 1997; Schneider et al., 1996c; Tonnesen et al., 1993). The results indicated that the inhaler is an efficacious aid to short-term smoking cessation when used in conjunction with minimal behavioral therapy and inhaler use instructions. However, the long-term efficacy results were contradictory. While two of the studies failed to find a significant difference in abstinence rates between the nicotine and the placebo inhaler beyond 3 months (Schneider et al., 1996c) and beyond 6 months (Leischow et al., 1996), the other two studies reported that the inhaler was successful in enhancing the 3-, 6-, and 12-month abstinence rates (Hjalmarson et al., 1997; Tonnesen et al., 1993).

Nicotine Sublingual Tablet and Oral Transmucasal Nicotine. The sublingual nicotine tablet (2 mg) and oral transmucosal nicotine (OT-nicotine) are being tested as alternative nicotine replacement systems. The nicotine in both products is absorbed bucally. Although clinical trials testing these two nicotine products have been conducted, the results have not yet been published.

Cotinine. Recently several researchers have been turning their attention to the use of nicotine metabolites such as cotinine. The results of animal and human studies on whether cotinine produces behavioral, subjective, and physiological effects have been mixed. While it appears that the use of high doses of cotinine results in either little (Benowitz, 1983; Keenen et al., 1994) or no physiological and behavioral effects (Hatsukami et al., 1997), it does appear that pretreatment with high doses of cotinine may antagonize some of the effects of nicotine and may therefore be suitable for the treatment of nicotine addiction. Animal studies have demonstrated these antagonistic effects of cotinine toward nicotine (Chanine et al., 1990, 1996; Kim et al., 1968).

The investigation into the safety and efficacy of cotinine is in its infancy. One study has reported that the short-term administration of high doses of cotinine is safe and produces no acute or withdrawal effects (Hatsukami et al., 1997).

Does Combining NRTs or Providing Higher Nicotine Doses Increase Efficacy?

The efficacy of increased nicotine dosing has been investigated by either combining two or more nicotine medications or by increasing the dose above the currently recommended levels. The rationale for combining NRTs to enhance efficacy is that the combined use of one of the self-administered delivery systems to provide rapid relief of acute withdrawal symptoms and the use of the slow, steady transdermal nicotine system to maintain plasma nicotine levels would result in greater abstinence rates than either system alone (Fagerstom, 1994; Schneider, 1992). In fact, several studies reported that the combination of patch and gum resulted in greater relief from withdrawal symptoms (Fagerstrom, 1994) and higher abstinence rates than when either medication is used alone (Fagerstrom, 1994; Fagerstrom et al., 1993; Kornitzer et al., 1995; Pushka et al., 1995).

The second hypothesis being tested is whether increasing the nicotine dose will result in greater replacement and therefore higher abstinence rates. Two studies compared the efficacy of a 22 mg and 44 mg nicotine patch (Dale et al., 1995; Jorenby et al., 1995), and the results are mixed. Dale and his colleagues reported a dose-response relationship for withdrawal symptom relief and for smoking cessation, and that there was no difference in the percentage of subjects that reported adverse events associated with nicotine toxicity. On the other hand, Jorenby and his colleagues reported that doubling the transdermal nicotine dose did not enhance abstinence rates but did increase the frequency of some nicotine toxicity side effects.

NON-NICOTINE FORMULATIONS FOR SMOKING CESSATION

There is a growing list of non-nicotine medications under investigation for smoking cessation. Some have been shown to be an effective means of helping a smoker quit, while others have not. However, few of these medications have been thoroughly tested in large, randomized, placebo-controlled, long-term clinical trials. It is proposed that these medications promote smoking cessation by either reducing specific symptoms of nicotine withdrawal, by replacing or blocking the reinforcing effects of nicotine, by making smoking aversive, or by treating the comorbid psychopathology.

Buspirone

Buspirone is a nonbenzodiazepine anxiolytic (serotonergic agonist). It has a relatively benign side effect profile (Hughes, 1996). Buspirone was evaluated as a smoking cessation medication because it was hypothesized to serve as a symptomatic treatment for nicotine withdrawal symptoms, for any general anxiety responses, and for mood regulation based on the notion that if the affective symptoms were not treated then the chances of relapse would be increased in certain smokers (Jarvik and Henningfield, 1988, 1993).

Two short-term studies reported that buspirone lowered withdrawal (Robinson et al., 1991) and lowered specific symptoms such as craving, irritability, sadness, anxiety, and restlessness (Hilleman et al., 1992). However, other short- and long-term studies reported that buspirone did not relieve withdrawal symptoms (Cinciripani et al., 1993; Gawin et al., 1989; Robinson et al., 1992; Schneider et al., 1996a; West et al., 1991).

The preliminary efficacy results have also been contradictory. Two placebo-controlled, short-term trials demonstrated improved cessation rates with the use of buspirone (Hilleman et al., 1992; West et al., 1991). In the most recent placebo-controlled, long-term trial, there was no difference in abstinence rates between the subjects receiving buspirone and placebo at both short- and long-term time points (Schneider et al., 1996a). Although buspirone was shown to improve short-term cessation rates in high-anxiety smokers (Circiripini et al., 1995), another study reported no differences between high- and low-anxiety smokers in abstinence rates (Schneider et al., 1996a).

Bupropion

Bupropion is the active ingredient in an antidepressant (Wellbutrin). In early 1997 the sustained-release form of bupropion was approved by the FDA as a prescription smoking cessation medication. There are three rationales for the use of bupropion and other antidepressant medications for smoking cessation. First, it has been hypothesized that it is necessary to pharmacologically treat the anxiety and dysphoria that occurs in some smokers trying to quit or they might relapse (Covey et al., 1990; Hall et al., 1994; Jarvik and Henningfield, 1988, 1993; Shiffman, 1982).

Second, it appears that there is a relationship between cigarette smoking and major depression. It has been shown that smokers are more likely to have a history of major depression than nonsmokers (Breslau et al., 1993; Breslau, 1995; Glassman et al., 1990; Hall et al., 1991) and that nicotine may act as an antidepressant in some smokers (Glass, 1990; Hughes, 1988). Since smokers with a history of major depression experience more significant symptoms of depression with nicotine withdrawal and lower abstinence rates than other smokers (Anda et al., 1990; Covey et al., 1990; Glassman et al., 1990; Hughes et al., 1986), there is a need to replace the antidepressant effect that nicotine has with another medication in order to prevent relapses.

A third reason that bupropion may have efficacy for smoking cessation is that bupropion appears to inhibit norepinephrine and dopamine uptake, resulting in elevated levels of these neurotransmitters. This then would compensate for the decreased levels of these neurotransmitters that occur during the withdrawal process when nicotine is no longer stimulating norepinephrine and dopamine release (Ascher et al., 1995).

Three double-blind, placebo-controlled trials have investigated the safety and efficacy of bupropion for smoking cessation (Ferry and Burchette, 1994; Hurt et al., 1997), and others have been conducted but not yet published. The first two studies found that, compared to placebo-treated smokers, bupropion use by depressed smokers resulted in increased abstinence rates (Ferry and Burchette, 1994). More recently the first three-site, long-term clinical trial of bupropion for smoking cessation was completed (Hurt et al., 1997). In smokers without current depression, the sustained-release form of bupropion was shown to be effective for smoking cessation. Specifically, at the end of treatment and at 1 year there was a significant dose response. At the end of treatment the abstinence rates were 19% in the placebo group, 28.8% in the 100 mg/day group, 38.6% in the 150 mg/day group, and 44.2% in the 300 mg/day group. At 1 year the success rates were 12.4%, 19.6%, 22.9%, and 23.1% for the respective groups. At 1 year, only the rates for the 150 mg and the 300 mg group were significantly higher than for the placebo group. There were minimal adverse events reported. The withdrawal efficacy results of this study were mixed. There was an inverse relationship between dose and weight gain, but there was no consistent dose effect on the suppression of other withdrawal symptoms.

Doxepine and Fluoxetine

Doxepine and fluoxetine are marketed as antidepressants (Sinequan and Prozac, respectively). Several small studies investigated whether using doxepine (Edwards et al., 1989; Murphy et al., 1990) or fluoxetine (Dalack et al., 1995; Niaura et al., 1995; Pomerleau et al., 1991) may be effective for smoking cessation. While Edwards and colleagues reported that doxepine enhanced abstinence rates (Edwards et al., 1989), the results of the study conducted by Murphy and colleagues indicated that doxepine was not very efficacious (Murphy et al., 1990). The results from one of the fluoxetine studies indicate that it may be more effective for smoking cessation with smokers with depression than general smokers (Niaura et al., 1995). The use of fluoxetine has been shown to ameliorate some of the symptoms of depression that are associated with nicotine withdrawal (Dalack et al., 1995) and to reduce the increase in appetite and weight gain withdrawal symptoms (Pomerleau et al., 1991). To date, there have not been any large clinical trials of either medication reported.

Nortriptyline

Nortriptyline is a tricyclic antidepressant. The preliminary results from only one nortriptyline smoking cessation trial have been presented to date. This study reported that the use of nortriptyline doubled abstinence rates, regardless of whether the subject had a history of depression (Humfleet et al., 1996). The mechanisms of action are not known, although the mechanisms that have been proposed are the same mechanisms proposed for bupropion.

Mecamylamine

Mecamylamine is a noncompetitive cholinergic blocking agent. It consistently reverses the central nervous system (CNS) effects of nicotine (Henningfield et al., 1983; Pickworth et al., 1988) and does not cause nicotine withdrawal (Clarke, 1991; Stolerman, 1986). Since mecamylamine is a nicotine blocker, it inhibits many of the behavioral, physiologic, and reinforcing effects of nicotine (Clarke, 1991; Collins et al., 1986; Corrigall and Coen, 1989; Martin et al., 1989; Rose et al., 1989; Stolerman, 1986). It also has been shown to increase nicotine preference and amount of smoking (Nemeth-Coslett et al., 1986; Pomerleau et al., 1987; Rose et al., 1989). Preliminary dosing studies reported that high doses of mecamylamine inhibited the subjective effects of smoking; however, it also produced significant adverse events and did not enhance success rates (Tennant et al., 1984; Tennant and Tarver, 1985). After these discouraging results, other strategies were developed and tested. The first strategy proposed was to modify the blocking treatment by lowering the dose and by beginning the treatment prior to quitting. It was thought that the smoker would experience fewer side effects yet would still realize the same reduction in smoking satisfaction. The other suggested strategy was to combine nicotine and low doses of mecamylamine, either concomitantly (Rose and Levin, 1991) or using nicotine (the agonist) first and then using mecamylamine (the antagonist) (Hughes, 1996). The rationale behind combining mecamylamine with transdermal nicotine is that the nicotine from the nicotine replacement and the mecamylamine would both occupy the nicotinic receptors that would normally be occupied by the nicotine from the cigarettes. Thus, the rewarding effects of nicotine would eventually be limited (Rose and Levin, 1991).

The combination of medications has been shown to be an effective means of reducing side effects and smoking satisfaction (Rose et al., 1994, 1996a). In addition, the smokers who used the combination treatment were more successful short and long term than those who used the patch alone (Rose et al., 1994, 1996a). Continuous quit rates at 7 weeks, 6 months, and 1 year when the mecamylamine pill (10 mg/day) and transdermal nicotine (21 mg/day) were administered concurrently versus transdermal nicotine alone were 50% versus 16.7%, 37.5% versus 12.5%, and 37.5% versus 4.2%, respectively (Rose et al., 1994).

Naltrexone

Naltrexone is an opioid antagonist. It is currently only approved for the treatment of alcohol and opioid addiction. Nicotine, like alcohol, activates the endogenous opioid system (Karras, 1982; Pomerleau and Pomerleau, 1984). Since naltrexone is an opioid blocker, it has been hypothesized that naltrexone will block the nicotine receptors and thus reduce the reinforcing effects of nicotine (Hughes, 1996). There has only been one placebo-controlled, randomized clinical trial of naltrexone. However, the results have not been published yet.

Clonidine

Clonidine is an alpha-2 adrenergic antagonist. Thus, it has been proposed that clonidine reduces nicotine withdrawal and craving because it reduces adrenergic overactivity; however, it does not appear that nicotine withdrawal is due to adrenergic overactivity (Hughes et al., 1990).

Clonidine reduces craving and nicotine withdrawal symptoms (Franks et al., 1989; Glassman et al., 1988; Gourlay et al., 1994; Ornish et al., 1988; Prochazka et al., 1992). However, as an aid for smoking cessation the results of subsequent short- and long-term trials have been mixed (Franks et al., 1989; Glassman et al., 1993; Gourlay et al., 1994; Niaura et al., 1993; Procazka et al., 1992; Zikos et al., 1991). It appears that clonidine is only effective with certain subgroups of smokers, such as women, highly dependent smokers, women smokers with a current history of depression, and smokers with multiple prior quit attempts (Covey and Glassman, 1991; Glassman et al., 1988, 1993; Hilleman et al., 1993). It is not known why clonidine appears to be more effective in these groups.

Lobeline

In the last 20 years there have been very few studies conducted to evaluate the safety and efficacy of lobeline as a smoking cessation medication (Davison and Rosen, 1972; Department of Health and Human Services, 1982; Glover et al., 1998; Kozlewski, 1984). The results provided little support for lobeline's smoking cessation efficacy. However, the latest study reported that, although there was no difference in the abstinence rates between lobeline and placebo, there might be possible benefits and that further research is warranted because of the trend for smoking reduction found in this study (Glover et al., 1998).

SUMMARY

Nicotine replacement therapy has become the most widely studied and used pharmacological treatment for nicotine dependence. There is no question that when nicotine replacement is used as recommended, the chances that a smoker will successfully quit are increased. With the recent FDA approval of transdermal nicotine and nicotine polacrilex for OTC use, one can expect that the use of NRT will continue to eclipse the use of the other pharmacotherapies available. However, it remains to be seen whether the use of OTC nicotine replacement, with and without concomitant behavioral therapy, will result in the same abstinence rates.

There are newer nicotine medications available that have proven to be just as safe and effective at relieving nicotine withdrawal and enhancing abstinence rates as the nicotine patch and gum. However, several novel medications that have either been approved or are currently under investigation are exciting advances because they point to new, and badly needed, ways of thinking about pharmacological treatment for smoking cessation. The line of research that has shown the most promise thus far is the dopaminergic enhancer bupropion, since the efficacy data have been consistent and robust. A mixed agonist and antagonist approach (combined mecamylamine and nicotine) looks promising as well, but judgment must be withheld until large clinical trials have been completed and reported. However, while nicotine replacement will continue to play an important role in smoking cessation treatment, it seems increasingly likely that the future of pharmacological treatment rests upon the further exploration and understanding of the mechanisms of nicotine addiction so that more refined and mechanism-specific medications can be formulated. Indeed, the continued development of new medications that affect specific areas of the brain associated with nicotine reward and cognitive enhancement have the potential to lead to dramatic advances in treatment efficacy with resulting increases in improved public health and quality of life.

REFERENCES

Anda RF, Williamson DF, Escobedo LG, Mast EE, Giocino GA, Remington PL (1990): Depression and the dynamics of smoking. JAMA 264: 1541–1545.

Ascher JA, Cole JO, Colin JN, et al. (1995): Bupropion: a review of its mechanism of antidepressant activity. J. Clin. Psychiatry 56: 395–401.

Benowitz NL (1983): The use of biologic fluid samples in assessing tobacco smoke consumption. In: Grabowski J, Bell CS, editors. Measurement in the analysis and treatment of smoking behavior. NIDA Research Monograph, Washington, DC: U.S. Government Printing Office, pp. 6–26.

Benowitz NL (1988): Pharmacologic aspects of cigarette smoking and nicotine addition. N. Engl. J. Med. 319: 1318–1330.

Breslau N (1995): Psychiatric comorbidity of smoking and nicotine dependence. Behav. Genet. 25: 95–101.

Breslau N, Fenn N, Peterson EL (1993): Early smoking initiation and nicotine dependence in a cohort of young adults. Drug Alcohol Depend. 33:129–137.

Cepeda-Benito A (1993): Meta-analytic review of the efficacy of nicotine chewing gum in smoking treatment programs. J. Consult. Clin. Psychol. 1: 822–830.

Chanine R, Calderone A, Navarro-Delmasure C (1990): The in vitro effects of nicotine and cotinine on prostacyclin and thromboxane biosynthesis. Prostanglandins Leukot. Essent. Fatty Acids 40: 261–266.

Chanine R, Aftimos G, Wainberg MC, Navarro-Delmasure C, Abou Khailil K, Chaboud B (1996): Cotinine modulates the cardiovascular effects of nicotine. Med. Sci. Res. 24: 21–23.

Cincirpini PM, Lapitsky L, Seay S, Wallfisch A, Meyer WJ, Vunakis H (1993): A placebo-controlled evaluation of the effects of buspirone on smoking cessation: differences between high and low anxiety smokers. J. Clin. Psychopharmacol. 15: 182–191.

Clark PBS (1991): Nicotinic receptor blockade therapy and smoking cessation. Br. J. Addict. 86: 501–505.

Cohen S, Lichtenstein E, Prochaska JO, Rossi JS, Gritz ER, Carr CR, Orleans CT, Schoenbach VJ, Biener L, Abrams D, DeClemente C, Curry S, Marlatt GA, Cumming KM, Emont SL, Giovino G, Ossip-Klein D (1989): Debunking myths about self quitting. Am. Psychol. 989: 44:1355–1365.

Collins AC, Evans CB, Mines LL, Marks MJ (1986): Mecamylamine blockade of nicotine responses: evidence for two brain nicotinic receptors. Pharmacol. Biochem. Behav. 24: 1767–1773.

Corrigall WA, Coen KM (1989): Nicotine maintains self-administration in rats on a limited-access schedule. Psychopharmacology 99: 473–478.

Covey LS, Glassman AH (1991): A meta-analysis of double-blinded placebo-controlled trials of clonidine for smoking cessation. Br. J. Addict. 86: 991–998.

Covey LS, Glassman AH, Stetner F (1990): Depression and depressive symptoms in smoking cessation. Compr. Psychiatry 31: 350–354.

Dalack GW, Glassman AH, Rivelli S, Lirio C (1995): Mood major depression, and fluoxetine response in cigarette smokers. Am. J. Psychiatry 152: 398–403.

Dale LC, Hurt RD, Offord KP, Lawson GM, Croghan IT, Schroeder DR (1995): High-dose nicotine patch therapy: percentage of replacement and smoking cessation. JAMA 274: 1353–1358.

Daughton DM, Heatley SA, Predergast JJ, Causey D, Knowles M, Rolf CN, Cheney RA, Hattelid K, Thompson AB, Rennard SI (1991): Effect of transdermal nicotine delivery as an adjunct to low-intervention smoking cessation therapy: a randomized, placebo-controlled, double-blind study. Arch. Intern. Med. 151: 749–752.

Davison GC, Rosen RC (1972): Lobeline and reduction of cigarette smoking. Psychol. Rep. 31: 443–456.

Department of Health and Human Services (1982): Smoking deterrent drug products for over-the-counter use: establishment of a monograph. Federal Register, Docket No. 81N-0027, pp. 490–500.

Edwards NB, Murphy JK, Downs AD, Ackerman BJ, Rosenthal TL (1989): Doxepin as an adjunct to smoking cessation: a double-blinded pilot study. Am. J. Psychiatry 146: 373–376.

Fagerstrom K-O (1994): Combined use of nicotine replacement products. Health Values 18: 15–20.

Fagerstrom K-O, Sawe U, Hurt, RD, Tonnesen P (1992): Therapeutic use of nicotine patches: efficacy and safety. J. Smoking Related Dis. 3: 247–261.

Fagerstrom K-O, Schneider NG, Lunell E (1993): Effectiveness of nicotine patch and nicotine gum as individual versus combined treatments for tobacco withdrawal symptoms. Psychopharmacology 111: 271–277.

Ferry LH & Burchette RJ (1994): Efficacy of bupropion for smoking cessation in non-depressed smokers. J. Addict. Dis. 13: 249.

Fiore MC, Novotny TE, Pierce JP, Giovino GA, Hatziandreu EJ, Newcomb PA, Surawicz TS, Davis RM (1990): Methods used to quit smoking in the United States: Do cessation programs help? JAMA 263: 2760–2765.

Fiore MC, Smith SS, Jorenby DE, Baker TB (1994): The effectiveness of the nicotine patch for smoking cessation: a meta-analysis. JAMA 271: 1940–1947.

Franks P, Harp J, Bell B (1989): Randomized, controlled trial of clonidine for smoking cessation in a primary care setting. JAMA 262: 3011–3013.

Gawin F, Compton M, Byck R (1989): Buspirone reduces smoking (letter). Arch. Gen. Psychiatry 46: 288.

Glass RM (1990): Blue mood, blackened lungs: depression and smoking. JAMA 264: 1583–1584.

Glassman AH, Stetner F, Walsh BT, Raizman P, Fleiss JL, Cooper TB, Covey LS (1988): Heavy smokers, smoking cessation, and clonidine: results of a double-blind, randomized trial. JAMA 259: 2863–2866.

Glassman AH, Helzer JE, Covey LS, Cotter LB, Stetner F, Tipp JE, Johnson J (1990): Smoking, smoking cessation and major depression. JAMA 264: 1546–1549.

Glassman AH, Covey LS, Dalack GW (1993): Smoking cessation, clonidine and vulnerability to nicotine among dependent smokers. Clin. Pharmacol. Ther. 54: 670–679.

Glover ED, Leischow SJ, Rennard SI, Glover PN, Daughton D, Quiring J, Schneider FH, Mione PJ (1998): A smoking cessation trial with lobeline sulfate: a pilot study. Am. J. Health Behav. 22(1): 62–75.

Gourlay SG, McNeil JJ (1990): Anti-smoking products. Med. J. Aust. 153: 699–707.

Gourlay S, Forbes A, Marriner T, Kutin J, McNeil J (1994): A placebo-controlled study of three clonidine doses for smoking cessation. Clin. Pharmacol. Ther. 55: 6469.

Hall SM, Munoz R, Reus V (1991): Smoking cessation, depression, and dysphoria. In: Harris LS, editor. Problems of drug dependence. NIDA Research Monograph 105: 312–313.

Hall SM, Munoz R, Reus V (1994): Cognitive-behavioral intervention increases abstinence rates for depressive-history smokers. J. Consult. Clin. Psychol. 62: 141–146.

Hatsukami DK, Grillo M, Pentel PR, Oncken C, Bliss R (1997): Safety of cotinine in humans: physiology, subjective, and cognitive effects. Pharm. Biochem. Beh. 57: 643–650.

Henningfield JE (1995): Nicotine medications for smoking cessation. N. Engl. J. Med. 333: 1196–1203.

Henningfield JE, Miyasato K, Johnson RE, Jasinski DR (1983): Rapid physiologic effects of nicotine in humans and selective blockade of behavioral effects by mecamylamine. In: Harris LS, editor. Problems of drug dependence. NIDA Research Monograph. Washington, DC: U.S. Government Printing Office, 43: 259–265.

Henningfield JE, London ED, Benowitz NL (1990a): Arterial-venous differences in plasma concentrations of nicotine after cigarette smoking. JAMA 263: 2049–2050.

Henningfield JE, Stapleton JM, Benowitz NL, Grayson RF, London ED (1990b): Higher levels of nicotine in arterial than in venous blood after cigarette smoking. Drug Alcohol Depend. 33: 23–29.

Henningfield JE, Stapleton JM, Benowitz NL, Grayson RF, London ED (1993): Higher levels of nicotine in arterial than in venous blood cigarette smoking. Drug Alcohol Depend. 33: 23–29.

Herrera N, Franco R, Herrera L, Partidas A, Rolando R, Fagerstrom KO (1995): Nicotine gum, 2 and 4 mg, for nicotine dependence. Chest 108: 447–451.

Hillemann DE, Mohiuddin SM, Del Core MG, Sketch MH (1992): Effect of buspirone on withdrawal symptoms associated with smoking cessation. Arch. Intern. Med. 152: 350–352.

Hilleman DE, Mohiuddin SM, Del Core MG, Lucas BD (1993): Randomized, controlled trial of transdermal clonidine for smoking cessation. Ann. Pharmacother. 27: 1025–1028.

Hjalmarson A, Franson M, Westin A, Wiklund O (1994): Effect of nicotine nasal spray on smoking cessation: a randomized placebo-controlled, double-blind study. Arch. Intern. Med. 154: 2567–2572.

Hjalmarsen A, Nilssen F, Sjostrom L, Wiklund O (1997): The nicotine inhaler in smoking cessation. Arch. Int. Med. 157: 1721–1728.

Hughes JR (1986): Problems of nicotine gum. In: Ockene JK, editor. Pharmacologic treatment to tobacco dependence: proceedings of the World Congress. Institute for the Study of Smoking Behavior and Policy. November, pp. 141–147.

Hughes JR (1988): Dependence potential and abuse liability of nicotine replacement therapies. In: Pomerleau O, Pomerleau C, editors. Progress in clinical evaluation. New York: Alan R. Liss, Inc., pp. 261–277.

Hughes JR (1991): Combined psychological and nicotine gum treatment for smoking: a critical review. J. Subst. Abuse 3: 337–350.

Hughes JR (1993a): Pharmacotherapy for smoking cessation: invalidated assumptions, anomalies, and suggestions for future research. J. Consult. Clin. Psychol. 61: 751–760.

Hughes JR (1993b): Risk/benefit of nicotine replacement in smoking cessation. Drug Safety 8: 49–56.

Hughes JR (1995): Combining behavioral therapy and pharmacotherapy for smoking cessation: an update. In: Blaine J, Onken L, editors. Integrating behavior therapies with medication in the treatment of drug dependence. NIDA Research Monograph. Washington, DC: U.S. Government Printing Office, vol. 150, pp. 92–109.

Hughes JR (1996): Pharmacotherapy of Nicotine Dependence. In: Schuster CK, Kuher MJ, editors. Pharmacological aspects of drug dependence toward an integrative neurobehavioral approach handbook of experimental pharmacology series. New York, Springer Verlag, pp. 599–626.

Hughes JR, Hatsukami DK (1992): The nicotine withdrawal syndrome: a brief review and update. Int. J. Smoking Cessation 1: 21–26.

Hughes JR, Hatsukami DK, Mitchell JE, Dahlgren LA (1986): Prevalence of smoking among psychiatric outpatients. Am. J. Psychiatry 143: 993–997.

Hughes JR, Higgins ST, Hatsukami DK (1990): Effects of abstinence from tobacco: a critical review. In: Kozlowski LT, Annis H, Cappell HD, Glaser F, Goodstadt M, Israel Y, Kalant H, Sellers EM, Vingilis J, editors. Research advances in alcohol and drug problems, New York: Wiley, vol. 10, pp. 317–398.

Hughes JR, Gust SW, Skoog K, Keenan RM, Fenwick JW (1991): Symptoms of tobacco withdraw: a replication and extension. Arch. Gen. Psychiatry 48: 52–59.

Humfleet G, Hall S, Reus V, Sees K, Munoz R, Triffleman E (1995): The efficacy of nortriptyline as an adjunct to the psychological treatment for smokers with and without depressive smokers. In: Harris LS, editor. Problems of drug dependence. 1995: Proceedings of the 57th Annual Scientific Meeting, The College on Problems of Drug Dependence, Inc. NIDA Research Monograph 162: 334 (abstract).

Hurt RD, Dale LC, Fredrickson PA, Caldwell CC, Lee GA, Offord KP, Lauger GG, Marusic Z, Neese LW, Lundberg TG (1994): Nicotine patch therapy for smoking cessation combined with physician advice and nurse follow-up. JAMA 271: 595–600.

Hurt RD, Sachs DPL, Glover ED, Offord KP, Johnston JA, Dale LC, Khayrallah MA, Schroeder DR, Glover PN, Sullivan CR, Croghan IT, Sullivan PM (1997): A comparison of sustained-release bupropion and placebo for smoking cessation. N. Engl. J. Med. 337: 1195–202.

Jarvik ME, Henningfield JE (1988): Pharmacological treatment of tobacco dependence. Pharmacol. Biochem. Behav. 30: 279–294.

Jarvik ME, Henningfield JE (1993): Pharmacological adjuncts for the treatment of tobacco dependence. In: Orleans CT, Slade J, editors. Nicotine addiction: principles and management. New York: Oxford University Press, pp. 245–261.

Jorenby DE, Smith SS, Fiore MC, Hurt RD, Offord Kp, Croghan IT, Hays JT, Lewis SF, Baker TB (1995): Varying nicotine patch dose and type of smoking cessation counseling. JAMA 274: 1347–1352.

Karras A (1982): Neurotransmitter and neuropeptide correlates of smoking cessation. In: Essman WR, Valvelli L, editors. Neuropharmacology: clinical applications. New York: Springer, pp. 41–66.

Keenen RM, Hatsukami D, Pentel PR, Thompson T, Grillo MA (1994): Pharmacodynamic effects of cotinine in abstinent smokers. Clin. Pharmacol. Ther. 55: 581–590.

Kim KS, Borzelleca JF, Bowman ER, McKennis H (1968): Effects of some nicotine metabolites and related compounds on isolated smooth muscle. Pharmacol. Exp. Ther. 161: 59–69.

Kornitzer M, Kittel F, Dramaix M, Bourdoux P (1987): A double-blind study of 2 mg versus 4 mg nicotine-gum in an industrial setting. J. Psychosomat. Res. 31: 171–176.

Kornitzer M, Boutsen M, Dramaix M, Thijs J, Gustavsson G (1995): Combined use of nicotine patch and gum in smoking cessation: a placebo-controlled clinical trial. Prev. Med. 24: 41–47.

Kozlowski LT (1984): Pharmacological approaches to smoking modification. In: Matarazzo JD, Weiss SM, Herd JA, Miller NE, editors. (1987) Behavioral health: a handbook of health enhancement and disease prevention. New York: Wiley, pp. 713–728.

Lam W, Sze PC, Sacks HS, Chalmers TC (1987): Meta-analysis of randomized controlled trials of nicotine chewing-gum. Lancet 2(8549): 27–30.

Leischow SJ (1994b): The nicotine vaporizer. Health Values 18: 4–9.

Leischow SJ, Valente SN, Hill AL, Otte PS, Aicken M, Kligman EW (1994): Nicotine patch and gum: withdrawal symptoms, side effects, and medication preference. In: Harris LS, editor. Problems of drug dependence. NIDA Research Monograph. Washington, DC: U.S. Government Printing Office. Vol. 154, pages 53–78..

Leischow SJ, Nilsson F, Franzon M, Hill A, Otte P, Merikle EP (1996): Efficacy of the nicotine inhaler as an adjunct to smoking cessation. Am. J. Health Behav. 20(5): 364–371.

Martin BR, Onaivi ES, Martin TJ (1989): What is the nature of mecamylamine's antagonism of the central effects of nicotine? Biochem. Pharmacol. 38: 3391–3397.

Murphy JK, Edwards NB, Downs AD (1990): Reduction of nicotine withdrawal symptoms with doxepin. Am. J. Psych. 147: 1353–1357.

Nemeth-Coslett R, Henningfield JE, O'Keefe MD, Griffiths RR (1986): Effects of mecamylamine on human cigarette smoking and subjective ratings. Psychopharmacology 88: 420–425.

Niaura R, Goldstein MG, Brown R, Murphy J, Abrams D (1993): A double-blind randomized dose-response trial of transdermal clonidine for smoking cessation. Presented at Society of Behavioral Medicine Meeting, San Francisco, March.

Niaura R, Goldstein MG, Depue J, Keuthen N, Kristeller J, Abrams D (1995): Fluoxetine, symptoms of depression, and smoking cessation. Ann. Behav. Med. 17(suppl. S061).

Ornish SA, Zisook S, McAdams LA (1988): Effects of transdermal clonidine treatment on withdrawal symptoms associated with smoking cessation: a randomized, controlled trial. Arch. Intern. Med. 148: 2027–2031.

Palmer KJ, Faulds D (1992): Transdermal nicotine: a review of its pharmacodynamic and pharmacokinetic properties, and therapeutic use as an aid to smoking cessation. Drugs 44: 498–529.

Pickworth WB, Herning RI, Henningfield JE (1988): Mecamylamine reduces some EEG effects of nicotine chewing gum in humans. Pharmacol. Biochem. Behav. 30: 149–153.

Pomerleau CS, Pomerleau OF (1984): Neuroregulators and the reinforcement of smoking: towards a biobehavioral explanation. Neurosci. Biobeh. Rev. 8: 503–513.

Pomerleau CS, Pomerleau OF, Majchrzak MJ (1987): Mecamylamine pretreatment increases subsequent nicotine self-administration as indicated by changes in plasma nicotine level. Psychopharmacology 91: 391–393.

Pomerleau CS, Pomerleau OF, Morrell EM, Lowenbergh JM (1991): Effects of fluoxetine on weight gain and food intake in smokers who reduce nicotine intake. Psychoneuroendocrinology 16: 433–440.

Prochazka AV, Petty TL, Nett L, Sivers GE, Sachs DPL, Rennard SI, Daughton DM, Grimm RH, Heim C (1992): Transdermal clonidine reduced some withdrawal symptoms but did not increase smoking cessation. Arch. Intern. Med. 152: 2065–2069.

Puska P, Korhonen HI, Vartiainer E, Urjanheimo E-L, Gustavsson G, Westin A (1995): Combined use of nicotine patch and gum compared with gum alone in smoking cessation: a clinical trial in North Karelia. Tobacco Control 4: 231–235.

Resnick MP (1993): Treating nicotine addiction in patients with psychiatric comorbidity. In Orleans CT, Slade J, editors. Nicotine addiction: principles and management. New York: Oxford University Press, pp. 327–336.

Robinson MD, Smith WA, Cederstrom EA, Sutherland DE (1991): Buspirone effect on tobacco withdrawal symptoms: a pilot study. J. Am. Board Fam. Pract. 4: 89–94.

Robinson MD, Pettice YL, Smith WA, Cederstrom EA, Sutherland DE, Davis H (1992): Buspirone effect on tobacco withdrawal symptoms: a randomized placebo-controlled trial. J. Am. Board Fam. Pract. 5: 1–4.

Rose JE (1988): The role of upper airway stimulation in smoking. In: Pomerleau OF, Pomerleau CS, editors. Nicotine replacement: a critical evaluation. New York: Han R. Liss, Inc., pp. 95–108.

Rose JE, Levin ED (1991): Concurrent agonist–antagonist administration for the analysis and treatment of drug dependence. Pharmacol. Biochem. Behav. 41: 219–226.

Rose JE, Sampson A, Levin ED, Henningfield JE (1989): Mecamylamine increases nicotine preference and attenuates nicotine discrimination. Pharmacol. Biochem. Behav. 32: 933–938.

Rose JE, Behm FM, Westman EC, Levin ED, Stein RM, Ripka GV (1994): Mecamylamine combined with nicotine skin patch facilitates smoking cessation beyond nicotine patch treatment alone. Clin. Pharmacol. Ther. 56: 86–99.

Rose JE, Westman EC, Behm FM (1996): Nicotine/mecamylamine combination treatment for smoking cessation. Drug Dev. Res. 38: 243–256.

Russell MAH, Feyerabend C (1978): Cigarette smoking: a dependence on high-nicotine boli. Drug Metabol. Rev. 80: 29–57.

Sachs DPL (1989): Nicotine polacrilex: practical use requirements. Curr. Pulmonol. 10: 141–159.

Sachs DPL (1995): Effectiveness of the 4 mg dose of nicotine polacrilex for the initial treatment of high-dependent smokers. Arch. Intern. Med. 155: 1973–1980.

Sachs DPL, Sawe U, Leischow SJ (1993): Effectiveness of a 16 hour transdermal nicotine patch in a medical practice setting, without intensive group counseling. Arch. Intern. Med. 153: 1881–1890.

Schneider NG (1992): Nicotine therapy in smoking cessation: Pharmacokinetic considerations. Clin. Pharmacokinet. 23: 169–172.

Schneider NG (1994a): Clinical signs and symptoms of withdraw during smoking cessation. In: Future directions in nicotine replacement therapy: proceedings. Paris: Adis International.

Schneider NG (1994b): Nicotine nasal spray. Health Values 18: 4–9.

Schneider NG, Jarvik ME, Forsythe AB (1984): Nicotine versus placebo gum in the alleviation of withdraws during smoking cessation. Addict. Behav. 9: 149–156.

Schneider NG, Olmstead R, Mody FV, Doan K, Franzon M, Jarvik ME, Steinberg C (1995): Efficacy of a nicotine nasal spray in smoking cessation: a placebo-controlled, double-blind trial. Addiction 90: 1671–1682.

Schneider NG, Olmstead RE, Sloan K, Steinberg C, Daims R, Brown HV (1996a): Efficacy of buspirone in smoking cessation: a placebo-controlled trial. Clin. Pharmacol. Ther. 60(5): 568–575.

Schneider NG, Lunell E, Olmstead RE, Fagerstrom KO (1996b): Clinical pharmacokinetics of nasal nicotine delivery: a review and comparison to other nicotine systems. Clin. Pharmacokinet. 31: 65–80.

Schneider NG, Olmstead R, Nilsson F, Mody F, Franzon M, Doan K (1996c): Efficacy of a nicotine inhaler in smoking cessation: double-blind, placebo-controlled trial. Addiction 91: 1293–1306.

Shiffman S (1982): Relapse following smoking cessation: a situational analysis. J. Consult. Clin. Psychol. 50: 71–86.

Silagy C, Mant D, Fowler G, Lodge M (1994): Meta-analysis on efficacy of nicotine replacement therapies in smoking cessation. Lancet 343: 139–142.

Stolerman IP (1986): Could nicotine antagonists be used in smoking cessation? Br. J. Addict. 81: 47–53.

Sutherland G, Stapleton JA, Russell MAH, Jarvis MJ, Hajek P, Belcher M, Feyeraband C (1992): Randomised controlled trial of nasal nicotine spray in smoking cessation. Lancet 340: 324–329.

Tang JL, Law M, Wald N (1994): How effective is nicotine replacement therapy in helping people to stop smoking? Br. Med. J. 308: 21–26.

Tennant FS, Tarver AL (1985): Withdrawal from nicotine dependence using mecamylamine: Comparison of three-week and six-week dosage schedules. In: Harris LS, editor. Problems of drug dependence. NIDA Research Monograph. Washington, DC: U.S. Government Printing Office, 55: 291–297.

Tennant FS, Tarver AL, Rawson RA (1984): Clinical evaluation of mecamylamine for withdrawal from nicotine dependence. In: Harris LS, editor. Problems of drug dependence. NIDA Research Monograph. Washington, DC: U.S. Government Printing Office, 43: 291–297.

Tonnesen P, Fryd V, Hansen M, Helsted J, Gunnersen A, Forchammer H, Stockner M (1988): Effect of nicotine chewing gum in combination with group counseling in the cessation of smoking. N. Engl. J. Med. 318: 15–27.

Tonnesen P, Norregaard J, Mikkelsen K, Jorgensen S, Nilsson F (1993): A double-blind trial of a nicotine inhaler for smoking cessation. JAMA 269: 1268–1271.

Tonnesen P, Mikkelsen K, Norregaard J, Jorgensen S (1996): Recycling of hardcore smokers with nicotine nasal spray. Eur. Resp. J. 9(8): 1619–1623.

West R, Hajek P, McNeill A (1991): Effects of buspirone on cigarette withdrawal symptoms and short-term abstinence rates in a smokers clinic. Psychopharmacology 104: 91–96.

Zikos A, Molter G, Romovacek M, Nellis K, Gaudio R (1991): Comparison of nicotine gum, clonidine and no medication in a multidisciplinary smoking cessation program. Chest 100: 3S.

19

Preclinical Evidence on the Neuroprotective Effects of Nicotinic Ligands

Diana L. Donnelly-Roberts, PhD, and Jorge D. Brioni, PhD
Neurological and Urological Diseases Research (NUDR)
Pharmaceutical Products Division
Abbott Laboratories
Abbott Park, Illinois

Neuronal cell death accompanies several neurodegenerative diseases that include Alzheimer's disease (AD), Parkinson's disease (PD), AIDS dementia complex (ADC), epilepsy, Huntington's disease, ischemia, amyotrophic lateral sclerosis (ALS), and various spinal neuropathies. The causes of cell death during these various neurodegenerative diseases is still being investigated. However, among the several hypotheses being actively researched is one involving an initial triggering step for the cascade of cellular destruction due to excess intracellular calcium levels $[Ca^{2+}]_i$ that leads to cell death either by impaired energy metabolism, mitochondrial dysfunction, free radical formation, or oxidative stress.

One common theme in these neurodegenerative diseases during the cascade of cellular destruction is the presence of excess $[Ca^{2+}]_i$, and some of these diseases involve areas rich in cholinergic synapses that express the neuronal nicotinic acetylcholine receptor (nAChR).

Excitotoxic neurodegenerative damage is thought to be involved in the etiology of both AD and PD (Lipton and Rosenberg, 1994). AD possesses two major hallmarks that involve β-amy-

Neuronal Nicotinic Receptors: Pharmacology and Therapeutic Opportunities, Edited by S. P. Arneric and J. D. Brioni
ISBN 0-471-24743-x, pages 337–348. Copyright © 1998 by Wiley-Liss, Inc.

loid (Aβ) protein deposition and a severe cholinergic deficit. The deposition of $A\beta_{1-42}$ peptide in senile plaques represents a major pathogenic event during disease progression and this peptide can be neurotoxic both *in vitro* and *in vivo* in rodents. Aβ-induced neurotoxicity involves the dysregulation of $[Ca^{2+}]$ homeostasis as the possible initial triggering step of cellular destruction (Mattson, 1994).

ADC shares many downstream pathological events with AD. ADC involves a set of neurological dysfunctions, e.g., dementia, loss of concentration, blindness, and motor deficits that affect greater than 60% of AIDS patients. Infection with HIV-1 leads to the coat protein of the virus, gp120, being shed. *In vitro,* gp120 has neurotoxic effects in picomolar concentrations on rodent primary cortical or hippocampal cultures, regions rich in cholinergic neurons. One putative mechanism for this effect is a disruption of $[Ca^{2+}]_i$ homeostasis as the triggering step for the cascade of cell destruction (Lipton and Rosenberg, 1994).

ALS, or Lou Gehrig's disease, is a progressive, fatal disease characterized by the loss of motoneurons in both the brain stem and spinal cord. There are three major hypotheses for the cause of ALS: (1) an autoimmune hypothesis based on antibodies to calcium channels; (2) an oxidative stress hypothesis based on mutations in the Cu/Zn superoxide dismustase (SOD-1) gene; and (3) an excitatory amino acid hypothesis, which is the focus here, based on the observation of increased levels of glutamate (GLU) in patients with sporadic ALS (Rothstein and Kuncl, 1995).

There has been accumulating evidence to suggest a role for nAChRs in neuroprotection. First, there is a reduced population of nAChRs in the cortex and hippocampus of AD patients. Second, there is a negative correlation between smokers and the onset of PD and more controversially AD (van Duijn, 1995), thereby suggesting a potential therapeutic role for nAChRs. Third, the α-bungarotoxin (α-Btx)-sensitive α7 nAChR subtype is highly permeable to Ca^{2+}, suggestive of a potential role in either intracellular signaling, neurite outgrowth, synaptic transmission, modulation of neurotrophin levels such as nerve growth factor (NGF), or brain-derived neurotrophic factor (BDNF) as well as excitotoxic processes (Role and Berg, 1996; Albuquerque et al., 1997). Finally, some nAChR ligands exhibit neuroprotective properties both *in vivo* and *in vitro,* as discussed below. However, the side-effect liabilities associated with the use of (-)-nicotine, which include gastrointestinal and cardiovascular disturbances, have limited the clinical use of this compound.

This obstacle may be overcome by identifying compounds that selectively interact with the emerging diversity of neuronal nAChRs recently identified (Sargent, 1993). To date, 11 gene products (α2–α9 and β2–β4) have been isolated from neuronal avian, rodent, or human tissues. The α2–α6 subunits require the presence of a β subunit to form a pentameric functional receptor. In contrast, the α7–α9 subunits are capable of forming functional channels as homo-oligomers when expressed in *Xenopus* oocytes or in stably transfected cell lines (Albuquerque et al., 1997). This recent evolution of the molecular diversity of subunits equates to a diverse number of potential nAChR subtype combinations that exhibit different modulatory roles for various therapeutic targets (Role and Berg, 1996; Arneric et al., 1996).

Efforts to identity cholinergic channel ligands with diminished side-effect liabilities as compared to (-)-nicotine have resulted in the identification of several novel cholinergic channel modulators (ChCMs) that include ABT-418, GTS-21, RJR-2043, SIB 1508Y, SIB 1553A, ABT-089 (see Chapter 15 of this book; Holladay and McDonald, 1998), and ABT-594 (Bannon et al., 1998). These novel ChCMs are analogs of (-)-nicotine with behavioral-enhancing properties in addition to a substantially reduced side-effect profile compared to (-)-nicotine, possibly due to a differential potency and efficacy to activate the various nAChR subtypes (Arneric et al., 1996; Brioni et al., 1997).

TABLE 19-1. Summary of the Experimental Studies Investigating the *In Vitro* Neuroprotective Efficacy of nACHR Ligands

Compound	Effect
(-)-Nicotine	Increased viability of rat cortical neurons against GLU but not kainic acid toxicity (Akaike et al., 1994; Kaneko et al., 1997)
GTS-21	Increased neurite extension and reduced cell loss in PC12 cells induced by NGF removal (Martin et al., 1994)
(-)-Nicotine	Reduced NMDA neurotoxicity in mice striatal cultures as well as ACh, nornicotine, and DMPP (Marin, et al., 1994)
ABT-418	Attenuated GLU neurotoxicity in rat cortical cultures and human IMR-32 cells (Donnelly-Roberts et al., 1996a)
(-)-Nicotine	Protected against dexamethasone-enhanced kainic acid neurotoxicity in rat hippocampal cells (Semba et al., 1996)
(-)-Nicotine	Modulated the neurotoxic effect of Aβ (25–35) in hippocampal cultures (Zamani et al., 1997)
(-)-Nicotine	Attenuated Aβ (25–35)-induced neurotoxicity in rat cortical cultures, also attenuation with GTS-21 (Kihara et al., 1997)
ABT-089	Attenuated GLU-induced neurotoxicity and possessed enhanced potency following subacute exposure in rat cortical cells (Sullivan et al., 1997)
ABT-418 ABT-089	Attenuated Aβ (1–42) and gp 120-induced neurotoxicity in rat cortical cells (Donnelly-Roberts et al., 1996b)
ABT-594	Attenuated GLU or substance P-induced neurotoxicity in rat spinal cord cultures (Donnelly-Roberts et al., 1997)

IN VITRO NEUROPROTECTION WITH nAChR LIGANDS

nAChR ligands with neuroprotective activity are listed in Table 19-1. The different *in vitro* models utilized in these studies mimic some of the neurodegenerative diseases previously discusssed such as ischemia, AD, ADC, and ALS (Lipton and Rosenberg, 1994).

In a GLU-induced model of neurotoxicity, three laboratories have investigated the neuroprotective properties of (-)-nicotine as well as other ChCMs. All reports demonstrate that nAChR ligands are neuroprotective against GLU-mediated toxicity in primary rat cultures, either cortical or striatal (Akaike et al., 1994; Marin et al., 1994; Donnelly-Roberts et al., 1996a; Kaneko et al., 1997). These neuroprotective effects of (-)-nicotine were nAChR mediated since the nAChR antagonists, mecamylamine, a noncompetitive channel blocker, or hexamethonium, attenuated the neuroprotection. ABT-418 was similar in potency (EC_{50} = 8 μM) to (-)-nicotine in protecting against GLU-induced toxicity in rat cortical cells (Donnelly-Roberts et al., 1996a). The potency reported in this study for (-)-nicotine and ABT-418 is in agreement with the results of Akaike and co-workers (1994) but differs from Marin and co-workers (1994), who showed a 26-fold lower potency for (-)-nicotine. The latter finding may be related to the fact that (-)-nicotine was only coapplied at the time of the GLU exposure, whereas the previous studies pretreated the cells that conceivably would permit changes in immediate early gene (IEG) expression. This nAChR-induced neuroprotection was shown to be both concentration and time dependent with a minimum 2-hour preincubation for ChCMs (Akaike et al., 1994; Donnelly-Roberts et al., 1996a). Additionally, the role of Ca^{2+} was critical to this nAChR-mediated neuroprotection since the protection was reversed upon removal of extracellular Ca^{2+} during pretreatment with ABT-418 (Donnelly-Roberts et al., 1996a).

Recently, ABT-089 demonstrated similar acute neuroprotective properties to ABT-418 in the GLU-induced neurotoxicity model in rat cortical cultures but an enhancement in potency upon subacute ABT-089 (0.01–10 μM) exposure. This increase in neuroprotective potency parallels the behavioral data, which also showed an enhanced potency in cognition following a 7-day exposure of rodents to ABT-089 (Sullivan et al., 1997).

GTS-21, a novel anabaseine analog, is a partial agonist for the α7 nAChR subtype versus the α4β2 or α3β4 subtypes (Brioni et al., 1997) and has been evaluated for neuroprotective activity (Martin et al., 1994). Using NGF-dependent PC12 cells, GTS-21 was shown to protect cells from death upon NGF removal as measured by an increase in neurite extension and reduced cell loss. This GTS-21–induced neuroprotection occurs via an nAChR mechanism, since mecamylamine attenuated this effect.

$A\beta_{1-42}$ exhibits both *in vitro* and *in vivo* neurotoxicity and is thought to be one of the first peptides formed in AD plaques (Roch and Puttfarcken, 1996). Aggregated $A\beta_{1-42}$ was shown to induce significant neurotoxicity above basal levels in cultures of rat cortical neurons. ABT-418 or ABT-089 exhibited neuroprotective properties that are nAChR mediated using the $A\beta_{1-42}$ model (Donnelly-Roberts et al., 1996b). (-)-Nicotine has been shown to reduce the neurotoxicity of the $A\beta_{25-35}$ peptide (Zamani et al., 1997; Kihara et al., 1997), a segment of $A\beta_{1-42}$ thought to represent the toxic part of the 1–42 amino acid peptide. This peptide has not been identified in the plaques within rat brains and therefore the pathophysiological relevance is controversial.

In an *in vitro* model of gp120-induced neurotoxicity using rat cortical cultures, ABT-418 and ABT-089 were also found to be neuroprotective, an effect blocked by mecamylamine. Again, the role of Ca^{2+} was critical to this nAChR-mediated neuroprotection since the protection was prevented upon removal of extracellular Ca^{2+} during the preincubation step with ABT-418 (Donnelly-Roberts et al., 1996b).

More recently, *in vitro* studies of ALS using either a chick spinal cord motoneuron survival model or an excess GLU insult to rat spinal cord motoneurons demonstrated the role nAChRs play in spinal cord motoneuron survival. The chick spinal cord motoneuron model showed that pulsed (-)-nicotinic activation of the α7 nAChR rescued a significant number of the spinal motoneurons from undergoing NGF removal–induced cell death (Messi et al., 1997). The novel analgesic ChCM (Bannon et al., 1998), ABT-594, could also prevent GLU-induced cell death in rat spinal cord motoneurons (Table 19-1), an effect blocked by mecamylamine (Donnelly-Roberts et al., 1997).

In further exploring potential nAChR subtype(s) that mediated the neuroprotective actions of ABT-418, it was shown that the neuroprotection may be dependent upon interactions with subtypes of nAChRs sensitive to α-Btx such as α7 (Donnelly-Roberts et al., 1996a). As shown in Fig. 19.1, Methyllycaconitine (MLA) and α-Btx, nAChR antagonists with reported neuronal selectivity toward the α7 subtype relative to the α4β2 and ganglionic nAChRs, attenuated the neuroprotective effects of ABT-418 and ABT-089 in rat cortical cultures (Donnelly-Roberts et al., 1996a; Sullivan et al., 1997) and ABT-594 (Donnelly-Roberts et al., 1997). These results with α7 selective antagonists were recently supported in a GLU neurotoxicity model showing attenuation of (-)-nicotine-induced neuroprotection (Kaneko et al., 1997). In contrast, atropine did not block the effects of ChCMs in these studies, indicating that muscarinic receptors are not involved.

IN VIVO NEUROPROTECTION WITH nAChR LIGANDS

Subchronic treatment with (-)-nicotine administered via minipumps (2 weeks) can reduce the loss of tyrosine-hydroxylase immunoreactive neurons after partial transection of the mesostri-

TABLE 19.2 Summary of the Experimental Studies Investigating the *In Vivo* Neuroprotective Efficacy of nACHR Ligands

Compound	Effect
(-)-Nicotine	Subchronic (-)-nicotine (2 weeks) reduced TH-IR cell body loss in the nigrostriatal pathway after partial hemitransection (Janson et al., 1988)
(-)-Nicotine	Subchronic (-)-nicotine (2 weeks) restored glucose utilization in the nigrostriatal hemitransection model (Owman et al., 1989)
(-)-Nicotine	Chronic (-)-nicotine (3 months) reduced neuronal cortical loss in ibotenic-lesioned rats in the nucleus basalis (Sjak-Shie and Meyer, 1993)
(-)-Nicotine	Subchronic (-)-nicotine (4 weeks) increased neuronal loss in the septum after fimbria-fornix transection (Fuxe et al., 1994)
(-)-Nicotine	(-)-Nicotine counteracted the increase in TH neurons following perinatal asphyxia in 4-week old rats (Chen et al., 1995)
(-)-Nicotine	Subchronic (-)-nicotine (2 weeks) protected against quinolinic-induced neurodegeneration in the rat hippocampus (O'Neill et al., 1997)
GTS-21	GTS-21 increased AChE-stained neurons after fimbria-fornix transection (Martin et al., 1994)
GTS-21	GTS-21 protected cortical neurons from degeneration after ibotenic acid sions in the nucleus basalis (Nanri et al., 1997)

atal dopaminergic pathway (Janson et al., 1988), leading to the suggestion that the protective effect of (-)-nicotine may be due to desensitization of excitatory nAChRs in dopaminergic terminals that could lead to a reduction in firing rates and the metabolic demands of the dopamine neurons. The demonstration that nAChR antagonists, e.g., d-tubocararine and mecamylamine, also shared the neuroprotective action of (-)-nicotine by enhancing neurite outgrowth (Lipton et al., 1988) may also indicate the importance of nAChR desensitization. In a hemitransected rat study of animals receiving a 2-week (-)-nicotine treatment (Owman et al., 1989), increased glucose utilization in the lesioned caudate putamen was observed (Table 19-2). Although (-)-nicotine "normalized" the activity of the lesioned side, this study fell short of providing any direct evidence on the neuroprotective effect of (-)-nicotine.

Following asphyxia for 20 min, rat pups were treated with (-)-nicotine via suckling (surrogate mothers implanted with minipumps delivering 0.2 μmol/kg/h nicotine for 4 weeks). Control pups (no(-)-nicotine) showed a 50% reduction in exploratory activity, a 50% reduction on the turnover rate of dopamine, and an increase in the number of TH-IR neurons in the substantia nigra compared to (-)-nicotine-treated rats (Chen et al., 1995). It is unclear from these data how to correlate the reduction in turnover rate of dopamine with the increase in neuronal counts in the nigra, even if there is a causal relationship between these events. (-)-Nicotine reversed the increase in the TH-IR neurons in the nigra in pups that suffered asphyxia, but there was no correlation between the behavioral, biochemical, and morpholocial changes induced by asphyxia. Thus, it is difficult to reach any conclusion on the neuroprotective effects of (-)-nicotine in this model.

Fuxe and co-workers (1994) also studied the effect of nicotine in fimbria-lesioned rats. A 4-week infusion of (-)-nicotine via minipumps (0.125 mg/kg/h) did not exert any protective action in rats with fimbria-fornix transection and actually increased the disappearance of cholinergic bodies in the septum (Fuxe et al., 1994). These data are not consistent with the effect of a similar treatment with (-)-nicotine on the survival of dopaminergic neurons after a mechanical lesion (Janson et al., 1988) and indicate that different neurons might exhibit a differential sensitivity to (-)-nicotine depending on the nAChR subtype expressed. Thus, (-)-nicotine might be able to selectively protect some neurons (dopaminergic) but not others (cholinergic).

The reactivity of hippocampal neurons to nicotine's neuroprotective effects has been different compared to the septal counterparts. In a quinolinic-induced model of neurodegeneration, hippocampal pyramidal and granule cells were protected after a subchronic treatment with (-)-nicotine (O'Neill et al., 1997). Quinolinic acid is an endogenous substance that activates the NMDA subtype of glutamatergic receptors, and it is significantly increased in the cerebrospinal fluid of patients suffering Huntington's disease and AIDS dementia (Heyes et al., 1993; Lipton and Gendelman, 1995). The finding that a subchronic treatment with (-)-nicotine provided protection against quinolinic-mediated neurotoxicity indicates that activation of nAChRs could be useful to ameliorate the neurodegenerative process associated to NMDA excitotoxicity (O'Neill et al., 1997). Rats were implanted with subcutaneous minipumps delivering saline or (-)-nicotine 62 μmol/kg/day, and quinolinic acid injections were made in the dorsal hippocampus 7 days later. Quinolinic-injected animals were significantly impaired in the water maze test, but those animals receiving (-)-nicotine exhibited a robust cognitive improvement as they performed as well as control rats. The neuroprotective effect of (-)-nicotine was also observed at the histological level by the staining of neurons with hematoxilin and eosin and showed agreement with the behavioral data.

GTS-21 is a novel nAChR ligand not only with *in vitro* neuroprotective properties but with *in vivo* efficacy as well. Based on the previous discussion of the potential neuroprotective effects of α7 activation, it is not surprising that neuroprotective properties have been reported for GTS-21. With regards to the *in vivo* efficacy, it has been reported that twice daily administration of GTS-21 reduces cell death in the septal area in rats with partial unilateral transection of the fimbria fornix (Martin et al., 1994). In a recent study, daily oral administration of GTS-21 for 20 weeks significantly attenuated the neuronal loss in layers II and III of the parietal cortex induced by ibotenic acid lesions in the nucleus basalis magnocellularis (Nanri et al., 1997).

As it has been summarized in this section, the evidence on the neuroprotective effect of (-)-nicotine in experimental animals *in vivo* indicates that subchronic administration of (-)-nicotine provides protection to certain neuronal populations but not all, and that desensitization of nAChRs may play an important role. However, much work remains to be done to fully understand the biochemical basis of this effect. The selective expression of immediate early genes and growth factors, as well as the hormonal effects induced by (-)-nicotine, may be related to the long-lasting neuroprotective actions by the stimulation of nAChRs.

MECHANISTIC BASIS FOR NEUROPROTECTION BY nAChR LIGANDS

Although very few physiological effects have been attributed to the α7 subtype of nAChRs despite its cloning and expression over the last few years, increasing data suggest these α-Btx–sensitive receptors have a role in modulating Ca^{2+} homeostasis. We propose here that α7 nAChRs play an important role in neuronal viability and are an amenable target for neuroprotection.

The homo-oligomeric α7 nAChRs have a high Ca^{2+} to Na^{+} permeability, which may be equivalent or higher than some of the NMDA receptors (Role and Berg, 1996). It has also been demonstrated that an α-Btx-sensitive nAChR subtype raises intracellular Ca^{2+} in cultures of chick ciliary ganglion neurons and mediates retraction of pseudoneurites (Oppenheim et al., 1992). Regulation of neurotrophin levels in hippocampus may also be linked to an α-Btx-sensitive nAChR subtype, since blockade of these receptors with α-Btx induces mRNA for NGF and BDNF (Freedman et al., 1993). More recently, a link between the α7 subtype and pathophysiological neurodegeneration has been revealed. A novel degenerative-inducing mutation in *C. elegans,* deg-3 (U662), was recently described and demonstrated to result in a gain of

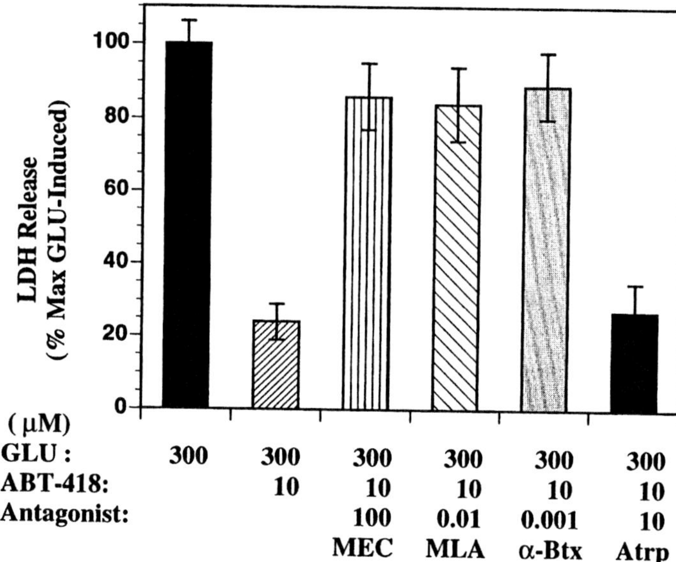

Figure 19.1. *The attenuation of ABT-418-induced neuroprotection against an insult by excess GLU is mediated via the nAChRs since it was reversed by mecamylamine (MEC) as well as α7-selective antagonists such as methyllycaconitine (MLA) or α-bungarotoxin (α-Btx) but not reversed with atropine (Atrp), a muscarinic antagonist.*

function that ultimately leads to vacuolated cell death (Treinin and Chalfie, 1995). This *C. elegans* mutation is equivalent to a mutation in the chick α7 transmembrane domain II (V251T), which also results in a gain of function as evidenced by increases in Ca^{2+} permeability and altered desensitization rates (Bertrand et al., 1993).

The mode of entry and levels of intracellular Ca^{2+} achieved appears to be decisive to the outcome of the intracellular signaling event. Ca^{2+} influx by alternate routes, either voltage-gated Ca^{2+} channels or chronic depolarization, can result in sustained increases in intracellular calcium to subtoxic levels, thereby allowing for neuroprotection from a toxic insult (Franklin and Johnson, 1992). This prompted Franklin and Johnson (1992) to propose the Ca^{2+} set point theory of cell survival (see Fig. 19-2). In addition, a role for Ca^{2+} entry through the α7 nAChR in the neuroprotective effects of ChCMs against different insults is suggested by the findings that removal of extracellular Ca^{2+} from the media during the preincubation period, but not during the time of insult, completely attenuated the nAChR-induced neuroprotection (Donnelly-Roberts et al., 1996a). The studies reported in this overview have documented the role of the α7 nAChR subtype in modulating $[Ca^{2+}]_i$, suggesting that alterations in Ca^{2+} influx might be a critical step underlying the observed neuroprotection. In addition, it was recently reported that ABT-418 can raise the intracellular Ca^{2+} levels to concentrations in the 300–390 nM range, which are considered trophic but not toxic in cells expressing the α7 subtype (Reganthan et al., 1997). This modulation of the $[Ca^{2+}]_i$ by ChCMs supports the Ca^{2+} set point theory, which suggests that nAChR ligands modulate Ca^{2+} dynamics to a subtoxic level that could potentially buffer incoming insults and thereby promote cell survival. Recently, a calcium hypothesis of ageing was postulated (Verkhratsky and Toescu, 1998) that implies changes in the regulation of $[Ca^{2+}]_i$ is the major cause of neurodegeneration. Future work in aged rat brains treated with and without novel ChCMs may reveal if any alterations in $[Ca^{2+}]_i$ occur.

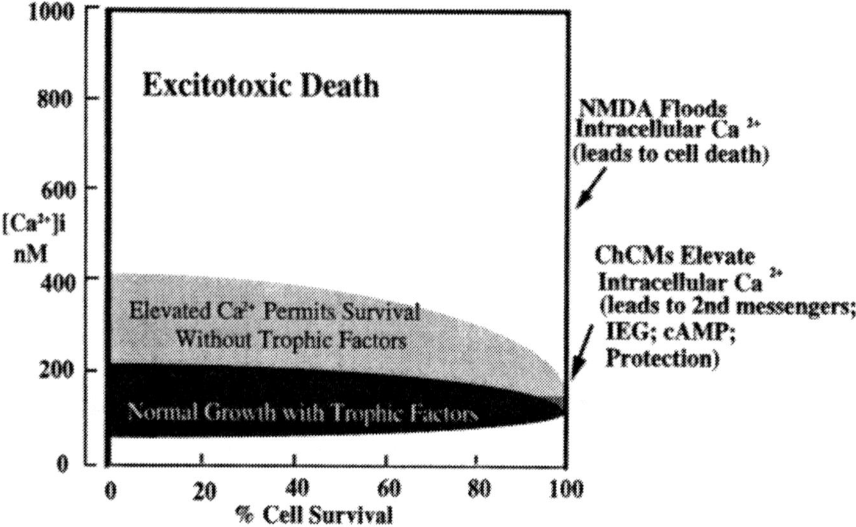

Figure 19.2. *The illustration of the Ca^{2+} set-point theory of cell survival (Franklin and Johnson, 1992). Excess $[Ca^{2+}]_i$, shown in white area, substantially above the normal conditions (greater than 3-fold) leads to excitotoxic cell death. In contrast, the normal conditions (area in black) for cells contain $[Ca^{2+}]_i$ of 200 nM or below but is dependent on trophic factors for growth, whereas a slightly elevated range of $[Ca^{2+}]_i$) (between 200 and 400 nM) above normal growing conditions permits cell survival without the addition of trophic factors (shaded area). This "subtoxic" $[Ca^{2+}]_i$, which is also elevated by not only chronic depolarization but ChCMs, could promote cell survival by the activation of several second-messenger systems such as IEGs or cAMP.*

The expression of bcl-2, a proto-oncogene, has been reported to overcome cell death induced by a variety of insults and in several cell types in vitro (Davies, 1995). Bcl-2 is only one of a large family of related genes controlling various aspects of apoptosis. Bcl_{xL} is a member of that family with the highest homology to bcl-2 and is also an anti-apoptotic gene. mRNA for bcl_{xL} is expressed at much higher levels in the developing and mature rodent brain than bcl-2 mRNA. It was demonstrated that nAChR ligands such as ABT-418 and ABT-594 were able to upregulate bcl_{xL} protein expression 2- to 3-fold over levels in control lysates (Donnelly-Roberts et al., 1996b, 1997). These results with nAChR ligands are consistent with past work with transforming growth factor β (Prehn et al., 1994) that indicates a minimum time of 2 hours and a 5-fold increase in bcl-2 is required to prevent apoptosis in rat hippocampal cultures against GLU insult. Studies are underway to determine the ratio between bcl and bax, a pro-apoptotic gene, since it has been suggested that it is the ratio between these two regulators that determines whether a cell undergoes death or survival. Nonetheless, this is evidence that a small molecule, such as a nAChR ligand, cannot only activate the opening of a multimeric ion channel but can possibly regulate the downstream synthesis of important cell-promoting proteins.

The finding that nAChR-mediated neuroprotection requires a minimum of 2 hours preincubation indicates that a downstream event is occurring. Although all the steps involved in mediating the neuroprotective effects of nAChR ligands are unclear, it is possible that the influx of Ca^{2+} through the α7 subtype triggers a cascade of intracellular events such as protein phosphorylation, induction of IEG expression, alteration in the levels of neurotrophic factors such as amyloid precursor protein (APP) [(-)-nicotine was shown to enhance the release of APP

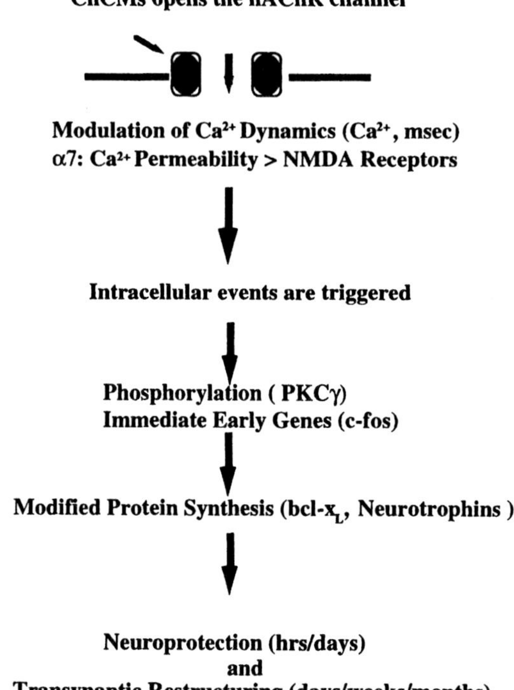

Figure 19.3 Proposed biochemical events leading to nAChR-mediated neuroprotection.

(Kim et al., 1997)], NGF, BDNF, neurotrophin-3, or induction of inhibitors of apoptosis (bcl-2). Other potential mechanisms are still under investigation such as modulation of nitric oxide levels (Shimohana et al., 1996) or effects on the release of arachidonic acid levels (Marin et al., 1997). However, it should be noted that the mechanism underlying the neuroprotection observed during acute (2-hour) treatment with ChCMs may be quite different from the mechanism underlying the subacute (7-day) exposure. It has been shown that chronic treatment with (-)-nicotine either *in vitro* or *in vivo* upregulates nAChRs (Flores et al., 1991). Whether upregulation of nAChRs is involved in the neuroprotection observed following subacute treatment has yet to be determined.

Most of the mechanistic studies for the rationale of nAChRs as neuroprotectants (Fig. 19-3) stem from *in vitro* studies. While not directly reflecting the complexity of the *in vivo* situation, the merits of the *in vitro* studies have enabled a number of important advances. *In vitro* studies make it possible to dissect each step involved in the neuroprotection pathway and help direct the longer and more complex *in vivo* studies. Naturally, both are needed to reach a greater understanding of the events involved. Most of these neurodegenerative diseases could possibly have a common triggering step in cell death and therefore nAChR ligands could also have a common pathway as neuroprotectants. Therefore, both the *in vivo* and *in vitro* neuroprotective effects of ChCMs raise the possibility that they may have clinical utility slowing the progression of several neurodegenerative diseases.

REFERENCES

Akaike A, Tamura Y, Yokota T, Shimohama S, Kimura J (1994): Nicotine-induced protection of cultured cortical neurons against N-methyl-D-aspartate receptor-mediated glutamate cytotoxicity. Brain Res. 644: 181–187.

Albuquerque EX, Manickavasagom A, Pereira EFR, Castro NG, Schrattenholz A, Barbosa CTF, Bonfante-Cabarcas R, Aracava Y, Eisenberg HM, Maelicke A (1997): Properties of neuronal nicotinic acetylcholine receptors: pharmacological characterization and modulation of synaptic function. J. Pharmacol. Exp. Ther. 280: 1117–1136.

Arneric SP, Holladay MW, Sullivan JP (1996): Cholinergic channel modulators as a novel therapeutic strategy for Alzheimer's disease. Exp. Opin. Invest. Drugs 5: 79–100.

Bannon AW, Decker MW, Holladay MW, Curzon P, Donnelly-Roberts DL, Puttfarcken PS, Bitner RS, Diaz A, Dickenson AH, Porsolt RD, Williams M, Arneric SP (1998): Broad spectrum, non-opioid analgesic activity by selective modulation of neuronal nicotinic acetylcholine receptors. Science 279: 77–81.

Bertrand D, Galzi JL, Devillers-Thiery A, Betrand S, Changeux JP (1993): Mutations at two distinct sites within the channel domain M2 alter calcium permeability of neuronal α7 nicotinic receptor. Proc. Natl. Acad. Sci. USA 90: 6971–6975.

Brioni JD, Decker, MW, Sullivan JD, Arneric SP (1997): The pharmacology of (-)-nicotine and novel cholinergic channel modulators. Adv. Pharmacol. 37: 153–214.

Chen Y, Ogren S, Bjelke B, Bolme P, Eneroth P, Gross J, Liodi F, Herrera-Marschitz M, Anderson K (1995): Nicotine treatment counteracts perinatal asphyxia-induced changes in the mesostriatal-limbic dopamine systems and in motor behaviour in the four-week-old male rat. Neuroscience 68: 531–538.

Davies AM (1995): The bcl-2 family of proteins and the regulation of neuronal survival. Trends Neurosci. 18: 335–358.

Donnelly-Roberts DL, Xue IC, Arneric SP, Sullivan JP (1996a): In vitro neuroprotective properties of the novel cholinergic channel activator (ChCA), ABT-418. Brain Res. 719: 36–44.

Donnelly-Roberts DL, Jacobs I, Arneric SP, Sullivan JP (1996b): The cholinergic channel activator (ChCA) ABT-418 protects against gp120 and Aβ(1–42)-induced neurotoxicity. Soc. Neurol. Abstr. 749.6.

Donnelly-Roberts DL, Jacobs I, Puttfarcken P, Kuntzweiler TA, Arneric SP (1997): ABT-594: neuroprotective effects in spinal cord (sc) cultures by a broad spectrum analgesic acting via nicotinic acetylcholine receptors (nAChRs). Soc. Neorol. Abstr. 477.12.

Flores CM, Rogers SW, Pabreza LA, Wolfe BB, Kellar KJ (1991): A subtype of nicotinic cholinergic receptor in rat brain is composed of α4 and β2 subunits and is up-regulated by chronic nicotine treatment. Mol. Pharmacol. 41: 31–37.

Franklin JL, Johnson EM Jr. (1992): Suppression of programmed neuronal death by sustained elevation of cytoplasmic calcium. Trends Neurosci. 15: 501–508.

Freedman R, Wetmore C, Stromberg I, Leonard S, Olson L (1993): α-Bungarotoxin binding to hippocampal interneurons: immunocytochemical characterization and effects on growth factor expression. J. Neurosci. 13: 1965–1975.

Fuxe K, Rosen L, Lippoldt P, Andbjer B, Hasselrot U, Agnati L (1994): Chronic continuous infusion of nicotine increases the disappearance of choline acetyltransferase immunoreactivity in the cholinergic cell bodies of the medial septal nucleus following a partial unilateral transection of the fimbria fornix. Clin. Investig. 72: 262–268.

Heyes MP, Saito K, Major EO, Milstein S, Markey SP, Vickers JH (1993): A mechanism of quinolinic acid formation by brain in inflammatory neurological disease. Brain 116: 1425–1450.

Janson A, Fuxe K, Agnati L, Katayama I, Harfstrand A, Anderson K, Goldstein M (1988): Chronic nicotine treatment counteracts the disappearance of tyrosine-hydroxylase-immunoreactive nerve cell bodies, dendrites and terminals in the mesostriatal dopamine system of the male rat after partial hemitransection. Brain Res. 455: 332–345.

Kaneko S, Maeda T, Kume T, Kochiy AH, Akaike A, Shimoshama S (1997): Nicotine protects cultured cortical-neurons against glutamate-induced cytotoxicity via alpha (7)-neuronal receptors and neuronal CNS receptors. Brain Res. 765: 135–140.

Kihara T, Shimohana S, Kimura J, Kume T, Kochiyama H, Maedao T, Akaike A (1997): Nicotinic receptor stimulation protects neurons against beta-amyloid toxicity. Ann. Neurol. 42: 159–163.

Kim SH, Kim YK, Jeong SJ, Haass C, Kim YH, Suh YH (1997): Enhanced release of secreted form of Alzheimer's amyloid presursor protein from PC12 cells by nicotine. Mol. Pharmacol. 52: 430–436.

Lipton SA, Rosenberg PA (1994): Exicitatory amino acids as a final common pathway for neurologic disorders. New Eng. J. Med. 330: 934–940.

Lipton S, Frosch M, Phillips M, Tauck D, Aizenmann E (1988): Nicotinic antagonists enhance process outgrowth by retinal ganglion cells in culture. Science 239: 1293–1296.

Marin P, Maus M, Desagher S, Glowinski J, Premont J (1994): Nicotine protects cultured striatal neurones against N-methyl-D-aspartate receptor-mediated neurotoxicity. Neuroreport 5: 1977–1980.

Marin P, Hamon B, Glowinski J, Premont J (1997): Nicotine-induced inhibition of neuronal phospholipase A2. J. Pharmacol. Exp. Ther. 280: 1277–1283.

Martin EJ, Panickar KS, King MA, Deyrup M, Hunter BE, Wang G, Meyer EM (1994): Cytoprotective actions of 2,4-dimethoxybenzylidene anabaseine in differentiated PC12 cells and septal cholinergic neurons. Drug Dev. Res. 31: 135–141.

Mattson MP (1994): β-amyloid precursor protein metabolites, metabolic compromise, loss of neuronal calcium homeostasis in Alzheimer's disease. Ann. NY Acad. Sci. 747: 50–76.

Messi ML, Renganathan M, Grigorenko E, Delbono O (1997): Activation of α7-nicotinic acetylcholine receptor promotes survival of spinal cord motoneurons. FEBS Lett. 411: 32–38.

Nanri M, Kasahara N, Kasahara N, Yamamoto J, Miyake H, Watanabe H (1997): GTS-21, a nicotinic agonist, protects against neocortical neuronal cell loss induced by the nucleus basalis magnocellularis lesion in rats. Jap. J. Pharmacol. 74: 285–289.

Newhouse P, Sunderland T, Tariot P, Blumhardt CL, Weingartner H, Mellow A (1988): Intravenous nicotine in Alzheimer's disease: a pilot study. Psychopharmacology 95: 171–175.

O'Neill AB, Morgan SB, Brioni JD (1997): Histological and behavioral protection by (-)-nicotine against quinolinic acid-induced neurodegeneration in the hippocampus. Neurobiology of Learning and Memory, 69: 46–64 (1998).

Oppenheim RW, Yin QW, Prevette D, Yan Q (1992): Brain-derived neurotrophic factor rescues developing avian motoneurons from cell death. Nature 360: 755–757.

Owman C, Fuxe K, Jason AM, Kahrstrom J (1989): Studies of protective actions of nicotine on neuronal and vascular functions in the brain of rats: comparison between sympathetic noradrenergic and mesostriatal dopaminergic fiber system, and the effects of a dopamine agonist. Prog. Brain Res. 79: 267–276.

Prehn JHM, Bindokas VP, Marcuccilli CJ, Krajewski S, Reed JC, Miller RJ (1994): Regulation of neuronal Bc12 protein expression and calcium homestasis by transforming growth factor type β confers wide-ranging protection on rat hippocampal neurons. Proc. Natl. Acad. Sci. USA 91: 12,599–12,603.

Reganthan M, Gopalakrishnan M, Messi ML, Arneric SP, Sullivan JP, Delbono O (1998): Regulation of intracellular calcium by ABT-418: involvement of the human α7 neuronal nicotinic acetylcholine receptor. Brain Res. submitted.

Roch J-M, Puttfarcken P (1996): Biological actions of the β-amyloid protein and its precursor. Current Drugs 1: 9–16.

Role LW, Berg DK (1996): Nicotinic receptors in the development and modulation of CNS synapses. Neuron 16: 1077–1985.

Rothstein JD, Kuncl RW (1995): Neuroprotective strategies in a model of chronic glutamate-mediated motor neuron toxicity. J. Neurochem. 65: 643–651.

Sargent PB (1993): The diversity of neuronal nicotinic acetylcholine receptor. Annu. Rev. Neurosci. 16: 403–443.

Semba J, Miyoshi R, Kito S (1996): Nicotine protects against the dexamethasone potentiation of kainic acid-induced neurotoxicity in cultured hippocampal neurons. Brain Res. 735: 335–338.

Shimohana S, Akaike A, Kimura J (1996): Nicotine-induced protection against glutamate cytotoxicity. Nicotinic cholinergic receptor-mediated inhibition of nitric oxide formation. Ann. NY Acad. Sci. 777: 356–361.

Sjak-Shie N, Meyer E (1993): Effects of chronic nicotine and pilocarpine administration on neocortical neuronal density and [^{3}H]GABA uptake in nucleus basalis lesioned rats. Brain Res. 624: 295–298.

Sullivan JP, Donnelly-Roberts D, Briggs CA, Anderson DJ, Gopalakrishnan M, Xue IC, Piattoni-Kaplan M, Molinari E, Campbell JE, McKenna DG, Gunn DE, Lin N-H, Ryther K, He Y, Holladay MW, Wonnacott S, Williams M, Arneric SP (1997): ABT-089 [2-methyl-3-(2-(S)-pyrrolidinylmethoxy)pyridine]: I. A potent and selective cholinergic channel modulator with neuroprotective properties. J. Pharmacol. Exp. Ther. 283: 235–246.

Treinin M, Chalfie M (1995): A mutated acetylcholine receptor subunit causes neuronal degeneration in *C. elegans*. Neuron 14: 871–877.

van Duijn CM, Havekes LM, Van Broeckhoven C, de Knijff P, Hofman A (1995): Apolipoprotein E genotype and association between smoking and early onset Alzheimer's disease. Brit. Med. J. 310: 627–631.

Verkhratsky A, Toescu EC (1998): Calcium and neuronal ageing. Trends Neurosci. 21: 2–7.

Zamani MR, Allen YS, Owen GP, Gray J (1997): Nicotine modulates the neurotoxic effect of β-amyloid protein (25–35) in cultured hippocampal cultures. Neuroreport 8: 513–517.

20

Transdermal Nicotine Treatment of Attention Deficit Hyperactivity Disorder

Edward D. Levin PhD, Barbara B. Simon PhD, and C. Keith Conners PhD
Department of Psychiatry and Department of Pharmacology
Duke University Medical Center
Durham, North Carolina

Nicotine has long been known to improve aspects of cognitive function in humans and laboratory animals (for reviews, see Brioni et al., 1997; Decker et al., 1995; Levin, 1992). Nicotine-induced improvements in attentional performance in humans have been repeatedly seen (for review, see Warburton, 1992). Nicotine-induced attentional improvement suggests the possible use of nicotine or other nicotinic ligands as therapy for attention deficit hyperactivity disorder (ADHD).

ADHD is a cognitive disorder that typically becomes manifest during the preschool years and can persist into adulthood. The symptomology includes increased impulsivity, hyperactivity, and most prominently attentional deficit (DSM-IV, 1994). Recently, the persistence of ADHD into adulthood has been better characterized. Approximately 60% of juveniles with ADHD have persisting symptoms of attentional deficit into adulthood (Wender, 1995).

NICOTINE EFFECTS ON LEARNING AND MEMORY

The best characterized effect of nicotine on cognitive function in experimental animal studies is its effect on memory. The most profound effect is improving working memory, while reference memory is much less affected (Levin et al., 1996a, 1997a, 1997c). Nicotine-induced

Neuronal Nicotinic Receptors: Pharmacology and Therapeutic Opportunities, Edited by S. P. Arneric and J. D. Brioni
ISBN 0-471-24743-x, pages 349–357. Copyright © 1998 by Wiley-Liss, Inc.

memory improvement has been seen in several different species in a variety of behavioral test procedures. We have documented working memory improvements in rats on the radial arm maze after acute (Levin, 1996, 1996c, 1997a; Levin and Torry, 1995) or chronic (Levin et al., 1993, 1996a; Levin and Rose, 1995; Levin and Torry, 1996) nicotine administration. Others have found similar improvements following nicotine treatment in rodents on passive avoidance (Brioni and Arneric, 1993; Decker et al., 1994a; Zarrindast et al., 1996) and Morris water maze tasks (Socci et al., 1995) and in monkeys on delayed or nondelayed matching to sample tasks (Buccafusco et al., 1995, 1996; Elrod et al., 1988; Jackson and Buccafusco, 1989).

Selective nicotinic agonists such as ABT-418, RJR 2403, DMAE, epibatidine, and SIB1765-F have also been shown to significantly improve working memory performance in experimental animals. For example, the $\alpha4\beta2$ agonist ABT-418 significantly increased retention of a passive avoidance task in rodents (Decker et al., 1994b) and enhanced delayed-match-to-sample performance in monkeys (Buccafusco et al., 1996) and rats (Terry, 1997). the nicotinic agonists dimethylethanolamine (DMAE) (Levin et al., 1995), epibatidine (Levin et al., 1996d), and lobeline (Decker et al., 1993; Terry, 1997) also increase working memory in rats on radial arm maze and delayed discrimination tasks. The $\alpha7$ agonist GTS-21 facilitates learning and performance of a variety of tasks including delayed match to sample in monkeys (Briggs, 1997), passive avoidance (Meyer et al., 1994), active avoidance, Lashley III maze, and 17-arm radial maze tasks (Arendash et al., 1995). GTS-21 also enhanced learning in older rabbits, suggesting that this and related compounds might have beneficial effects in aged or deficient organisms (Woodruff-Pak et al., 1994).

As with all nootropics, the positive effects of nicotine are critically dependent on dose and requirements of the test procedure. There is evidence for the typical inverted U–shaped dose-response curve, with moderate doses causing improvements and higher doses being less effective or actually causing decrements. Nicotine-induced increases in proactive interference can cause choice accuracy deficits in situations where proactive interference is prominent (Dunnett and Martel, 1990; Levin et al., 1997b).

In contrast, nicotinic antagonists have been shown to significantly impair working memory performance. The most commonly examined antagonist is mecamylamine, which significantly impairs working memory performance in the radial arm maze after acute administration (Andrews et al., 1994; Levin et al., 1997a; Zarrindast et al., 1996). Interestingly, chronic administration of low doses of mecamylamine has been shown to significantly improve working memory performance on the radial arm maze (Levin et al., 1993) and T-maze spatial alternation (Levin et al., 1997b). ICV infusion of mecamylamine (Brucato et al., 1994; Decker and Majchrzak, 1992) or DHβE (Curzon et al., 1996) causes memory impairments. Local infusion of nicotinic antagonists into the brain significantly impairs working memory performance on the radial arm maze. Infusion of mecamylamine in the VTA, substantia nigra, or ventral hippocampus significantly impaired working memory performance in the radial arm maze (Levin et al., 1994). Infusion of the same dose range of mecamylamine into the nucleus accumbens was without apparent effect (Kim and Levin, 1996). Infusion of the more specific nicotinic antagonists MLA and DHβE into the ventral hippocampus also significantly impaired working memory in the radial arm maze (Kim and Levin, 1996). However, the high doses of these drugs needed to invoke the effect suggests that there was probably less receptor subtype–selective effects.

NICOTINE EFFECTS ON ATTENTION

Human Studies

Much of the early work concerning nicotine effects on cognitive function studied smokers as a group and cigarette smoking as a delivery form. Warburton and his colleagues have found that smoking improves attentional performance (Rusted and Warburton, 1992). However, nicotine-induced memory improvement can also be seen even when nicotine is administered after initial encoding (Warburton et al., 1992a). In particular, nicotine can prevent the decrement typically found in visual (Wesnes and Warburton, 1984a, 1984b) and auditory vigilance tasks (Mangan, 1983). Initially, there was concern over the relationship of nicotine effects separate from the relief of withdrawal (see Heishman et al., 1993, for a more complete review). However, more recently this has been resolved. Although there are identifiable cognitive deficits associated with smoking withdrawal (Hatsukami et al., 1989), nicotine seems to improve attentiveness even under minimal or nonexistent withdrawal states.

Warburton and Arnall (1994) found that smoking in minimally deprived smokers significantly improved attentiveness. In addition, Rusted and colleagues have been able to show that performance can be significantly improved in "light" or infrequent smokers (Rusted and Eaton-Williams, 1991; Rusted et al., 1995). Other studies found that nicotine reduced decision time on choice tasks of varying complexity irrespective of the subjects' level of smoking (Bates et al., 1994). We recently studied the effects of nicotine skin patch effects in nonsmoking normal adults (Levin et al. 1998). Nicotine skin patch administration caused a significant improvement on the continuous performance task. Other research has also found that subcutaneous nicotine in nonsmokers can reliably improve performance of an information processing task (Le Houezec et al., 1994). These findings demonstrate that nicotine can improve attentiveness even in people who do not smoke and thus are not either already under the influence of nicotine or in the state of nicotine withdrawal.

There was at one time some controversy concerning the degree to which this reflected merely an attenuation of a nicotine withdrawal–induced attentional deficit. More recently, it has been demonstrated that nicotine will improve attentiveness in nondeprived smokers and "chippers," smokers who only smoke occasionally and do not show pronounced withdrawal symptoms (Warburton and Arnall, 1994). Thus, nicotine seems to have attentional-improving effects apart from alleviation of withdrawal symptoms.

While the majority of studies find enhanced performance in both smokers and nonsmokers following nicotine, some have found either no facilitation or even performance decrements following treatment (Spilich et al., 1992). It may be that nicotine can enhance attention and consolidation processes while not affecting retrieval from long-term memory or integration and processing of complex material. These conclusions are based on results from studies failing to find nicotine facilitation on the Wechsler Memory Scale and text comprehension tasks (Spilich et al., 1992). However, nicotine appears to enhance processing resources as well, since oral treatment in smokers improved recall on a 32-item list (Warburton et al., 1992b) and also increased scores on the Raven IQ Advanced Progressive Matrices, a measure of information processing, in smokers deprived for 2 hours (Stough, 1994).

While nicotine can facilitate performance on a variety of cognitive tasks, cognitive impairment can result following nicotinic antagonists in humans. The nicotine antagonist mecamylamine caused a significant increase in errors during learning of a repeated acquisition task and also increased the number of false alarms and reaction time during a recognition task (Newhouse et al., 1992).

Animal Studies

Nicotine-induced attentional improvements have been much more difficult to demonstrate in animal subjects, especially rodents. In fact, nicotine had only moderate effects at best in studies of attention in rats (Bushnell, 1997; Evenden et al., 1993). Other studies have found no effects (Turchi et al., 1995) or even a trend toward detrimental effects following nicotine (Turchi et al., 1996). However, failures to show improvements with nicotine in animal studies are probably the result of factors including dose, the nature of the task, and level of training (see Levin, 1992, for a more complete review). For example, in studies by Turchi et al. (1995, 1996), rats were required to perform with a baseline accuracy of only 59%. As this is only slightly above chance level, it is difficult to ascertain that the animals learned the task contingencies well enough to overcome the disruptive effects of stimulant administration (Weiner et al., 1986).

In contrast, other studies have shown that nicotine can provide protection in rats against the vigilance decrement induced by intracerebroventricular (ICV) administration of ethylcholine mustard aziridinium ion toxin (Miyata, 1991). Moreover, we have recently found a dose-dependent effect of nicotine in a two-choice light discrimination task that maintained baseline performance between 75% and 87% correct. While lower doses (0.1–0.2 mg/kg sc) had minimal effects in aged (24 months) rats, higher doses (up to 0.4 mg/kg) of nicotine facilitated performance in this population (Simon et al., 1998).

ATTENTION DEFICIT HYPERACTIVITY DISORDER

The idea that nicotine may be useful in the treatment of ADHD came from three different lines of evidence. First was the previously described attentional improving effects of nicotine in smokers. Second was the finding that nicotine potently simulated the release of dopamine (Wonnacott et al., 1989), a primary mechanism of action of the stimulants such as methylphenidate, amphetamine, and pemoline currently used to treat ADHD. Third was the finding that adults with ADHD showed a smoking rate about twice the societal background rate (Pomerleau et al., 1995). Might these adults with ADHD be self-medicating with nicotine to attenuate their symptoms of ADHD?

Given the attentional improvements caused by nicotine (see above), it is possible that people with attention deficits might use nicotine as a form of self-medication. There is recent evidence that this may be true. Smoking rates in adults with ADHD are substantially higher than the general population (Milberger et al., 1997; Pomerleau et al., 1995). However, this is only suggestive, because it has also been found that adults with ADHD also use a variety of drugs at higher rates than the general population (Biederman et al., 1997). Further evidence supporting the possible self-medication of adults with ADHD with nicotine from cigarette smoking is the finding that nicotine, like the standard treatments for ADHD, methylphenidate and amphetamine, acts as an indirect agonist of dopamine and norepinepherine (Wonnacott et al., 1989). These findings led to the hypothesis that nicotine administered via skin patch would attenuate symptoms of inattention in adults with ADHD.

We conducted several studies of adults with ADHD to determine in placebo-controlled double-blind studies whether nicotine administered via skin patches would attenuate symptoms in adults with ADHD (Conners et al., 1996; Levin et al., 1996b). Adults ages 19–51 years with a positive diagnosis for ADHD residual type (DSM-IV, 1994) were administered nicotine skin patches or placebos (Nicoderm). Six subjects were cigarette smokers. They smoked an average of 31.4 ± 15.5 (mean ± SD) cigarettes/day for an average of 18.7 ± 7.3 years. They were administered 21 mg/day nicotine skin patches or placebos after overnight abstinence. Eleven subjects were nonsmokers verified by end-tidal CO < 8 ppm. They were administered 7 mg/day nicotine skin patches or placebos. The subjects were tested for a baseline session and counterbalanced order sessions with either nicotine or placebo patches. Patches were administered 3 hours prior to the Clinical Global Impressions (CGI) rating. Ten minutes later the Conners Continuous Performance Test (CPT) (Conners, 1994, 1995) was given. Finally, an interval timing test was given 3 hours and 45 minutes after initial patch administration.

The CGI showed significant nicotine-induced effects of enhancing the improvement ($p <$

0.005) and the therapeutic effect ($p < 0.01$) subscales over all subjects. Importantly, these significant beneficial effects ($p < 0.001$ for improvement and $p < 0.001$ for therapeutic effect) were also seen when the analysis was restricted to the nonsmoking subjects. This demonstrates that the improvement in clinical symptoms was not just an effect of reducing the adverse effects of nicotine withdrawal. Nicotine caused a significant effect even when the subjects were nonsmokers. The effect was also seen in terms of subjective self-perceived vigor as measured on the Profile of Mood States (POMS) test. Again this significant effect of nicotine was also seen when the analysis was restricted to nonsmokers. A significant reduction in variability of response speed across the blocks of a CPT session was seen with the smokers ($p < 0.05$). A nearly significant reduction was found in variability of response speed over different interstimulus delays in the nonsmokers ($p < 0.06$). The nonsmokers were given an additional test of interval timing at the end of the session. They showed a significant nicotine-induced reduction in spread of response times and increase in mean accuracy. This acute study with nicotine administered over a 4.5-hour session showed promising results concerning the potential use of nicotinic treatment for ADHD. A study of chronic effects of nicotine in comparison with methylphenidate is currently underway.

Nicotine has also been found by Sanberg and his coworkers (Sanberg et al., 1997) to be effective in attenuating the motor symptoms of Tourette's syndrome, a disorder characterized by twitches and explosive movement of the limbs and coprolalia. Nicotine given via either chewing gum or skin patch is effective in attenuating motor symptoms when given alone or in combination with the usual treatment of dopamine antagonists (McConville et al., 1991; Sanberg et al., 1988; Shytle et al., 1996; Silver and Sanberg, 1993; Silver et al., 1995). Attention deficit disorder is seen at a higher incidence in children with Tourette's disorder than in the general population. Sanberg et al. found that nicotine attenuated symptoms of inattention in children with Tourette's syndrome (Sanberg et al., 1997).

CONCLUSIONS

Nicotine or other nicotinic ligands hold promise as treatments for attention deficit hyperactivity disorder. Nicotine given via skin patches has been found to significantly attenuate symptoms of inattention in adults with ADHD and children with Tourette's disorder. Nicotine given via skin patches is likely to be substantially safer than cigarette smoking because none of the over 4,000 compounds present in tar are delivered. The risk of administering a drug must also be rated in comparison to the current line of therapy.

The stimulants methylphenidate and amphetamine are currently widely used. They have side effects of appetite suppression, sleep disturbance, and abuse liability. This is a similar profile as nicotine. Nicotine's abuse liability seems to be related to the kinetics of self-administration. The abuse liability of the nicotine patch has been found to be low in a study by Henningfield and co-workers (Pickworth et al., 1994). This may be due to the zero-order pharmacokinetics achieved by the patch. Nicotinic sister drugs are being developed by a variety of pharmaceutical companies and academic labs. Some may prove to be useful for ADHD. Certainly the working memory improvement caused by ABT-418, ABT-089, DMAE, nornicotine, norisonicotine, RJR 2403, and SIB 1765F show that these compounds hold promise worthy of further investigation.

REFERENCES

Andrews JS, Jansen JHM, Linders S, Princen A (1994): Effects of disrupting the cholinergic system on short-term spatial memory in rats. Psychopharmacology 115: 485–494.

Arendash GW, Sengstock GJ, Sanberg PR, Kem WR (1995): Improved learning and memory in aged rats with chronic administration of the nicotinic receptor agonist GTS-21. Brain Res. 674: 252–259.

Bates T, Pellett O, Stough C, Mangan G (1994): Effects of smoking on simple and choice reaction time. Psychopharmacology 114: 365–368.

Biederman J, Wilens T, Mick E, Faraone SV, Weber W, Curtis S, Thornell A, Pfister K, Jetton JG, Soriano J (1997): Is ADHD a risk factor for psychoactive substance use disorders? Findings from a four-year prospective follow-up study. J. Am. Acad. Child Adolesc. Psych. 36: 21–29.

Briggs CAA, Brioni JG, Buccafusco JJ, Buckley MJ, Campbell JE, Decker MW, Donnelly-Roberts D, Elliott RL, Gopalakrishnan M, Holladay MW, Hui Y-H, Jackson WJ, Kim DJB, Marsh KC, O'Neill A, Prendergast MA, Ryther KB, Sullivan JP, Arneric SP (1997): Functional characterization of the novel neuronal nicotinic acetylcholine receptor ligand GTS-21 in vitro and in vivo. Pharmacol. Biochem. Behav. 57: 231–241.

Brioni JD, Arneric SP (1993): Nicotinic receptor agonists facilitate retention of avoidance training—participation of dopaminergic mechanisms. Behav. Neural Biol. 59: 57–62.

Brioni JD, Decker MW, Sullivan JP, Arneric SP (1997): The pharmacology of (-)-nicotine and novel cholinergic channel modulators. Adv. Pharmacol. 37: 153–214.

Brucato FH, Levin ED, Rose JE, Swartzwelder HS (1994): Interocerebroventricular nicotine and mecamylamine alters radial-arm maze performance in rats. Drug Dev. Res. 31: 18–23.

Buccafusco JJ, Jackson WJ, Terry AV, Marsh KC, Decker MW, Arneric SP (1995): Improvement in performance of a delayed matching-to-sample task by monkeys following ABT-418: a novel cholinergic channel activator for memory enhancement. Psychopharmacology 120: 256–266.

Buccafusco JJ, Prendergast MA, Terry AV, Jackson WJ (1996): Cognitive effects of nicotinic cholinergic receptor agonists in nonhuman primates. Drug Dev. Res. 38: 196–203.

Bushnell PJ, Oshiro, WM, Padnos BK (1997): Detection of visual signals by rats: effects of chlordiazepoxide and cholinergic and adrenergic drugs on sustained attention. Psychopharmacology, 134: 230–241.

Conners CK (1994): The Continuous Performance Test (CPT): use as a diagnostic tool and measure of treatment outcome. Annual Convention of the American Psychological Association. Los Angeles, California.

Conners CK (1995): The Continuous Performance Test. Toronto: Multi-Health Systems.

Conners CK, Levin ED, Sparrow E, Hinton S, Ernhardt D, Meck WH, Rose JE, March J (1996): Nicotine and attention in adult ADHD. Psychopharmacol. Bull. 32: 67–73.

Curzon P, Brioni JD, Decker MW (1996): Effect of intraventricular injections of dihydro-beta-erythroidine (DHβE) on spatial memory in the rat. Brain Res. 714: 185–191.

Decker MW, Majchrzak MJ (1992): Effects of systemic and intracerebroventricular administration of mecamylamine, a nicotine cholinergic antagonist, on spatial memory in rats. Psychopharmacology 107: 530–534.

Decker MW, Majchrzak MJ, Arneric SP (1993): Effects of lobeline, a nicotinic receptor agonist, on learning and memory. Pharmacol. Biochem. Behav. 45: 571–576.

Decker MW, Buckley MJ, Brioni JD (1994a): Differential effects of pretreatment with nicotine and lobeline on nicotine-induced changes in body temperature and locomotor activity in mice. Drug Dev. Res. 31: 52–58.

Decker MW, Curzon P, Brioni JD, Arneric SP (1994b): Effects of ABT-418, a novel cholinergic channel ligand, on place learning in septal-lesioned rats. Eur. J. Pharmacol. 261: 217–222.

Decker MW, Brioni JD, Bannon AW, Arneric SP (1995): Diversity of neuronal nicotinic acetylcholine receptors: lessons from behavior and implications for CNS therapeutics—minireview. Life Sci. 56: 545–570.

DSM-IV (1994): Diagnostic and statistical manual for mental disorders: DSM-IV. Washington, DC: American Psychiatric Association.

Dunnett SB, Martel FL (1990): Proactive interference effects on short-term memory in rats: 1. Basic parameters and drug effects. Behav. Neurosci. 104: 655–665.

Elrod K, Buccafusco JJ, Jackson WJ (1988): Nicotine enhances delayed matching-to-sample performance by primates. Life Sci. 43: 277–287.

Evenden JL, Turpin M, Oliver L, Jennings C (1993): Caffeine and nicotine improve visual tracking by rats—a comparison with amphetamine, cocaine and apomorphine. Psychopharmacology 110: 169–176.

Hatsukami D, Fletcher L, Morgan S, Keenan R, Amble P (1989): The effects of varying cigarette deprivation duration on cognitive and performance tasks. J. Sub. Abuse 1: 407–416.

Heishman SJ, Snyder FR, Henningfield JE (1993): Performance, subjective, and physiological effects of nicotine in non-smokers. Drug Alcohol Depend. 34: 11–18.

Jackson WJ, Buccafusco JJ (1989): Nicotinic enhancement of delayed matching by monkeys is specific to the most difficult problems. Soc. Neurosci. Abstr. 15: 731.

Kim J, Levin ED (1996): Nicotinic, muscarinic and dopaminergic actions in the ventral hippocampus and the nucleus accumbens: effects on spatial working memory in rats. Brain Res. 725: 231–240.

Le Houezec J, Halliday R, Benowitz NL, Callaway E, Naylor H, Herzig K (1994): A low dose of subcutaneous nicotine improves information processing in non-smokers. Psychopharmacology 114: 628–634.

Levin ED (1992): Nicotinic systems and cognitive function. Psychopharmacology 108: 417–431.

Levin ED (1996): Nicotinic agonist and antagonist effects on memory. Drug Dev. Res. 38: 188–195.

Levin ED, Rose JE (1995): Acute and chronic nicotinic interactions with dopamine systems and working memory performance. In: Lajtha A, Abood L, editors. Functional diversity of interacting receptors. New York: The New York Academy of Sciences, pp. 218–221.

Levin ED, Torry D (1995): Nicotine effects on memory performance. In: Clarke PBS, Quik M, Thurau K, Adlkofer F, editors. International Symposium on Nicotine: The effects of nicotine on biological systems II. Boston: Birkhäuser, pp. 329–335.

Levin ED, Torry D (1996): Acute and chronic nicotine effects on working memory in aged rats. Psychopharmacology 123: 88–97.

Levin ED, Briggs SJ, Christopher NC, Rose JE (1993): Chronic nicotinic stimulation and blockade effects on working memory. Behav. Pharmacol. 4: 179–182.

Levin ED, Briggs SJ, Christopher NC, Auman JT (1994): Working memory performance and cholinergic effects in the ventral tegmental area or substantia nigra. Brain Res. 657:165–170.

Levin ED, Rose JE, Abood L (1995): Effects of nicotinic dimethylaminoethyl esters on working memory performance of rats in the radial-arm maze. Pharmacol. Biochem. Behav. 51: 369–373.

Levin ED, Kim P, Meray R (1996a): Chronic nicotine effects on working and reference memory in the 16-arm radial maze: interactions with D_1 agonist and antagonist drugs. Psychopharmacology 127: 25–30.

Levin ED, Conners CK, Sparrow E, Hinton S, Meck W, Rose JE, Ernhardt D, March J (1996b): Nicotine effects on adults with attention-deficit/hyperactivity disorder. Psychopharmacology 123: 55–63.

Levin ED, Lippiello P, Robinson J, Rose JE (1996c): Nicotine and neurobehavioral function. Drug Dev. Res. 38: 135.

Levin ED, Toll K, Chang G, Christopher NC, Briggs SJ (1996d): Epibatidine, a potent nicotinic agonist: effects on learning and memory in the radial-arm maze. Medicinal Chem. Res. 6: 543–554.

Levin ED, Conners C, Silva D, Hinton S, Meck W, March J, Rose J (1998): Transdermal nicotine patch effects on attention. Psychopharmacology in press.

Levin ED, Kaplan S, Boardman A (1997a): Acute nicotine interactions with nicotinic and muscarinic antagonists: working and reference memory effects in the 16-arm radial maze. Behav. Pharmacol. 8: 236–242.

Levin ED, Christopher NC, Briggs SJ (1997b): Chronic nicotinic agonist and antagonist effects on T-maze alternation. Physiol. Behav. 61: 863–866.

Levin ED, Torry D, Christopher NC, Yu X, Einstein G, Schwartz-Bloom R (1997c): Is binding to nicotinic acetylcholine and dopamine receptors related to working memory in rats? Brain Res. Bull. 43: 295–304.

Mangan GL (1983): The effects of cigarette smoking on vilance performance. J. Gen. Psychol. 108: 203–210.

McConville BJ, Fogelson MH, Norman AB, Klykylo WM, Manderschied PZ, Parker KW, Sanberg PR (1991): Nicotine potentiation of haloperidol in reducing tic frequency in Tourette's disorder. Am. J. Psych. 148: 793–794.

Meyer EM, de Fiebre CM, Hunter BE, Simpkins CE, Frauworth N, de Fiebre NE (1994): Effects of anabaseine-related analogs on rat brain nicotinic receptor binding and on avoidance behaviors. Drug Dev. Res. 31: 127–134.

Milberger S, Biederman J, Faraone SV, Chen L, Jones J (1997): Further evidence of an association between attention-deficit/hyperactivity disorder and cigarette smoking. Findings from a high-risk sample of siblings. Am. J. Addict. 6: 205–217.

Miyata H, Hironaka N, Ando K (1991): Effects of cholinergic drugs on delayed discrimination disruption by AF64A in rats. Nippon Yakurigaku Zasshi 97: 259–266.

Newhouse PA, Corwin J, Lenox R (1992): Acute nicotinic blockade produces cognitive impairment in normal humans. Psychopharmacology 108: 480–484.

Pickworth WB, Bunker EB, Henningfield JE (1994): Transdermal nicotine: reduction of smoking with minimal abuse liability. Psychopharmacology 115: 9–14.

Pomerleau OF, Downey KK, Stelson FW, Pomerleau CS (1995): Cigarette smoking in adult patients diagnosed with attention deficit hyperactivity disorder. J. Subst. Abuse 7: 373–378.

Rusted J, Eaton-Williams P (1991): Distinguishing between attentional and amnestic effects in information processing: the separate and combined effects of scopolamine and nicotine on verbal free recall. Psychopharmacology 104: 363–366.

Rusted JM, Warburton DM (1992): Facilitation of memory by post-trial administration of nicotine: evidence for an attentional explanation. Psychopharmacology 108: 452–455.

Rusted J, Graupner L, Warburton D (1995): Effects of post-trial administration of nicotine on human memory: evaluating the conditions for improving memory. Psychopharmacology 119: 405–413.

Sanberg PR, Fogelson HM, Manderscheid PZ, Parker KW, Norman AB, McConville BJ (1988): Nicotine gum and haloperidol in Tourette's syndrome. Lancet 592.

Sanberg PR, Silver AA, Shytle RD, Philipp MK, Cahill DW, Fogelson HM, McConville BJ (1997): Nicotine for the treatment of Tourette's syndrome. Pharmacol. Ther. 74: 21–25.

Shytle RD, Silver AA, Philipp MK, McConville BJ, Sanberg PR (1996): Transdermal nicotine for Tourette's syndrome. Drug Dev. Res. 38: 290–298.

Silver AA, Sanberg PR (1993): Transdermal nicotine patch and potentiation of haloperidol in Tourette's syndrome. Lancet 342: 182.

Silver AA, Shytle RD, Philipp MK, Sanberg PR (1995): Transdermal nicotine in Tourette's syndrome. In: Clarke PBS, Quik M, Thurau K, editor. The effects of nicotine on biological systems. Basal: Burkhäuser, pp. 293–299.

Simon BB, Levin ED, Grilly DM (1998): Nicotine enhances vigilance task performance in aged rats. in preparation.

Socci DJ, Sanberg PR, Arendash GW (1995): Nicotine enhances morris water maze performance of young and aged rats. Neurobiol. Aging 16: 857–860.

Spilich GJ, June L, Renner J (1992): Cigarette smoking and cognitive performance. Br. J. Addict. 87: 1313–1326.

Stough C, Mangan G, Bates T, Pellett O (1994): Smoking and Raven I.Q. Psychopharmacology 116: 382–384.

Terry AV, Buccafusco JJ, Decker MW (1997): Cholinergic channel activator, ABT-418, enhances delayed-response accuracy in rats. Drug Dev. Res. 40: 304–312.

Turchi J, Holley LA, Sarter M (1995): Effects of nicotinic acetylcholine receptor ligands on behavioral vigilance in rats. Psychopharmacology 118: 195–205.

Turchi J, Holley LA, Sarter M (1996): Effects of benzodiazepine receptor inverse agonists and nicotine on behavioral vigilance in senescent rats. J. Gerontol. A Biol. Sci. Med. Sci. 51: B225–31.

Warburton DM (1992): Nicotine as a cognitive enhancer. Prog. Neuropsychopharmacol. Biol. Psych. 16: 181–919.

Warburton DM, Arnall C (1994): Improvements in performance without nicotine withdrawal. Psychopharmacology 115: 539–542.

Warburton DM, Rusted JM, Fowler J (1992a): A comparison of the attentional and consolidation hypotheses for the facilitation of memory by nicotine. Psychopharmacology 108: 443–447.

Warburton DM, Rusted JM, Muller C (1992b): Patterns of facilitation of memory by nicotine. Behav. Pharmacol. 3: 375–378.

Weiner I, Horin EB, Feldon J (1986): Amphetamine and overtraining reversal effect. Pharmacol. Biochem. Behav. 24: 1539–1542.

Wender P (1995): Attention-deficit hyperactivity disorder in adults. New York: Oxford University Press.

Wesnes K, Warburton DM (1984a): The effects of cigarettes of varying yield on rapid information processing performance. Psychopharmacology 82: 338–342.

Wesnes K, Warburton DM (1984b): Effects of scopolamine and nicotine on human rapid information information processing performance. Psychopharmacology 82: 147–150.

Wonnacott S, Irons J, Rapier C, Thorne B, Lunt GG (1989): Presynaptic modulation of transmitter release by nicotinic receptors. In: Nordberg A, Fuxe K, Holmstedt B, Sundwall A, editors. Progress in brain research. Elsevier Science Publishers B.V., pp. 157–163.

Woodruff-Pak DS, Li YT, Kem WR (1994): A nicotinic agonist (GTS-21), eyeblink classical conditioning, and nicotinic receptor binding in rabbit brain. Brain Res. 645: 309–317.

Zarrindast MR, Sadegh M, Shafaghi B (1996): Effects of nicotine on memory retrieval in mice. Eur. J. Pharmacol. 295: 1–6.

21

Neuronal Nicotinic Receptors: New Targets in the Treatment of Pain

Christopher M. Flores, Ph.D., and Kenneth M. Hargreaves, DDS, Ph.D.

Departments of Endodontics and Pharmacology
University of Texas Health Science Center at San Antonio
San Antonio, Texas

Pain is a major problem of both biomedical and societal importance. It is the foremost indication for which humans seek medical attention and, in the United States, costs an estimated $100 billion annually in treatment expenses, lost productivity, and income. The negative impact of chronic and debilitating pain states on quality of life in terms of human suffering is profound and immeasurable. Although the opioid narcotics and nonsteroidal anti-inflammatory drugs (NSAIDs) have been used successfully in the treatment of pain for millennia, they are at times inadequate, especially in many types of severe or chronic pain, due to side effects (e.g., tolerance, respiratory depression, gastrointestinal distress) and/or a lack of efficacy (especially with respect to neuropathic pain syndromes). Thus, the development of safer, more efficacious analgesic therapeutics is and should be a primary research objective.

In 1992, Spande and colleagues, working in the NIH laboratory of John Daly, reported the isolation of a novel alkaloid from the skin of *Epipedobates tricolor*, an Ecuadoran frog of the Dendrobatidae family. The substance, which they called epibatidine, was originally targeted for purification from extracts containing over 200 trace dendrobatid compounds, because it elicited the Straub-tail response, a hallmark of the opiate alkaloids. In addition, epibatidine exhibited antinociceptive activity in a mouse hot plate model with a potency at least two orders of magnitude greater than that of morphine. Remarkably, neither of these responses was

Neuronal Nicotinic Receptors: Pharmacology and Therapeutic Opportunities, Edited by S. P. Arneric and J. D. Brioni
ISBN 0-471-24743-x, pages 359–378. Copyright © 1998 by Wiley-Liss, Inc.

blocked by the opiate receptor antagonist naloxone; moreover, epibatidine only weakly displaced [^{3}H]dihydromorphine binding from guinea pig brain homogenates. Of key interest, however, were the subsequent demonstrations that epibatidine possessed extremely high affinity for [^{3}H]cytisine and [^{3}H]nicotine binding sites in rat brain, with a Ki of approximately 40 pM, and that epibatidine's antinociceptive effects were blocked by mecamylamine, a noncompetitive antagonist at neuronal nicotinic acetylcholine receptors (nAChRs; Qian et al., 1993; Badio and Daly, 1994). Although the concept of nicotinic analgesia was nothing new, the reigning prototype of this drug class had been nicotine, the pharmaceutical development of which has always been precluded by an extremely poor therapeutic index and, more recently, a politically undesirable reputation. Thus, epibatidine has served to jumpstart research on the potential role of nAChR ligands in the treatment of pain and to renew optimism about eventually generating a novel class of nAChR-based analgesics. The creation of a new class of efficacious, analgesic drugs holds the promise of mitigating, at least in part, some of the above-mentioned drawbacks associated with the use of currently available therapeutic agents.

This chapter (1) reviews and summarizes the literature on the ability of nAChR agonists to modulate the processing of nociceptive information in animals, emphasizing mechanisms and sites of action; (2) highlights some emerging information about novel targets for these compounds in the peripheral nervous system; (3) integrates the potential contribution from the clinical literature on human smokers; and (4) discusses the prospects for the development and approval of clinically useful nicotinic analgesics.

ANTINOCICEPTIVE EFFECTS OF NICOTINE AND nAChR AGONISTS

There can be little doubt that the discovery of epibatidine and its potent antinociceptive properties has had a profound impact upon research relating nAChRs and pain. However, these findings were just the latest contributions to a field whose roots can be traced to the early 1930s, when Davis et al. first showed that nicotine produced antinociception in a cat model of visceral pain. Since then, several groups of investigators have elegantly and systematically elucidated many of the mechanisms and pathways involved in nAChR-mediated antinociception (Table 21-1). From this impressive body of work, several themes have emerged. It is a primary objective of this review to identify those that will provide some insight into endogenous pain pathways and the mechanisms by which nicotinic drugs may modulate them. However, it is important to remember that no discrete neurotransmitter system is singular in its function nor does it exist and operate in isolation, and so the reader is entreated to bear in mind two equally important and relevant areas of research that are not discussed at length here. First, acetycholine has been shown to evoke pain responses in humans and animals via an action on cutaneous nAChRs (Armstrong et al., 1953; Jancsó et al., 1961; Keele and Armstrong, 1964). More recent studies have demonstrated that acetylcholine and carbachol activate peripheral nociceptors by a nAChR-dependent mechanism (Tanelian, 1991; Steen and Rhee, 1993). In addition to this peripheral site, Martin and colleagues have identified a medullary hyperalgesic center that may be stimulated by nicotinic agonists (Martin et al., 1988; Hamann and Martin, 1992; Parvini et al., 1993). Thus, under certain conditions, nicotine and its congeners may be algogenic via both peripheral and central mechanisms. Second, muscarinic cholinergic receptors also play a role in modulating nociceptive systems. Indeed, these receptors likely mediate many of the effects on measures of pain observed in response to the administration of choline esters or cholinesterase inhibitors (Pedigo et al., 1975; for reviews, see Green and Kitchen, 1986; Pert, 1987). Hence, it must be appreciated that endogenous cholinergic systems and their modulation by exogenous pharmacological agents will manifest the composite, potentially indepen-

TABLE 21-1. Antinociceptive Activity of Acute nACuR Agonist Administration in Animals

Drug Species	Model	Dose	Route	Antagonism	Site	Mechanism of Action	Reference
ABT-418							
Mouse	Tail-flick	35–50 μg	it	Mec it	Spinal	nAChR agonism	Damaj et al., 1995b
Mouse	Tail-flick	1–20 mg/kg	sc	Mec sc	and		
Mouse	Hot plate	1–20 mg/kg	sc	Mec sc	supraspinal		
ABT-594							
Rat	Paw withdrawal	.03–.3 μmol/kg	ip, po	Mec ip, Chl icv	CNS	nAChR agonism	Bannon et al., 1997, 1998
Rat	Formalin	.1–.3 μmol/kg	ip	Mec ip, Chl icv	CNS	nAChR agonism	Bannon et al., 1997, 1998
Rat	Paw withdrawal	.04–.13 nmol	mi NRM	Mec ip/NRM, 5,7-DHT	NRM	nAChR activation of NRM 5-HT	Curzon et al., 1997
Mouse	Hot plate	.62 μmol/kg	ip	Mec	Systemic	nAChR agonism	Decker et al., 1997
Rat	Chung	.1–.3 μmol/kg	ip	Mec	Systemic	nAChR agonism	Kim et al. 1997; Bannon et al., 1998
Carbamylcholine							
Rat	Tail-flick	.2 μg	mi PAG	Mec mi PAG/ip, Pob ip	PAG	nAChR/α-adrenoceptor	Guimarães and Prado, 1994
(-)-Cytisine							
Rat	Paw withdrawal	~.002–2 μg	icv (IVth vent.)	ND	CNS	nAChR agonism	Martin et al. 1988
Rat	Tail-flick	6.25–100 μg	icv	Mec sc	CNS	nAChR agonism	Rao et al., 1996
Dimethylphenylpiperazinium (DMPP)							
Rabbit	Tooth pulp stimulation	5 mg/kg	sc	ND	PNS	Ganglionic nAChR agonism	Mattila et al., 1968
Rat	Tail-flick	150 μg	icv	Mec ip	CNS	nAChR agonism	Phan et al., 1973
Rat	Tail-flick	.35 μg	mi PAG	ND	PAG	nAChR agonism	Guimarães and Prado, 1994
Rat	Tail-flick	1–100 μg	icv	Mec sc	CNS	nAChR agonism	Rao et al., 1996
Epibatidine							
Mouse	Tail-flick	5–50 μg/kg	sc	ND	Systemic	Unknown high-affinity receptors	Li et al., 1993
Mouse	Tail-flick	13.6 μg/kg (ED_{50})	?	Mec	Systemic	nAChR agonism	Qian et al., 1993
Rat	Tail-flick	4.3 μg/kg (ED_{50})	?	Mec	Systemic	nAChR agonism	
Mouse	Hot plate	.625–5 μg/kg	ip	Mec, DβE ip	Systemic	Ganglionic-type nAChR	Badio and Daly, 1994
Mouse	Tail-flick	~2.5–25 μg/kg	sc	Mec sc	CNS	nAChR agonism	Damaj et al., 1994a
		~.625–.625 μg	it	ND	Spinal	nAChR agonism	
Mouse	Hot plate	1–20 μg/kg	sc	Mec, DβE sc	Systemic	nAChR agonism	Rupniak et al., 1994
Rat	Carrageenan	5 ug/kg	sc	DβE sc	Systemic	nAChR agonism	
Mouse	Hot plate	.05–.1 μmol/kg	ip	Mec ip	Systemic	nAChR agonism	Sulivan et al., 1994

(*continued*)

TABLE 21-1. (*Continued*)

Drug Species	Model	Dose	Route	Antagonism	Site	Mechanism of Action	Reference
Mouse	Hot plate	.1 μmol/kg	ip	Chl ip	CNS	nAChR agonism	Bannon et al., 1995a
Mouse	Hot plate	.3–30 μg/kg	sc	Mec	Systemic	nAChR agonism	Bonhaus et al., 1995
Epibatidine							
Mouse	Hot plate	5 μg/kg	ip	Ibo ip	CNS	α4β2 or α3β4 nAChR agonism	Badio et al., 1997b
Rat	Carrageenan	1–100 nmol	ipl	ND	Systemic	nAChR agonism	Kilo et al., 1996
Rat	Tail-flick	.3–3 μg/kg	sc	ND	Systemic	nAChR agonism	Rao et al., 1996
		.125–1 μg	icv	Mec, Pob sc DSP-4 ip/icv	CNS	Noradrenergic NT and/or α-adrenergic receptor	
Epiboxidine							
Mouse	Hot plate	20–100 μg/kg	ip	Mec ip	CNS	α4β2 nAChR agonism	Badio et al., 1997
Nicotine							
Cat	Gall bladder distension	.1 mg/kg	injection	ND	Systemic	nAChR agonism	Davis et al., 1932
Mouse	Hot plate	2.5 mg/kg	sc	Adi	Systemic	Cholinergic receptor	Mattila et al., 1968
Rabbit	Tooth pulp stimulation	50–200 μg/kg 2.5 mg/kg	iv sc	Mec iv	MRF	Depolarizing blockade	
Mouse	Hot plate	2.5 mg/kg	sc	ND	CNS	nAChR agonism	Mansner et al., 1972
Mouse	Hot plate	10–20 mg/kg	ip	Mec sc/ip, Trem, Phy ip	Systemic	Depolarizing blockade	Phan et al., 1973
Mouse	Acetic acid	5–15 mg/kg	sc	ND	Systemic	nAChR agonism	
Rat	Hot plate	5 mg/kg	ip	Mec sc	Systemic	nAChR agonism	
Rat	Tail-flick	150 μg	icv	Mec ip	CNS	nAChR agonism	
Rat	Tail-flick	.16–.50 mg/kg	ip	Mec, Scop sc	CNS	ACh release	Sahley and Berntson, 1979
Rat	Tail-flick	5 μg	mi PAG	Mec ip/mi PAG	PAG	nAChR agonism	Llewelyn et al., 1981
Dog	Skin twitch	40 μg/kg/min	iv	ND	Systemic	nAChR agonism	Kamerling et al., 1982
Rat	Tail-flick	~2–2 mg/kg	sc	Mec, Yoh	CNS	nAChR agonism; α-adrenoceptor	Tripathi et al., 1982
Mouse	Tail-flick	~1–5 mg/kg	sc	Mec, Yoh, Nal(p.ant.)	Systemic	nAChR agonism; opiate receptor	
Rat (an)	Tail-flick	1500 μg/kg 50 μg/kg 400 μg/kg 300 μg/kg 10 μg/kg	sc ia (vertebral) ia (carotid) icv sub-arach.	ND	Thoracic spinal cord	nAChR agonism	Aceto et al., 1986
Mouse	Tail-flick	8 μmol/kg	sc	NSTX, Mec, Pem sc	CNS	nAChR agonism	Yamada et al., 1986

Rat (sp;dc)	Paw withdrawal	.01–.8 μg	icv (IIVth vent.)	Mec, Nalt ip	CNS	Nicotinic and opiatergic	Martin et al., 1988
Rat	Tail-flick	5–50 μg	icv	Mec, Nal sc; SP icv	CNS	nAChR agonism, opioids	Molinero and Del Rio, 1987
Rat	Hot plate	0.77–12.35 nmol	mi PPTN	Mec, Scop, Pzp,	PPTN	nAChR agonism, ACh release,	Iwamoto, 1989
Rat	Tail-flick			Ves mi PPTN		M_1 receptor	
Rat(sp)	Tail-flick	43 nmol	it	Mec, Yoh ip	Spinal	noradrenergic NT	Christensen and Smith, 1990
Rat	Tail-flick	.8 mg/kg	sc	Mec ip	CNS	nAChR agonism	Cooley et al. 1990
Rat	Hot plate	.77–30.85 nmol	mi brain, 185	HC-3, Mec	PPTN	NRM ACh release, M_2 receptor,	Iwamoto, 1991
Rat	Tail-flick		discrete sites	Pzp mi NRM		descending spinal inhibition	
Mouse	Tail-flick	1–30 mg/kg	sc	Mec, Pem sc	Systemic	nAChR agonism	Martin et al., 1990
Rat	Tail-flick	.001–20 μg	mi	ND	Rostral medulla	nAChR agonism	Hamann and Martin, 1992
Rat	Tail-flick	.4 μg	icv (IVth vent.)	Mec mi PDMT	PDMT	nAChR agonism	Parvini et al., 1992
Mouse	Tail-flick	1–10 mg/kg	sc	Nal sc	Systemic	Opioidergic component	Aceto et al., 1993
Rat	Tail-flick	1 mg/kg	sc	$CaCl_2$ sc	Systemic	Ca^{2+} involvement	Block et al., 1993
Rat	Tail-flick	1 mg/kg	sc	$CACl_2$ ip	Systemic	Ca^{2+} involvement	Chin et al., 1993
Mouse	Tail-flick	1.5 mg/kg	sc	Nif, Nim ip/sc	Systemic	L-type Ca^{2+} channel	Damaj and Martin, 1993a
Mouse	Tail-flick	.15–.625 mg/kg	sc	EGTA, CGRP it	Systemic	Ca^{2+} channel;nAChR desensitization	Damaj et al., 1993
Rat (an)	C-fiber stimulation	.01–.5 mg/kg	iv	Mec iv	Spinal	nAChR agonism	Jurna et al., 1993
		5–30 μg	it				
Rat	Hot plate	.125–.5 mg/kg	sc or it	HM-3 (RVM), Scop, Met	RVM	Spinal α2-adrenergic,	Rogers and Iwamoto, 1993
Rat	Tail-flick			Yoh, Ida, Mec, Atr it		muscarinic, serotonergic NT	
Mouse	Tail-flick	~.625–25 mg/kg	sc	ND	CNS	nAChR agonism	Damaj et al., 1994a
		~6.25–25 μg	it	ND	Spinal	nAChR agonism	
Mouse	Tail-flick	1.5 mg/kg	sc	DPAT, Busp sc	Systemic	5-HT_{1A} receptor	Damaj et al., 1994b
Mouse	Tail-flick	1.5 mg/kg	sc	PTX, Fsk, 8CPTcAMP it	Spinal	Ca^{2+}, G-protein, PKA	Damaj et al., 1994c
Mouse	Hot plate	3 mg/kg	sc	Mec sc	Systemic	nAChR agonism	Rupniak et al., 1994
Rat	Carrageenan	1.5 mg/kg	sc	DβE sc	Systemic	nAChR agonism	
Mouse	Hot plate	.3–30 μg/kg	sc	Mec	Systemic	nAChR agonism	Bonhaus et al., 1995
Rat	Tail withdrawal	.75 mg/kg	sc	Mec, Chl	PNS	nAChR agonism	Caggiula et al., 1995
Rat	Hot plate	.125–.75 mg/kg	sc	Mec	CNS	nAChR agonism; assay-dependent	
Mouse	Tail-flick	.5–12.5 μmol/kg	sc	Mec, DβE sc and it	(Supra)spinal	nAChR agonism	Damaj et al., 1995a
Rat	Capsaicin	6.6–660 nmol	ipl	ND	PNS	nAChR agonism	Kilo et al., 1996
Rat	Tail-flick	.1–1.2 mg/kg	sc	Mec, Hex sc	CNS and PNS	nAChR agonism	Rao et al., 1996
		3–200 μg	icv	Mec, Atr, Hex, Pob (+) Nic sc, DSP-4 ip/icv	CNS	Noradrenergic NT and/or α-adrenoceptor	

Table 21.1 (*continued*)

TABLE 21-1. *(Continued)*

Drug Species	Model	Dose	Route	Antagonism	Site	Mechanism of Action	Reference
N-Methyl-Carbamylcholine							
Rat	Hot plate	40 nmol	mi	Ida, Pro, LY,	PPTN, NRM	Spinal $\alpha 2$, 5–$HT_{1c/2}$,	Iwamoto and Marion, 1993
Rat	Tail-flick		PPTN, NRM	Zac, Pzp it		5-HT_3 and M_2 receptors	
Rat	Tail-flick	6.25–200 μg	icv	Mec, Atr sc	CNS	nAChR agonism	Rao et al., 1996

Abbreviations: ACh, acetylcholine; Adi, adiphenine; an, anesthetized; Atr, atropine; Busp, buspirone; CGRP, calcitonin gene-related peptide; Chl, chlorisondamine; CNS, central nervous system; 8CPTcAMP, 8-(-4-chlorophenylthio) adenosine-3′, 5′ monophosphate, cyclic; DβE, dihydro-β-erythroidine; dc, decerebrated; 5,7-DHT,5,7-dihydroxytryptamine; DPAT, 8-hydroxy-2-(di-n-propylamino) tetralin; DSP-4, [N-(2-chloroethyl)-N-ethyl-2-bromobenzylamine]; EGTA, ethylenediaminetetraacetic acid; Fsk, forskolin; HC-3, hemicholinium-3; Hex, hexamethonium; 5-HT, 5-hydroxytryptamine; ia, intra-arterial; Ibo, ibogaine; icv intracerebroventricular; Ida, idazoxan; ip, intraperitoneal; ipl, intraplantar; it, intrathecal; iv, intravenous; IVth vent., fourth ventricle; LY, LY53857; Mec, mecamylamine; Met, methysergide; mi, microinjection; MRF, mesencephalic reticular formation; nAChR, nicotinic acetylcholine receptor; Nal, naloxone; Nalt, naltrexone; ND, not determined; Nic, nicotine; Nif, nifedipine; Nim, nimodipine; NRM, nucleus raphe magnus; NSTX, neosurugatoxin; NT, neurotransmission; PAG, periaqueductal gray; p.ant, partial antagonist; PDMT, posterior dorsal mesencephalic tegmentum; Pem, pempidine; Phy, physostigmine; PKA, cAMP-dependent protein kinase (protein kinase A); PNS, peripheral nervous system; po, per os; Pob, phenoxybenzamine; PPTN, pedunculopontine tegmental nucleus; Pro, propranolol; PTX, pertussis toxin; Pzp, pirenzepine; RVM, rostal ventral medulla; sc, subcutaneous; Scop, scopolamine; SP, substance P; sp, spinalized; sub-arach, sub-arachnoid space; Trem, tremorine; Ves, vesamicol; Yoh, yohimbine; Zac, zacopride.

dent, contribution of both muscarinic as well as nicotinic receptor subtypes. However, to the extent that muscarinic acetylcholine receptors may mediate some of the antinociceptive effects of nAChR activation (i.e., via the release of ACh), they should be regarded in the same context as any other receptor class and are discussed below.

There is strong evidence that the antinociception observed following nicotinic agonist administration can occur via activation of receptors expressed in a variety of central nervous system (CNS) loci. Specifically, based on studies utilizing a microinjection approach, there appear to be several brain regions that contain nAChRs capable of producing antinociception. These include the rostral (Hamann and Martin, 1992; Rogers and Iwamoto, 1993) and ventral (Iwamoto and Marion, 1993; Curzon et al., 1997) medulla, the periaqueductal gray (Llewelyn et al., 1981; Guimarães and Prado, 1994), as well as certain areas of the midbrain/pons, specifically the mesencephalic reticular formation (Mattila, 1968), the pedunculopontine tegmental nucleus (Iwamoto, 1989, 1991; Iwamoto and Marion, 1993), and the posterior dorsal mesencephalic tegmentum (Parvini et al., 1992). Agonist activation of at least some of these nAChR sites are thought to engage multiple, descending, pain-modifying pathways involving multiple neurotransmitter systems that terminate in the dorsal spinal cord. Whether the spinal cord represents a primary site of action for nAChR agonists, on the other hand, is controversial. Thus, while several investigators could demonstrate antinociception following the intrathecal administration of nicotine (Christensen and Smith, 1990; Jurna et al., 1993; Damaj et al., 1994a), others could not (Yaksh et al., 1985; Smith et al., 1989; Gillberg et al., 1990). The implication by Aceto et al. (1986), that nicotine may be acting at the level of the thoracic spinal cord, suggests that it may be exerting its effects by modulating preganglionic sympathetic outflow from the mediolateral cell column, although there is no direct evidence to support this possibility. Antinociceptive effects of intrathecally administered acetylcholinesterase inhibitors or nonselective cholinergic receptor agonists are well documented. However, these effects are almost always attenuated by muscarinic, but not nAChR, antagonists (see Green and Kitchen, 1985, and Pert, 1987), suggesting that the cholinergic modulation of pain at the level of the spinal cord is principally muscarinic in nature.

The receptor subtypes at which nAChR agonists exert their antinociceptive actions have received only moderate attention. Based on a wealth of pharmacological data, however, it is clearly evident that (−)-nicotine, epibatidine, etc., initially engage neuronal-type nAChRs to produce their antinociceptive effects, because these effects are blocked by several biochemically and pharmacologically distinct nicotinic receptor antagonists (e.g., mecamylamine, chlorisondamine, and dihydro-β-erythroidine; see Table 21-1). Specifically, α4β2 (Badio et al., 1997a,b) as well as α3β4 (Badio et al., 1997b) neuronal nAChR subtypes have been implicated. Apparently, α7-containing subtypes that are sensitive to α-bungarotoxin do not play a role in either the tail-flick response itself or in the antinociceptive effects of (−)-nicotine, (−)-cytisine, epibatidine, DMPP, or N-methylcarbamylcholine, since the intracerbroventricular coadministration of methyllycaconitine failed to block the increased tail-flick latencies to any of these compounds (Rao et al., 1996). Thus, it is still unclear which nAChR subtypes subserve the antinociceptive effects of nAChR agonists. Until such determinations are made, the reasonably attractive proposition of designing subtype-selective analgesic compounds should be pursued with caution. Finally, it has been suggested that nicotine and other pharmacologically related compounds may exert centrally mediated antinociceptive actions as a result of depolarizing blockade rather than agonist stimulatory effects, possibly at the level of the mesencephalic reticular formation (Mattila et al., 1968; Phan et al., 1973). If this were the case, it would be consistent with an action of (−)-nicotine at a tonically active, nicotinic hyperalgesic locus such as that mentioned above.

What happens after neuronal nAChR activation in the present context is much less clear. It is generally assumed that agonist binding to nAChRs, which are ligand-gated cation channels, results in a depolarizing, excitatory event. Moreover, these receptors have been localized to both pre- and postsynaptic neural elements, where they may mediate the local regulation of transmitter release and the trans-synaptic propagation of electrochemical neurotransmission, respectively. Appreciating these roles for nAChRs and the extent to which such processes depend upon calcium ions, several groups sought to evaluate the influence of calcium-modifying agents on nicotinic antinociception. Insofar as nAChRs can themselves conduct calcium ions and the activation of these receptors may lead to the opening of voltage-gated calcium channels, one might predict a priori that agents that facilitate calcium entry into cells will enhance nicotinic antinociception while those that inhibit calcium entry will antagonize it. Indeed, it was shown that an intraperitoneal injection of the calcium channel agonist (+/−)-Bay K 8644 significantly potentiated the antinociceptive effect of epibatidine in mice, while (+)-Bay K 8644, a calcium channel antagonist, blocked epibatidine's effect (Bannon et al., 1995b). Similarly, Damaj and Martin (1993a) reported that the ability of (−)-nicotine to produce antinociception in mice was blocked by the calcium channel antagonists nifedipine and nimodipine but not by verapamil, indicating the involvement of dihydropyridine-sensitive, L-type calcium channels. Moreover, Damaj et al. (1993) showed that the antinociceptive effect of subcutaneously administered (−)-nicotine was enhanced by the intrathecal administration of either (+/-)-Bay K 8644, thapsigargin, glyburide, A23187, or calcium itself and blocked by EGTA, suggesting that intracellular calcium concentrations in the spinal cord play an important role in mediating nAChR agonist–induced antinociception. It is significant that these drugs were administered by the intrathecal route, given the evidence cited against a contribution of spinal nAChRs in mediating the antinociceptive effects of nAChR agonists. Thus, the mechanism of action of these calcium-modifying compounds is unlikely to involve a direct action on spinal nAChRs and will require further investigation.

In contrast to these studies, systemic administration of the calcium chelator EDTA potentiated, whereas calcium chloride inhibited, nicotine-induced antinociception in rats (Block et al., 1993; Chin et al., 1993). Similarly, Wong et al. (1994) showed that nifedipine, but not verapamil, dose-dependently potentiated nicotine-induced antinociception. It is immediately difficult to reconcile these apparently paradoxial results with respect to the direction of calcium manipulation and its subsequent influence on nAChR-mediated antinociception. In fact, the potential role of calcium on nociceptive signaling in animals made tolerant to nicotine is similarly confounding (Damaj, 1997; Zbuzek et al., 1997); however, a dependence on the species of animal under study is possible.

As mentioned above, numerous studies have been conducted to ascertain which neurotransmitter systems are activated in mediating the antinociceptive effects of nAChR agonists. So far, there is evidence indicating the involvement of muscarinic cholinergic (Sahley and Berntson, 1979; Iwamoto, 1989, 1991; Rogers and Iwamoto, 1993; Iwamoto and Marion, 1993), (nor)adrenergic (Guimarães and Prado, 1994; Rao et al., 1996; Tripathi et al., 1982; Christensen and Smith, 1990; Rogers and Iwamoto, 1993); serotonergic (Curzon et al., 1997; Rogers and Iwamoto, 1993; Damaj et al., 1994b; Iwamoto and Marion, 1993), and opioidergic (Tripathi et al., 1982; Martin et al., 1988; Molinero and Del Rio, 1987; Aceto et al., 1993) synapses in nAChR-mediated antinociception. Dopaminergic neurotransmission, at least that mediated by D1 or D2 dopamine receptors, apparently is not involved (Damaj and Martin, 1993b).

The demonstration that antimuscarinic agents, notably atropine, scopolamine, and pirenzepine, blocked nicotine's antinociceptive effects has led to the suggestion that these effects are due, at least in part, to the ability of (−)-nicotine to release acetylcholine (a similar effect of nicotine on acetylcholine release in rat cerebral cortex is well documented). There is further evidence that this or a similar pathway may be tonically active, as the M2 muscarinic receptor antagonist methoctramine exhibited antinociceptive effects by itself when injected into either the mesopontine tegmentum (Iwamoto, 1989) or the nucleus raphe magnus (Iwamoto, 1991), and these effects were abolished by pretreatment with vesamicol or hemicholinium, respectively. On the other hand, it was also shown that mecamylamine either injected intraperitoneally (approximately 3–30 mg/kg) or microinjected into hyperalgesic brain stem loci (Hamann and Martin, 1992) or into the fourth ventricle (Parvini et al., 1993) produced antinociception in the rat tail-flick assay, indicating the existence of central, nAChR-mediated hyperalgesic tone. In contrast, the intravenous administration of mecamylamine had no effect on nociceptive thresholds to pulpal electrical stimulation in the rabbit (Mattila et al., 1968). Similarly, 2–10 mg/kg mecamylamine injected intraperitonially had no effect on nociceptive thresholds in the mouse hot plate model (Phan et al., 1973). The reason for these differences, at least where the route and dose of antagonist are comparable, is unclear and further points to the complexity of cholinergic regulation of nociception. Interestingly, Onaivi et al. (1993) showed that chronic nicotine administration (for 2 years, po), restored the age-related loss of nociceptive sensitivity in older rats, suggesting that nAChR-mediated function, which is known to be diminished in older mammals including humans, may indeed play a role in establishing nociceptive thresholds, at least in older individuals.

Based on a series of elegant studies, Iwamoto and colleagues have proposed a pathway that initially involves the activation of nAChRs in the pedunculopontine tegmental nucleus, which then projects to the nucleus raphe magnus (NRM), resulting in the release of acetylcholine and the stimulation of both nicotinic as well as muscarinic receptors. This leads to the activation of multiple, descending inhibitory pathways originating in the NRM and involving adrenergic, cholinergic, and serotonergic neurons synapsing on spinal $\alpha 2$ adrenoceptors, M2 muscarinic receptors, and $5\text{-HT}_{1c/2/3}$ receptors, respectively. The concept of descending spinal inhibition would seem particularly tenable to explain the nAChR-mediated antinociceptive effects observed in the tail-flick assay in which a spinal reflex is initiated that does not require the involvement of higher CNS regions. As mentioned above, Martin and co-workers have demonstrated both antinociceptive as well as hyperalgesic nicotinic loci in different parts of the medulla (Martin et al., 1988; Hamann and Martin, 1992; Parvini et al., 1993), and Rao et al. (1996) have suggested that the algesic and analgesic responses to nAChR agonists may be subserved by distinct receptor subtypes. Collectively, these studies imply that the effect of nAChR agonists on nociceptive behaviors will depend both on the neuroanatomical location and the subtype of receptors being activated, similar to the opioid systems that modulate nociceptive processing (Wang and Nakai, 1994).

Several investigators have demonstrated that naloxone (Tripathi et al., 1982; Molinero and Del Rio, 1987; Aceto et al., 1993) or naltrexone (Martin et al., 1988) can antagonize, at least partially, nicotine-induced antinociception, suggesting that the phenomenon has an opioidergic component. It is curious, therefore, that the initial demonstrations of epibatidine-mediated antinociception were not naloxone reversible (Qian et al., 1993; Badio and Daly, 1994). The well-known ability of nicotine to release endogenous opioids (pituitary β-endorphin, for example) would reasonably be expected to lead to antinociception that is at least partially dependent on opioid receptor activation. Such differences may be species dependent, as Tripathi et al. (1982) have shown that nicotine-induced antinociception was naloxone reversible in the mouse but not in the rat. However, the preponderance of data indicates that the cited descrepancies cannot be accounted for principally by differences in the species or assay under study (see Table 21-1). It is possible, on the other hand, that nicotine and epibatidine have overlapping as well as distinct pharmacological activities. However, there appears to be little difference in the selectivity of either compound for neuronal nAChR subtypes, albeit epibatidine exhibits an approximately 100-fold greater affinity than (−)-nicotine for α4β2 and α3β4 nAChRs. Alternatively, these differences may be related to the distribution and/or receptor subtype selectivity of the opioid agonist/antagonist under study, as Suh and colleagues (1996a,b) have shown that (−)-nicotine is able to enhance morphine- or β-endorphin-induced antinociception but not that induced by DPDPE or U50,488H. These studies suggest that supraspinal nAChR activation is capable of modulating the antinociceptive activity of μ- but not δ- or κ-opioid receptor agonists. Obviously, the validity of each of these putative mechanisms will require further investigation.

While the potential dependence of experimental results on the species of animal being tested and the assay being used cannot largely explain many of the differences noted previously and in Table 21-1, there are several salient cases in which such differences do come to bear. For example, Sahley and Berntson (1979) reported that nicotine produced antinociception in the tail-flick but not in the tail-pinch assay, suggesting that nicotine was effective on only certain types of noxious stimuli. Similarly, it was observed that (−)-nicotine had no effect on nociception produced by mechanical (Kamat et al., 1972) or electrical stimulation (Oelssner and Andreas, 1969; Pert, 1975; Rogers, 1979). Collectively, these data might indicate that nAChR agonists are selectively efficacious for only thermal types of pain. The fortuitous aspect of these results, however, is that this differential antinociceptive property of (−)-nicotine indicates that it is not acting merely to depress motoric function, a legitimate concern given the ability of some nAChR agonists to stimulate inhibitory Renshaw cells in the spinal cord. In addition, while multiple investigators have demonstrated in the rat tail-flick assay that nAChR agonists produced antinociception when microinjected into the PAG (Llewelyn et al., 1981; Guimarães and Prado, 1994), no such effect of (−)-nicotine in the PAG was observed for the skin twitch reflex in the dog (Kamerling et al., 1982). In another example of assay dependence within a species, it was shown that chronic nicotine-induced antinociception as well as the hyperalgesia induced by nicotine withdrawal could be demonstrated in the hot plate but not in the tail-flick assay, leading the authors to suggest that these phenomena involved supraspinal sites of action (Yang et al., 1992). Interestingly, in addition to being antinociceptive, nAChR agonists appear to be antihyperalgesic, as these compounds have proved efficacious in both the carrageenan (Rupniak et al., 1994; Kilo et al., 1996) and capsaicin (Kilo et al., 1996) models of inflammation, and these studies will be discussed in more detail below with respect to possible peripheral sites of action.

Obviously, no review of the effects of (−)-nicotine, or any other potential drug of abuse for that matter, would be complete without a discussion of tolerance and dependence. Most investigators who have attempted to address the effects of chronic or repeated administration of nAChR agonists have observed some degree of tolerance to the antinociceptive effects. Here again, however, there were some notable exceptions. For example, Phan et al. (1973) reported the development of tolerance to multiple injections of nicotine chronically over several days but not acutely over 2 hours. Highlighting the differences between drugs, in this regard, epibatidine was shown to produce less tolerance than nicotine (Damaj and Martin, 1996), and the novel cholinergic channel modulator ABT-594 was shown to produce less tolerance than epibatidine (Bannon et al., 1997; Kim et al., 1997). Such findings are significant insofar as they underlie the potential for developing nAChR analgesics that may be relatively devoid of certain side effects, notably tolerance and dependence. Moreover, in a clever experiment that paralleled what was known to occur in the opioid field, Epstein et al. (1989) reported that the tolerance that developed to the antinociceptive effect of (−)-nicotine in the rat tail withdrawal assay could be reversed by altering the environment in which the drug was administered to the animal, suggesting that the tolerance to (−)-nicotine was learned or conditioned.

It is of particular interest that the development of tolerance to (−)-nicotine in the rat tail-flick test could be replicated with tobacco smoke and that the latter exhibited cross-tolerance to restraint stress-induced antinociception (Mousa et al., 1988), suggesting that a common mechanism may be involved. Clearly, the potential development of tolerance and dependence in response to chronic, as well as acute, administration of nAChR analgesics will be an important issue to resolve before such compounds would have widespread clinical utility paralleling that of NSAIDs. Related to this point, it should be emphasized that in virtually every study reported to date, the duration of nicotinic-mediated antinociception is extremely short, with peak effects typically occurring within 5 to 30 minutes and completely abating within 1 hour, even at the maximally effective dose. Given the half-life of nicotine in rodents (~1 hr), it is reasonable that the short duration of effect is related to nAChR desensitization, for which there is substantial precedent. Here too, however, differences may exist with respect to duration of effect and the model of pain under study, as Iwamoto (1989) has shown that microinjection of (−)-nicotine into the mesopontine tegmentum leads to a longer antinociception in the tail flick as compared to the hot plate assay. Whether this is the result of pharmacokinetic issues related to the distribution of drug to its site(s) of action or pharmacodynamic issues related to the functional state of receptor subtypes will require further study.

PERIPHERAL SENSORY NEURONS AS TARGETS FOR nAChR ANALGESICS

There is no question that antinociception in response to nAChR agonist administration can be evoked by activating nAChRs in the CNS, albeit at multiple potential sites (see above). This fact was derived empirically using drugs, primarily nicotine and, more recently, epibatidine, that were extracted from the leaves of the tobacco plant and the skin of a poison arrow frog, respectively. As such, these agents are naturally occurring and do not represent the product of a rational drug design strategy. Bearing in mind the potential for centrally mediated nAChR side effects, the demonstration of a peripheral site of action for these agents offers the possibility of developing a peripherally active and/or subtype-selective nAChR analgesic. As mentioned above, many of the limiting side effects of currently available, and often very effective, analgesic drugs such as the opioids arise from their untoward activity in the CNS. It is ironic that the opioids, whose CNS-mediated abuse liability is well appreciated, have found utility outside of the brain. First proposed in the early 1980s and directly demonstrated more recently, opioid receptors exist on neurons in the periphery where they are capable of mediating antinociception. Thus, the questions arise: Are there nAChRs on peripheral sensory neurons and, if so, can they favorably modulate nociceptive processing?

While there is substantial evidence indicating the existence of nAChRs on sensory neurons (e.g., see Flores et al., 1996), at first such inquiries with respect to their function in modulating nociception might appear misguided. After all, Armstrong (1953) clearly demonstrated that acetylcholine was algogenic when applied to the skin of humans, and supporting data for the direct activation of pain-conveying C-fibers by nAChR agonists have come from several other investigators (Jansco et al., 1961; Juan, 1982; Tanelian, 1991; Steen and Rhee, 1993). As mentioned earlier, the discussion of the potentially pain-producing properties of nAChR agonists, however well founded, are beyond the scope of this chapter. Nonetheless, anyone interested in afferent nAChR pharmacology is referred to the enlightening dissertation of Ginzel (1975) on "The Importance of Sensory Nerve Endings as Sites of Drug Action."

In 1983, Aceto et al. directly tested the hypothesis that (−)-nicotine has an antinociceptive site of action in the periphery using quaternary methiodide derivatives of (−)-nicotine in three animal models of pain. They showed that several (−)-nicotine methiodides were completely ineffective following peripheral administration in the mouse and rat tail-flick assays of thermal nociceptive thresholds and were only weakly antinociceptive in the mouse phenylquinone test of hyperalgesia, leading them to conclude that nicotine's antinociceptive site of action was restricted to the CNS. However, there are several pharmacokinetic and pharmacodynamic explanations for their results. Indeed, Caggiula et al. (1995) reported that nicotine-induced increases in the latency to tail withdrawal but not to paw withdrawal to a noxious thermal stimulas were blocked by the peripherally active nAChR antagonist chlorisondamine. Thus, in addition to suggesting that nicotine was capable of modulating nociceptive behavior in a manner that was mediated at or dependent upon the integrity of peripheral nAChRs, their data also suggested that the extent to which this is the case may be assay dependent. Moreover, a complex involvement of both the CNS and PNS cannot be ruled out, as Hajós and Engberg (1988) have shown that (−)-nicotine is capable of increasing dose dependently the firing of neurons located in the locus coeruleus. Importantly, this effect was shown to be mediated by the direct activation of primary sensory C-fiber neurons, since it was abolished by the peripherally active, quaternary nAChR antagonists hexamethonium and chlorisondamine as well as by neonatal treatment with the C-fiber neurotoxin capsaicin. Finally, Kilo et al. (1996), using an experimental paradigm specifically designed to distinguish between local and systemic drug effects, found that (−)-nicotine almost completely reversed the decrease in paw withdrawal latency induced by peripherally administered capsaicin. Importantly, these data suggested that in addition to being antinociceptive (i.e., able to increase nociceptive thresholds), (−)-nicotine was also antihyperalgesic (i.e., able to reverse the decrease in threshold caused by an inflammatory substance), consistent with the findings of Rupniak et al. (1994). It was also demonstrated that epibatidine was capable of reducing edema caused by an intraplantar injection of carrageenan (Kilo et al., 1996). Although, it was not ascertained whether epibatidine's effect was mediated locally or systemically, this provides compelling evidence of additional anti-inflammatory properties of nAChR agonists.

The recent evidence indicating that sensory neurons of the trigeminal ganglion express multiple nAChR subtypes and that at least some of these receptors are localized to nociceptive neurons (Flores et al., 1995) provides a molecular rationale for the ability of nAChR agonists to directly influence the function of neurons important in nociceptive processing and inflammation. It is still a matter of debate whether the consequence of sensory neuronal nAChR activation is to enhance or inhibit nociceptive neurotransmission and whether these disparate effects are mediated by distinct receptor subtypes. Nonetheless, the prospect of developing peripherally selective nAChR analgesics, based on the evidence presented, would have widespread appeal.

TOBACCO USE AND THE PERCEPTION OF PAIN

Humans have self-administered nicotine for centuries in the form of smoked and smokeless tobacco products and, more recently, by itself in the form of gum or transdermal patches (ironically, to overcome their addictions to the former). However, the data in the human clinical literature on the relationship between nicotine and pain are inconclusive at best. As early as 1921, Mendenhall observed the effect of tobacco smoking on human sensory thresholds. He concluded that the effect of smoking on pain threshold was conditioned upon the state of the sensory mechanism, leading to a return toward the normal condition. Since then there have been a dozen or so reports attempting to link nicotine, usually through the use of tobacco products, to alterations in pain sensitivity, threshold, and/or endurance. The difficulty in interpreting such studies derives from several issues. First, very few of these investigations examined the effects of nicotine proper, in the form of nicotine-containing gum for example, as opposed to smoked or smokeless tobacco products, which contain a variety of other components that may or may not contribute to the sensation of pain. Second, most studies were performed in smokers at various stages of deprivation, and thus there is relatively little information on naive individuals who are not experiencing some sort of withdrawal and some of whose nAChRs are not, presumably, in a desensitized or inactivated state. That is not to say that valuable information cannot be obtained from these individuals, only that such data do not primarily address the issue of whether nicotine administration to a naive individual will have pain modulating activity. Finally, it is often difficult to separate nicotine's effects on arousal, cognition, and anxiety associated with withdrawal aversion from those on the dependent measure under study.

Indeed, it has been suggested that the crucial outcome of cigarette smoking is the alleviation of withdrawal symptoms rather than a direct effect of reducing pain and/or anxiety, the evidence for which led Schachter (1978) to conclude that "the sensory and manipulative gratifications of smoking are illusory. Serious smokers smoke to prevent withdrawal." And this view was reiterated by Silverstein (1982). In contrast to this hypothesis, Milgrom-Friedman et al. (1983) reported that cigarette deprivation diminished the sensitivity to pain of smokers compared to nonsmokers and, furthermore, that smoking per se rather than nicotine absorption increased sensitivity to pain. Taken together, it is hardly surprising that the potential analgesic effects of nicotine in humans, at least to the extent that it is addressed by the current literature, is extremely controversial. Thus, while Nesbitt (1973) found a positive correlation between increased pain endurance thresholds and the nicotine content of cigarettes smoked by smokers but not nonsmokers (which effects Waller et al., 1983, suggested were the result of peripheral rather than central activation), Seltzer (1974) reported that deep pain tolerance, as measured by the Achilles tendon pressure test, was reduced in white smokers compared to nonsmokers. Moreover, while several studies found that smoking was associated with reduced pain responses in the cold pressor test (Pomerleau et al., 1984; Fertig and Pomerleau, 1986; Pomerleau, 1986), others reported no effect of cigarette smoking on response measures for the cold pressor (Sult and Moss, 1986) or electrical stimulation tasks (Waller et al., 1983; Sult and Moss, 1986).

Collectively, these studies do little to build a consensus on the potential role of nicotine or nAChRs in modulating nociceptive systems in humans, and it is unlikely that similar such studies will shed considerably more light on the subject. Instead, it is most probable that the answers will have to wait for controlled clinical trials evaluating the analgesic properties of late-generation nAChR agonists on various forms of clinical or experimentally induced pain in human subjects.

OUTLOOK FOR A NEW CLASS OF nAChR ANALGESICS

With the concept of nAChR-mediated analgesia more than a half century old, what are the current prospects that it will lead to a clinically viable drug for the treatment of pain? Clearly, there is an overwhelming body of evidence (see Table 21-1) documenting the antinociceptive activity of nAChR agonists in animal models of pain, including substantial information on mechanisms and sites of action. These contributions, therefore, not only strongly support the further research and development of a nAChR analgesic drug class but also provide some important clues regarding where and how such drugs should act. Indeed, some of the most recent evidence presages a generation of compounds that, at therapeutic doses, are not only relatively devoid of many of the undesirable effects that have plagued their predecessors but also are endowed with additionally desirable therapeutic efficacies. For example, Donnelly-Roberts et al. (1997) recently presented evidence that, similar to ABT-418, the orally effective antinociceptive agent ABT-594 exerted neuroprotective effects in spinal cord cultures against substance P- or glutamate-induced toxicity. Moreover, Kim et al. (1997) have shown that ABT-594 was active in an animal model of neuropathic pain, one of the most intractable types of pain to treat in humans. Also, it was reported that the time courses of epibatidine-induced changes in locomotor activity, temperature, jump latency, and rotorod performance were different and that the antinociception, but not the other dependent measures under study, could be potentiated by a subsequent injection of mecamylamine (Bannon et al., 1995a). This same group also showed that chlorisondamine pretreatment blocked the effects of epibatidine on antinociception but not on hypothermia or activity reduction. Taken together, these studies suggest that certain effects of nAChR agonists, notably antinociception, can be dissociated from others at a variety of levels. Similarly, Damaj and Martin (1996) showed that compared to (−)-nicotine, epibatidine-induced antinociception did not exhibit tolerance after acute or chronic administration and that there was no cross-tolerance to epibatidine in nicotine-tolerant mice (tail-flick).

Interestingly, Damaj and Martin (1993b) also showed that neither D1 nor D2 dopamine receptors are involved in nicotine-mediated antinociception in a mouse tail-flick assay, making it one of the few neurotransmitter systems not involved in nAChR-mediated antinociception. This may be fortuitous, to the extent that dopamine release in response to nicotine (as well as several other drugs of abuse) is thought to underlie some of the reinforcing properties of the drug. Thus, it might be reasonable that the coadministration of either a subtype-selective nAChR agonist that does not cause dopamine release or a dopamine receptor antagonist, which has been shown to reduce nicotine self-administration in the rat, would constitute a viable pharmacophore. Similarly, intrathecal nimodipine blocked, while calcium and thapsigargin enhanced, the acute tolerance to nicotine-induced antinociception (Damaj et al., 1996). This evidence provides a rationale for the combined use of calcium channel blockers with nAChR agonists to help prevent the development of tolerance. An additional possibility is that nAChR agonists might be used as opioid-sparing adjuncts, as suggested by the work of Suh et al. (1996a,b).

Given all of the data gathered to date, it appears likely that there will be substantial effort to develop a new class of nAChR analgesic drugs to put into the pain-fighting arsenal. Based on many of the studies cited, such drugs may have a variety of indications, including neuropathic as well as inflammatory types of pain. The stage has also been set for the advent of nAChR ligands as adjunctive therapy in combination with opioids or other drug classes, serving to potentially reduce the effective doses of these drugs and thereby decrease side-effect liabilities. Conversely, nAChR analgesics might themselves make use of adjuncts, such as antidopaminergics or calcium modifiers, the application of which would seek to mitigate some of the potentially untoward effects of nAChR agonists, including dependence and tolerance. In any case, there seems sufficient promise that we may be on the threshold of having a new and powerful weapon that may offer improved efficacy and toxicity profiles in the fight against various types of pain.

ACKNOWLEDGMENTS

The authors gratefully acknowledge Ms. Gilda Castano, Mr. Anthony Leong, and Ms. Catherine Harding-Rose for their generous assistence in the preparation of this manuscript. Dr. Flores is supported by R01 DA10510 from the National Institute on Drug Abuse.

REFERENCES

Aceto MD, Awaya H, Martion BR, May EL (1983): Antinociceptive action of nicotine and its methiodide derivatives in mice and rats. Br. J. Pharm. 79: 869–876.

Aceto MD, Bagley RS, Dewey WL, Fu TC, Martin BR (1986): The spinal cord as a major site for the antinociceptive action of nicotine in the rat. Neuropharmacology 25: 1031–1036.

Aceto MD, Scates SM, Ji Z, Bowman ER (1993): Nicotine's opioid and anti-opioid interactions: proposed role in smoking behavior. Eur. J. Pharmacol. 248: 333–335.

Armstrong D, Dry RML, Keele CA, Markham JW (1953): Observations on chemical excitants of cutaneous pain in man. J. Physiol. 120: 326–351.

Badio B, Daly JW (1994): Epibatidine, a potent analgetic and nicotinic agonist. Mol. Pharmacol. 45: 563–569.

Badio B, Garraffo HM, Plummer CV, Padgett WL, Daly JW (1997a): Synthesis and nicotinic activity of epiboxidine: an isoxazole analogue of epibatidine. Eur. J. Pharmacol. 321: 189–194.

Badio B, Padgett WL, Daly JW (1997b): Ibogaine: a potent noncompetitive blocker of ganglionic/neuronal nicotinic receptors. Mol. Pharmacol. 51: 1–5.

Bannon AW, Gunther KL, Decker MW (1995a): Is epibatidine really analgesic? Dissociation of the activity, temperature, and analgesic effects of (±)-epibatidine. Pharmacol. Biochem. Behav. 51: 693–698.

Bannon AW, Gunther KL, Decker MW, Arneric SP (1995b): The influence of Bay K 8644 treatment on (±)-epibatidine-induced analgesia. Brain Res. 678: 244–250.

Bannon AW, Decker MW, Curzon P, Buckley MJ, Kim DJB, Lynch JK, Wasicak JT, Arnold WH, Holladay MW, Arneric SP (1997): Characterization of ABT-594 in rats: a novel, orally effective antinociceptive agent acting via nicotinic acetylcholine receptors. Soc. Neurosci. Abstr. 22: 1199.

Bannon AW, Decker MW, Holladay MW, Curzon P, Donnelley-Roberts D, Puttfarcken PS, Bitner S, Diaz A, Dickenson AH, Porsolt RD, Williams M, Arneric SP (1998): Broad-spectrum, non-opoid analgesic activity by selective modulation of neuronal nicotinic acetylcholine receptors. Science. 279: 77–80.

Block RC, Chin CWY, Wu WH, Zbuzek VK (1993): Nicotine-induced analgesia in rats: the role of calcium and the diversity of responders and nonresponders. Life Sci. 53: 195–200.

Bonhaus DW, Bley KR, Broka CA, Fontana DJ, Leung E, Lewis R, Shieh A, Wong EHF (1995): Characterization of the electrophysiological, biochemical and behavioral actions of epibatidine. J. Pharmacol. Exp. Ther. 272: 1199–1203.

Caggiula AR, Epstein LH, Perkins KA, Saylor S (1995): Different methods of assessing nicotine-induced antinociception may engage different neural mechanisms. Psychopharmacology 122: 301–306.

Chin CWY, Block RC, Wu WH, Zbuzek VK (1993): Modulation of nicotine-induced analgesia by calcium agonist and antagonist in adult rats. Psychopharmacology 110: 497–499.

Christensen MK, Smith DF (1990): Antinociceptive effects of the stereoisomers of nicotine given intrathecally in spinal rats. J. Neural Transm. 80: 189–194.

Cooley JE, Villarosa GA, Lombardo TW, Moss RA, Fowler SC, Sult S (1990): Effect of pCPA on nicotine-induced analgesia. Pharmacol. Biochem. Behav. 36: 413–415.

Curzon P, Bitner RS, Nikkel AL, Bannon AW, Arneric SP, Decker MW (1997): Central mechanism involved in ABT-594 antinociception. Soc. Neurosci. Abstr. 22: 1200.

Damaj MI (1997): Altered behavioral sensitivity of Ca^{2+}-modulating drugs after chronic nicotine administration in mice. Eur. J. Pharmacol. 322: 129–135.

Damaj MI, Martin BR (1993a): Calcium agonists and antagonists of the dibydropyridine type: effect on nicotine-induced antinociception and hypomotility. Drug Alcohol Dependence 32: 73–79.

Damaj MI, Martin BR (1993b): Is the dopaminergic system involved in the central effects of nicotine in mice? Psychopharmacology 111: 106–108.

Damaj MI, Martin BR (1996): Tolerance to the antinociceptive effect of epibatidine after acute and chronic administration in mice. Eur. J. Pharmacol. 300: 51–57.

Damaj MI, Welch SP, Martin BR (1993): Involvement of calcium and L-type channels in nicotine-induced antinociception. J. Pharmacol. Exp. Ther. 266: 1330–1338.

Damaj MI, Creasy KR, Grove AD, Rosecrans JA, Martin BR (1994a): Pharmacological effects of epibatidine optical enantiomers. Brain Res. 664: 34–40.

Damaj MI, Glennon RA, Martin BR (1994b): Involvement of the serotonergic system in the hypoactive and antinociceptive effects of nicotine in mice. Brain Res. Bull. 33: 199–203.

Damaj MI, Welch SP, Martin BR (1994c): Nicotine-induced antinociception in mice: role of G-proteins and adenylate cyclase. Pharmacol. Bio. Behav. 48: 37–42.

Damaj MI, Welch SP, Martin BR (1995a): In vivo pharmacological effects of dihydro-b-erythroidine, a nicotinic antagonist, in mice. Psychopharmacology 117: 67–73.

Damaj MI, Creasy KR, Welch SP, Rosecrans JA, Aceto MD, Martin BR (1995b): Comparative pharmacology of nicotine and ABT-418, a new nicotinic agonist. Psychopharmacology 120: 483–490.

Damaj MI, Welch SP, Martin BR (1996): Characterization and modulation of acute tolerance to nicotine in mice. J. Pharmacol. Exp. Ther. 277: 454–461.

Davis L, Pollock LJ, Stone TT (1932): Visceral pain. Surg. Gynecol. Obstet. 55: 418–426.

Donnelly-Roberts D, Puttfarcken P, Jacobs IX, Kuntzweiller T, Arneric SP (1997): ABT-594: neuroprotective effects in spinal cord (SC) cultures by a broad spectrum analgesic acting via nicotinic acetylcholine receptors (nAChRs). Soc. Neurosci. Abstr. 22: 1199.

Epstein LH, Caggiula AR, Stiller R (1989): Environment-specified tolerance to nicotine. Psychopharmacology 97: 235–237.

Fertig JB, Pomerleau OF, Sanders B (1986): Nicotine-produced antinociception in minimally deprived smokers and ex-smokers. Addict. Behav. 11: 239–248.

Flores CM, Kilo S, DeCamp RM, Hargreaves KM (1995): Neuronal nicotinic receptor expression in the trigeminal ganglion of the rat. Soc. Neurosci. Abstra. 21: 1414.

Flores CM, DeCamp RM, Kilo S, Rogers SW, Hargreaves KM (1996): Neuronal nicotinic receptor expression in sensory neurons of the rat trigeminal ganglion: demonstration of $\alpha3\beta4$, a novel subtype in the mammalian nervous system. J. Neurosci. 16(24): 7892–7901.

Gillberg PG, Hartvig P, Gordh T, Sottile A, Jansson I, Archer T, Post C (1990): Behavioral effects after intrathecal administration of cholinergic receptor agonists in the rat. Psychopharmacology 100: 464–469.

Ginzel KH (1975): The importance of sensory nerve endings as sites of drug action. Naunyn-Schmiedegerg's Arch. Pharmacol. 288: 29–56.

Green PG, Kitchen I (1986): Antinociception opioids and the cholinergic system. Prog. Neurobiol. 26: 119–146.

Guimarães APC, Prado WA (1994): Antinociceptive effects of carbachol microinjected into different portions of the mesencephalic periaqueductal gray matter of the rat. Brain Res. 647: 220–230.

Hajós M, Engberg G (1988): Role of primary sensory neurons in the central effects of nicotine. Psychopharmacology 94: 468–470.

Hamann SR, Martin WR (1992): Opioid and nicotinic analgesic and hyperalgesic loci in the rat brain stem. J. Pharmacol. Exp. Ther. 261: 707–715.

Iwamoto ET (1989): Antinociception after nicotine administration ino the mesopontine tegmentum of rats: evidence for muscarinic actions. J. Pharmacol. Exp. Ther. 251: 412–421.

Iwamoto ET (1991): Characterization of the antinociception induced by nicotine in the pedunculopontine tegmental nucleus and the nucleus raphe magnus. J. Pharmacol. Exp. Ther. 257: 120–133.

Iwamoto ET, Marion L (1993): Adrenergic, serotonergic and cholinergic components of nicotinic antinociception in rats. J. Pharmacol. Exp. Ther. 265: 777–789.

Jancsó N, Jancsó-Gábor S, Takáts I (1961): Pain and inflammation induced by nicotine, acetylcholine and structurally related compounds and their prevention by desensitizing agents. Acta Physiol. 19: 113–132.

Juan H (1982): Nicotinic nociceptors on perivascular sensory nerve endings. Pain 12: 259–264.

Jurna I, Kraub P, Baldauf J (1993): Depression by nicotine of pain-related nociceptive activity in the rat thalamus and spinal cord. Clin. Investig. 72: 65–73.

Kamat UG, Pradhan RJ, Sheth UK (1972): Potentiation of a non-narcotic analgesic, dipyrone, by cholinomimetic drugs. Psychopharmacologia 23: 180–186.

Kamerling SG, Wettstein JG, Sloan JW, Su TP, Martin WR (1982): Interaction between nicotine and endogenous opioid mechanisms I the unanesthetized dog. Pharmacol. Biochem. Behav. 17: 733–740.

Keele CA, Armstrong D (1964): Substances producing pain and itch. London: Edward Arnold Ltd., pp. 107–123.

Kilo S, Hargreaves KM, Flores CM (1996): Thermal nociception is attenuated by nicotinic agonists in two models of peripheral inflammation in the rat. Soc. Neurosci. Abstr. 22: 879.

Kim DJB, Bannon AW, Campbell JE, Lynch JK, Bai H, Holladay MW, Arneric SP, Decker MW (1997): Efficacy of ABT-594, a novel anti-nociceptive agent acting via nicotinic acetylcholine receptors, in a model of neuropathic pain. Soc. Neurosci. Abstr. 22: 1199.

Li T, Qian C, Eckman J, Huang DF, Shen TY (1993): The analgesic effect of epidatidine and isomers. Bioorgan. Med. Chem. Lett. 3: 2759–2764.

Llewelyn MB, Azami J, Grant CM, Roberts MHT (1981): Analgesia following microinjection of nicotine into the periaqueductal gray matter. Neurosci. Lett. Suppl. 7: S277.

Mansner R (1972): Relation between some central effects of nicotine and its brain levels in the mouse. Ann. Med. Exp. Biol. Fenn. 50: 205–212.

Martin WR, Kumar S, Sloan JW (1988): Opioid and nicotinic medullary hyperalgesic influences in the decerebrated rat. Pharmacol. Bio. Behav. 29: 725–731.

Martin TJ, Suchocki J, May EL, Martin BR (1990): Pharmacological evaluation of the antagonism of nicotine's central effects by mecamylamine and pempidine. J. Pharmacol. Exp. Ther. 254: 45–51.

Mattila MJ, Ahtee L, Saarnivaara L (1968): The analgesic and sedative effects of nicotine in white mice, rabbits and golden hamsters. Ann. Med. Exp. Biol. Fenn. 46: 78–84.

Mendenhall WL (1921): Effect of tobacco smoking on human sensory thresholds. J. Pharmacol. Exp. Ther. 17: 333–335.

Milgrom-Friedman J, Penman R, Meares R (1983): A preliminary study on pain perception and tobacco smoking. Clin. Exp. Pharmacol. Physiol. 10: 161–169.

Molinero MT, Del Rio J (1987): Sustance P, nicotinic acetylcholine receptors and antinociception in the rat. Neuropharmacology 26: 1715–1720.

Mousa SA, Aloyo VJ, Van Loon GR (1988): Tolerance to tobacco smoke- and nicotine-induced analgesia in rats. Pharmacol. Bio. Behav. 31: 265–268.

Nesbitt PD (1973): Smoking, physiological arousal, and emotional response. J. Pers. Soc. Psychol. 25: 137–144.

Oelssner W, Andreas K (1969): The inhibition of pain reactions of the mouse by cholinomimetics and their antagonistic modification. Acta Biol. Med. Germ. 22: 369–385.

Onaivi ES, Payne S, Brock JW, Hamdi A, Faroouqui S, Prasad C (1993): Chronic nicotine reverses age-associated increases in tail-flick latency and anxiety in rats. Life Sci. 54: 193–202.

Parvini S, Hamann SR, Martin WR (1993): Pharmacologic characteristics of a medullary hyperalgesic center. J. Pharmacol. Exp. Ther. 265: 286–293.

Pedigo NW, Dewey WL, Harris LS (1975): Determination and characterization of the antinociceptive activity of intraventricularly administered acetylcholine in mice. J. Pharmacol. Exp. Ther. 193: 845–852.

Pert A (1975): The colinergic system and nociception in the primate: interactions with morphine. Psychopharmacologia 44: 131–137.

Pert A (1987): Cholinergic and catecholaminergic modulation of nociceptive reactions. Pain and Headache 9: 1–63.

Phan DV, Dóda M, Bite A, György L (1973). Antinociceptive activity of nicotine. Acta Physiol. Acad. Sci. Hung 44: 85–93.

Pomerleau OF (1986): Nicotine as a psychoactive drug: anxiety and pain reduction. Psychopharmacol. Bull. 22: 865–869.

Pomerleau OF, Turk DC, Fertig JB (1984): The effects of cigarette smoking on pain and anxiety. Addictive Behav. 9: 265–271.

Qian C, Li T, Shen TY, Libertine-Garahan L, Eckman J, Biftu T, Ip S (1993): Epibatidine is a nicotinic analgesic. Eur. J. Pharmacol. 250: R13–R14.

Rao TS, Correa LD, Reid RT, Lloyd GK (1996): Evaluation of anti-nociceptive effects of neuronal nicotinic acetylcholine receptor (NAChR) ligands in the rat tail-flick assay. Neuropharmacology 35: 393–405.

Rogers RJ (1979): Effects of nicotine, mecamylamine, and hexamethonium on shock-induced fighting, pain reactivity, and locomotor behaviour in rats. Psychopharmacology 66: 93–98.

Rogers DT, Iwamoto ET (1993): Multiple spinal mediators in parenteral nicotine-induced antinociception. J. Pharmacol. Exp. Ther. 267: 341–349.

Rupniak NMJ, Patel S, Marwood R, Webb J, Traynor JR, Elliott J, Freedman SB, Fletcher SR, Hill RG (1994): Antinociceptive and toxic effects of (+)-epibatidine oxalate attributable to nicotinic agonist activity. Br. J. Pharmacol. 113: 1487–1493.

Sahley TL, Berntson GG (1979): Antinociceptive effects of central and systemic administrations of nicotine in the rat. Psychopharmacology 65: 279–283.

Schachter S (1978): Pharmacological and psychological determinants of smoking. Ann. Int. Med. 88: 104–114.

Seltzer CC, Friedman GD, Siegelaub AB, Collen MF (1974): Smoking habits and pain tolerance. Arch. Environ. Health 29: 170–172.

Silverstein B (1982): Cigarette smoking, nicotine addiction, and relaxation. J. Pers. Soc. Psychol. 42: 946–950.

Smith MD, Yang X, Nha J-Y, Buccafusco JJ (1989): Antinociceptive effect of spinal cholinergic stimulation: interaction with substance P. Life Sci. 45: 1255–1261.

Spande TF, Garraffo HM, Edwards MW, Yeh HJC, Pannel L, Daly JW (1992): Epibatidine: a novel (chloropyridyl)azabicycloheptane with potent analgesic activity from an ecuadoran poison frog. J. Am. Chem. Soc. 114: 3475–3478.

Steen KH, Reeh PW (1993): Actions of cholinergic agonists and antagonists on sensory nerve endings in rat skin, in vitro. J. Neurophysiol. 70: 397–405.

Suh HW, Song DK, Choi SR, Chung KM, Kim YH (1996a): Nicotine enhances morphine- and beta-endorphin-induced antinociception at the supraspinal level in the mouse. Neuropeptides 30: 479–484.

Suh HW, Song DK, Lee KJ, Choi SR, Kim YH (1996b): Intrathecally injected nicotine enhances the antinociception induced by morphine but not beta-endorphin, D-Pen^{2,5}-enkephalin and U50,488H administered intrathecally in the mouse. Neuropeptides 30: 373–378.

Sullivan JP, Decker MW, Brioni JD, Donnelly-Roberts D, Anderson DJ, Bannon AW, Kang CH, Adams P, Piattoni-Kaplan M, Buckley MJ, Gopalakrishnan M, Williams M, Arneric SP (1994): (±)-Epibatidine elicits a diversity of in vitro and in vivo effects mediated by nicotinic acetylcholine receptors. J. Pharmacol. Exp. Ther. 271: 624–631.

Sult SC, Moss RA (1986): The effects of cigarette smoking on the perception of electrical stimulation and cold pressor pain. Addict. Behav. 11: 447–451.

Tanelian DL (1991): Cholinergic anctivation of a population of corneal afferent nerves. Exp. Brain Res. 86: 414–420.

Tanelian DL (1991): Cholinergic activation of a population of corneal afferent nerves. Exp. Brain Res. 86: 414–420.

Tripathi HL, Martin BR, Aceto MD (1982): Nicotine-induced antinociception in rats and mice: correlation with nicotine brain levels. J. Pharmacol. Exp. Ther. 221: 91–96.

Waller D, Schalling D, Levander S, Edman G (1983): Smoking, pain tolerance, and physiological activation. Psychopharmacology 79: 193–198.

Wang Q-P, Nakai Y (1994): The dorsal raphe: an important nucleus in pain modulation. Brain Res. Bull. 34: 575–585.

Wong JON, Chin CWY, Wu H, Zbuzek VK (1994): The effect of nifedipine and verapamil on nicotine-induced antinociception in rats. Life Sci. 54: 1711–1718.

Yaksh TI, Dirksen R, Harty GJ (1985): Antinociceptive effects of intrathecally injected cholinomimetic drugs in the rat and cat. Eur. J. Pharmacol. 117: 81–88.

Yamada S, Kagawa Y, Takayanagi N, Hayashi E, Tsuji K, Kosuge T (1986): Inhibition by neosurugatoxin of nicotine-induced antinociception. Brain Res. 375: 360–362.

Yang CY, Hu WH, Zbuzek VK (1992): Antinociceptive effect of chronic nicotine and nociceptive effect of its withdrawal measured by hot-plate and tail-flick in rats. Psychopharmacology 106: 417–420.

Zbuzek VK, Cohen B, Wu WH (1997): Antinociceptive effect of nifedipine and verapamil tested on rats chronically exposed to nicotine and after its withdrawal. Life Sci. 60: 1651–1658.

22

Subtype-Selective nAChR Agonists for the Treatment of Neurological Disorders: SIB-1508Y and SIB-1553A

Frédérique Menzaghi, Ph.D., David E. McClure, Ph.D. and G. Kenneth Lloyd, Ph.D.
SIBIA Neurosciences Inc.
La Jolla, California

Neuronal nicotinic acetylcholine receptors (nAChRs) exist in a variety of forms composed of alpha ($\alpha 2$–$\alpha 9$) and beta ($\beta 2$–24) subunits, which combine to form a pentameric structure. Pentameric alpha homomers ($\alpha 7$, $\alpha 8$, $\alpha 9$) may also exist. NAChRs mediate synaptic transmission directly by enhancing neurotransmitter release via a presynaptic action and indirectly by increasing the neuronal firing rate. While it is known that different nAChRs are responsible for modulating the release of different neurotransmitters (Sacaan et al., 1995), the structural identity for the specific nAChR(s) releasing a given neurotransmitter (e.g., dopamine, norepinephrine, or acetylcholine) in a given region (e.g., striatal vs. limbic vs. cortical) is as yet undetermined. The observation that specific neurotransmitter pathways are modulated by different nAChRs is the basis for the hypothesis that subtype-selective nAChR agonists will have therapeutic potential in different neurological and psychiatric conditions. Thus, striatal dopamine (DA) release modulated by nAChR agonists in appropriate brain regions should relate to the treatment of extrapyramidal motor disorders, whereas the modulation of hippocampal and cortical acetylcholine, dopamine, and norepinephrine release should be useful in the treatment of cognitive disorders.

Neuronal Nicotinic Receptors: Pharmacology and Therapeutic Opportunities, Edited by S. P. Arneric and J. D. Brioni
ISBN 0-471-24743-x, pages 379–394. Copyright © 1998 by Wiley-Liss, Inc.

SIB-1508Y
[(*S*)-]

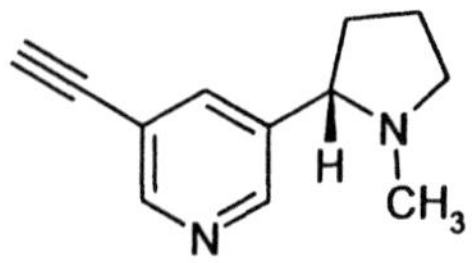

SIB-1553A
[Racemic]

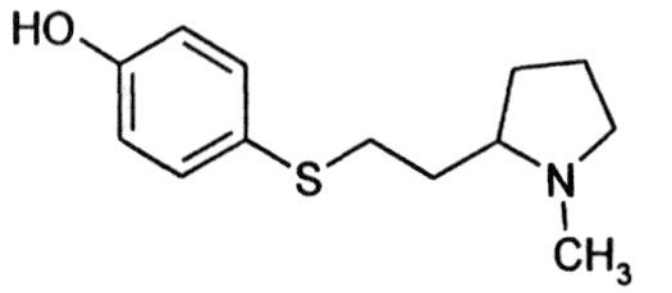

Figure 22.1. Structures of SIB-1508Y (S)-(−)-5-ethynyl-3-(1-methyl-2-pyrrolidinyl)pyridine, maleate salt) and SIB-1553A {(±)-2-[2-(4-hydroxyphenylthio)ethyl]-1-methylpyrrolidine, hydrochloride}. The suffix letters indicate the acid used to provide the corresponding salt for each compound. Y, maleic acid salt; A, hydrochloride salt.

The present chapter summarizes the activity of SIB-1508Y and SIB-1553A, two structurally different nAChR agonists (Fig. 22-1) with different selectivity for nAChR subtypes and differential activity on neurotransmitter release. From their *in vivo* profiles in animal models, it is envisaged that SIB-1508Y should be of therapeutic benefit for the motor, cognitive, and affective symptoms of Parkinson's disease and that SIB-1553A should be of therapeutic benefit for the symptomatic treatment of Alzheimer's disease.

SIB-1508Y

Subtype Selectivity

SIB-1508Y is highly selective for nAChRs with a ratio (Kd's) of at least 1000 separating the displacement of ^{3}H-nicotine binding to rat brain cortex as compared to the displacement of ^{3}H-ligands to more than 55 other neurotransmitter and ligand binding sites. This includes the muscarinic cholinergic site and also other sites within the same receptor superfamily (GABA-A, 5-HT3) (Sacaan et al., 1997).

When the functional activity of SIB-1508Y is assessed on recombinant human nAChRs expressed in stable cell lines or in *Xenopus* oocytes, the compound shows a high degree of preference for the $\alpha4\beta2$-containing cell line as compared to other nAChR-containing preparations (Sacaan et al., 1997; Chavez-Noriega et al., 1997). Notably, SIB-1508Y exhibits very little activity at the human neuromuscular junction nAChR ($\alpha1\beta1\gamma\delta$) or the proposed ganglionic receptor ($\alpha3\beta4$). In all cases, SIB-1508Y is a full agonist.

Neurotransmitter Release

SIB-1508Y enhances the release of several neurotransmitters *in vitro* and *in vivo*. Dopamine is released by SIB-1508Y from the striatum, limbic system, and frontal cortex to a greater extent than that induced by nicotine; norepinephrine is released from the hippocampus, thalamus, and frontal cortex to a lesser extent than by nicotine (Sacaan et al., 1997); acetylcholine is released from the hippocampus and frontal cortex to a much greater extent than by nicotine (hippocampus: 3- to 4-fold over baseline for SIB-1508Y vs. 2-fold over baseline for nicotine; frontal cortex: 6- to 7-fold over baseline for SIB-1508Y vs. 2- to 3-fold over baseline for nicotine). SIB-1508Y did not induce the release of acetylcholine from the striatum (Reid et al., unpublished results).

Motor Activity

Parkinson's disease is classically viewed as a disease involving the loss of nigrostriatal dopamine neurons with subsequent motor dysfunction (rigidity, akinesia, tremor, and loss of balance). In addition to the loss of dopaminergic neurons, other neurotransmitter systems are altered in Parkinson's disease and certainly play a role in the symptomatology observed. These additional systems include noradrenergic, serotoninergic, and cholinergic pathways, all of which fall under the mediation of nAChRs. In Parkinson's disease the concentration of nAChRs decreases by up to 50% in some brain regions, probably due to their presynaptic localization on degenerated neurons. Administration of L-DOPA, a precursor of DA that increases striatal DA levels, is currently the most effective treatment of Parkinson's disease.

Rodents. SIB-1508Y and its racemate, but not the optical antipode, increase locomotor activity without evidence of stereotypies or other abnormal movements in young adult male rats (Fig. 22-2). This enhanced locomotor activity is much greater than that induced by nicotine (Fig. 22-3) or other nAChR agonists (Menzaghi et al., 1997a) and is maintained after 8 weeks of daily injections of the racemate (unpublished observations). This activity depends upon activation of nAChR receptors (blocked by the ion channel blocker mecamylamine and the competitive nAChR antagonist DHβE) and stimulation of dopamine receptors (antagonized by D-1 or D-2 receptor antagonists) (Menzaghi et al., 1997a).

In rats with unilateral 6-OH dopamine lesions of the nigrostriatal dopamine neurons, SIB-1508Y induces ipsilateral rotations, which is indicative of dopamine release from the unlesioned dopamine tract (Cosford et al., 1996). In reserpinized rats, SIB-1508Y exhibits a weak intrinsic activity; however, the effects of levodopa (Fig. 22-4) and amantadine are markedly potentiated (Menzaghi, et al., 1997b).

Primates. SIB-1508Y has been administered to different strains of primates with MPTP-induced lesions of their dopamine neurons. The different primates studied provide an index of the different stages of Parkinson's disease based on the degree of MPTP-induced striatal dopamine loss needed to provoke Parkinsonian motor symptoms. The squirrel monkey exhibits a marked response to SIB-1508Y (Langston et al., personal communication). In this regard, the squirrel monkey has the least severe lesion (~75% of dopamine neurons lesioned), corresponding to stage 2–3 of Parkinson's disease (Lloyd et al., 1975). In macaques, which need an almost total destruction of dopamine neurons before Parkinson motor symptoms appear (Hornykiewicz and Pifl, 1994), SIB-1508Y exerts a significant but weak effect. Coadministration of minimal doses of L-DOPA and inactive doses of SIB-1508Y provides an effect equivalent to the maximum benefit observed with L-DOPA alone (Fig. 22-5) (Lloyd et al., 1995; Menzaghi et al., 1996; Schneider et al., 1998a,b). Thus, the effect of L-DOPA is markedly potentiated by SIB-1508Y. This applies to the total rating score, to object retrieval (fine motor control), and to initiation of movement.

These data suggest that SIB-1508Y may provide clinical benefit for the motor symptoms of Parkinson's disease. This benefit should be maximal for the early stages of the disease (stages 1 through 3) when a significant proportion of dopamine neurons remains intact. In

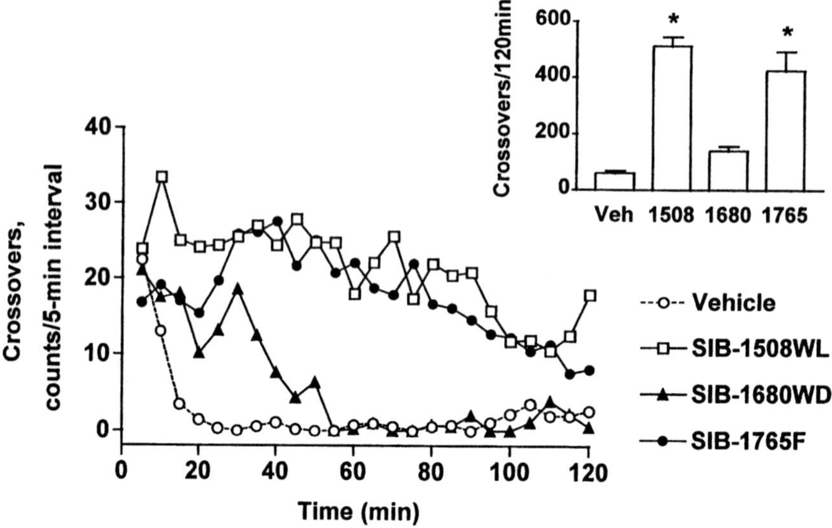

Figure 22.2. *Effect of SIB-1765F, and its isomers SIB-1508WL and SIB-1680WD, on locomotor activity in young adult Sprague-Dawley rats habituated to the test cages. Locomotor activity was assessed in photocell activity cages according to the method described in Menzaghi et al., 1997a. Data are presented as average number of cage crosses (crossovers, i.e., consecutive interruptions of one photocell followed immediately by interruption of an adjacent photocell) taken at 5 min intervals beginning immediately after the subcutaneous administration of SIB-1765F (24 mg/kg, free base) (F, fumaric acid salt), SIB-1508WL (12 mg/kg, free base), and SIB-1680WD (12 mg/kg, free base) (WD, di-p-toluoyl-D-tartaric acid salt) (n = 7–8/group). SEMs were not shown for the clarity of the graphs. Inset: Locomotor activity presented as total crossovers over a 120 min period beginning immediately after the subcutaneous administration of the test compounds (mean ± SEM). SIB-1508WL and its racemate, but not the optical antipode (SIB-1680WD), increased locomotor activity in rats. *p $<$ 0.05 vs. vehicle-treated group, Dunnett's test.*

cotherapy with L-DOPA, SIB-1508Y should provide marked benefit with much lower doses of L-DOPA than would be necessary without the nAChR agonist.

Cognitive Activity

Although Parkinson's disease is classically viewed as a motor disorder, cognitive changes are a frequent occurrence in this disorder. Cognitive deficits associated with Parkinson's disease may exist early in the course of the disease and may occur in the absence of dementia (up to 40% of patients) (Lieberman, 1997). In general, the cognitive deficits described in Parkinson's disease patients cover a wide range of cognitive functions. Primarily though, they resemble problems associated with frontal lobe and frontal cortico-striatal circuit dysfunction including deficits in attention set shifting, deficits in executive function in planning and sequencing, impairment of memory (spatial memory deficits, impaired immediate and delayed recall), bradyphrenia, and increased distractibility (Levin and Katzen, 1995). Many of these cognitive deficits can be found at the earliest stages of Parkinson's disease, and some are thought to actually predate the detection of parkinsonian motor deficits. Thus, as motor symptoms progress, increased motor disability may be superimposed on earlier-appearing cognitive deficits. It is probable that the degeneration of several neurotransmitter systems contribute to these cognitive alterations, including cholinergic, dopaminergic, and noradrenergic pathways, all of which

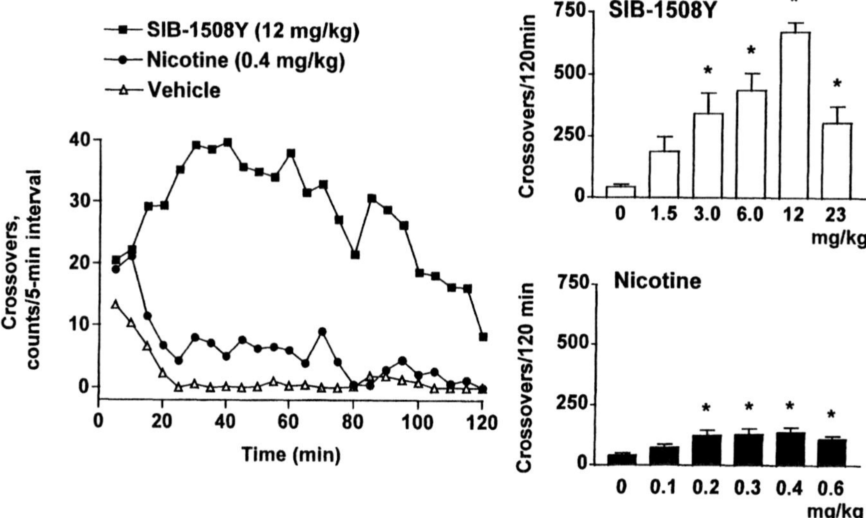

*Figure 22.3. Increases in locomotor activity induced by SIB-1508Y and (−)-nicotine in young adult Sprague-Dawley rats habituated to the test cages. Left panel: Time course of effects presented as crossovers taken at 5 min intervals beginning immediately after the subcutaneous administration of SIB-1508Y (12 mg/kg, free base) and nicotine (0.4 mg/kg, free base) (n = 7–8/group). SEMs were not shown for the clarity of the graphs. Right panels: Dose-related effects presented as total crossovers over a 120 min period beginning immediately after the subcutaneous administration of the test compounds (mean ± SEM). SIB-1508Y was more efficacious than nicotine at increasing locomotor activity in young adult rats. *p < 0.05 vs. vehicle-treated group, Dunnett's test.*

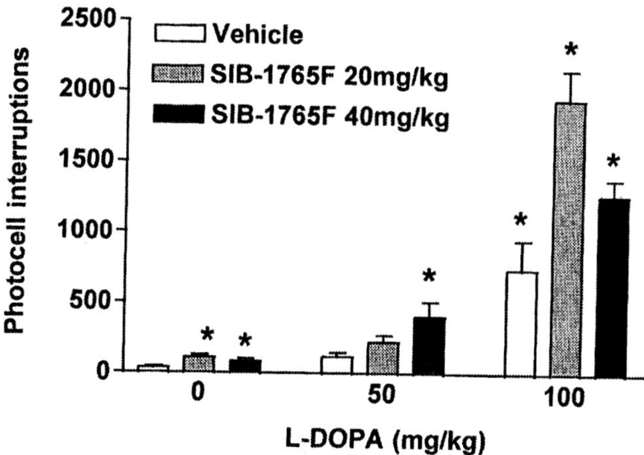

*Figure 22.4. Dose-related effects of subcutaneous administration of SIB-1765F in combination with L-DOPA on locomotor activity in young adult Sprague-Dawley rats pretreated with reserpine (1 mg/kg, intravenous 24 hr before the experiment). Benserazide (50 mg/kg, i.p.) was administered 15 min before L-DOPA (50 and 100 mg/kg, i.p.) according to methods described in Menzaghi et al., 1997b. SIB-1765F was administered 30 min after L-DOPA. Data are presented as total photocell interruptions taken over a 120 min period after the administration of the test compounds (mean ± SEM, n = 8/group). The combination of SIB-1765F and L-DOPA resulted in a significant greater locomotor stimulant effect than that produced by either treatment alone. *p < 0.05 vs. respective control, Newman Keul's test. (From Menzaghi et al., 1997b.)*

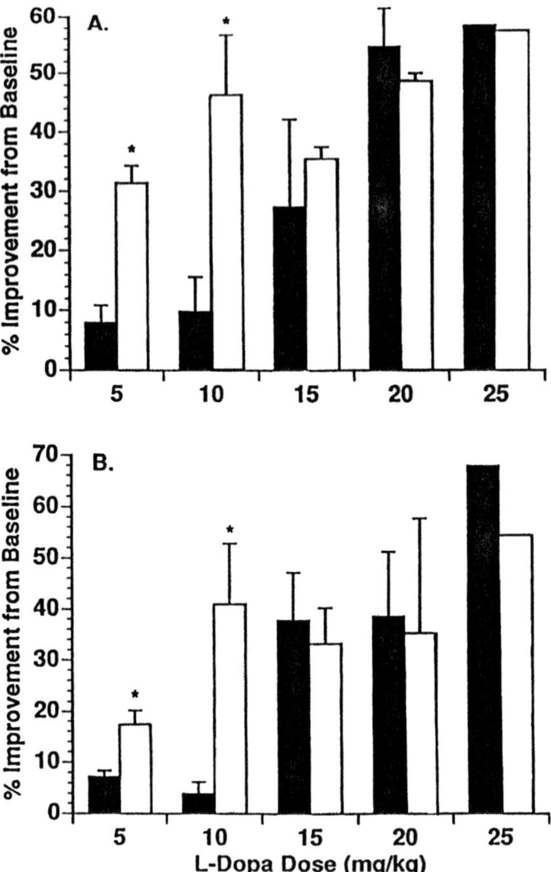

*Figure 22.5. Effects of L-DOPA (dark bars) and combined SIB-1508Y and L-DOPA (light bars) on behavioral ratings of parkinsonian monkeys obtained 20 min (**A**) and 60 min (**B**) after drug administration. Data are presented as percent (±SEM) of motor improvement (use of a motor/behavioral rating scale) over predrug baseline. A single dose of SIB-1508Y (1.0 mg/kg, i.m.), minimally effective on its own, was given in combination with various doses of L-DOPA (administered intramuscularly 30 min after administration of 10.0 mg/kg benserazide). The combination of SIB-1508Y and L-DOPA caused significant functional improvements at doses that were essentially ineffective on their own. *p < 0.01 compared to L-DOPA alone. (From Schneider et al., 1998b.)*

are modulated by nAChRs. Although there is some disagreement in the literature, the consensus is that present antiparkinsonian medications provide minimal if any benefit for the cognitive symptoms of Parkinson's disease.

Rodents. SIB-1508Y consistently enhances the speed of performance (with maintained accuracy) of rats exposed to the working memory version of the 8-arm radial maze (Menzaghi et al., unpublished results); however, SIB-1508Y did not significantly improve working and reference memory performances in young and aged male rats (Menzaghi et al., unpublished results).

Primates. In agreement with the results obtained in rodents, SIB-1508Y did not exhibit a statistically significant cognitive activity in aged (20–23 years) female rhesus monkey in a work-

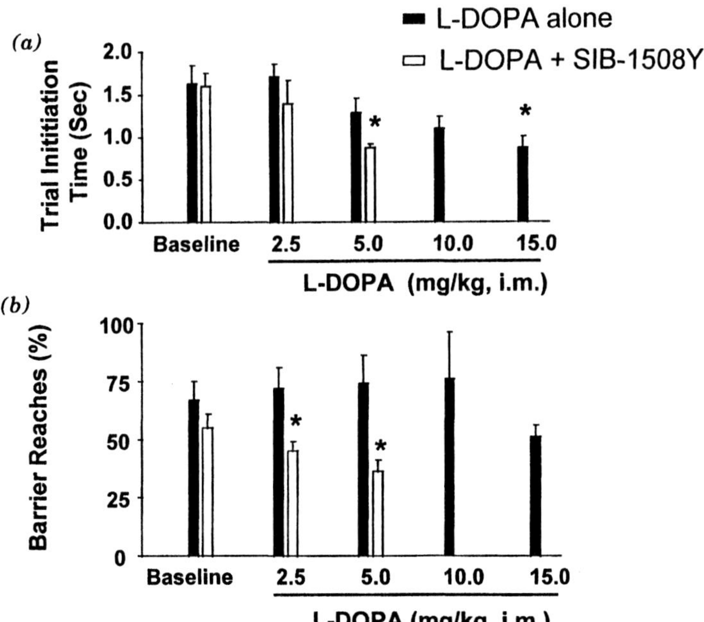

Figure 22.6. *Effects of L-DOPA and SIB-1508Y on reversal of cognitive and motor deficits in MPTP-treated monkeys performing a detour reaching/object retrieval task. Three adult male* Macaca fascicularis *monkeys were trained to perform an object retrieval task according to the method described by Schneider et al., 1993. After training, animals received chronic MPTP treatment to the point where the animals developed motor deficits superimposed on their cognitive deficits. L-DOPA was administered intramuscularly 30 minutes after administration of benserazide (10.0 mg/kg, i.m.). Testing took place 30 minutes after administration of L-DOPA. The response to SIB-1508Y in combination with L-DOPA was assessed using a standard dose of SIB-1508Y (1 mg/kg, i.m.) (Schneider et al., 1998a) in combination with either 2.5 or 5.0 mg/kg L-DOPA. SIB-1508Y and L-DOPA were administered simultaneously 30 minutes after benserazide administration, and testing took place 30 minutes after administration of the combined drugs. The combination of SIB-1508Y and L-DOPA caused significant improvements in both motor (object retrieval trial initiation times [panel A]) and cognitive (number of barrier reaches or cognitive errors made per trial [panel B]) performances. *p < 0.05 vs. respective L-DOPA alone or baseline, student* t *test.*

ing memory task (delayed matching to sample). However, acute intramuscular administration of SIB-1508Y produced an overall improvement of 12.8 ± 4.9% of performance at long delays (J. Buccafusco, unpublished results). SIB-1508Y exerts a marked, significant, cognitive effect in MPTP-treated monkeys. In marmosets with a marked parkinsonian symptomatology, SIB-1508Y exerts its most marked effects on alertness and attention, as compared to the locomotor activity (P. Jenner, unpublished results). In severely MPTP-lesioned cynomolgus monkeys, the combination of SIB-1508Y with L-DOPA reduce akinesia (decreased initiation time), reduce motor slowness (decreased task completion time), and improve cognitive function (reduced barrier failures), whereas L-DOPA alone improves motor functions without significantly improving cognitive functions (Fig. 22-6) (Schneider et al., unpublished results).

Cynomolgus monkeys that have received a regimen of low-dose MPTP show marked cognitive deficits without motor impairment (Schneider and Kovelowski, 1990). In this model, SIB-1508Y increases performance of a visual memory task (variable delayed response) with both attentional and short-term memory components. This effect lasts for at least 48 hours following a single administration of SIB-1508Y (Fig. 22-7) (Pope-Coleman et al., 1996). Neither

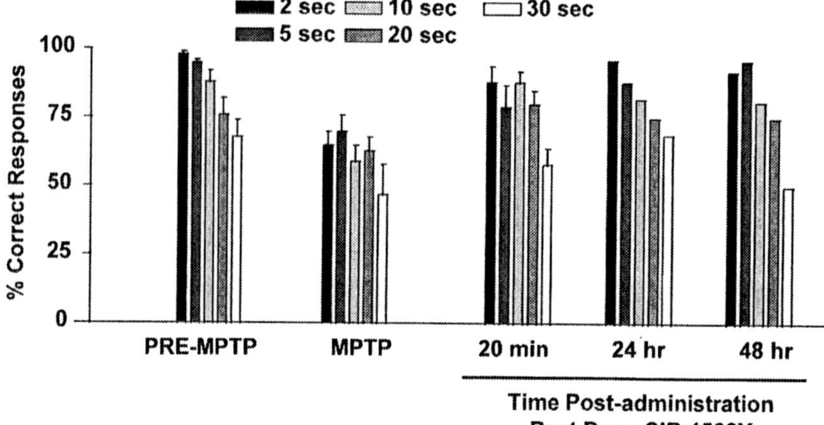

Figure 22.7. *Best dose effect of SIB-1508Y on reversal of spatial working memory deficits in chronic low-dose MPTP-treated monkeys. Four adult male* Macaca fascicularis *monkeys and one adult female* Macaca nemistrina *monkey were trained to perform a variable spatial delayed-response task according to the method described by Schneider et al., 1998b. After training, animals received chronic low-dose MPTP treatment to the point where the animals developed cognitive deficits but no motor deficits. Overall, monkeys performed the task at an 83.6% correct level (± 7.0) during the pre-MPTP baseline period. After chronic MPTP exposure, overall performance deteriorated to 66% (± 3.1). In contrast to the normal performance of this task, monkeys exhibited a delay-independent performance deficit after chronic exposure to MPTP. This delay-independent pattern of responding is classically observed in patients with Parkinson's disease. SIB-1508Y was administered intramuscularly 30 minutes prior to testing. Overall performance on the task improved in a dose-dependent fashion with administration of SIB-1508Y and performance tended to revert back to a normal (pre-MPTP) delay-dependent pattern of responding. There were individual differences in the dose response to SIB-1508Y such that animals had a "best dose" effect at different doses. The most facilitating dose of SIB-1508Y (i.e., best dose) was selected for each animal as shown in this figure. The average "best dose" was 1.7 mg/kg (range = 1.0–2.5 mg/kg). Immediate (20 min) and long-lasting effects (at least 24–48 hrs after drug administration) were observed.*

L-DOPA nor nicotine significantly improved the cognitive performances of monkeys treated with low doses of MPTP (Schneider et al., unpublished results). These results strongly suggest that SIB-1508Y may be useful in the treatment of the cognitive deficits observed in Parkinson's disease.

Affect and Mood

Patients with Parkinson's disease also frequently exhibit changes in affect, notably apathy, anergia, and anhedonia with or without accompanying depression. Approximately 40% of patients with idiopathic Parkinson's disease exhibit affective disorders. These symptoms are dissociable from the motor disability, and dopamine replacement therapy is not associated with marked changes in affect even when good motor responses are attained (Lieberman, 1998). The rationale for the use of nAChR agonists for the affective symptoms of Parkinson's disease is appealing, since nAChRs mediate the release of neurotransmitters (norepinephrine, dopamine, serotonin) in brain regions (limbic areas, frontal cortex) relevant for the modulation of affect and mood.

SIB-1508Y releases dopamine and norepinephrine in limbic areas and frontal cortex (Sacaan et al., 1997), actions highly relevant to arousal, attentional processes, and reward sys-

TABLE 22-1. Antidepressant-like Activity of SIB-1508Y in the Learned Helplessness Test in Rats

	Number of Escape Failures per Session (mean ± SEM)		
Group (n = 10)	Session 1	Session 2	Session 3
Nonhelpless controls	8 ± 2**	9 ± 4**	11 ± 4**
Helpless controls	20 ± 3	23 ± 2	25 ± 2
SIB-1508Y (20 mg/kg)	11 ± 4	9 ± 4**	9 ± 4**
Nicotine (0.4 mg/kg)	16 ± 4	17 ± 4	17 ± 4
Imipramine (32 mg/kg)	9 ± 3**	6 ± 2***	7 ± 3***

According to the model of "learned helplessness" (Martin et al., 1990), animals that have been exposed to inescapable aversive stimuli (e.g., electric shocks) are unable on a subsequent occasion to learn the responses necessary to escape from similar aversive stimulation ("learned helplessness"). Repeated administration of antidepressants (such as imipramine) after exposure to shocks decreases the escape deficits (e.g., the number of escape failures per session) induced in rats by exposure to inescapable shocks. SIB-1508Y reversed the helpless condition to the same extent as imipramine. (**$p < 0.01$, ***$p < 0.001$ one-way ANOVA). (Data collected by R. Porsolt (I.T.E.M. Labo), Paris, France, and P. Lacroix [Biotral], Rennes, France.)

tems. In the learned helplessness model of depression in rats, SIB-1508Y reversed the helpless condition to the same extent as imipramine (Table 22-1), a highly effective antidepressant medication. These data suggest that SIB-1508Y holds promise for the management of the affective components of Parkinson's disease in addition to addressing the cognitive and motor symptoms of the disorder.

Phase I Clinical Studies

SIB-1508Y has been administered in double-blind placebo-controlled studies to healthy volunteers of either sex within the age range of 40 to 60 years as both a single oral administration and in a multiple-dose regimen over 7 days. SIB-1508Y was generally well tolerated with the dose-limiting effects noted to be nausea and vomiting. These are common effects of virtually all dopamine-related antiparkinsonian medications. Other adverse events noted included lightheadedness, headache, salivation, and sweating. No cardiovascular events, changes in vital signs, EKG, or standard laboratory parameters were noted. The adverse event profile was similar for both single- and multiple-dose regimens, and side effects diminished or disappeared upon repeated dosing. Female subjects appeared to be more susceptible to nausea and vomiting than the males; females were able to attain similar maximal repeated doses to those achieved with direct dosing in males when a dose-escalation regimen was employed for the female subjects. The bioavailability of SIB-1508Y appears to be considerably greater in humans than in rats or dogs based upon pharmacokinetics (AUC, Cmax) and pharmacodynamics (emesis). The half-life of SIB-1508Y is considerably longer in humans (approximately 10 hours) as compared to laboratory species (about 1.5 hours).

SIB-1508Y for Parkinson's Disease

The data summarized above provide a sound basis for the testing of SIB-1508Y as a therapeutic agent for the broad spectrum of the symptomatology of Parkinson's disease, including cognitive, motor, and affective aspects. The data also suggest that SIB-1508Y should be an effective monotherapy in mild to moderate cases of Parkinson's disease and should be useful in all stages of Parkinson's disease when used in cotherapy with levodopa.

SIB-1553A

Alzheimer's disease is a neurodegenerative disorder leading to the dementia of Alzheimer's type. The use of nAChR agonists for the treatment of this disease has become an active target for novel drug discovery and development (Decker and Brioni, 1997). The impetus for this activity is based on several observations:

1. The cholinergic deficits occurring in Alzheimer's disease and the hypothesis that increasing synaptic availability of acetylcholine in relevant brain regions will ameliorate cognitive function. This reasoning has led to the development and approval of various acetylcholinesterase inhibitors, which at the time of writing this chapter are the only approved treatments for Alzheimer's disease in the United States.
2. The fact that activation of specific nAChRs induce the release of neurotransmitters other than ACh (such as dopamine and norepinephrine), which have also been shown to be depleted in specific brain regions of patients with Alzheimer's disease and are known to play an important role in the mediation of cognitive processes.
3. The activity of nicotine and other nAChR agonists in animal models of cognition and the decrements in cognitive performances following administration of nAChR antagonists.
4. The reports on the activity of nicotine and ABT-418 in Alzheimer's disease patients, suggesting that better-tolerated and more efficacious nAChR agonists will be useful in this disorder.

Subtype Selectivity

SIB-1553A (Fig. 22-1) is a nAChR agonist that exhibits a subtype selectivity for β4 subunit–containing nAChRs as demonstrated by profiling the compound against different human recombinant nAChRs in stable cells lines and in transfected *Xenopus* oocytes using either calcium fluorescent dyes or electrophysiological techniques (Washburn et al., 1997). This profile is unlike that observed for SIB-1508Y (see above) or that published for other nAChR agonists (see relevant chapters in this volume).

Neurotransmitter Release

In vitro and *in vivo* studies have shown that SIB-1553A stimulates the release of different neurotransmitters (acetylcholine, norepinephrine, dopamine) from different rat brain regions (limbic areas, frontal cortex, striatum) known to play an important role in cognitive processes (Reid et al., 1997). The differentiating aspect of SIB-1553A-induced neurotransmitter release, in comparison with SIB-1508Y and other nAChR agonists, is its high efficacy in inducing the release of acetylcholine from the hippocampus and frontal cortex as assessed by *in vivo* microdialysis technique (hippocampus: 20-fold over baseline for SIB-1553A vs. 2-fold over baseline for nicotine; frontal cortex: 10-fold over baseline for SIB-1553A vs. 2- to 3-fold over baseline for nicotine) (Figs. 22-8 and 22-9). To the authors' knowledge, this is the most efficacious compound reported in terms of enhanced synaptic release of acetylcholine.

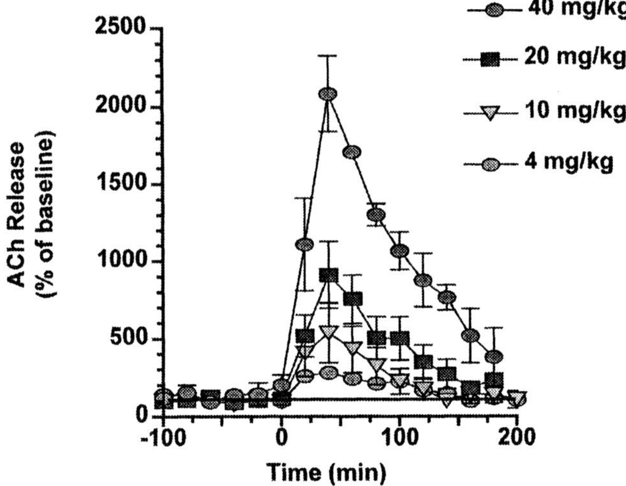

Figure 22.8. Dose-related effect of SIB-1553A on extracellular levels of acetylcholine (ACh) in the hippocampus as measured by microdialysis in freely moving rats. SIB-1553A was administered subcutaneously at time = 0 in young adult Sprague-Dawley rats. Samples were collected every 20 min and directly injected onto an HPLC with electrochemical detection (2 mm dialysis probe, perfusion rate 1 μl/min). Data represent mean ± SEM. *$p < 0.05$ versus saline (two-way repeated measures ANOVA followed by Newman Keul's post hoc test) ($n = 3–4$/group).

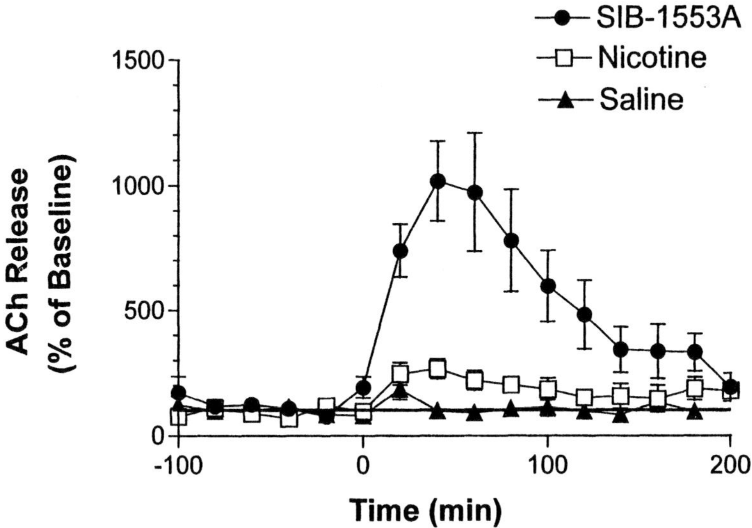

Figure 22.9. Comparison of the effect between the maximum tested dose of SIB-1553A and a maximum tolerated dose of nicotine on extracellular levels of acetylcholine (ACh) in the prefrontal cortex as measured by microdialysis in freely moving rats. SIB-1553A (40 mg/kg free base, s.c.) and (−)-nicotine (0.4 mg/kg free base, s.c.) were administered subcutaneously at time = 0 in young adult Sprague-Dawley rats. Samples were collected every 20 min and directly injected onto an HPLC with electrochemical detection (2 mm dialysis probe, perfusion rate 1 μl/min). Data represent mean ± SEM. *$p < 0.05$ versus saline (two-way repeated measures ANOVA followed by Newman Keuls post hoc test ($n = 4–6$/group).

Cognitive Activity

By inducing the release of dopamine, norepinephrine, and predominantly acetylcholine, SIB-1553A may be beneficial for the symptomatic treatment of Alzheimer's disease. Alzheimer's disease is characterized by a dysfunction of several forms of memory including spatial learning and memory, and short-term memory according to the DSM-IV definition. Three stages of Alzheimer's disease can be grossly defined:

1. An early stage characterized by an inability to remember novel information. At this stage, the initial encoding of information is affected but not the retrieval of previously learned information.
2. A middle stage where semantic memory, working memory, spatial orientation, and attentional deficits are especially observed.
3. A late stage (or dementia) characterized by a global intellectual breakdown associated with changes in personality and behavior.

No animal model exists that mimics the different memory deficits or different stages associated with Alzheimer's disease. Therefore, in order to determine the potential usefulness of SIB-1553A in the symptomatic treatment of this disease, the effects of this compound were studied in three animal models of a specific subset of memory impairment seen in early- and middle-stage Alzheimer's disease. SIB-1553A was evaluated in two animal models where the cholinergic system was disrupted by pharmacological manipulation (administration of scopolamine, a muscarinic antagonist) or by selective destruction of cholinergic neurons (central administration of the immunotoxin 192 IgG saporin). Finally, SIB-1553A was evaluated in a monkey model of catecholaminergic dysfunction (low-dose MPTP) and a more physiological model with multiple-system dysfunction, the aging model.

As shown on Table 22-2, SIB-1553A was investigated using different paradigms looking at different types of behavior (exploratory behavior in a T- or Y-maze, escape behavior in a water maze, and reinforced behavior in a radial maze and delayed-response task) in three different species: mice, rats, and monkeys. These paradigms were tailored to target specific memory forms: spatial and nonspatial, working, and reference memory. Working memory refers to the temporary storage of information; reference memory refers to the long-term storage of information through consolidation processes. Acute and repeated administration of SIB-1553A improved spatial and nonspatial working and reference memory (Table 22-2). In all models tested, SIB-1553A was markedly better than nicotine and was active across the three different species and different models studied (Tables 22-2 and 22-3) (Bontempi et al., 1997; Menzaghi et al., 1997c). At the range of doses active in these models, SIB-1553A did not alter motor activity and was without observable adverse effects. These data suggest that SIB-1553A should be useful in the treatment of cognitive disorders.

DISCUSSION AND CONCLUSION

The data presently available for SIB-1508Y and SIB-1553A together with published information on other nAChR agonists such as ABT-418 and ABT-089 (see the relevant chapters in this volume) provide a firm basis for the hypothesis that this class of drugs will be useful in a wide range of neurological and psychiatric disorders. The therapeutic indication for a specific compound will be determined by its selectivity for different subtypes of nAChRs and subsequently its profile on neurotransmitter release, which translates into specific types of behavioral activity.

TABLE 22-2. Summary of Cognitive Effects of SIB-1553A in Rodents and Nonhuman Primates

Animal Model	Cognitive Function Tested	Cognitive Effect
Pharmacological model		
Adult mice treated with scoplolamine	Working memory[a]	+++
	T maze	
Lesioned animals		
a. Rats with selective destruction of	Spatial reference memory[b]	++
cholinergic neurons (192 IgG/saporin)	Water maze	
75–80% depletion of ChAT in	Spatial working memory[b]	+++
hippocampus and cortical areas)	Water maze/Y maze	
b. Low-dose MPTP-treated monkeys	Working memory[c]	++
	Delayed response (DR)/Delayed matching to sample (DMTS)	
Aged animals		
a. Male C57BL/6 mice	Working memory	+++
(24–26 months old)	T maze*/Radial maze[c]	
	Reference memory[a]	
	T maze	0
b. Male Fisher 344/Nhsd rats	Spatial reference memory[b]	
(22–23 years old)	Water maze	++
	Spatial working memory[b]	
	Water maze	0
c. Female Rhesus monkeys	Working memory[c]	
(20–23 months old)	DMTS	+++
Intact animals		
Adult rats	Spatial refrc./working memory[b]	0
	Water maze	
	Spatial working memory[b]	+
	Y maze	

[a]Subucutaneous acute administration before task.
[b]Repeated subcutaneous administration once daily before task.
[c]Ascending dose response in the same animals after subcutaneous administration in mice and intramuscular or oral administration in nonhuman primates.
0 No improvement.
\+ Mild improvement but not statistically significant.
++ Moderate improvement (statistically significant).
+++ Major improvement (statistically significant).

SIB-1508Y enhances the release of dopamine and acetylcholine and, to a lesser extent, norepinephrine. The behavioral profile is that of a marked enhancement of locomotor activity without stereotyped behavior, which is due to the release of dopamine from vesicular storage sites as demonstrated by the weak activity in reserpinized rats. SIB-1508Y increases the speed of performance in cognitive tasks while maintaining accuracy. All of these activities are in marked contrast to non-nAChR-related locomotor stimulants such as amphetamines, which induce stereotyped behavior, release dopamine from cytoplasmic pools, and increase the speed of performance at the cost of decreased accuracy. In primate models of the motor and cognitive deficits found in Parkinson's disease, SIB-1508Y reverses both types of deficits and, in animals with severe motor deficits, is synergistic with levodopa. The effect of SIB-1508Y on cognitive functions in monkeys was greater in the MPTP model of Parkinson's disease (i.e., catecholaminergic dysfunctions) as compared to the aging model (Table 22-3). In a rodent model of depression, SIB-1508Y was as active as imipramine, an effective antidepressant in humans. Together with the good tolerability and safety profile demonstrated in phase 1 clinical studies, these data strongly support the testing of SIB-1508Y as a therapeutic agent for the motor, cognitive, and affective symptomatology of Parkinson's disease.

TABLE 22-3. Summary of *In Vivo* Pharmacology of SIB-1508Y and SIB-1553A: Comparison with Nicotine

		SIB-1508Y	SIB-1553A	(-)Nicotine
Human nAChRs	$\alpha4\beta2$	+++	++	++
	$\alpha\times\beta2$	++	+	++
	$\alpha\times\beta4$	++	+++	++
	$\alpha7$	0	0	++
Motor behavior	Locomotor activity	+++	++	+
	6-OHDA	+++	+	+
	MPTP-treated monkeys	++	Not tested	0
	Potentiation L-Dopa	+++	Not tested	0
Cognition	Working memory			
	Scopolamine	Not tested	+++	+++
	Lesion immunotoxin	Not tested	+++	0
	Lesion MPTP	+++	++	0
	Aging	+	+++	++
	Reference memory			
	Lesion immunotoxin	Not tested	++	0
	Aging	0	++	+

[a]Agonist activity on recombinant human nAChRs expressed in *Xenopus* oocytes.
($\alpha x\beta2$: $\alpha2\beta2$, $\alpha3\beta2$ $\alpha x\beta4$: $\alpha2\beta4$, $\alpha4\beta4$)

0 No activity.
\+ Mild activity but not statisically significant.
++ Moderate activity (statisically significant).
+++ Marked activity (statisically significant).

SIB-1553A demonstrates a broad activity in reversing deficits in working and reference memory induced by severe lesions of the central cholinergic system or aging in rodents and nonhuman primates (Table 22-3). SIB-1553A, in a dose-dependent manner, also induces the release of a large amount of acetylcholine in hippocampal and cortical areas. This effect is greater than that of other compounds (including nAChR agonists) described so far and may be beneficial in the case of progressive loss of cholinergic neurons. It is therefore projected that this compound will be efficacious in ameliorating the symptoms of Alzheimer's disease. In addition, SIB-1553A induces the release of neurotransmitters, other than acetylcholine, that are known to play an important role in various cognitive processes. SIB-1553A may therefore offer advantages for the symptomatic treatment of Alzheimer's disease over current, more limited therapies.

REFERENCES

Bontempi B, Whelan KT, Risbrough VB, Joppa MA, Kille N, Buccafusco JJ, Menzaghi F, Lloyd GK (1997): SIB-1553A, a novel nAChR agonist with cognitive enhancing properties in aged rodents and non-human primates: effects on various memory forms. Soc. Neurosci. Abstr. 23: 477.15.

Chavez-Noriega LE, Washburn MS, Crona JH, Johnson EC (1997): The novel nicotinic agonist SIB-1508Y and its racemate display subtype-selective activity on recombinant human neuronal nicotinic acetylcholine receptors expressed in *Xenopus* oocytes. Soc. Neurosci. Abstr. 23: 477.4.

Cosford NDP, Bleicher L, Herbaut A, McCallum S, Vernier J-M, Dawson H, Whitten JP, Adams P, Chavez-Noriega L, Correa LD, Crona JH, Mahaffy LS, Menzaghi F, Rao TS, Reid RT, Sacaan AI, Santori EM, Stauderman KA, Whelan KD, Lloyd GK, McDonald IA (1996): (S)-(−)-5-Ethynyl-3-(1-methyl-2-

pyrrolidinyl)pyridine maleate (SIB-1508Y): a novel antiparkinsonian agent with selectivity for neuronal nicotinic acetylcholine receptors. J. Med. Chem. 39(17): 3235–3237.

Decker MW, Brioni JD (1997): Neuronal nicotinic receptors: potential treatment of Alzheimer's disease with novel cholinergic channel modulators. In Brioni JD, Decker MW, editors. Pharmacological treatment of Alzheimer's disease: molecular and neurobiological foundations. New York: Wiley-Liss, Inc., pp. 433–459.

Hornykiewicz O, Pifl C (1994): The validity of the MPTP primate model for the neurochemical pathology of idiopathic Parkinson's disease. In Briley M, Marien M, editors. Noradrenergic mechanisms in Parkinson's disease. Boca Raton: CRC Press, Inc., pp. 11–23.

Levin BE, Katzen HL (1995): Early cognitive changes and nondementing behavioral abnormalities in Parkinson's disease. In Weiner WJ, Lang AE, editors. Behavioral neurology of movement disorders. Advances in Neurology, vol. 65. New York: Raven Press, Ltd., pp. 85–95.

Lieberman AN (1997): Point of view: dementia in Parkinson's disease. Parkinson and Related Disorders. 2: 151–458.

Lieberman AN (1998): Mental disorders in Parkinson's disease: dementia, depression and the cognitive disorder. (In press).

Lloyd KG, Davidson L, Hornykiewicz O (1975): The neurochemistry of Parkinson's disease: effect of L-DOPA therapy. J. Pharmacol. Exp. Ther. 195: 453–464.

Lloyd GK, Rao TS, Sacaan A, Reid RT, Correa LD, Whelan K, Risbrough V, Menzaghi F (1995): SIB-1765F, a novel nicotinic agonist: profile in models of extrapyramidal motor dysfunction. Soc. Neurosci. Abstr. 21: 11.10.

Martin P, Soubrie P, Puech AJ (1990): Reversal of helplessness behavior by serotonin uptake blockers in rats. Psychopharmacology 101: 403–407.

Menzaghi F, Sacaan AI, Reid RT, Santori E, Correa LD, Adams PB, Whelan KT, Risbrough VB, Rao TS, Schneider J, Lloyd GK (1996): Characterization of SIB-1508Y, the active enantiomer of a novel nicotinic acetylcholine receptor (nAChR) agonist, SIB-1765F. Soc. Neurosci. Abstr. 22: 602.11.

Menzaghi F, Whelan KT, Risbrough VB, Rao TS, Lloyd GK (1997a): Effects of a novel cholinergic ion channel agonist SIB-1765F on locomotor activity in rats. J. Pharmacol. Exp. Ther. 280: 384–392.

Menzaghi F, Whelan KT, Risbrough VB, Rao TS, Lloyd GK (1997b): Interactions between a novel cholinergic ion channel agonist, SIB-1765F and L-DOPA in the reserpine model of Parkinson's disease in rats. J. Pharmacol. Exp. Ther. 280: 393–401.

Menzaghi F, Joppa MA, Reid RT, Sacaan AI, Bontempi B, Whelan KT, Santori EM, Risbrough VB, Lloyd GK (1997c): Memory deficits associated with cholinergic dysfunctions: reversal by SIB-1553A, a novel nAChR agonist. Soc. Neurosci. Abstr. 23: 477.16.

Pope-Coleman A, Lloyd GK, Schneider JS (1996): The nicotinic receptor agonist SIB-1508Y potentiates L-DOPA responses in parkinsonian monkeys. Soc. Neurosci. Abstr. 22: 89.8.

Reid RT, Saccan AI, Adams PB, Correa LD, Santoru EM, Mcdonald IA, Lloyd GK, Rao TS (1997): Pharmacological characterization of SIB-1553A, a novel, subtype-selective neuronal nicotinic acetylcholine receptor agonist. Soc. Neurosci. Abstr. 23: 477.17.

Sacaan AI, Dunlop JL, Lloyd GK (1995): Pharmacological characterization of neuronal acetylcholine-gated ion channel receptor-mediated hippocampal norepinephrine and striatal dopamine release from rat brain slices. J. Pharmacol. Exp. Ther. 274: 224–230.

Sacaan AI, Reid JT, Santori EM, Adams P, Corres LD, Mahaffy LS, Bleicher L, Cosford NPD, Stauderman KA, McDonald IA, Rao TS, Lloyd GK (1997): Pharmacological characterization of SIB-1765F: a novel cholinergic ion channel agonist. J. Pharmacol. Exp. Ther. 280: 373–383.

Schneider JS, Kovelowski CJ (1990): Chronic exposure to low doses of MPTP. I. Cognitive deficits in motor asymptomatic monkeys. Brain Res. 519: 122–128.

Schneider JS, Roeltgen DP (1993): Delayed matching-to-sample, object retrieval, and discrimination reversal deficits in chronic low dose MPTP-treated monkeys. Brain Res. 615: 351–354.

Schneider JS, Van Velson M, Menzaghi F, Lloyd GK (1998a): Treatment with the nicotinic acetylcholine receptor agonist SIB-1508Y improves object retrieval performance in MPTP-treated monkeys: comparison with Levodopa treatment. Ann. Neurol. 43: 311–317.

Schneider JS, Pope-Coleman A, Van Velson M, Menzaghi F, Lloyd GK (1998b): Effects of the novel neuronal nicotinic acetylcholine receptor agonist SIB-1508Y on motor behavior in parkinsonian monkeys. Movement Disorders, (in press).

Washburn MS, Chavez-Noriega LE, Crona JH, Vernier JM, Johnson EC (1997): Electrophysiological characterization of SIB-1553A activity on recombinant human neuronal nicotinic receptors expressed in *Xenopus* oocyte and HEK293 cells. Soc. Neurosci. Abstr. 23: 477.3.

23

Nicotinic Acetylcholine Receptor-Targeted Compounds: A Summary of the Development Pipeline and Therapeutic Potential

Michael W. Decker, PhD,
Neurological and Urological Diseases Research
Pharmaceutical Products Division
Abbott Laboratories
Abbott Park, Illinois

and Stephen P. Arneric, PhD
CNS Diseases Research
DuPont Pharmaceuticals Company
Wilmington, Delaware

Nicotine is generally regarded as a toxic and addictive substance. This negative perception, of course, stems from nicotine's presence in tobacco and the many adverse consequences associated with cigarette smoking. The adverse effects of smoking, however, are not necessarily related to nicotine. Nicotine, for example, is not generally believed to be the basis for the increased risk of lung cancer among smokers, and cigarette smoke contains many bioactive substances that may be more harmful than nicotine. Moreover, nicotine may actually have some

Neuronal Nicotinic Receptors: Pharmacology and Therapeutic Opportunities, Edited by S. P. Arneric and J. D. Brioni
ISBN 0-471-24743-x, pages 395–411. Copyright © 1998 by Wiley-Liss, Inc.

beneficial effects (for additional reviews, see Balfour and Fagerström, 1996; Baron, 1996; Birtwistle, 1996; Jarvik, 1991), some of which have been addressed in chapters in this volume. (−)-Nicotine has been reported to be cognition enhancing (Chapters 20 and 22), to ameliorate sensorimotor gating deficits characteristic of schizophrenia (Chapter 17), and to be antinociceptive (Chapter 21). There is clinical evidence that (−)-nicotine may be a useful adjunct for treating Tourette's syndrome (Sanberg et al., 1997), and agents acting at neuronal nicotinic acetylcholine receptors (nAChRs) may have potential for treating parkinsonian symptoms (Chapter 22). In addition, it has been suggested that the high degree of nicotine dependence among smokers with a history of major depression may be the result of antidepressant effects of (−)-nicotine (Gilbert and Gilbert, 1995; Glassman et al., 1990). Moreover, nicotine use (smoking) has been associated with reduced risk for developing Alzheimer's disease (Baron, 1996; Lee, 1994; van Duijn and Hofman, 1991), Parkinson's disease (Baron, 1996; Smith and Giacobini, 1992), and ulcerative colitis (Calkins, 1989). Thus, there is reason to believe that there may be some therapeutic potential for compounds that target nAChRs.

There is no doubt, however, that nicotine has some major liabilities. (−)-Nicotine is addictive and clearly has adverse effects on the cardiovascular and gastrointestinal systems, but much of drug discovery is based on investigation of the potential therapeutic properties of toxic substances. Isolation of the toxic component from spoiled clover lethal to livestock resulted in the discovery of warfarin, an anticoagulant that is still widely used. Similarly, cocaine served as the structural lead for the development of currently available local anesthetics, despite its stimulant and cardiovascular effects and obvious abuse potential. These examples serve as illustrations that the toxicities associated with nicotine would not necessarily preclude the development of therapeutic agents targeting nAChRs. Thus, ignoring the potential benefits of a nAChR approach because of negative perceptions about nicotine is both shortsighted and inconsistent with the historical realities of drug discovery.

The molecular biology of neuronal nAChRs has seen dramatic advances over the last decade (see Chapter 1 and Boyd, 1997; Changeux, 1990; Luetje et al., 1990; Sargent, 1993, for reviews). It has been established that these nAChRs are ligand-gated ion channels composed of five subunits. At least eight α and three β subunits have been found in vertebrate nervous tissue, and various combinations of these subunits form functional receptors in artificial systems. Importantly, the subunits that form nAChRs found in the central and autonomic nervous systems differ from those found at the neuromuscular junction in striated muscle. The discovery of distinct neuronal nAChR subtypes suggests the possibility that nicotine's diverse effects might be mediated by actions at discrete nAChR subtypes. Unfortunately, the data currently available do not permit the identification of the specific nAChR subtypes responsible for mediating each of the effects of nicotine. Moreover, it has not been clearly established that nAChR subunit combinations expressed in artificial systems, such as oocytes or transfected cell lines, are isomorphic with native receptors. Adding to the complexity of the approach, of course, is the possibility that there may be species differences in the pharmacology of nAChR subtypes. Despite our current inability to reliably relate the various physiological and behavioral effects of nAChR stimulation to actions at distinct subtypes, nAChR diversity suggests that it may be possible to design molecules that would produce only a subset of these effects.

POTENTIAL THERAPEUTIC ACTIONS OF AGENTS TARGETING nAChRs

Tourette's Syndrome

Tourette's syndrome (TS) is characterized by severe involuntary motor movements and vocalizations. Sometimes these vocalizations take the form of obscenities. This condition is frequently treated with antipsychotics such as the dopamine antagonist haloperidol. These medications are ineffective in about 30% of the cases, and even when effective they can produce a variety of side effects, including sedation and attentional impairments that can affect compliance (Erenberg et al., 1987; Shapiro et al., 1989). Moreover, long-term haloperidol use can lead to the development of tardive dyskinesia.

The observation that (−)-nicotine potentiates the behavioral effects of neuroleptics in a number of animal models (Emerich et al., 1991; Sanberg et al., 1989) led to the idea that (−)-nicotine could be used clinically as a neuroleptic-sparing agent. By potentiating the therapeutic effects of neuroleptics, (−)-nicotine might allow for reduced neuroleptic doses in TS patients and might therefore decrease side-effect liabilities. Initial open trials with (−)-nicotine gum and patches in TS patients receiving neuroleptics were positive (McConville et al., 1991) with the beneficial effects of (−)-nicotine lasting days to weeks following discontinuation of treatment (Silver and Sanberg, 1993). The finding that the beneficial effects are present long after (−)-nicotine has been cleared from the system raises the possibility that a longer-lasting metabolite of (−)-nicotine, such as cotinine, could have activity (Crooks and Dwoskin, 1997), but even this would not necessarily explain such an extended duration of action. Alternatively, since the prolonged activity is more impressive with (−)-nicotine patches than with (−)-nicotine gum, the extended duration of action may be related to prolonged desensitization or inactivation following continuous exposure of nAChRs to (−)-nicotine (Sanberg et al., 1997). This would be consistent with preliminary reports that the nAChR antagonist, mecamylamine, produces similar beneficial effects.

Schizophrenia

It has long been recognized that smoking is unusually common among schizophrenics and that schizophrenic smokers show unusually high levels of cigarette (nicotine) consumption. One possible explanation of this observation is that nicotine-induced dopamine release may overcome some of the anhedonic effects of dopamine antagonists used to treat schizophrenia. However, it is also possible that schizophrenics smoke as a form of self-treatment. Support for this notion comes from studies on sensory gating (see Chapter 17). Sensory gating is an involuntary process important for selective attention to sensory stimuli and is significantly impaired in schizophrenics. Sensory gating impairments can also be induced in experimental animals by the nAChR antagonist α-bungarotoxin, suggesting that α-bungarotoxin-sensitive receptors (α7) may play an important role in this process (Luntz-Leybman et al., 1992). Consistent with this interpretation is the finding of reduced numbers of α-bungarotoxin-sensitive receptors in the temporal lobes of schizophrenics (Freedman et al., 1993).

(−)-Nicotine ameliorates sensory gating deficits in both humans and experimental animals (Adler et al., 1992; Stevens and Wear, 1997). The novel nAChR activator, ABT-418, can also restore normal gating in animals displaying gating deficits and is 10-fold more potent than (−)-nicotine (Stevens and Wear, 1997). However, the therapeutic utility of ABT-418 and (−)-nicotine appears to be limited since both compounds lose efficacy with rapid, repeated administration. This tachyphylaxis may reflect the very rapid desensitization rate for α7 nAChRs. Interestingly, compounds with partial agonist activities at this nAChR subtype, such as GTS-21, may not desensitize with repeated administration (R. Freedman, personal communication) and could represent an alternative therapeutic approach. Another possible approach to avoid the nAChR desensitization effect of (−)-nicotine would be to develop a positive allosteric modulator of these receptors. More extensive discussion of the therapeutic potential of nAChR agents in schizophrenia can be found in Chapter 17.

Parkinson's Disease

Parkinson's disease (PD) is characterized by tremors at rest, rigidity, bradykinesia, and a loss of postural reflexes. These extrapyramidal effects result from the loss of dopaminergic neurons in the substantia nigra of the brain (German et al., 1989).

In rats subjected to partial dimesencephalic hemitransection, (−)-nicotine can increase DA levels in the substantia nigra (Janson et al., 1989) and attenuate lesion-induced loss of nigral tyrosine hydroxylase-like immunoreactive neurons (Janson and Moller, 1993). This neuroprotective effect of nicotine may account for the inverse relationship between smoking and PD that has been demonstrated in epidemiological studies (Baron, 1996; Smith and Giacobini, 1992).

Another explanation of the inverse relationship between smoking and PD is that (−)-nicotine facilitates DA release in the striatum (Chapter 8), which would help to alleviate the extrapyramidal effects found in PD. nAChR agonists, then, might be useful for treating the motor symtomatology of PD. SIB-1508Y (SIBIA; see Chapter 22) is currently under clinical development for this indication. SIB-1508Y releases DA from striatal slices and elicits ipsilateral rotations in rats with unilateral 6-hydroxydopamine lesions, providing evidence for increased striatal dopamine release in vivo. This compound can also reverse haloperidol-induced catalepsy in rats, a model of extrapyramidal motor dysfunction, suggesting that SIB-1508Y may be useful for treating the motor aspects of PD.

Cognitive Function

Experimental studies in normal humans have demonstrated the beneficial effects of nicotine on cognitive function, particularly with respect to vigilance and attention. A potential confound in these studies is that most have been conducted with subjects who were smokers. Thus, what appears to be cognitive enhancement could merely be the reversal of cognitive-impairing withdrawal symptoms (Rusted et al., 1994). However, positive effects of nicotine have been shown in studies with minimally deprived smokers (Pritchard et al., 1992; Warburton and Arnall, 1994) and in studies with rodents and nonhuman primates (Arendash et al., 1995; Brioni and Arneric, 1993; Buccafusco and Jackson, 1991; Decker et al., 1992; Elrod et al., 1988; Levin, 1992). Therefore, it would appear that amelioration of withdrawal effects does not provide a complete account of the positive effects of nicotine on cognitive function.

Several small clinical studies have also documented positive effects of nicotine in conditions marked by impaired cognitive function. Notably, (−)-nicotine appears to benefit adults with attention deficit disorder (ADD; Levin et al., 1996; Chapter 20) and patients with Alzheimer's disease (AD). Because the most effective current treatments for ADD (e.g., d-amphetamine, methylphenidate, and pemoline) act on catecholamines, the prevailing dogma has been that dysfunction of catecholaminergic systems may underlie ADD (Ernst and Zemetkin, 1995). (−)-Nicotine could have a similar mechanism of action, since it enhances dopamine and norepinephrine release (Benwell and Balfour, 1992; Grady et al., 1992; Mitchell, 1993). However, d-amphetamine and (−)-nicotine also enhance ACh release, which could contribute their beneficial effects (Day and Fibiger, 1992, 1994; Wonnacott, 1997). A clear role for basal forebrain cholinergic systems in attentional processing has been established in the last few years, and (−)-nicotine can ameliorate some of the attentional deficits observed in rats with lesions of the cholinergic basal forebrain (Muir et al., 1995).

The beneficial effects of nicotine that have been reported in studies with AD patients thus far have been modest (Jones et al., 1992; Newhouse et al., 1988; Wilson et al., 1995). It is possible, though, that some of these studies may underestimate the efficacy of the approach given that nAChR-mediated side effects, such as sympathetic activation in nicotine-naive patients, may have blunted cognitive performance. Moreover, these side effects may be particularly pronounced in the elderly. Thus, an agent with reduced side-effect liabilities may allow for a more

TABLE 23-1. nAChR Ligand Competitive Dynamics: Compounds Filling the Pipeline

Compound	Company	Phase	Indication	Comments	References
(-)-Nicotine Patches (Nicotrol, Nicoderm, and Habitrol)	Warner-Lambert, HMR, Novartis	Launched	SC	Potential for ADHD, TS, AD	Levin et al., 1996; Sanberg et al., 1997; Wilson et al., 1995; Chapter 18
(-)-Lobeline	Dynagen	Phase 3	SC		Glover et al., 1998
ABT-418	Abbott	Phase 2	SC/AD	AD discontinued/smoking cessation and ADHD ongoing	Arneric et al.,. 1995; Prendergast et al., 1998
ABT-594	Abbott	Phase 2	Pain	Potential to treat acute and neuropathic pain	Bannon et al., 1998
RJR-2403	Targacept-RJR	Phase 2	AD	Potential for ADHD, pain, and TS	Bencherif et al., 1996; Lippiello et al., 1996
SIB-1508Y	Sibia	Phase 2	PD	Given in conjunction with L-DOPA; potential as antidepressant	Chapter 22
SIB-1553A	Sibia	Phase 2	SC		Chapter 22
GTS-21	Taiho	Phase 1	AD		Meyer et al. 1994; Meyer et al., 1997
ABT-089	Abbott	Phase 1	ADHD		Decker et al., 1997; Prendergast et al., in press; Sullivan et al., 1997
RJR-2557	Targacept-RJR	Preclinical	AD	Potential for ADHD, pain, and TS	Bencherif et al., 1996; Lippiello et al., 1996
CMI-980	CytoMed	Preclinical	Pain	Epibatidine analogs	iDdb
DBO-083	Univ. Milan/Abbott	Preclinical	Pain		Ghelardini et al., 1997
Epibatidine analogs	Astra	Preclinical	Pain		

AD (Alzheimer's disease); PD (Parkinson's disease); SC (smoking cessation); ADHD (attention deficit hyperactivity disorder); TS (Tourette's syndrome); RJR (R. J. Reynolds Tobacco); HMR (Hoechst-Marion-Roussel); iDdb (Investigational Drug Database).

accurate estimate of the potential of this approach. Several other compounds targeting nAChRs (e.g., SIB-1553, ABT-418, ABT-089, RJR-2403, GTS-21, and lobeline) have positive effects on cognitive function in preclinical studies (Arendash et al., 1995; Briggs et al., 1997; Buccafusco et al., 1995; Decker et al., 1993, 1994a, 1994c, 1997; Lippiello et al., 1996; Meyer et al., 1997), some of which are in clinical development (see Table 23-1). The viability of the approach will be determined from large-scale, pivotal clinical trials.

Although not as well established as with PD, there is a controversial negative relationship between AD and smoking (Baron, 1996; Lee, 1994; van Duijn and Hofman, 1991). The neuroprotective effects of nAChR activators (see Chapter 19) may underlie this relationship and could provide an approach for arresting neurodegenerative processes occurring in AD. The combination of neuroprotective and cognitive-enhancing effects in a single compound would be particularly appealing. Agents that treat AD only by slowing degenerative processes may have some utility, but since AD is currently diagnosed behaviorally significant disease progression typically occurs before a diagnosis can be made. This clearly limits the potential of purely neuroprotective approaches, since marked cognitive decline present at the time treatment is started would not be addressed. A neuroprotective agent with cognition-enhancing properties, on the other hand, would improve cognitive performance, as opposed to merely stabilizing or slowing disease progression.

Anxiety

Smokers often make reference to the "calming" effects of smoking, and there is experimental support for mild anxiolytic effects of smoking. For example, smoking has been shown to decrease anxiety produced by either a stressful movie or social interaction (Gilbert et al., 1989; Gilbert and Spielberger, 1987). The act of smoking, apart from delivery of nicotine, does not provide a complete account for the anxiolytic effects in humans (Pomerleau, 1986). Again, the smoking literature is difficult to interpret since apparent anxiolytic effects of nicotine could actually result from the reversal of withdrawal effects in smokers. There is, however, some evidence from animal studies demonstrating anxiolytic-like effects of nAChR agents, including (−)-nicotine, lobeline, ABT-418, and ABT-594 (Brioni et al., 1993, 1994; Cao et al., 1993; Costall et al., 1989; Decker et al., 1998; Onaivi et al., 1994). However, negative results have been reported with several compounds that activate nAChRs, including cytisine, GTS-21, and epibatidine (Briggs et al., 1997; Brioni et al., 1993; Sullivan et al., 1994). Anxiolytic-like effects of nAChR agonists are more readily demonstrated in ethologically based assays than in the conflict tests more traditionally used as anxiolytic screening assays, and the preclinical efficacy of nAChR ligands is generally less robust than that of the benzodiazepines. In contrast to the benzodiazepines, however, nAChR agents are less likely to disrupt motor coordination and cognitive performance.

Balfour and Fagerström (1996) raise the interesting possibility that desensitization of the stress response may form the basis for the anxiolytic effects of (−)-nicotine. They point to the similarity between the physiological effects of (−)-nicotine and those of stressors and suggest that (−)-nicotine might desensitize this response system. This proposal is consistent with the finding that continuous infusion of doses of (−)-nicotine that would be expected to desensitize nAChRs has an anxiolytic-like effect (Brioni et al., 1994). Although this would not explain why nAChR blockade prevents the effects of acute (−)-nicotine, it is possible that the anxiolytic-like effects of acute and chronic (−)-nicotine are mediated by different mechanisms.

Depression

Interest in the possibility that nicotine might have antidepressant activity is based, in part, on the association between major depression and nicotine dependence (Glassman et al., 1990).

Nicotine-dependent smokers are more likely to have a history of major depression than are nonsmokers or nondependent smokers. This relationship is between major depression and nicotine dependence rather than being between major depression and smoking per se (Breslau et al., 1991). This contrasts with the connection between smoking and alcohol and illicit drug use, where the relationship is with smoking without regard to the degree of nicotine dependence (Breslau, 1995). Thus, it does not appear that the correlation between depression and smoking is merely a reflection of lifestyle.

One interpretation of the relationship between smoking and major depression is that nicotine use may be a form of self-medication. This idea is supported by the finding that, for people with a history of major depression, depressive episodes become more likely during attempts to stop smoking (Covey et al., 1997). In addition, smokers with a history of major depression have an unusually difficult time giving up smoking (Covey et al., 1997), which would be expected if mood enhancement and alleviation of depressive symptoms associated with smoking were particularly reinforcing in depressives (Gilbert and Gilbert, 1995). Moreover, there are limited clinical data suggesting that (−)-nicotine itself may have antidepressant effects. In an open-label trial assessing the effects of transdermal (−)-nicotine over a 4-day period, significant improvements were noted—effects that disappeared when treatment was discontinued (Salin-Pascual et al., 1996). In a single-day, double-blind, placebo-controlled trial, similar improvements were observed (Salin-Pascual et al., 1995). In addition, in this latter study, transdermal (−)-nicotine increased REM sleep parameters in depressed patients while decreasing these same parameters in normal controls. This is consistent with the idea that (−)-nicotine may have different effects on EEG in normals and depressives (Gilbert et al., 1994). Preclinical data on the novel nAChR ligand, SIB-1508Y, suggests that this compound may have antidepressant potential (Chapter 22). Whether a nAChR-targeted antidepressant would have advantages relative to the selective serotonin reuptake inhibitors that dominate the market is an open question, but it is interesting that (−)-nicotine effects appear to have a fairly rapid onset, compared to the delayed onset of action characteristic of available antidepressants.

Smoking Cessation

The addictive properties of tobacco appear to be due to (−)-nicotine (Benowitz, 1996; Corrigall, 1993; Rose, 1996). Experimental animals (Cox et al., 1984) will self-administer (−)-nicotine, although self-administration is not as readily established with (−)-nicotine as it is with highly abused drugs such as cocaine and heroin. As with most drugs that are self-administered, the ability of (−)-nicotine to support self-administration depends on activation of the mesolimbic dopaminergic system (Corrigall et al., 1992, 1994). Nicotine addiction is complex and multidimensional, which may explain the difficulty in developing broadly effective smoking cessation treatments (Oates et al., 1988; Stolerman and Jarvis, 1995).

One approach to smoking cessation, of course, has been to provide (−)-nicotine using an alternative delivery system while patients are attempting to quit smoking (see Chapter 18). (−)-Nicotine is administered in gum (Nicorette) or transdermal patches (Nicotrol, Habitrol, NicoDerm). After the regulatory approval of these delivery systems in 1989, sales were initially brisk but softened when it became clear that the approach was only modestly successful in the absence of behavioral support therapy.

An alternative therapeutic strategy for smoking cessation would be nAChR blockade (Clarke, 1994). nAChR antagonists, such as mecamylamine, attenuate nicotine self-administration in animals, and mecamylamine can reverse an acquired preference for cigarette smoke in monkeys (Glick et al., 1990). A related approach combines both replacement (agonist) and blockade of reinforcement (antagonist). Proof of principle for this approach comes from a recent randomized, placebo-controlled, double-blind trial in which oral mecamylamine administration was combined with a (−)-nicotine transdermal patch (Rose et al., 1994). Perhaps a more practical alternative to this polypharmacy approach would be the use of an nAChR partial agonist (Rose and Levin, 1991); and (−)-lobeline, a nAChR ligand with full agonist, partial agonist, or full antagonist properties, depending on the nAChR subtype studied, is currently in phase 3 clinical trials for smoking cessation (Glover et al., 1998).

Analgesia

(−)-Nicotine is modestly antinoceptive in both experimental animals and humans (Pomerleau, 1986; Tripathi et al., 1982; see Chapter 21). The precise site of action for (−)-nicotine is not entirely clear, but antinociception can be observed after direct injection of (−)-nicotine into several brain-stem locations (Iwamoto, 1989, 1990). Perhaps the most consistent effects are obtained by injecting into the lateral tegmental peducular pontine and nucleus raphe magnus (Iwamoto, 1990). (−)-Nicotine injected into these sites stimulates release of ACh, resulting in activation of descending pain inhibitory pathways. Alternatively, nAChR stimulation in the spinal cord can also directly activate antinociceptive mechanisms by a process that appears to involve an increase in intracellular calcium (Damaj et al., 1993).

The recent discovery that the potent antinociceptive agent epibatidine (Badio and Daly, 1994), which is approximately 100 times more potent than morphine, produces its effects via nAChR activation renewed interest in the possible development of nAChR agents as analgesics. The compound is toxic at doses at or only slightly higher than those producing antinociception (Sullivan et al., 1994), so it is not a good candidate for development. To develop a nAChR-targeted analgesic, then, the toxicities associated with epibatidine must be separated from its beneficial effects. Evidence that different nAChR subtypes might underlie these undesirable and desirable effects comes from studies demonstrating that the calcium channel antagonist (+)-Bay K 8644 can block the antinociceptive effect of epibatidine (Bannon et al., 1995) but not its nicotine-like discriminative stimulus properties or its effects on locomotion (J. Brioni and A. Bannon, unpublished observations). Thus, the nAChRs involved in antinociception may act through second-messenger and neurotransmitter systems that are distinct from those underlying other nAChR-mediated behaviors. Analgesics acting at nAChRs have the potential to overcome two of the four major limitations of opioid analgesics, which include constipation, central respiratory depression, tolerance, and addiction, since (−)-nicotine facilitates gastrointestinal motility and enhances central respiratory drive (Taylor, 1990). An example of a novel antinociceptive nAChR ligand with promising preclinical data is ABT-594 (Bannon et al., 1998), which is currently in phase 2 clinical trials.

Inflammatory Bowel Disorders

As with PD and AD, epidemiological data suggest that smokers have a reduced risk of developing ulcerative colitis (UC; Calkins, 1989). The possible therapeutic utility of a nAChR-based approach to treating UC is supported by clinical trials demonstrating positive effects of (−)-nicotine patches. The effects of (−)-nicotine, however, appear to be more pronounced in the active stage of the disease than in remission (Pullan et al., 1994; Thomas et al., 1995). Several mechanisms may contribute to the therapeutic effects of (−)-nicotine in active UC. (−)-Nicotine can downregulate the expression of E-selectin in intestinal tissue biopsies, an action that could modify the properties of inflammatory cell infiltrates (Bhatti et al., 1997). (−)-Nicotine can also reduce basal- and mitogen-stimulated IL-8 production (Bhatti and Hodgson, 1997).

Side effects associated with systemic administration of (−)-nicotine (e.g., by transdermal patches) have led to attempts to develop topical therapies. Pilot clinical studies have been conducted with nicotine enemas in an attempt to minimize systemic exposure. Early results are promising (Green et al., 1997) and randomized controlled studies using such topical formulations of (−)-nicotine are in progress. Interestingly, smoking has an adverse effect on Crohn's disease, a transmural inflammatory disease of the distal ileum and colon, and differences in the effects of nicotine on Crohn's disease and UC would be consistent with differences in the etiology of these inflammatory bowel disorders (Rhodes et al., 1997).

CHALLENGES FOR DRUG DISCOVERY

Recently, several pharmaceutical and biotechnology companies, including SIBIA Neurosciences, Taiho, Abbott, CytoMed, Targacept (R. J. Reynolds), and Dynagen, have attempted to identify novel ligands for neuronal nAChRs (see Table 23-1 for a summary of compound pipeline). These efforts have focused on therapeutic indications such as AD, PD, ADD, and chronic pain management. The most advanced compounds are GTS-21, SIB-1508Y, SIB-1553A, ABT-418, ABT-089, ABT-594, and RJR-2403 (see Fig. 23-1 for structures). Other companies that have shown notable patent activity in the area include Bayer, Astra Arcus, Lundbeck, Pfizer, and Merck. There are, however, several challenges that must be addressed in developing nAChR-targeted compounds as therapeutic agents.

Separation of Undesirable from Desirable Effects of Nicotine

Despite the current lack of understanding of the specific role of nAChR subtypes in the various beneficial and adverse effects of nicotine, there is clear empirical evidence that some degree of functional separation between these effects can be achieved. Some of this evidence comes from the observed differences in the rate at which tolerance develops to different nAChR-mediated effects. For example, the anxiolytic-like effects and cognition-enhancing effects of (−)-nicotine in rodents are maintained during a period of continuous infusion, whereas the locomotor and hypothermic effects are not (Brioni et al., 1994; Decker et al., 1997). Although differences in the development of tolerance here may be related to effects downstream from the activation of nAChRs, these data would be consistent with the idea that these effects are mediated through different nAChR subtypes.

Additional support for the position that the various effects of (−)-nicotine are mediated by different nAChR subtypes comes from recent studies with novel nAChR-targeted compounds. One of these compounds, ABT-418, is at least as potent as (−)-nicotine in improving cognitive performance in experimental animals but is less potent in producing hypothermia, emesis, seizures, and cardiovascular effects (Decker et al., 1994a, 1994c). Similar separations between desirable and adverse effects of nAChR activation have been achieved with other compounds, including GTS-21, SIB-1508Y, ABT-089, ABT-594, and RJR-2403 (Briggs et al., 1997; Decker et al., 1997; Lin et al., 1997; Lippiello et al., 1996; Meyer et al., 1994, 1997; Chapter 22). In general, the patterns of biochemical effects produced by these compounds in preclinical studies are distinct from those observed with (−)-nicotine. While the preclinical functional selectivity of these compounds has yet to be confirmed fully in humans, the preclinical findings suggest that it may be possible to develop compounds that selectively target nAChRs involved in mediating the beneficial effects of nicotine. Furthermore, the pattern of beneficial effects observed with even this limited set of compounds differ, which suggests that beneficial effects mediated via actions at nAChRs also have distinct underlying mechanisms. GTS-21, for example, shares the cognition-enhancing the cytoprotective effects of ABT-418 (Decker et al.,

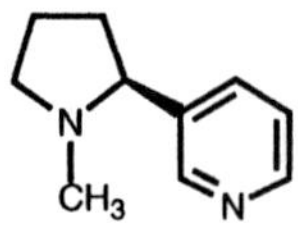

(-)-Nicotine

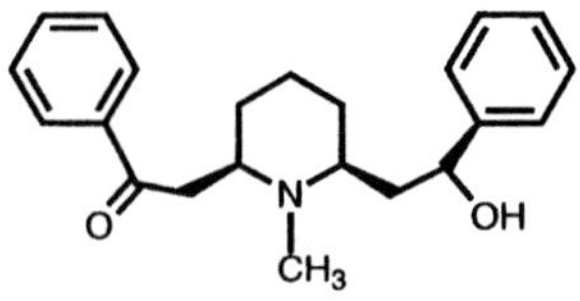

Lobeline

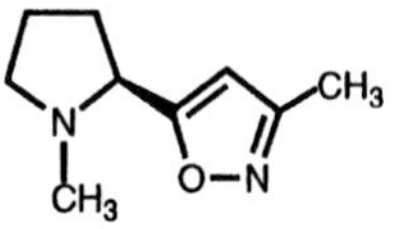

ABT-418

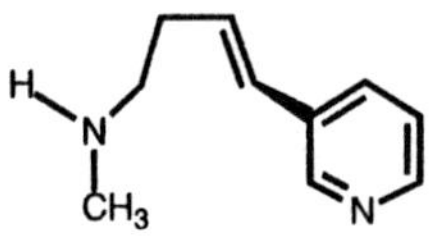

RJR-2403

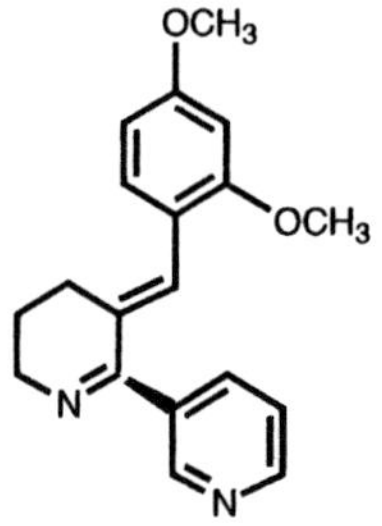

GTS-21

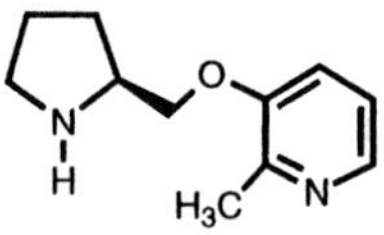

ABT-089

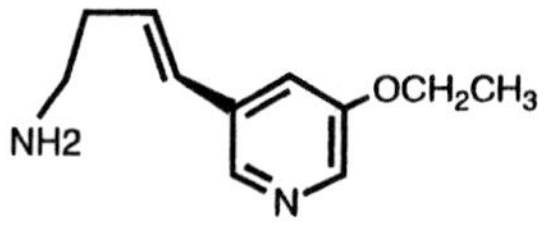

RJR-2557

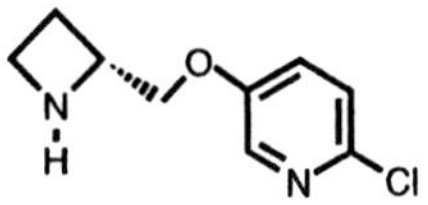

ABT-594

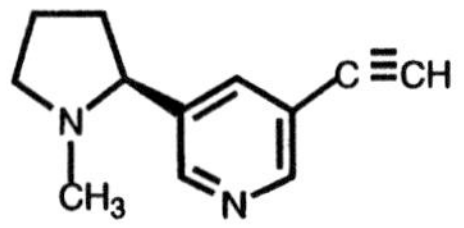

SIB-1508Y

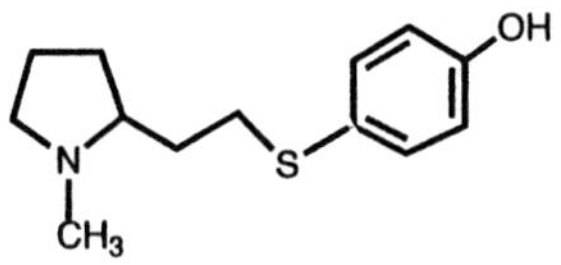

SIB-1553A

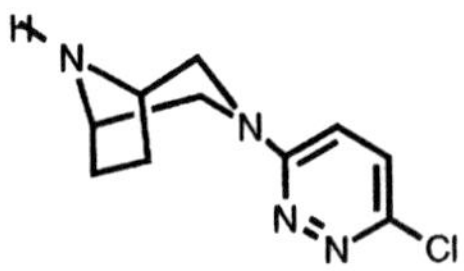

DBO-083

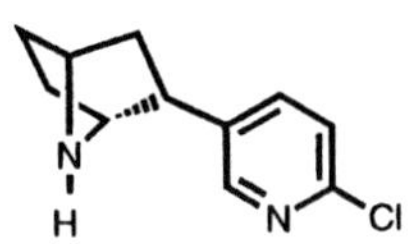

(±)-Epibatidine

Figure 23.1. *Structures of nAChR-Targeted Compounds*

1994a; Meyer et al., 1997; Nanri et al., 1997; Chapter 19) but does not have the anxiolytic-like effects that have been obtained preclinically with ABT-418 (Briggs et al., 1997; Brioni et al., 1994). Similarly, ABT-594 has potent, broad-spectrum antinociceptive properties (Bannon et al., 1998) that are not observed with ABT-089 (unpublished observations). Again, these different activity profiles may be related to actions at distinct subtypes or combinations of subtypes.

Agonism Versus Antagonism

Because (−)-nicotine is a cholinergic channel activator, it is frequently assumed that the physiological and behavioral effects of nicotinic are mediated by nAChR activation. This view is supported by the observation that many effects of (−)-nicotine can be prevented by nAChR antagonists, such as mecamylamine. Yet many of the beneficial effects of (−)-nicotine are observed with continuous administration that might be expected to desensitize nAChRs. For example, antidepressant effects in humans are observed with continuous administration using (−)-nicotine patches (Salin-Pascual et al., 1995, 1996). Similarly, anxiolytic-like and cognitive-enhancing effects can be obtained in rats with continuous infusion via subcutaneous osmotic pumps (Brioni et al., 1994; Decker et al., 1997; Levin, 1992). Thus, it is at least possible that some of the effects of (−)-nicotine are related to receptor desensitization. Other effects, however, are markedly decreased by repeated administration. The hypothermic and the drug discrimination effects in rodents, for example, can be attenuated by pre-exposure to even a low dose of (−)-nicotine given shortly before the (−)-nicotine challenge is administered (Decker et al., 1994b; Rosecrans and James, 1993). A similar reduction in potency by previous exposure to (−)-nicotine can be demonstrated for the amelioration of sensory gating deficits (Stevens and Wear, 1997). Thus, some, but not all, effects of (−)-nicotine could actually be related to its ability to desensitize nAChRs.

In their discussion of the possible role of desensitization in some of the behavioral actions of (−)-nicotine, Balfour and Fagerström (Balfour and Fagerström, 1996) note that the effects of continuous and acute (−)-nicotine on catecholamine release differ. However, the effects of continuous infusion are not the same as those of nAChR blockade. Thus, effects are not simply based on whether receptors are turned on or off. Instead, it may be that activated, desensitized, and inactivated nAChRs have distinct signaling properties. Clearly, an important question that must be addressed when developing nAChR-based therapies is: What kind of activity at the receptor is responsible for the desired (and undesired) effects?

Assessment of Competitive Position of the Approach

Evaluation of the competitive position of a nAChR-based approach varies as a function of the proposed therapeutic indication. Identification of the degree of unmet medical need and a realistic appraisal of the likely efficacy of nAChR-targeted compounds are required to evaluate the likelihood of successfully penetrating the marketplace. The nAChR approach is novel, and novel mechanisms of action are risky. For patients who do not respond adequately to currently available therapies, however, compounds with novel mechanisms of action are likely to provide more realistic alternatives than are compounds representing incremental improvements within a class.

With nAChR ligands, it may be particularly important to consider the full range of potential beneficial effects when evaluating potential. For example, one could argue that the anxiolytic-like properties of these compounds are modest and that this limits the potential for development of this class of compounds as anxiolytics. Similarly, the availability of effective antidepressants might be viewed as an impediment to the successful development of nAChR-targeted antidepressants. However, anxiolytic and/or antidepressant properties might be important adjunctive properties for compounds targeting other therapeutic indications. Cognitive enhancers with anxiolytic and/or antidepressant activity might be particularly useful in AD, where agitation and depression can be major problems. Similarly, it is conceivable that anxiolytic and/or antidepressant properties would be useful in an analgesic agent targeting chronic pain or that antipsychotic (Chapter 17) and/or antiparkinsonian (Chapter 22) properties would be useful in a cognition enhancer targeting Lewy body dementia, which is accompanied by hallucinations and motor impairment (Hansen et al., 1990; Perry et al., 1990). Thus, a complete evaluation of the market potential of nAChR agents would include a consideration of the possibility of complementary therapeutic actions of a single molecule.

CONCLUSION

The notion that (−)-nicotine may have therapeutic potential is not new (Jarvik, 1991), but the side effects and negative perceptions associated with (−)-nicotine have limited the development of nicotine as a therapeutic agent. The emerging molecular neurobiology of neuronal nAChRs, however, suggests the possibility of separating desirable and undesirable effects mediated by nAChRs and may provide specific targets for the development of selective compounds. The possibility of targeting nAChR subtypes would be consistent with the distinct clinical pharmacology of other ligand-gated ion channels that display subtype diversity, such as the GABA receptor. Moreover, dissimilarities in the behavioral effects of currently available nAChR ligands (Decker et al., 1995) indicate that selectivity is possible, but targeted compound development awaits an improved understanding of the pharmacology and function of the putative nAChR subtypes. Once this information is available, it may become possible to predict in vivo effects based on in vitro profiles, facilitating the design and development of safer, more efficacious compounds and allowing clinical validation of the therapeutic potential of the nAChR-based approach.

REFERENCES

Adler LE, Hoffer LJ, Griffith J, Waldo MC, Freedman R (1992): Normalization by nicotine of deficient auditory sensory gating in the relatives of schizophrenics. Biol. Psych. 32: 607–616.

Arendash GW, Sengstock GJ, Sanberg PR, Kem WR (1995): Improved learning and memory in aged rats with chronic administration of the nicotinic receptor agonist GTS-21. Brain Res. 674: 252–259.

Arneric SP, Sullivan JP, Decker MW, Brioni JD, Bannon AW, Briggs CA, Donnelly-Roberts D, Radek RJ, Marsh KC, Kyncl J, Williams M, Buccafusco JJ (1995): Potential treatment of Alzheimer disease using cholinergic channel activators (chCAs) with cognitive enhancement, anxiolytic-like, and cytoprotective properties. Alzheimer Dis. Assoc. Disorders. 9 (suppl.2): 50–61.

Badio B, Daly JW (1994): Epibatidine, a potent analgetic and nicotinic agonist. Molec. Pharmacol. 45: 563–569.

Balfour DJK, Fagerström KO (1996): Pharmacology of nicotine and its therapeutic use in smoking cessation and neurodegernative disorders. Pharmacol. Ther. 72: 51–81.

Bannon AW, Gunther KL, Decker MW, Arneric SP (1995): The influence of Bay K 8644 treatment on (±)-epibatidine-induced analgesia. Brain Res. 678: 244–250.

Bannon AW, Decker MW, Holladay MW, Curzon P, Donnelly-Roberts D, Puttfarcken PS, Bitner RS, Diaz A, Dickenson AH, Porsolt RD, Williams M, Arneric SP (1998): Broad-spectrum, non-opoid analgesic activity by selective modulation of neuronal nicotinic acetylcholine receptors. Science 279: 77–81.

Baron JA (1996): Beneficial effects of nicotine and cigarette smoking: the real, the possible and the spurious. Br. Med. Bull. 52: 58–73.

Bencherif M, Lovette ME, Fowler KW, Arrington S, Reeves L, Caldwell WS, Lippiello PM (1996): RJR-2403: a nicotinic agonist with CNS selectivity I. In vitro characterization. J. Pharmacol. Exp. Ther. 279(3): 1413–1421.

Benowitz NL (1996): Pharmacology of nicotine: addiction and therapeutics. Annu. Rev. Pharmacol. Toxicol. 36: 597–613.

Benwell MEM, Balfour DJK (1992): The effects of acute and repeated nicotine treatment on nucleus accumbens dopamine and locomotor activity. Br. J. Pharmacol. 105: 849–856.

Bhatti MA, Hodgson HJF (1997): Nicotine down-regulates IL-8 production and tissue expression in inflammatory bowel disorder (IBD). Gastroenterology 112: A934.

Bhatti MA, Haskard DO, Hodgson HJF (1997): In intestinal tissue biopsies expression of E-selectin is modulated by nicotine and crude tobacco extract. Gastroenterology 112: A934.

Birtwistle J (1996): Does nicotine have beneficial effects in the treatment of certain diseases? Br. J. Nursing 5: 1195–1202.

Boyd RT (1997): The molecular biology of neuronal nicotinic acetylcholine receptors. Crit. Rev. Toxicol. 27: 299–318.

Breslau N (1995): Psychiatric comorbidity of smoking and nicotine dependence. Behav. Genet. 25(2): 95–101.

Breslau N, Kilbey M, Andreski P (1991): Nicotine dependence, major depression, and anxiety in young adults. Arch. Gen. Psychiatry 48(12): 1069–1074.

Briggs CA, Anderson DJ, Brioni JD, Buccafusco JJ, Buckley MJ, Campbell JE, Decker MW, Donnelly-Roberts D, Elliott RL, Gopalakrishnan M, Holladay MW, Hui YH, Jackson WJ, Kim DJB, Marsh KC, Oneill A, Prendergast MA, Ryther KB, Sullivan JP, Arneric SP (1997): Functional characterization of the novel neuronal nicotinic acetylcholine receptor ligand GTS-21 in-vitro and in-vivo. Pharmacol. Biochem. Behav. 57: 231–241.

Brioni JD, Arneric SP (1993): Nicotinic receptor agonists facilitate retention of avoidance training: participation of dopaminergic mechanisms. Behav. Neural. Biol. 59: 57–62.

Brioni JD, O'Neill AB, Kim DJB, Decker MW (1993): Nicotinic receptor agonists exhibit anxiolytic-like effects on the elevated plus-maze. Eur. J. Pharmacol. 238: 1–8.

Brioni JD, O'Neill AB, Kim DJB, Buckley MJ, Decker MW, Arneric SP (1994): Anxiolytic-like effects of the novel cholinergic channel activator ABT-418. J. Pharmacol. Exp. Ther. 271: 353–361.

Buccafusco JJ, Jackson WJ (1991): Beneficial effects of nicotine administered prior to a delayed matching-to-sample task in young and aged monkeys. Neurobiol. Aging 12: 233–238.

Buccafusco JJ, Jackson WJ, Terry AV Jr., Marsh KC, Decker MW, Arneric SP (1995): Improvement in performance of a delayed matching-to-sample task by monkeys following ABT-418: a novel cholinergic channel activator for memory enhancement. Psychopharmacology 120: 256–266.

Calkins BM (1989): A meta-analysis of the role of smoking in inflammatory bowel disease. Dig. Dis. Sci. 34: 1841–1854.

Cao W, Burkholder T, Wilkins L, Collins AC (1993): A genetic comparison of behavioral actions of ethanol and nicotine in the mirrored chamber. Pharmacol. Biochem. Behav. 45(4): 803–809.

Changeux J-P (1990): The nicotinic acetylcholine receptor: an allosteric protein prototype of ligand-gated ion channels. Trends Pharmacol. Sci. 11: 485–491.

Clarke PBS (1994): Nicotine dependence—mechanisms and therapeutic strategies. Biochem. Soc. Symp. 59: 83–95.

Corrigall W (1993): Nicotine addiction: considerations in the therapeutic use of nicotine. Med. Chem. Res. 2: 603–611.

Corrigall WA, Franklin KBJ, Coen KM, Clarke PBS (1992): The mesolimbic dopaminergic system is implicated in the reinforcing effects of nicotine. Psychopharmacology 107: 285–289.

Corrigall WA, Coen KM, Adamson KL (1994): Self-administered nicotine activates the mesolimbic dopamine system through the vental tegmental area. Brain Res. 653: 278–284.

Costall B, Kelly ME, Naylor RJ, Onaivi ES (1989): The actions of nicotine and cocaine in a mouse model of anxiety. Pharmacol. Biochem. Behav. 33(1): 197–203.

Covey LS, Glassman AH, Stetner F (1997): Major depression following smoking cessation. Am. J. Psychiatry 154(2): 263–265.

Cox BM, Goldstein A, Nelson WT (1984): Nicotine self-administration in rats. Br. J. Pharmacol. 83: 49–55.

Crooks PA, Dwoskin LP (1997): Contribution of CNS nicotine metabolites to the neuropharmacological effects of nicotine and tobacco smoking. Biochem. Pharmacol. 54(7): 743–753.

Damaj MI, Welch SP, Martin BR (1993): Involvement of calcium and L-type channels in nicotine-induced antinociception. J. Pharmacol. Exp. Ther. 266: 1330–1338.

Day J, Fibiger HC (1992): Dopaminergic regulation of cortical acetylcholine release. Synapse 12(4): 281–286.

Day JC, Fibiger HC (1994): Dopaminergic regulation of septohippocampal cholinergic neurons. J. Neurochem. 63(6): 2086–2092.

Decker MW, Majchrzak MJ, Anderson DJ (1992): Effects of nicotine on spatial memory deficits in rats with septal lesions. Brain Res. 572: 281–285.

Decker MW, Majchrzak MJ, Arneric SP (1993): Effects of lobeline, a nicotinic receptor agonist, on learning and memory. Pharmacol. Biochem. Behav. 45: 571–576.

Decker MW, Brioni JD, Sullivan JP, Buckley MJ, Radek RJ, Raskiewickz JL, Kang CH, Kim DJB, Giardina WJ, Williams M, Arneric SP (1994a): (S)-3-Methyl-5-(1-methyl-2-pyrrolidinyl) isoxazole (ABT 418): a novel cholinergic ligand with cognition enhancing and anxiolytic activities: II. In vivo characterization. J. Pharmacol. Exp. Ther. 270: 319–328.

Decker MW, Buckley MJ, Brioni JD (1994b): Differential effects of pretreatment with nicotine and lobeline on nicotine-induced changes in body temperature and locomotor activity in mice. Drug Dev. Res. 31: 52–58.

Decker MW, Curzon P, Brioni JD, Arneric SP (1994c): Effects of ABT 418, a novel cholinergic channel ligand, on learning and memory in septal-lesioned rats. Eur. J. Pharmacol. 261: 217–222.

Decker M, Brioni J, Bannon A, Arneric S (1995): Diversity of neuronal nicotinic acetylcholine receptors: lessons from behavior and implications for CNS therapeutics. Life Sci. 56: 545–570.

Decker MW, Bannon AW, Curzon P, Gunther KL, Brioni JD, Holladay MW, Lin N-H, Li Y, Daanen JF, Buccafusco JJ, Prendergast MA, Jackson WJ, Arneric SP (1997): ABT-089[2-methyl-3-(2-(s)-pyrrolidinylmethoxy)pyridine dihydrochloride]: II. A novel cholinergic channel modulator with effects on cognitive performance in rats and monkeys. J. Pharmacol. Exp. Ther. 283: 247–258.

Decker MW, Bannon AW, Buckley MJ, Kim DJB, Holladay MW, Ryther KB, Lin N-H, Wasicak JT, Williams M, Arneric SP (1998): Antinociceptive effects of the novel neuronal nicotinic acetylycholine receptor agonist, ABT-594, in mice. Eur. J. Pharmacol. 346: 23–33.

Elrod K, Buccafusco JJ, Jackson WJ (1988): Nicotine enhances delayed matching-to-sample performance by primates. Life Sci. 43: 277–287.

Emerich DF, Zanol MD, Norman AB, McConville BJ, Sanberg PR (1991): Nicotine potentiates haloperidol-induced catalepsy and locomotor hypoactivity. Pharmacol. Biochem. Behav. 38: 875–880.

Erenberg E, Cruse RP, Rothner AD (1987): The natural history of Tourette syndrome: a follow-up study. Ann. Neurol. 22: 383–385.

Ernst M, Zemetkin A (1995): The interface of genetics, neuroimaging, and neurochemistry in attention-deficit hyperactivity disorder. In: Bloom FE, Kupfer DJ, editors. Psychopharmacology: the fourth generation of progress. New York: Raven Press, Ltd., pp. 1643–1652.

Freedman R, Wetmore C, Strömberg I, Leonard S, Olson L (1993): α-Bungarotoxin binding to hippocampal interneurons: immunocytochemical characterization and effects on growth factor expression. J. Neurosci. 13: 1965–1975.

German DC, Manaye K, Smith WK, Woodward DJ, Saper CB (1989): Midbrain dopaminergic cell loss in Parkinson's disease: computer visualization. Ann. Neurol. 26: 507–514.

Ghelardini C, Galeotti N, Barlocco D, Bartolini A (1997): Antinociceptive profile of the new nicotinic agonist DBO-83. Drug Dev. Res. 40(3): 251–258.

Gilbert DG, Spielberger CD (1987): Effects of smoking on heart rate, anxiety, and feelings of success during social interaction. J. Behav. Med. 10: 629–638.

Gilbert DG, Gilbert BO (1995): Personality, psychopathology, and nicotine response as mediators of the genetics of smoking. Behav. Genet. 25: 133–147.

Gilbert DG, Robinson JH, Chamberlin CL, Spielberger CD (1989): Effects of smoking/nicotine on anxiety, heart rate, and lateralization of EEG during a stressful movie. Psychophysiology. 26(3): 311–320.

Gilbert DG, Meliska CJ, Welser R, Estes SL (1994): Depression, personality, and gender influence EEG, cortisol, beta-endorphin, heart rate, and subjective responses to smoking multiple cigarettes. Person Individ. Diff. 16: 247–264.

Glassman AH, Helzer JE, Covey LS, Cottler LB, Stetner F, Tipp JE, Johnson J (1990): Smoking, smoking cessation, and major depression. J. Amer. Med. Assoc. 264: 1546–1549.

Glick SD, Jarvik ME, Nakamura RK (1990): Inhibition of smoking behavior in monkeys. Nature 227: 573–575.

Glover ED, Leischow SJ, Rennard SI, Glover PN, Daughton D, Quiring J, Schneider FH, Mione PJ (1998): A smoking cessation trial with lobeline sulfate: a pilot study. Am. J. Health Behav. 22: 62–75.

Grady S, Marks MJ, Wonnacott S, Collins AC (1992): Characterization of nicotinic receptor-mediated [^{3}H]dopamine release from synaptosomes prepared from mouse striatum. J. Neurochem. 59: 848–856.

Green JT, Thomas GAO, Rhodes J, Williams GT, Evans BK, Russell MAH, Feyerbend C, Rhodes P, Sandborn WJ (1997): Nicotine enemas for ulcerative colitis. Gastroenterology 112: A984.

Hansen L, Salmon D, Galasko D, Masliah E, Katzman R, DeTeressa R, Thal L, Pay MM, Hofstetter R, Klauber M, Rice V, Butters N, Alford M (1990): The Lewy body variant of Alzheimer's disease: a clinical and pathologic entity. Neurology 40: 1–8.

Iwamoto ET (1989): Antinociception after nicotine administration into the mesopontine tegmentum of rats: evidence for muscarinic actions. J. Pharmacol. Exp. Ther. 251: 412–421.

Iwamoto ET (1990): Characterization of the aninociception induced by nicotine in the pedunculopontine tegmental nucleus and the nucleus raphe magnus. J. Pharmacol. Exp. Ther. 257: 120–133.

Janson AM, Moller A (1993): Chronic nicotine treatment counteracts nigral cell loss induced by a partial mesodiencephalic hemitransection: an analysis of the total number and mean volume of neurons and glia in substantia nigra of the male rat. Neuroscience 57: 931–941.

Janson AM, Fuxe K, Agnati LF, Jasson A, Bjeklke B, Sundstrom E, Anderson K, Harfstrand A, Goldstein M, Owman C (1989): Protective effects of chronic nicotine treatment on lesioned nigrostriatal dopamine neurons in the male rat. Prog. Brain Res. 79: 257–265.

Jarvik ME (1991): Beneficial effects of nicotine. Br. J. Addict. 86: 571–575.

Jones GMM, Sahakian BJ, Levy R, Warburton DM, Gray JA (1992): Effects of acute subcutaneous nicotine on attention, information processing short-term memory in Alzheimer's disease. Psychopharmacology 108: 485–494.

Lee PN (1994): Smoking and Alzheimer's disease: a review of the epidemiological evidence. Neuroepidemiology 13: 131–144.

Levin ED (1992): Nicotinic systems and cognitive function. Psychopharmacology 108: 417–431.

Levin ED, Conners CK, Sparrow E, Hinton SC, Erhardt D, Meck WH, Rose JE, March J (1996): Nicotine effects on adults with attention-deficit/hyperactivity disorder. Psychopharmacology 123(1): 55–63.

Lin N-H, Gunn DE, Ryther KB, Garvey DS, Donnelly-Roberts DL, Decker MW, Brioni JD, Buckley MJ, Rodrigues AD, Marsh KG, Anderson DJ, Buccafusco JJ, Prendergast MA, Sullivan JP, Williams M, Arneric SP, Holladay MW (1997): Structure-activity studies on 2-methyl-3-(2-(S)-pyrrolidinylmethoxy) pyridine (ABT-089): an orally bioavailable 3-pyridyl ether nicotinic acetylcholine receptor (nAChR) ligand with cognition enhancing properties. J. Med. Chem. 40: 385–390.

Lippiello PM, Bencherif M, Gray JA, Peters S, Grigoryan G, Hodges H, Collins AC (1996): RJR-2403: a nicotinic agonist with CNS selectivity II. In vivo characterization. J. Pharmacol. Exp. Ther. 279(3): 1422–1429.

Luetje CW, Patrick J, Séguéla P (1990): Nicotine receptors in the mammalian brain. FASEB J. 4: 2753–2760.

Luntz-Leybman V, Bickford PC, Freedman R (1992): Cholinergic gating of response to auditory stimuli in rat hippocampus. Brain Res. 587: 130–136.

McConville BJ, Fogelson MH, Klykylo WM, Manderscheid PZ, Parder KW, Sanberg PR (1991): Nicotine potentiation of haloperidol in reducing tic frequency in Tourette's disorder. Am. J. Psych. 148: 793–794.

Meyer EM, de Fiebre CM, Hunter BE, Simpkins CE, Frauworth N, de Fiebre NEC (1994): Effects of anabaseine-related analogs on rat brain nicotinic receptor binding and on avoidance behaviors. Drug Dev. Res. 31: 127–134.

Meyer EM, Tay ET, Papke RL, Meyers C, Huang GL, de Fiebre CM (1997): 3-[2,4-dimethoxybenzylidene]anabaseine (DMXB) selectively activates rat α7 receptors and improves memory-related behaviors in a mecamylamine-sensitive manner. Brain Res. 768(1–2): 49–56.

Mitchell SN (1993): Role of the locus coeruleus in the noradrenergic response to a systemic administration of nicotine. Neuropharmacology 32: 937–949.

Juir JL, Everitt BJ, Robbins TW (1995): Reversal of visual attentional dysfunction following lesions of the cholinergic basal forebrain by physostigmine and nicotine but not by the 5-HT_3 receptor antagonist, ondansetron. Psychopharmacology 118: 82–92.

Nanri M, Kasahara N, Yamamoto J, Miyake H, Watanabe H (1997): GTS-21, a nicotinic agonist, protects against neocortical neuronal cell loss induced by the nucleus basalis magnocellularis lesion in rats. Japan. J. Pharmacol. 74: 285–289.

Newhouse PA, Sunderland T, Tariot PN, Blumhardt CL, Weingartner H, Mellow A (1988): Intravenous nicotine in Alzheimer's disease: a pilot study. Psychopharmacology 95: 171–175.

Oates JA, Wood AJJ, Benowitz NL (1988): Pharmacologic aspects of cigarette smoking and nicotine addiction. N. Eng. J. Med. 319: 1318–1330.

Onaivi ES, Payne S, Brock JW, Hamdi A, Faroouqui S, Prasad C (1994): Chronic nicotine reverses age-associated increases in tail-flick latency and anxiety in rats. Life Sci. 54(3): 193–202.

Perry EK, Marshall E, Kerwin J, Smith CJ, Jabeen S, Cheng AV, Perry RH (1990): Evidence of a monoaminergic-cholinergic imbalance related to visual hallucinations in Lewy body dementia. J. Neurochem. 55: 1454–1456.

Pomerleau OF (1986): Nicotine as a psychoactive drug: anxiety and pain reduction. Psychopharm. Bull. 22: 865–869.

Prendergast MA, Jackson WJ, Terry AV Jr., Decker MW, Arneric SP, Buccafusco JJ (1998): Central nicotinic receptor agonists ABT-418, ABT-089 and (−)-nicotine reduce distractibility in young-adult monkeys. Psychopharmacology.

Pritchard WS, Robinson JH, Guy TD (1992): Enhancement of continuous performance task reaction time by smoking in non-deprived smokers. Psychopharmacology 108: 437–442.

Pullan RD, Rhodes J, Ganesh S, Mani V, Morris JS, Williams GT, Newcombe RG, Russell MA, Feyerabend C, Thomas GA, et al. (1994): Transdermal nicotine for active ulcerative colitis. N. Eng. J. Med. 330: 811–815.

Rhodes J, Thomas G, Evans BK (1997): Inflammatory bowel disease management. Current Opin. Drugs 53: 189–194.

Rose JE (1996): Nicotine addiction and treatment. Ann. Rev. Med. 47: 493–507.

Rose JE, Levin ED (1991): Concurrent agonist-antagonist administration for the analysis and treatment of drug dependence. Pharmacol. Biochem. Behav. 41: 219–226.

Rose JE, Behm FM, Westman EC, Levin ED, Stein RM, Ripka GV (1994): Mecamylamine combined with nicotine skin patch facilitates smoking cessation beyond nicotine patch treatment alone. Clin. Pharmacol. Ther. 56: 86–99.

Rosecrans JA, James JR (1993): The evaluation of central cholinergic neurons using an in vivo drug discrimination paradigm in rats. Med. Chem. Res. 2: 530–545.

Rusted J, Graupner L, O'Connell N, Nicholls C (1994). Does nicotine improve cognitive function? Psychopharmacology 115: 547–549.

Salin-Pascual RJ, de la Fuente JR, Galicia-Polo L, Drucker-Colin R (1995): Effects of transdermal nicotine on mood and sleep in nonsmoking major depressed patients. Psychopharmacology 121: 476–479.

Salin-Pascual RJ, Rosas M, Jimenez-Genchi A, Rivera-Meza BL, Delgado-Parra V (1996): Antidepressant effect of transdermal nicotine patches in nonsmoking patients with major depression. J. Clin. Psych. 57: 387–389.

Sanberg PR, McConville BJ, Fogelson HM, Manderscheid PZ, Parker KW, Blythe MM, Klykylo WM, Norman AB (1989): Nicotine potentiates the effects of haloperidol in animals and in patients with Tourette's syndrome. Biomed. Pharmacother. 43: 19–23.

Sanberg PR, Silver AA, Shytle RD, Philipp MK, Cahill DW, Fogelson HM, McConville BJ (1997): Nicotine for the treatment of tourette's syndrome. Pharmacol. Ther. 74: 21–25.

Sargent PB (1993): The diversity of neuronal nicotinic acetylcholine receptors. Ann. Rev. Neurosci. 16: 403–443.

Shapiro ES, Shapiro AK, Fulop G, Hubbard M, Mandeli J, Mordie J, Phillips RA (1989): Controlled study of haloperidol, pimozide, and placebo for the treatment of Gilles de la Tourette's syndrome. Arch. Gen. Psych. 46: 722–730.

Silver AA, Sanberg PR (1993): Transdermal nicotine patch and potentiation of haloperidol in Tourette's syndrome. Lancet 342: 182.

Smith CJ, Giacobini E (1992): Nicotine, Parkinson's and Alzheimer's disease. Rev. Neurosci. 3: 25–43.

Stevens KE, Wear KD (1997): Normalizing effects of nicotine and a novel nicotinic agonist on hippocampal auditory gating in two animal models. Pharmacol. Biochem. Behav. 57(4): 869–874.

Stolerman IP, Jarvis MJ (1995): The scientific case that nicotine is addictive. Psychopharmacology 117: 2–10.

Sullivan J, Decker MW, Brioni JD, Donnelly-Roberts D, Anderson DJ, Bannon AW, Kang C, Adams P, Piattoni-Kaplan M, Buckley MJ, Gopalakrishnan M, Williams M, Arneric SP (1994): (±)-Epibatidine elicits a diversity of in vitro and in vivo effects mediated by nicotinic acetylcholine receptors. J. Pharmacol. Exp. Ther. 271: 624–631.

Sullivan JP, Donnelly-Roberts D, Briggs CA, Anderson DJ, Gopalakrishnan M, Xue IC, Plattonikaplan M, Molinari E, Campbell JE, McKenna DG, Gunn DE, Lin NH, Ryther KB, He Y, Holladay MW, Wonnacott S, Williams M, Arneric SP (1997): ABT-089 [2-methyl-3(2-(S)-pyrrolidinylmethoxy)pyridine]. 1. A potent and selective cholinergic channel modulator with neuroprotective properties. J. Pharmacol. Exp. Ther. 283: 235–246.

Taylor P (1990): Agents acting at the neuromuscular junction and autonomic ganglia, In: Gilman AG, Rall TW, Nies AS, Taylor P, editors. Goodman and Gilman's the pharmacological basis of therapeutics, 8th ed. New York: Pergamon Press, pp. 166–186.

Thomas GAO, Rhodes J, Mani V, Williams GT, Newcombe R, Russell MA, Feyerabend C (1995): Transdermal nicotine as maintenance therapy for ulcerative colitis. N. Engl. J. Med. 332: 988–992.

Tripathi HL, Martin BR, Aceto MD (1982): Nicotine-induced antinociception in rats and mice: correlation with nicotine brain levels. J. Pharmacol. Exp. Ther. 221: 91–96.

van Duijn CM, Hofman A (1991): Relation between nicotine intake and Alzheimer's disease. Br. Med. J. 302: 1491–1494.

Warburton DM, Arnall C (1994): Improvements in performance without nicotine withdrawal. Psychopharmacology 115: 539–540.

Wilson AL, Langley LK, Monley J, Bauer T, Rottunda S, McFalls E, Kovera C, McCarten JR (1995): Nicotine patches in Alzheimer's disease: pilot study on learning, memory, and safety. Pharmacol. Biochem. Behav. 51(2–3): 509–514.

Wonnacott S (1997): Presynaptic nicotinic ACh receptors. Trends Neurosci. 20: 92–98.

Index

Printed in the United Kingdom
by Lightning Source UK Ltd.
109353UKS00001B/100